Essentials of Biostatistics in Public Health
Third Edition

Lisa M. Sullivan, PhD

Professor of Biostatistics
Associate Dean for Education
Boston University School of Public Health
Boston, Massachusetts

JONES & BARTLETT
LEARNING

World Headquarters
Jones & Bartlett Learning
5 Wall Street
Burlington, MA 01803
978-443-5000
info@jblearning.com
www.jblearning.com

Jones & Bartlett Learning books and products are available through most bookstores and online booksellers. To contact Jones & Bartlett Learning directly, call 800-832-0034, fax 978-443-8000, or visit our website, www.jblearning.com.

Substantial discounts on bulk quantities of Jones & Bartlett Learning publications are available to corporations, professional associations, and other qualified organizations. For details and specific discount information, contact the special sales department at Jones & Bartlett Learning via the above contact information or send an email to specialsales@jblearning.com.

Copyright © 2018 by Jones & Bartlett Learning, LLC, an Ascend Learning Company

All rights reserved. No part of the material protected by this copyright may be reproduced or utilized in any form, electronic or mechanical, including photocopying, recording, or by any information storage and retrieval system, without written permission from the copyright owner.

The content, statements, views, and opinions herein are the sole expression of the respective authors and not that of Jones & Bartlett Learning, LLC. Reference herein to any specific commercial product, process, or service by trade name, trademark, manufacturer, or otherwise does not constitute or imply its endorsement or recommendation by Jones & Bartlett Learning, LLC and such reference shall not be used for advertising or product endorsement purposes. All trademarks displayed are the trademarks of the parties noted herein. *Essentials of Biostatistics in Public Health, Third Edition* is an independent publication and has not been authorized, sponsored, or otherwise approved by the owners of the trademarks or service marks referenced in this product.

There may be images in this book that feature models; these models do not necessarily endorse, represent, or participate in the activities represented in the images. Any screenshots in this product are for educational and instructive purposes only. Any individuals and scenarios featured in the case studies throughout this product may be real or fictitious, but are used for instructional purposes only.

The author, editor, and publisher have made every effort to provide accurate information. However, they are not responsible for errors, omissions, or for any outcomes related to the use of the contents of this book and take no responsibility for the use of the products and procedures described. Treatments and side effects described in this book may not be applicable to all people; likewise, some people may require a dose or experience a side effect that is not described herein. Drugs and medical devices are discussed that may have limited availability controlled by the Food and Drug Administration (FDA) for use only in a research study or clinical trial. Research, clinical practice, and government regulations often change the accepted standard in this field. When consideration is being given to use of any drug in the clinical setting, the health care provider or reader is responsible for determining FDA status of the drug, reading the package insert, and reviewing prescribing information for the most up-to-date recommendations on dose, precautions, and contraindications, and determining the appropriate usage for the product. This is especially important in the case of drugs that are new or seldom used.

18513-3

Production Credits
VP, Executive Publisher: David D. Cella
Publisher: Michael Brown
Associate Editor: Lindsey M. Sousa
Senior Production Editor: Amanda Clerkin
Senior Marketing Manager: Sophie Fleck Teague
Manufacturing and Inventory Control Supervisor: Amy Bacus
Composition: Integra Software Services Pvt. Ltd.
Cover Design: Scott Moden
Rights & Media Specialist: Wes DeShano
Media Development Editor: Shannon Sheehan
Cover Image: © Sergey Nivens/Shutterstock
Printing and Binding: Edwards Brothers Malloy
Cover Printing: Edwards Brothers Malloy

Library of Congress Cataloging-in-Publication Data
Names: Sullivan, Lisa M. (Lisa Marie), 1961-author.
Title: Essentials of biostatistics in public health / Lisa Sullivan.
Description: Third edition. | Burlington, Massachusetts : Jones & Bartlett
 Learning, [2018] | Includes bibliographical references and index.
Identifiers: LCCN 2016048158 | ISBN 9781284108194
Subjects: | MESH: Biostatistics—methods | Biometry—methods | Public
 Health—methods
Classification: LCC RA409 | NLM WA 950 | DDC 610.72—dc23
LC record available at https://lccn.loc.gov/2016048158

6048

Printed in the United States of America
21 20 19 18   10 9 8 7 6 5 4

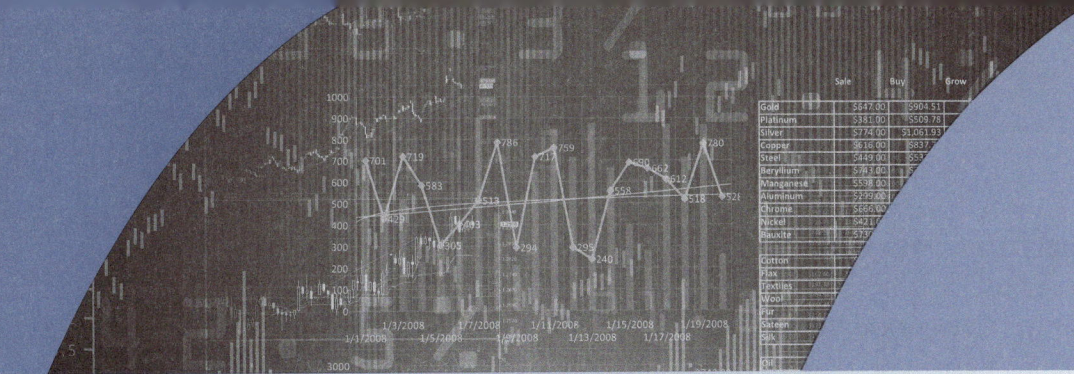

Contents

Acknowledgments — vii
Preface — ix
Prologue — xi
About the Author — xiii

Chapter 1	Introduction	1
1.1	What Is Biostatistics?	2
1.2	What Are the Issues?	2
1.3	Summary	4

Chapter 2	Study Designs	7
2.1	Vocabulary	8
2.2	Observational Study Designs	9
2.3	Randomized Study Designs	14
2.4	The Framingham Heart Study	17
2.5	More on Clinical Trials	18
2.6	Sample Size Implications	20
2.7	Summary	20
2.8	Practice Problems	20

Chapter 3	Quantifying the Extent of Disease	23
3.1	Prevalence	24
3.2	Incidence	24
3.3	Relationships Between Prevalence and Incidence	27
3.4	Comparing the Extent of Disease Between Groups	27
3.5	Summary	30
3.6	Practice Problems	30

Chapter 4	Summarizing Data Collected in the Sample	35
4.1	Dichotomous Variables	37
4.2	Ordinal and Categorical Variables	41
4.3	Continuous Variables	50
4.4	Summary	61
4.5	Practice Problems	63

Chapter 5	The Role of Probability	67
5.1	Sampling	68
5.2	Basic Concepts	70
5.3	Conditional Probability	71
5.4	Independence	74
5.5	Bayes' Theorem	74
5.6	Probability Models	75
5.7	Summary	96
5.8	Practice Problems	97

Chapter 6	Confidence Interval Estimates	101
6.1	Introduction to Estimation	102
6.2	Confidence Intervals for One Sample, Continuous Outcome	104
6.3	Confidence Intervals for One Sample, Dichotomous Outcome	105
6.4	Confidence Intervals for Two Independent Samples, Continuous Outcome	107
6.5	Confidence Intervals for Matched Samples, Continuous Outcome	110
6.6	Confidence Intervals for Two Independent Samples, Dichotomous Outcome	114
6.7	Summary	119
6.8	Practice Problems	120

Chapter 7	Hypothesis Testing Procedures	123
7.1	Introduction to Hypothesis Testing	124
7.2	Tests with One Sample, Continuous Outcome	131
7.3	Tests with One Sample, Dichotomous Outcome	133
7.4	Tests with One Sample, Categorical and Ordinal Outcomes	135
7.5	Tests with Two Independent Samples, Continuous Outcome	139
7.6	Tests with Matched Samples, Continuous Outcome	142
7.7	Tests with Two Independent Samples, Dichotomous Outcome	145
7.8	Tests with More Than Two Independent Samples, Continuous Outcome	149
7.9	Tests for Two or More Independent Samples, Categorical and Ordinal Outcomes	157
7.10	Summary	163
7.11	Practice Problems	164

Chapter 8	**Power and Sample Size Determination**	**171**
8.1	Issues in Estimating Sample Size for Confidence Intervals Estimates	172
8.2	Issues in Estimating Sample Size for Hypothesis Testing	179
8.3	Summary	188
8.4	Practice Problems	188
Chapter 9	**Multivariable Methods**	**193**
9.1	Confounding and Effect Modification	195
9.2	The Cochran–Mantel–Haenszel Method	201
9.3	Introduction to Correlation and Regression Analysis	203
9.4	Multiple Linear Regression Analysis	212
9.5	Multiple Logistic Regression Analysis	216
9.6	Summary	219
9.7	Practice Problems	219
Chapter 10	**Nonparametric Tests**	**223**
10.1	Introduction to Nonparametric Testing	226
10.2	Tests with Two Independent Samples	227
10.3	Tests with Matched Samples	232
10.4	Tests with More Than Two Independent Samples	241
10.5	Summary	245
10.6	Practice Problems	246
Chapter 11	**Survival Analysis**	**249**
11.1	Introduction to Survival Data	252
11.2	Estimating the Survival Function	253
11.3	Comparing Survival Curves	261
11.4	Cox Proportional Hazards Regression Analysis	267
11.5	Extensions	271
11.6	Summary	272
11.7	Practice Problems	273
Chapter 12	**Data Visualization**	**279**
12.1	Design Principles	280
12.2	When and How to Use Text, Tables, and Figures	281
12.3	Presenting Data and Statistical Results in Tables	285
12.4	Presenting Data and Statistical Results in Figures	295
12.5	Summary	328
12.6	Practice Problems	328
Appendix		**333**
Glossary		**351**
Index		**361**

Acknowledgments

I am very grateful to Dr. Richard Riegelman for his unending energy and enthusiasm for this project. I thank my family for their encouragement; Casey, Ryan, Aislinn, Abbi, and Gavin for their inspiration; most especially, Kimberly Dukes and Kevin Green for their unending love and support.

This textbook is dedicated to the memory of my cousin Catherine Render. Catherine was a remarkable woman. She was a loving daughter, sister, wife, mother, cousin, and friend who excelled in every role. She lost a long battle with breast cancer in January 2006. She fought through many setbacks, always pushing forward optimistically, taking care of everyone around her, never asking why. She was and always will be an inspiration to me and to so many others who were fortunate enough to know her.

Biostatisticians play an extremely important role in addressing important medical and public health problems. Unfortunately, there are many more problems than solutions. We must never lose sight of the fact that our work is important in improving health and well-being. We need qualified biostatisticians to work in research teams to address problems like breast cancer, cardiovascular disease, diabetes, and so many others.

Preface

Essentials of Biostatistics in Public Health, Third Edition provides a fundamental and engaging background for students learning to apply and appropriately interpret biostatistical applications in the field of public health. The examples are real, important, and represent timely public health problems. The author aims to make the material relevant, practical, and engaging for students. Throughout the textbook, the author uses data from the Framingham Heart Study and from observational studies and clinical trials in a variety of major areas. The author presents example applications involving important risk factors—such as blood pressure, cholesterol, smoking, and diabetes and their relationships to incident cardiovascular and cerebrovascular disease—throughout. Clinical trials investigating new drugs to lower cholesterol, to reduce pain, and to promote healing following surgery are also considered. The author presents examples with relatively few subjects to illustrate computations while minimizing the actual computation time, as a particular focus is mastery of "by-hand" computations. All of the techniques are then applied to and illustrated on real data from the Framingham Heart Study and large observational studies and clinical trials. For each topic, the author discusses methodology—including assumptions, statistical computations, and the appropriate interpretation of results. Key formulas are summarized at the end of each chapter.

Prologue

Understanding how to present and interpret data is the foundation for evidence-based public health. It is essential for public health practitioners, future clinicians, and health researchers to know how to use data and how to avoid being deceived by data. In *Essentials of Biostatistics in Public Health*, Lisa Sullivan ably guides students through this maze. To do so, she uses an abundance of real and relevant examples drawn from her own experience working on the Framingham Heart Study and clinical trials.

Essentials of Biostatistics in Public Health takes an intuitive, step-by-step, hands-on approach in walking students though statistical principles. It emphasizes understanding which questions to ask and knowing how to interpret statistical results appropriately.

The third edition of *Essentials of Biostatistics in Public Health* builds upon the success of the previous editions in presenting state-of-the-art biostatistical methods that are widely used in public health and clinical research. A new chapter on data visualization provides important insights about how to produce and interpret data presented as tables, figures, and newer forms of data visualization. Dr. Sullivan provides information on both good and bad data presentations, and teaches the reader how to recognize the difference. Her recommendations are based on sound biostatical principles presented in a way that can be appreciated without extensive statistical background.

The *Third Edition* also features a new series of integrative exercises based on data collected in the Framingham Heart Study. The data set includes real data on more than 4,000 participants and gives students an opportunity to "get their hands dirty" by using real data.

In addition, the *Third Edition* includes a set of key questions for each chapter to engage students—that is, it adopts an inquiry-based approach to teaching and learning. Dr. Sullivan provides links to recent "in the news" articles to encourage students to "dig in" and to critically think about how to draw conclusions from data.

The strategies used in *Essentials of Biostatistics in Public Health* represent a tried-and-true, classroom-tested approach. Lisa Sullivan has more than 2 decades of experience teaching biostatistics to both undergraduates and graduate students. As Assistant Dean for Undergraduate Programs in Public Health at Boston University, she has developed and taught undergraduate courses in biostatistics. She has also served as the chair of the Department of Biostatistics. Today she is the Associate Dean for Education at Boston University School of Public Health. Her background speaks to her unique ability to combine the skills of biostatistics with the skills of education.

Dr. Sullivan has won numerous teaching awards for her skills and commitment to education in biostatistics, including the Association of Schools of Public Health Award for Teaching Excellence. She possesses a unique combination of sophisticated biostatistics expertise and a clear and engaging writing style—one that can draw students in and help them understand even the most difficult topic. Even a quick glance through *Essentials of Biostatistics in Public Health* will convince you of her skills in communication and education.

I am delighted that Dr. Sullivan has included her book and workbook in our *Essential Public Health* series. There is no better book to recommend for the anxious student first confronting the field of biostatistics. Students will find the book and workbook engaging and relevant. Just take a look and see for yourself.

Richard Riegelman, MD, MPH, PhD
Editor, *Essential Public Health* series

About the Author

Lisa M. Sullivan has a PhD in statistics and is Professor and former Chair of the Department of Biostatistics at the Boston University School of Public Health. She is also Associate Dean for Education. She teaches biostatistics for MPH students and lectures in biostatistical methods for clinical researchers. From 2003 to 2015, Lisa was the principal investigator of the National Heart, Lung, and Blood Institute's *Summer Institute for Training in Biostatistics*, which was designed to promote interest in the field of biostatistics and to expose students to the many exciting career opportunities available to them. Lisa is the recipient of numerous teaching awards, including the Norman A. Scotch Award and the prestigious Metcalf Award, both for excellence in teaching at Boston University. In 2008 she won the Association of Schools of Public Health/Pfizer Excellence in Teaching Award. In 2011, she won the American Statistical Association's Section on Teaching Statistics in the Health Sciences Outstanding Teaching Award. In 2013, she won the Mostellar Statistician of the Year Award, presented by the Boston Chapter of the American Statistical Association. Also in 2013, she won the Massachusetts ACE National Network of Women Leaders Leadership Award. Lisa is also a biostatistician on the Framingham Heart Study, working primarily on developing and disseminating cardiovascular risk functions. She is active in several large-scale epidemiological studies for adverse pregnancy outcomes. Her work has resulted in more than 200 peer-reviewed publications.

CHAPTER 1

Introduction

LEARNING OBJECTIVES

By the end of this chapter, the reader will be able to

- Define biostatistical applications and their objectives
- Explain the limitations of biostatistical analysis
- Compare and contrast a population and a sample
- Explain the importance of random sampling
- Develop research questions and select appropriate outcome variables to address important public health problems
- Identify the general principles and explain the role and importance of biostatistical analysis in medical, public health, and biological research

Biostatistics is central to public health education and practice; it includes a set of principles and techniques that allows us to draw meaningful conclusions from information or data. Implementing and understanding biostatistical applications is a combination of art and science. Appropriately understanding statistics is important both professionally and personally, as we are faced with statistics every day.

For example, cardiovascular disease is the number one killer of men and women in the United States. The American Heart Association reports that more than 2600 Americans die every day of cardiovascular disease, which is approximately one American every 34 seconds. There are over 70 million adults in the United States living with cardiovascular disease, and the annual rates of development are estimated at 7 cases per 1000 in men aged 35–44 years and 68 cases per 1000 in men aged 85–94 years.[1] The rates in women are generally delayed about 10 years as compared to men.[2] Researchers have identified a number of risk factors for cardiovascular disease including blood pressure, cholesterol, diabetes, smoking, and weight. Smoking and weight (specifically, overweight and obesity) are considered the most and second-most, respectively, preventable causes of cardiovascular disease death in the United States.[3,4] Family history, nutrition, and physical activity are also important risk factors for cardiovascular disease.[5]

The previous example describes cardiovascular disease, but similar statistics are available for many other diseases including cancer, diabetes, asthma, and arthritis. Much of what we know about cardiovascular and many other diseases comes from newspapers, news reports, or the Internet. Reporters describe or write about research studies on a daily basis. Nightly newscasts almost always contain a report of at least one research study. The results from some studies seem quite obvious, such as the positive effects of exercise on health, whereas other studies describe breakthrough medications that cure disease or prolong a healthy life. Newsworthy topics can include conflicting or contradictory results in medical research. One study might report that a new medical therapy is effective, whereas another study might suggest this new therapy is ineffectual; other studies may show vitamin supplements thought to be effective as being ineffective or even harmful. One study might demonstrate the effectiveness of a drug, and years later it is determined to be harmful due to some serious side effect. To understand and interpret these results requires knowledge of statistical principles and statistical thinking.

How are these studies conducted in the first place? For example, how is the extent of disease in a group or region quantified? How is the rate of development of new disease estimated? How are risk factors or characteristics that might be related to development or progression of disease identified? How is the effectiveness of a new drug determined? What could explain contradictory results? These questions are the essence of biostatistics.

1.1 WHAT IS BIOSTATISTICS?

Biostatistics is defined as the application of statistical principles in medicine, public health, or biology. Statistical principles are based in applied mathematics and include tools and techniques for collecting information or data and then summarizing, analyzing, and interpreting those results. These principles extend to making inferences and drawing conclusions that appropriately take uncertainty into account.

Biostatistical techniques can be used to address each of the aforementioned questions. In applied biostatistics, the objective is usually to make an inference about a specific population. By definition, this population is the collection of all individuals about whom we would like to make a statement. The population of interest might be all adults living in the United States or all adults living in the city of Boston. The definition of the population depends on the investigator's study question, which is the objective of the analysis. Suppose the population of interest is all adults living in the United States and we want to estimate the proportion of all adults with cardiovascular disease. To answer this question completely, we would examine every adult in the United States and assess whether they have cardiovascular disease. This would be an impossible task! A better and more realistic option would be to use a statistical analysis to estimate the desired proportion.

In biostatistics, we study samples or subsets of the population of interest. In this example, we select a sample of adults living in the United States and assess whether each has cardiovascular disease or not. If the sample is representative of the population, then the proportion of adults in the sample with cardiovascular disease should be a good estimate of the proportion of adults in the population with cardiovascular disease. In biostatistics, we analyze samples and then make inferences about the population based on the analysis of the sample. This inference is quite a leap, especially if the population is large (e.g., the United States population of 300 million) and the sample is relatively small (for example, 5000 people). When we listen to news reports or read about studies, we often think about how results might apply to us personally. The vast majority of us have never been involved in a research study. We often wonder if we should believe results of research studies when we, or anyone we know, never participated in those studies.

1.2 WHAT ARE THE ISSUES?

Appropriately conducting and interpreting biostatistical applications require attention to a number of important issues. These include, but are not limited to, the following:

- Clearly defining the objective or research question
- Choosing an appropriate study design (i.e., the way in which data are collected)
- Selecting a representative sample, and ensuring that the sample is of sufficient size
- Carefully collecting and analyzing the data
- Producing appropriate summary measures or statistics
- Generating appropriate measures of effect or association
- Quantifying uncertainty
- Appropriately accounting for relationships among characteristics
- Limiting inferences to the appropriate population

In this book, each of the preceding points is addressed in turn. We describe how to collect and summarize data and how to make appropriate inferences. To achieve these, we use biostatistical principles that are grounded in mathematical and probability theory. A major goal is to understand and interpret a biostatistical analysis. Let us now revisit our original questions and think about some of the issues previously identified.

How Is the Extent of Disease in a Group or Region Quantified?

Ideally, a sample of individuals in the group or region of interest is selected. That sample should be sufficiently large so that the results of the analysis of the sample are adequately precise. (We discuss techniques to determine the appropriate sample size for analysis in Chapter 8.) In general, a larger sample for analysis is preferable; however, we never want to sample more participants than are needed, for both financial and ethical reasons. The sample should also be representative of the population. For example, if the population is 60% women, ideally we would like the sample to be approximately 60% women. Once the sample is selected, each participant is assessed with regard to disease status. The proportion of the sample with disease is computed by taking the ratio of the number with disease to the total sample size. This proportion is an estimate of the proportion of the population with disease. Suppose the sample proportion is computed as 0.17 (i.e., 17% of those sampled have the disease). We estimate the proportion of the population with disease to be approximately 0.17 (or 17%). Because this is an estimate based on one sample, we must account for uncertainty, and this is reflected in what is called a *margin of error*. This might result in our estimating the proportion of the population with disease to be anywhere from 0.13 to 0.21 (or 13% to 21%).

This study would likely be conducted at a single point in time; this type of study is commonly referred to as a cross-sectional study. Our estimate of the extent of disease refers only to the period under study. It would be inappropriate to make inferences about the extent of disease at future points based on this study. If we had selected adults living in Boston as our population, it would also be inappropriate to infer that the extent of disease in other cities or in other parts of Massachusetts would be the same as that observed in a sample of Bostonians. The task of estimating the extent of disease in a region or group seems straightforward on the surface. However, there are many issues that complicate things. For example, where do we get a list of the population, how do we decide who is in the sample, how do we ensure that specific groups are represented (e.g., women) in the sample, and how do we find the people we identify for the sample and convince them to participate? All of these questions must be addressed correctly to yield valid data and correct inferences.

How Is the Rate of Development of a New Disease Estimated?

To estimate the rate of development of a new disease—say, cardiovascular disease—we need a specific sampling strategy. For this analysis, we would sample only persons free of cardiovascular disease and follow them prospectively (going forward) in time to assess the development of the disease. A key issue in these types of studies is the follow-up period; the investigator must decide whether to follow participants for either 1, 5, or 10 years, or some other period, for the development of the disease. If it is of interest to estimate the development of disease over 10 years, it requires following each participant in the sample over 10 years to determine their disease status. The ratio of the number of new cases of disease to the total sample size reflects the proportion or cumulative incidence of new disease over the predetermined follow-up period. Suppose we follow each of the participants in our sample for 5 years and find that 2.4% develop disease. Again, it is generally of interest to provide a range of plausible values for the proportion of new cases of disease; this is achieved by incorporating a margin of error to reflect the precision in our estimate. Incorporating the margin of error might result in an estimate of the cumulative incidence of disease anywhere from 1.2% to 3.6% over 5 years.

Epidemiology is a field of study focused on the study of health and illness in human populations, patterns of health or disease, and the factors that influence these patterns. The study described here is an example of an epidemiological study. Readers interested in learning more about epidemiology should see Magnus.[6]

How Are Risk Factors or Characteristics That Might Be Related to the Development or Progression of Disease Identified?

Suppose we hypothesize that a particular risk factor or exposure is related to the development of a disease. There are several different study designs or ways in which we might collect information to assess the relationship between a potential risk factor and disease onset. The most appropriate study design depends, among other things, on the distribution of both the risk factor and the outcome in the population of interest (e.g., how many participants are likely to have a particular risk factor or not). (We discuss different study designs in Chapter 2 and which design is optimal in a specific situation.) Regardless of the specific design used, both the risk factor and the outcome must be measured on each member of the sample. If we are interested in the relationship between the risk factor and the development of disease, we would again involve participants free of disease at the study's start and follow all participants for the development of disease. To assess whether there is a relationship between a risk factor and the outcome, we estimate the proportion (or percentage) of participants with the risk factor who go on to develop disease and compare that to the proportion (or percentage) of participants who do not have the risk factor and go on to develop disease. There are several ways to make this comparison; it can be based on a difference in proportions or a ratio of proportions. (The details of these comparisons are discussed extensively in Chapter 6 and Chapter 7.)

Suppose that among those with the risk factor, 12% develop disease during the follow-up period, and among those free of the risk factor, 6% develop disease. The ratio of the proportions is called a **relative risk** and here it is equal to 0.12 / 0.06 = 2.0. The interpretation is that twice as many people with the risk factor develop disease as compared to people without the risk factor. The issue then is to determine whether this estimate, observed in one study sample, reflects an increased risk in the population. Accounting for uncertainty might result in an estimate of the relative risk anywhere from 1.1 to 3.2 times higher for persons with the risk factor. Because the range contains risk values greater than 1, the data reflect an increased risk (because a value of 1 suggests no increased risk).

Another issue in assessing the relationship between a particular risk factor and disease status involves understanding complex relationships among risk factors. Persons with the risk factor might be different from persons free of the risk factor; for example, they may be older and more likely to have other risk factors. There are methods that can be used to assess the association between the hypothesized risk factor and disease status while taking into account the impact of the other risk factors. These techniques involve statistical modeling. We discuss how these models are developed and, more importantly, how results are interpreted in Chapter 9.

How Is the Effectiveness of a New Drug Determined?

The ideal study design from a statistical point of view is the **randomized controlled trial** or the **clinical trial**. (The term *clinical* means that the study involves people.) For example, suppose we want to assess the effectiveness of a new drug designed to lower cholesterol. Most clinical trials involve specific inclusion and exclusion criteria. For example, we might want to include only persons with total cholesterol levels exceeding 200 or 220, because the new medication would likely have the best chance to show an effect in persons with elevated cholesterol levels. We might also exclude persons with a history of cardiovascular disease. Once the inclusion and exclusion criteria are determined, we recruit participants. Each participant is randomly assigned to receive either the new experimental drug or a control drug. The randomization component is the key feature in these studies. Randomization theoretically promotes balance between the comparison groups. The control drug could be a *placebo* (an inert substance) or a cholesterol-lowering medication that is considered the current standard of care.

The choice of the appropriate comparator depends on the nature of the disease. For example, with a life-threatening disease, it would be unethical to withhold treatment; thus a placebo comparator would never be appropriate. In this example, a placebo might be appropriate as long as participants' cholesterol levels were not so high as to necessitate treatment. When participants are enrolled and randomized to receive either the experimental treatment or the comparator, they are not told to which treatment they are assigned. This is called blinding or masking. Participants are then instructed on proper dosing and after a predetermined time, cholesterol levels are measured and compared between groups. (Again, there are several ways to make the comparison and we will discuss different options in Chapter 6 and Chapter 7.) Because participants are randomly assigned to treatment groups, the groups should be comparable on all characteristics except the treatment received. If we find that the cholesterol levels are different between groups, the difference can likely be attributed to treatment.

Again, we must interpret the observed difference after accounting for chance or uncertainty. If we observe a large difference in cholesterol levels between participants receiving the experimental drug and the comparator, we can infer that the experimental drug is effective. However, inferences about the effect of the drug are only able to be generalized to the population from which participants are drawn—specifically, to the population defined by the inclusion and exclusion criteria. Clinical trials must be carefully designed and analyzed. There exist a number of issues that are specific to clinical trials, and we discuss these in detail in Chapter 2.

Clinical trials are discussed extensively in the news, particularly recently. They are heavily regulated in the United States by the Food and Drug Administration (FDA).[7] Recent news reports discuss studies involving drugs that were granted approval for specific indications and later removed from the market due to safety concerns. We review these studies and assess how they were conducted and, more important, why they are being reevaluated. For evaluating drugs, randomized controlled trials are considered the gold standard. Still, they can lead to controversy. Studies other than clinical trials are less ideal and are often more controversial.

What Could Explain Contradictory Results Between Different Studies of the Same Disease?

All statistical studies are based on analyzing a sample from the population of interest. Sometimes, studies are not designed appropriately and results may therefore be questionable. Sometimes, too few participants are enrolled, which could lead to imprecise and even inaccurate results. There are also instances where studies are designed appropriately, yet two different replications produce different results. Throughout this book, we will discuss how and when this might occur.

1.3 SUMMARY

In this book, we investigate in detail each of the issues raised in this chapter. Understanding biostatistical principles is critical to public health education. Our approach will be through active learning: examples are taken from the Framingham Heart Study and from clinical trials, and used throughout the book to illustrate concepts. Example applications involving important risk factors such as blood pressure, cholesterol, smoking, and diabetes and their relationships to incident cardiovascular and cerebrovascular disease are discussed. Examples with relatively few subjects help to illustrate computations while minimizing the actual computation time; a particular focus is mastery of "by-hand" computations. All of the techniques are then applied to real data from the Framingham study and from clinical trials. For each topic, we discuss methodology—including assumptions, statistical formulas, and the appropriate interpretation of results. Key formulas are summarized at the end of each chapter. Examples are selected to represent important and timely public health problems.

REFERENCES

1. American Heart Association. Available at *http://www.americanheart.org*.
2. Sytkowski, P.A., D'Agostino, R.B., Belanger, A., and Kannel, W.B. "Sex and time trends in cardiovascular disease incidence and mortality: The Framingham Heart Study, 1950–1989." *American Journal of Epidemiology* 1996; 143(4): 338–350.

3. Wilson, P.W.F., D'Agostino, R.B., Levy, D., Belanger, A.M., Silbershatz, H., and Kannel, W.B. "Prediction of coronary heart disease using risk factor categories." *Circulation* 1998; 97: 1837–1847.
4. The Expert Panel. "Expert panel on detection, evaluation, and treatment of high blood cholesterol in adults: summary of the second report of the NCEP expert panel (Adult Treatment Panel II)." *Journal of the American Medical Association* 1993; 269: 3015–3023.
5. Kaikkonen, K.S., Kortelainen, M.L., Linna, E., and Huikuri, H.V. "Family history and the risk of sudden cardiac death as a manifestation of an acute coronary event." *Circulation* 2006; 114(4): 1462–1467.
6. Magnus, M. *Essentials of Infectious Disease Epidemiology*. Sudbury, MA: Jones and Bartlett, 2007.
7. United States Food and Drug Administration. Available at *http://www.fda.gov*.

CHAPTER 2

Study Designs

When and Why

Key Questions

- How do you know when the results of a study are credible?
- What makes a good study? How do you decide which kind of study is best?
- How do you know when something is a cause of something else?

In the News

The following are recent headlines of reports summarizing investigations conducted on important public health issues.

"Could going to college or being married give you brain cancer?"

Sharon Begle of *STAT* News reports on a study of more than 4 million residents of Sweden and finds that people with 3 years of college or more had a 20% higher risk of developing glioma (a brain cancer) than those with elementary school-level education, as did married men compared to their unmarried counterparts.[1]

"Does Monsanto's Roundup herbicide cause cancer or not? The controversy explained."

Sarah Zhang of *Wired* comments on conflicting reports from the United Nations and the World Health Organization about glyphosate (a weed killer) and whether it is carcinogenic.[2]

"Eating pasta does not cause obesity, Italian study finds."

Tara John of *Time* reports that a new study on more than 20,000 Italians found that pasta consumption is not associated with obesity, but rather with a reduction in body mass index. The author does note that the study was partially funded by a pasta company, Barilla, and the Italian government.[3]

---

[1]Begle S. *STAT News*. June 20, 2016. Available at https://www.statnews.com/2016/06/20/brain-cancer-college-marriage/.
[2]Zhang S. *Wired*. May 17, 2016. Available at http://www.wired.com/2016/05/monsantos-roundup-herbicide-cause-cancer-not-controversy-explained/.
[3]John T. *Time*. July 16, 2016. Available at http://time.com/4393040/pasta-fat-obesity-body-mass-index-good/.

Dig In

Choose any one of the studies mentioned previously and consider the following.

- What was the research question that investigators were asking?
- What was the outcome and how was it measured? What was the exposure or risk factor that they were trying to link to the outcome, and how was it measured?
- Is it appropriate to infer causality based on this study?

Learning Objectives

By the end of this chapter, the reader will be able to

- List and define the components of a good study design
- Compare and contrast observational and experimental study designs
- Summarize the advantages and disadvantages of alternative study designs
- Describe the key features of a randomized controlled trial
- Identify the study designs used in public health and medical studies

---

Once a study objective or research question has been refined—which is no easy task, as it usually involves extensive discussion among investigators, a review of the literature, and an assessment of ethical and practical issues—the next step is to choose the study design to most effectively and efficiently answer the question. The **study design** is the methodology that is used to collect the information to address the research question. In Chapter 1, we raised a number of questions that might be of interest, including: How is the extent of a disease in a group or region quantified? How is the rate of development of a new disease estimated? How are risk factors or characteristics that might be related to the development or progression of a disease identified? How is the effectiveness of a new drug determined? To answer each of these questions, a specific study design must be selected. In this chapter, we review a number of popular study designs. This review is not meant to be exhaustive but instead illustrative of some of the more popular designs for public health applications.

The studies we present can probably be best organized into two broad types: observational and randomized studies. In **observational studies**, we generally observe a phenomenon, whereas in randomized studies, we intervene and measure a response. Observational studies are sometimes called descriptive or associational studies, nonrandomized, or historical studies. In some cases, observational studies are used to alert the medical community to a specific issue, whereas in other instances, observational studies are used to generate hypotheses. We later elaborate on other instances where observational studies are used to assess specific associations. Randomized studies are sometimes called analytic or experimental studies. They are used to test specific hypotheses or to evaluate the effect of an intervention (e.g., a behavioral or pharmacologic intervention).

Another way to describe or distinguish study types is on the basis of the time sequence involved in data collection. Some studies are designed to collect information at a point in time, others to collect information on participants over time, and others to evaluate data that have already been collected.

In biostatistical and epidemiological research studies, we are often interested in the association between a particular exposure or risk factor (e.g., alcohol use, smoking) and an outcome (e.g., cardiovascular disease, lung cancer). In the following sections, we discuss several observational study designs and several randomized study designs. We describe each design, detail its advantages and disadvantages, and distinguish designs by the time sequence involved. We then describe in some detail the Framingham Heart Study, which is an observational study and one of the world's most important studies of risk factors for cardiovascular disease.[1] We then provide more detail on clinical trials, which are often considered the gold standard in terms of study design. At the end of this chapter, we summarize the issues in selecting the appropriate study design. Before describing the specific design types, we present some key vocabulary terms that are relevant to study design.

2.1 VOCABULARY

- **Bias**—A systematic error that introduces uncertainty in estimates of effect or association
- **Blind/double blind**—The state whereby a participant is unaware of his or her treatment status (e.g., experimental drug or placebo). A study is said to be double blind when both the participant and the outcome assessor are unaware of the treatment status (*masking* is used as an equivalent term to blinding).
- **Clinical trial**—A specific type of study involving human participants and randomization to the comparison groups
- **Cohort**—A group of participants who usually share some common characteristics and who are monitored or followed over time
- **Concurrent**—At the same time; optimally, comparison treatments are evaluated concurrently or in parallel
- **Confounding**—Complex relationships among variables that can distort relationships between the risk factors and the outcome
- **Cross-sectional**—At a single point in time
- **Incidence (of disease)**—The number of new cases (of disease) over a period of time
- **Intention-to-treat**—An analytic strategy whereby participants are analyzed in the treatment group they were assigned regardless of whether they followed the study procedures completely (e.g., regardless of whether they took all of the assigned medication)
- **Matching**—A process of organizing comparison groups by similar characteristics
- **Per protocol**—An analytic strategy whereby only participants who adhered to the study protocol (i.e., the specific procedures or treatments given to them) are analyzed (in other words, an analysis of only those assigned to a particular group who followed all procedures for that group)
- **Placebo**—An inert substance designed to look, feel, and taste like the active or experimental treatment (e.g., saline solution would be a suitable placebo for a clear, tasteless liquid medication)
- **Prevalence (of disease)**—The proportion of individuals with the condition (disease) at a single point in time
- **Prognostic factor**—A characteristic that is strongly associated with an outcome (e.g., disease) such that it could be used to reasonably predict whether a person is likely to develop a disease or not
- **Prospective**—A study in which information is collected looking forward in time
- **Protocol**—A step-by-step plan for a study that details every aspect of the study design and data collection plan
- **Quasi-experimental design**—A design in which subjects are not randomly assigned to treatments
- **Randomization**—A process by which participants are assigned to receive different treatments (this is usually based on a probability scheme)

- **Retrospective**—A study in which information is collected looking backward in time
- **Stratification**—A process whereby participants are partitioned or separated into mutually exclusive or non-overlapping groups

2.2 OBSERVATIONAL STUDY DESIGNS

There are a number of observational study designs. We describe some of the more popular designs, from the simplest to the more complex.

2.2.1 The Case Report/Case Series

A **case report** is a very detailed report of the specific features of a particular participant or case. A *case series* is a systematic review of the interesting and common features of a small collection, or series, of cases. These types of studies are important in the medical field as they have historically served to identify new diseases. The case series does not include a control or comparison group (e.g., a series of disease-free participants). These studies are relatively easy to conduct but can be criticized as they are unplanned, uncontrolled, and not designed to answer a specific research question. They are often used to generate specific hypotheses, which are then tested with other, larger studies. An example of an important case series was one published in 1981 by Gottlieb et al., who reported on five young homosexual men who sought medical care with a rare form of pneumonia and other unusual infections.[2] The initial report was followed by more series with similar presentations, and in 1982 the condition being described was termed Acquired Immune Deficiency Syndrome (AIDS).

2.2.2 The Cross-Sectional Survey

A **cross-sectional survey** is a study conducted at a single point in time. The cross-sectional survey is an appropriate design when the research question is focused on the prevalence of a disease, a present practice, or an opinion. The study is non-randomized and involves a group of participants who are identified at a point in time, and information is collected at that point in time. Cross-sectional surveys are useful for estimating the prevalence of specific risk factors or prevalence of disease at a point in time. In some instances, it is of interest to make comparisons between groups of participants (e.g., between men and women, between participants under age 40 and those 40 and older). However, inferences from the cross-sectional survey are limited to the time at which data are collected and do not generalize to future time points.

Cross-sectional surveys can be easy to conduct, are usually ethical, and are often large in size (i.e., involve many participants) to allow for estimates of risk factors, diseases, practices, or opinions in different subgroups of interest. However, a major limitation in cross-sectional surveys is the fact that both the exposure or development of a risk factor (e.g., hypertension) and the outcome have occurred. Because the study is conducted at a point in time (see **Figure 2–1**), it is not possible to assess temporal relationships, specifically whether the exposure or risk factor occurred prior to the outcome of interest. Another issue is related to non-response. While a large sample may be targeted, in some situations only a small fraction of participants approached agree to participate and complete the survey. Depending on the features of the participants and non-participants, non-response can introduce bias or limit generalizability.

In Figure 2–1, approximately one-third of the participants have the risk factor and two-thirds do not. Among those with the risk factor, almost half have the disease, as compared to a much smaller fraction of those without the risk factor. Is there an association between the risk factor and the disease?

2.2.3 The Cohort Study

A **cohort study** involves a group of individuals who usually meet a set of inclusion criteria at the start of the study. The cohort is followed and associations are made between a risk factor and a disease. For example, if we are studying risk factors for cardiovascular disease, we ideally enroll a cohort of individuals free of cardiovascular disease at the start of the study. In a prospective cohort study, participants are enrolled and followed going forward in time (see **Figure 2–2**). In some

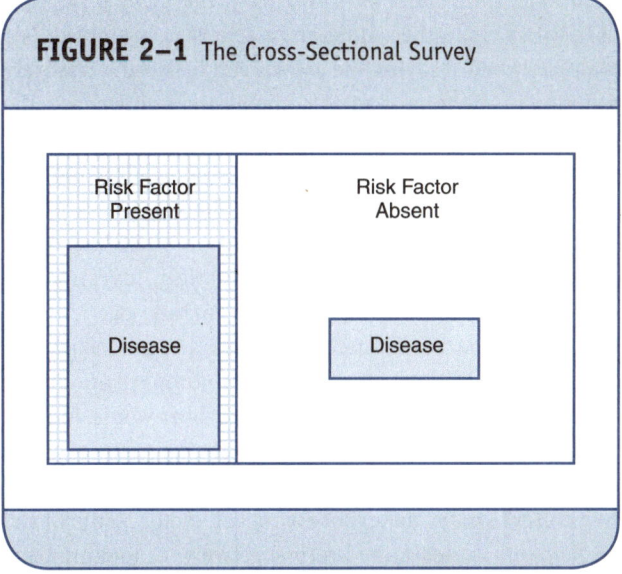

FIGURE 2–1 The Cross-Sectional Survey

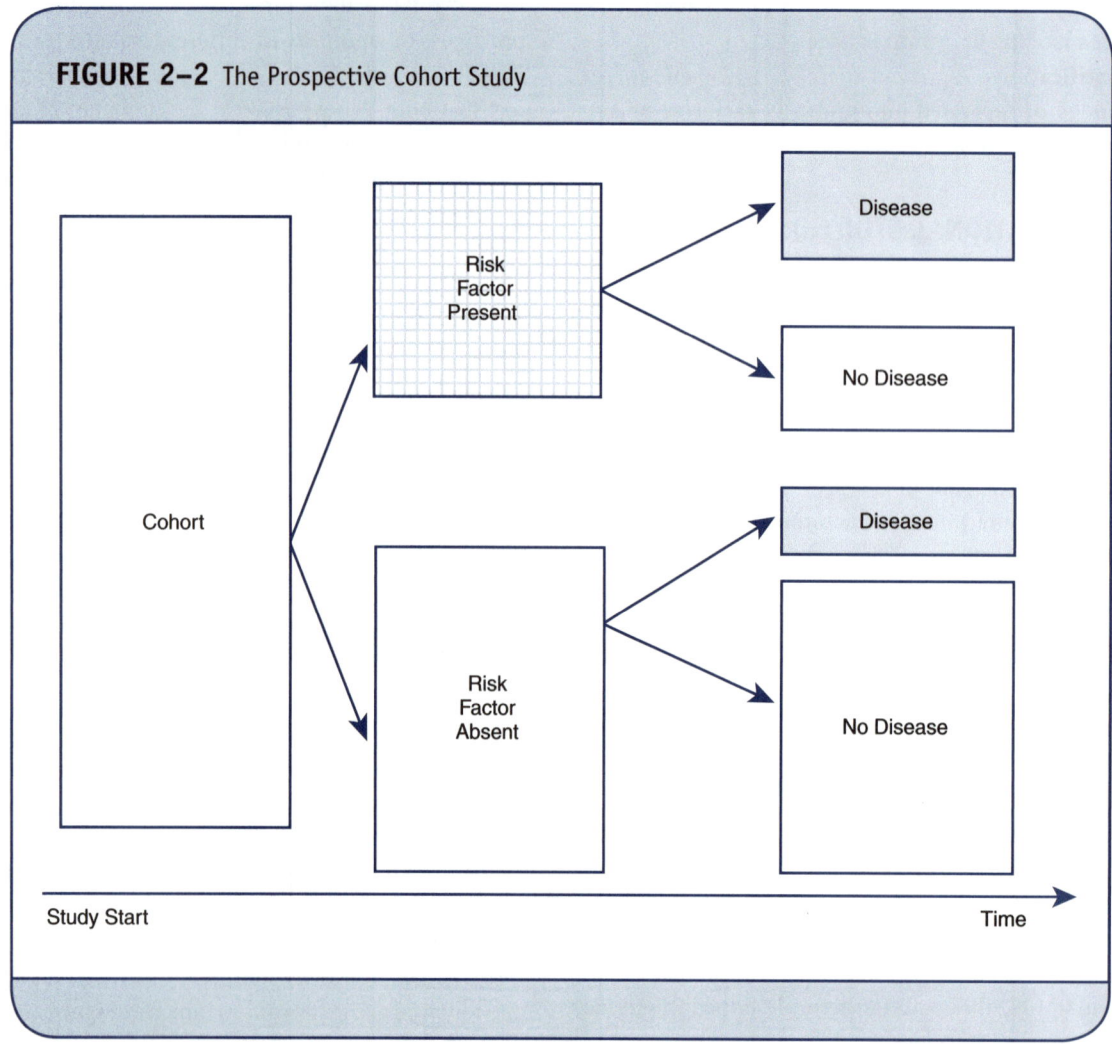

FIGURE 2-2 The Prospective Cohort Study

situations, the cohort is drawn from the general population, whereas in other situations a cohort is assembled. For example, when studying the association between a relatively common risk factor and an outcome, a cohort drawn from the general population will likely include sufficient numbers of individuals who have and do not have the risk factor of interest.

When studying the association between a rare risk factor and an outcome, special attention must be paid to constructing the cohort. In this situation, investigators might want to enrich the cohort to include participants with the risk factor (sometimes called a special exposure cohort). In addition, an appropriate comparison cohort would be included. The comparison cohort would include participants free of the risk factor but similar to the exposed cohort in other important characteristics. In a **retrospective cohort study,** the exposure or risk factor status of the participants is ascertained retrospectively, or looking back in time (see **Figure 2-3** and the time of study start). For example, suppose we wish to assess the association between multivitamin use and neural tube defects in newborns. We enroll a cohort of women who deliver live-born infants and ask each to report on their use of multivitamins before becoming pregnant. On the basis of these reports, we have an exposed and unexposed cohort. We then assess the outcome of pregnancy for each woman. Retrospective cohort studies are often based on data gathered from medical records where risk factors and outcomes have occurred and been documented. A study is mounted and records are reviewed to assess risk factor and outcome status, both of which have already occurred.

The prospective cohort study is the more common cohort study design. Cohort studies have a major advantage in that they allow investigators to assess temporal relationships. It is also possible to estimate the incidence of a disease (i.e., the rate at which participants who are free of a disease develop that disease). We can also compare incidence rates

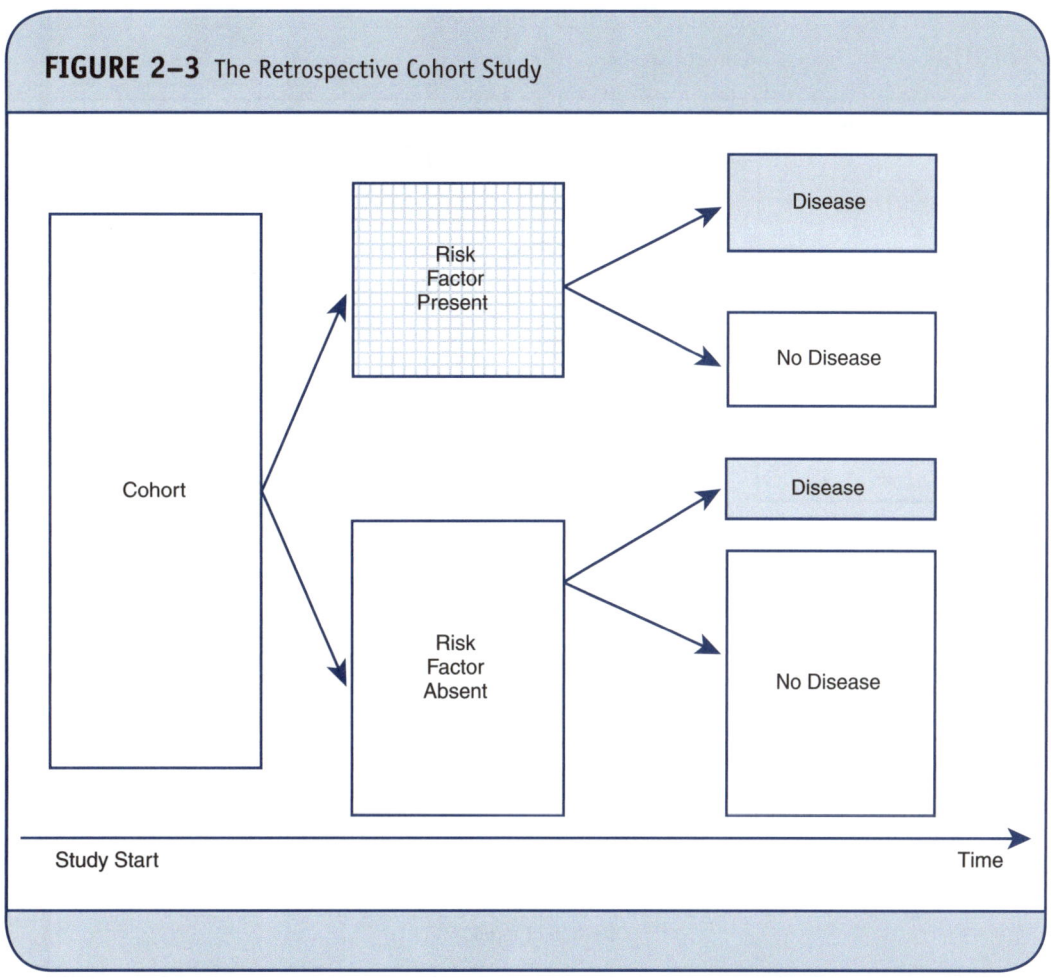

FIGURE 2-3 The Retrospective Cohort Study

between groups. For example, we might compare the incidence of cardiovascular disease between participants who smoke and participants who do not smoke as a means of quantifying the association between smoking and cardiovascular disease. Cohort studies can be difficult if the outcome or disease under study is rare or if there is a long latency period (i.e., it takes a long time for the disease to develop or be realized). When the disease is rare, the cohort must be sufficiently large so that adequate numbers of events (cases of disease) are observed. By "adequate numbers," we mean specifically that there are sufficient numbers of events to produce stable, precise inferences employing meaningful statistical analyses. When the disease under study has a long latency period, the study must be long enough in duration so that sufficient numbers of events are observed. However, this can introduce another difficulty, namely loss of participant follow-up over a longer study period.

Cohort studies can also be complicated by confounding. **Confounding** is a distortion of the effect of an exposure or risk factor on the outcome by other characteristics. For example, suppose we wish to assess the association between smoking and cardiovascular disease. We may find that smokers in our cohort are much more likely to develop cardiovascular disease. However, it may also be the case that the smokers are less likely to exercise, have higher cholesterol levels, and so on. These complex relationships among the variables must be reconciled by statistical analyses. In Chapter 9, we describe in detail the methods used to handle confounding.

2.2.4 The Case-Control Study

The **case-control study** is a study often used in epidemiologic research where again the question of interest is whether there is an association between a particular risk factor or exposure and an outcome. Case-control studies are particularly useful when the outcome of interest is rare. As noted previously, cohort studies are not efficient when the outcome of interest is rare as they require large numbers of participants to be enrolled in the study to realize a sufficient number of outcome events. In a case-control study, participants are identified on the basis of their outcome status. Specifically, we select a set of *cases*, or persons with the outcome of interest. We then select

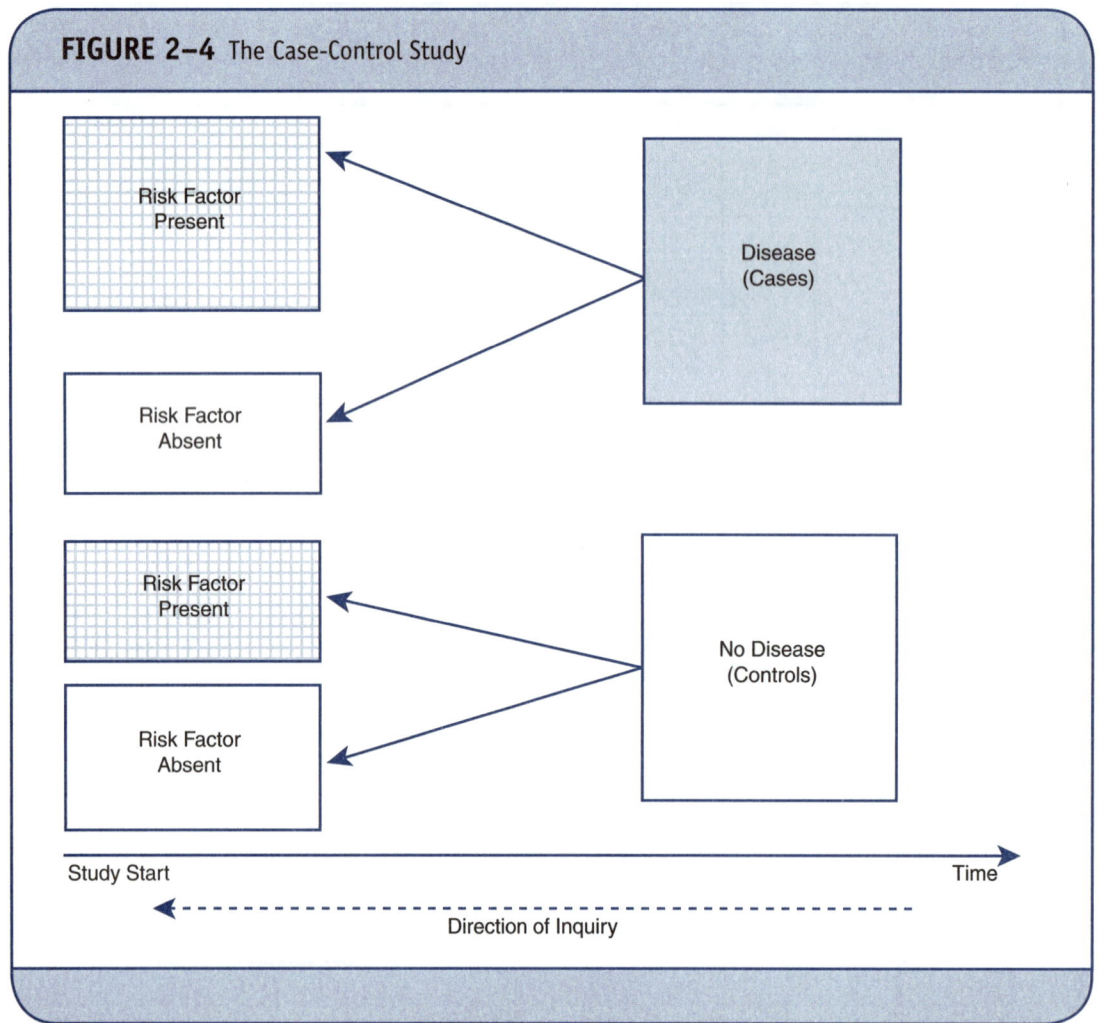

FIGURE 2-4 The Case-Control Study

a set of controls, who are persons similar to the cases except for the fact that they are free of the outcome of interest. We then assess exposure or risk factor status retrospectively (see **Figure 2-4**). We hypothesize that the exposure or risk factor is related to the disease and evaluate this by comparing the cases and controls with respect to the proportions that are exposed; that is, we draw inferences about the relationship between exposure or risk factor status and disease. There are a number of important issues that must be addressed in designing case-control studies. We detail some of the most important ones.

First, cases must be selected very carefully. An explicit definition is needed to identify cases so that the cases are as homogeneous as possible. The explicit definition of a case must be established before any participants are selected or data collected. Diagnostic tests to confirm disease status should be included whenever possible to minimize the possibility of incorrect classification.

Controls must also be selected carefully. The controls should be comparable to the cases in all respects except for the fact that they do not have the disease of interest. In fact, the controls should represent non-diseased participants who would have been included as cases if they had the disease. The same diagnostic tests used to confirm disease status in the cases should be applied to the controls to confirm non-disease status.

Usually, there are many more controls available for inclusion in a study than cases, so it is often possible to select several controls for each case, thereby increasing the sample size for analysis. Investigators have shown that taking more than four controls for each case does not substantially improve the precision of the analysis.[3] (This result will be discussed in subsequent chapters.) In many instances, two controls per case are selected, which is denoted as a 2:1 ("two to one") control to case ratio.

The next issue is to assess exposure or risk factor status, and this is done retrospectively. Because the exposure or risk factor might have occurred long ago, studies that can establish risk factor status based on documentation or records are preferred over those that rely on a participant's memory of past events. Sometimes, such data are not documented, so participants are queried with regard to risk factor status. This must be done in a careful and consistent manner for all participants, regardless of their outcome status—assessment of exposure or risk factor status must be performed according to the same procedures or protocol for cases and controls. In addition, the individual collecting exposure data should not be aware of the participant's outcome status (i.e., they should be blind to whether the participant is a case or a control).

Case-control studies have several positive features. They are cost- and time-efficient for studying rare diseases. With case-control studies, an investigator can ensure that a sufficient number of cases are included. Case-control studies are also efficient when studying diseases with long latency periods. Because the study starts after the disease has been diagnosed, investigators are not waiting for the disease to occur during the study period. Case-control studies are also useful when there are several potentially harmful exposures under consideration; data can be collected on each exposure and evaluated.

The challenges of the case-control study center mainly around bias. We discuss several of the more common sources of bias here; there are still other sources of bias to consider. **Misclassification bias** can be an issue in case-control studies and refers to the incorrect classification of outcome status (case or control) or the incorrect classification of exposure status. If misclassification occurs at random—meaning there is a similar extent of misclassification in both groups—then the association between the exposure and the outcome can be dampened (underestimated). If misclassification is not random—for example, if more cases are incorrectly classified as having the exposure or risk factor—then the association can be exaggerated (overestimated). Another source of bias is called selection bias, and it can result in a distortion of the association (over- or underestimation of the true association) between exposure and outcome status resulting from the selection of cases and controls. Specifically, the relationship between exposure status and disease may be different in those individuals who chose to participate in the study as compared to those who did not. Yet another source of bias is called **recall bias,** and again, it can result in a distortion of the association between exposure and outcome. It occurs when cases or controls differentially recall exposure status. It is possible that persons with a disease (cases) might be more likely to recall prior exposures than persons free of the disease. The latter might not recall the same information as readily. With case-control studies, it is also not always possible to establish a temporal relationship between exposure and outcome. For example, in the present example both the exposure and outcome are measured at the time of data collection. Finally, because of the way we select participants (on the basis of their outcome status) in case-control studies, we cannot estimate incidence (i.e., the rate at which a disease develops).

2.2.5 The Nested Case-Control Study

The **nested case-control study** is a specific type of case-control study that is usually designed from a cohort study. For example, suppose a cohort study involving 1000 participants is run to assess the relationship between smoking and cardiovascular disease. In the study, suppose that 20 participants develop myocardial infarction (MI, i.e., heart attack), and we are interested in assessing whether there is a relationship between body mass index (measured as the ratio of weight in kilograms to height in meters squared) and MI. With so few participants suffering this very specific outcome, it would be difficult analytically to assess the relationship between body mass index and MI because there are a number of confounding factors that would need to be taken into account. This process generally requires large samples (specifics are discussed in Chapter 9). A nested case-control study could be designed to select suitable controls for the 20 cases that are similar to the cases except that they are free of MI. To facilitate the analysis, we would carefully select the controls and might match the controls to cases on gender, age, and other risk factors known to affect MI, such as blood pressure and cholesterol. Matching is one way of handling confounding. The analysis would then focus specifically on the association between body mass index and MI.

Nested case-control studies are also used to assess new biomarkers (measures of biological processes) or to evaluate expensive tests or technologies. For example, suppose a large cohort study is run to assess risk factors for spontaneous preterm delivery. As part of the study, pregnant women provide demographic, medical, and behavioral information through self-administered questionnaires. In addition, each woman submits a blood sample at approximately 13 weeks gestation, and the samples are frozen and stored. Each woman is followed in the study through pregnancy outcome and is classified as having a spontaneous preterm delivery or not (e.g., induced preterm delivery, term delivery, etc.). A new test is developed to measure a hormone in the mother's blood that is hypothesized to be related to spontaneous preterm delivery. A nested case-control study is designed in which women who deliver prematurely and spontaneously (cases) are matched to women who do not (controls) on the basis of maternal age, race/ethnicity, and prior history of premature delivery. The hormone is measured in each case and control using the new test applied to the stored (unfrozen) serum samples. The analysis is focused on the association between hormone levels and spontaneous preterm delivery. In this situation the nested case-control study is an efficient way to evaluate whether the risk factor (i.e., hormone) is related to the outcome (i.e., spontaneous preterm delivery). The new test is applied to only those women who are selected into the nested case-control study and not to every woman enrolled in the cohort, thereby reducing cost.

2.3 RANDOMIZED STUDY DESIGNS

Cohort and case-control studies often address the question: Is there an association between a risk factor or exposure and an outcome (e.g., a disease)? Each of these observational study designs has its advantages and disadvantages. In the cohort studies, we compare incidence between the exposed and unexposed groups, whereas in the case-control study we compare exposure between those with and without a disease. These are different comparisons, but in both scenarios, we make inferences about associations. (In Chapter 6 and Chapter 7, we detail the statistical methods used to estimate associations and to make statistical inferences.) As we described, observational studies can be subject to bias and confounding. In contrast, randomized studies are considered to be the gold standard of study designs as they minimize bias and confounding. The key feature of randomized studies is the random assignment of participants to the comparison groups. In theory, randomizing makes the groups comparable in all respects except the way the participants are treated (e.g., treated with an experimental medication or a placebo, treated with a behavioral intervention or not). We describe two popular randomized designs in detail.

2.3.1 The Randomized Controlled Trial (RCT) or Clinical Trial

The **randomized controlled trial** (RCT) is a design with a key and distinguishing feature—the randomization of participants to one of several comparison treatments or groups. In pharmaceutical trials, there are often two comparison groups; one group gets an experimental drug and the other a control drug. If ethically feasible, the control might be a placebo. If a placebo is not ethically feasible (e.g., it is ethically inappropriate to use a placebo because participants need medication), then a medication currently available and considered the standard of care is an appropriate comparator. This is called an **active-controlled trial** as opposed to a **placebo-controlled trial**. In clinical trials, data are collected prospectively (see **Figure 2-5**).

The idea of randomization is to balance the groups in terms of known and unknown prognostic factors (i.e., characteristics that might affect the outcome), which minimizes confounding. Because of the randomization feature, the comparison groups—in theory—differ only in the treatment received. One group receives the experimental treatment and the other does not. With randomized studies, we can make much stronger inferences than we can with observational studies. Specifically, with clinical trials, inferences are made with regard to the effect of treatments on outcomes, whereas with observational studies, inferences are limited to associations between risk factors and outcomes.

It is important in clinical trials that the comparison treatments are evaluated concurrently. In the study depicted in Figure 2-5, the treatments are administered at the same point in time, generating parallel comparison groups. Consider a clinical trial evaluating an experimental treatment for allergies. If the experimental treatment is given during the spring and the control is administered during the winter, we might see very different results simply because allergies are highly dependent on the season or the time of year.

It is also important in clinical trials to include multiple study centers, often referred to as **multicenter trials**. The reason for including multiple centers is to promote generalizability. If a clinical trial is conducted in a single center and the experimental treatment is shown to be effective, there may be a question as to whether the same benefit would be seen in other centers. In multicenter trials, the homogeneity of the effect across centers can be analyzed directly.

Ideally, clinical trials should be double blind. Specifically, neither the investigator nor the participant should be aware of the treatment assignment. However, sometimes it is impossible or unethical to blind the participants. For example, consider a trial comparing a medical and a surgical procedure. In this situation, the participant would definitely know whether they underwent a surgical procedure. In some very rare situations, sham surgeries are performed, but these are highly unusual, as participant safety is always of the utmost concern. It is critical that the outcome assessor is blind to the treatment assignment.

There are many ways to randomize participants in clinical trials. Simple randomization involves essentially flipping a coin and assigning each participant to either the experimental or the control treatment on the basis of the coin toss. In multicenter trials, separate randomization schedules are usually developed for each center. This ensures a balance in the treatments

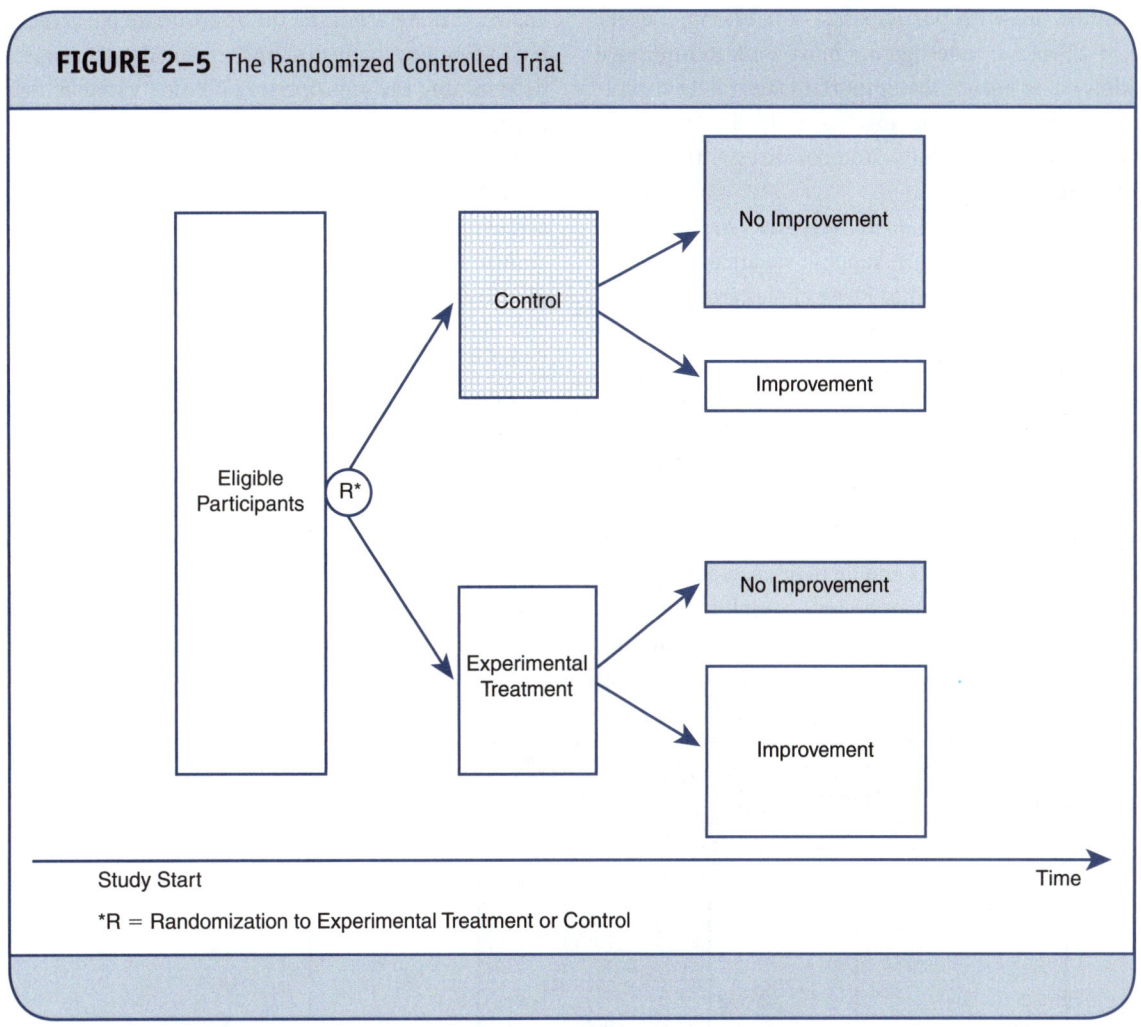

FIGURE 2-5 The Randomized Controlled Trial

*R = Randomization to Experimental Treatment or Control

within each center and does not allow for the possibility that all patients in one center get the same treatment. Sometimes it is important to minimize imbalance between groups with respect to other characteristics. For example, suppose we want to be sure we have participants of similar ages in each of the comparison groups. We could develop separate or stratified randomization schedules for participants less than 40 years of age and participants 40 years of age and older within each center. There are many ways to perform the randomization and the appropriate procedure depends on many factors, including the relationship between important prognostic factors and the outcome, the number of centers involved, and so on.

The major advantage of the clinical trial is that it is the cleanest design from an analytic point of view. Randomization minimizes bias and confounding so, theoretically, any benefit (or harm) that is observed can be attributed to the treatment. However, clinical trials are often expensive and very time-consuming. Clinical trials designed around outcomes that are relatively rare require large numbers of participants to demonstrate a significant effect. This increases the time and cost of conducting the trial. There are often a number of challenges in clinical trials that must be faced. First, clinical trials can be ethically challenging. Choosing the appropriate control group requires careful assessment of ethical issues. For example, in cancer trials it would never be possible to use a placebo comparator, as this would put participants at unnecessary risk. Next, clinical trials can be difficult to set up. Recruitment of centers and participants can be difficult. For example, participants might not be willing to participate in a trial because they cannot accept the possibility of being randomly assigned to the control group. Careful monitoring of participants is also a crucial aspect of clinical trials. For example, investigators must be sure that participants are taking the assigned drug as planned and are not taking other medications that might interfere with the study medications (called concomitant medications). Most clinical trials require frequent follow-up with participants—for example, every 2 weeks for 12 weeks. Investigators must work to minimize loss to follow-up to ensure that important study data are collected at every time point during the study. Subject retention and adherence to the study protocol are essential for the success of a clinical trial.

In some clinical trials, there are very strict inclusion and exclusion criteria. For example, suppose we are evaluating a new medication hypothesized to lower cholesterol. To allow the medication its best chance to demonstrate benefit, we might include only participants with very high total cholesterol levels. This means that inferences about the effect of the medication would then be limited to the population from which the participants were drawn. Clinical trials are sometimes criticized for being too narrow or restrictive. In designing trials, investigators must weigh the impact of the inclusion and exclusion criteria on the observed effects and on their generalizability.

Designing clinical trials can be very complex. There are a number of issues that need careful attention, including refining the study objective so that it is clear, concise, and answerable; determining the appropriate participants for the trial (detailing inclusion and exclusion criteria explicitly); determining the appropriate outcome variable; deciding on the appropriate control group; developing and implementing a strict monitoring plan; determining the number of participants to enroll; and detailing the randomization plan. While achieving these goals is challenging, a successful randomized clinical trial is considered the best means of establishing the effectiveness of a medical treatment.

2.3.2 The Crossover Trial

The **crossover trial** is a clinical trial where each participant is assigned to two or more treatments sequentially. When there are two treatments (e.g., an experimental and a control), each participant receives both treatments. For example, half of the participants are randomly assigned to receive the experimental treatment first and then the control; the other half receive the control first and then the experimental treatment. Outcomes are assessed following the administration of each treatment in each participant (see **Figure 2–6**). Participants receive the

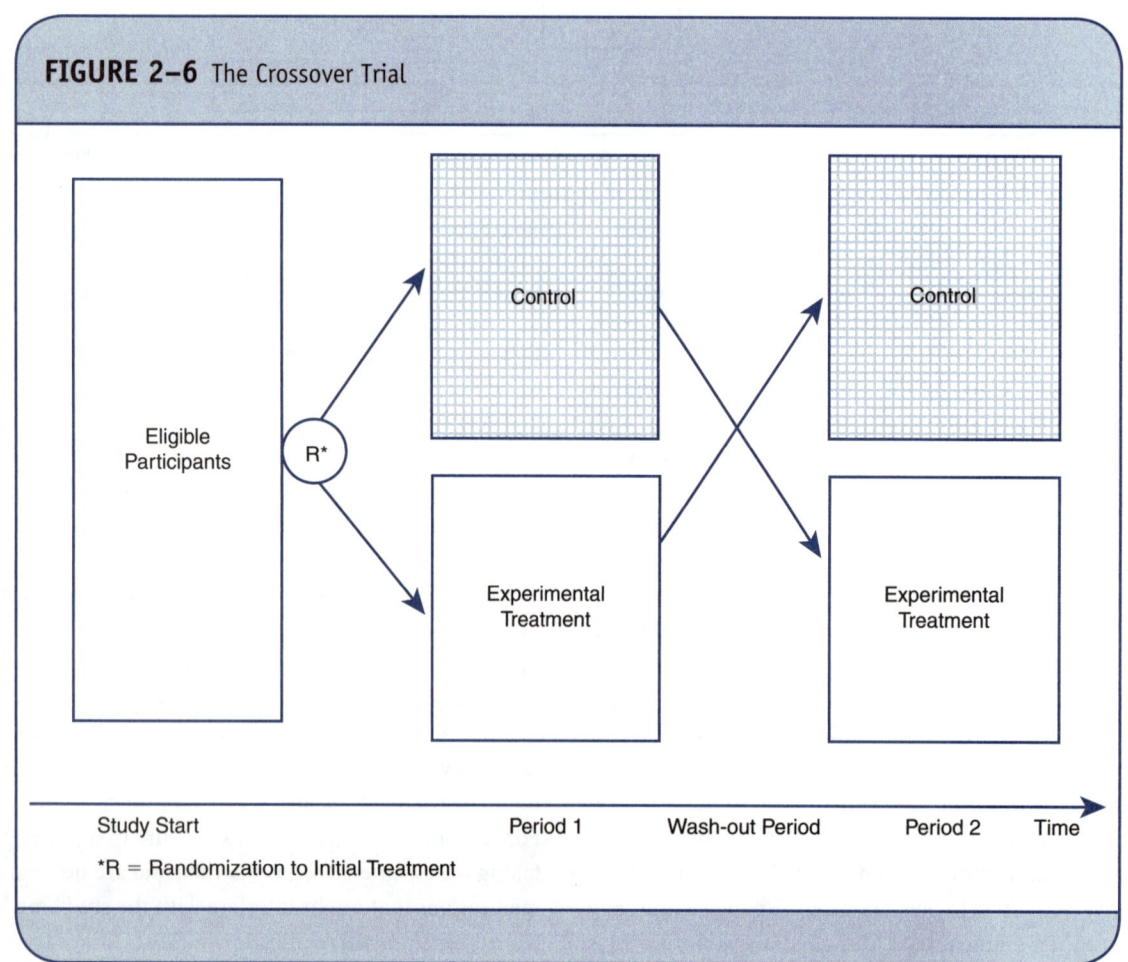

FIGURE 2–6 The Crossover Trial

*R = Randomization to Initial Treatment

randomly assigned treatment in Period 1. The outcome of interest is then recorded for the Period 1 treatment. In most crossover trials, there is then what is a called a **wash-out period** where no treatments are given. The wash-out period is included so that any therapeutic effects of the first treatment are removed prior to the administration of the second treatment in Period 2. In a trial with an experimental and a control treatment, participants who received the control treatment during Period 1 receive the experimental treatment in Period 2 and vice versa.

There are several ways in which participants can be assigned to treatments in a crossover trial. The two most popular schemes are called random and fixed assignment. In the random assignment scheme (already mentioned), participants are randomly assigned to the experimental treatment or the control in Period 1. Participants are then assigned the other treatment in Period 2. In a fixed assignment strategy, all participants are assigned the same treatment sequence. For example, everyone gets the experimental treatment first, followed by the control treatment or vice versa. There is an issue with the fixed scheme in that investigators must assume that the outcome observed on the second treatment (and subsequent treatments, if there are more than two) would be equivalent to the outcome that would be observed if that treatment were assigned first (i.e., that there are no carry-over effects). Randomly varying the order in which the treatments are given allows the investigators to assess whether there is any order effect.

The major advantage to the crossover trial is that each participant acts as his or her own control; therefore, we do not need to worry about the issue of treatment groups being comparable with respect to baseline characteristics. In this study design, fewer participants are required to demonstrate an effect. A disadvantage is that there may be carry-over effects such that the outcome assessed following the second treatment is affected by the first treatment. Investigators must be careful to include a wash-out period that is sufficiently long to minimize carry-over effects. A participant in Period 2 may not be at the same baseline as they were in Period 1, thus destroying the advantage of the crossover. In this situation, the only useful data may be from Period 1. The wash-out period must be short enough so that participants remain committed to completing the trial. Because participants in a crossover trial receive each treatment, loss to follow-up or dropout is critical because losing one participant means losing outcome data on both treatments.

Crossover trials are best suited for short-term treatments of chronic, relatively stable conditions. A crossover trial would not be efficient for diseases that have acute flare-ups because these could influence the outcomes that are observed yet have nothing to do with treatment. Crossover trials are also not suitable for studies with death or another serious condition considered as the outcome.

Similar to the clinical trial described previously, adherence or compliance to the study protocol and study medication in the crossover trial is critical. Participants are more likely to skip medication or drop out of a trial if the treatment is unpleasant or if the protocol is long or difficult to follow. Every effort must be made on the part of the investigators to maximize adherence and to minimize loss to follow-up.

2.4 THE FRAMINGHAM HEART STUDY

We now describe one of the world's most well-known studies of risk factors for cardiovascular disease. The Framingham Heart Study started in 1948 with the enrollment of a cohort of just over 5000 individuals free of cardiovascular disease who were living in the town of Framingham, Massachusetts.[1] The Framingham Heart Study is a longitudinal cohort study that involves repeated assessments of the participants approximately every 2 years. The study celebrated its fiftieth anniversary in 1998 and it still continues today. The original cohort has been assessed over 30 times. At each assessment, complete physical examinations are conducted (e.g., vital signs, blood pressure, medication history), blood samples are taken to measure lipid levels and novel risk factors, and participants also have echocardiograms in addition to other assessments of cardiovascular functioning. In the early 1970s, approximately 5000 offspring of the original cohort and their spouses were enrolled into what is called the Framingham Offspring cohort (the second generation of the original cohort). These participants have been followed approximately every 4 years and have been assessed over nine times. In the early 2000s, a third generation of over 4000 participants was enrolled and are being followed approximately every 4 years.

Over the past 50 years, hundreds of papers have been published from the Framingham Heart Study identifying important risk factors for cardiovascular disease, such as smoking, blood pressure, cholesterol, physical inactivity, and diabetes. The Framingham Heart Study also identified risk factors for stroke, heart failure, and peripheral artery disease. Researchers have identified psychosocial risk factors for heart disease, and now, with three generations of participants in the Framingham Study, investigators are assessing genetic risk factors for obesity, diabetes, and cardiovascular disease. More details on the Framingham Heart Study, its design, investigators, research milestones, and publications can be found at *http://www.nhlbi.nih.gov/about/framingham* and at *http://www.bu.edu/alumni/bostonia/2005/summer/pdfs/heart.pdf*.

2.5 MORE ON CLINICAL TRIALS

Clinical trials are extremely important, particularly in medical research. In Section 2.3, we outlined clinical trials from a design standpoint, but there are many more aspects of clinical trials that should be mentioned. First, clinical trials must be conducted at the correct time in the course of history. For example, suppose we ask the research question: Is the polio vaccine necessary today? To test this hypothesis, a clinical trial could be initiated in which some children receive the vaccine while others do not. The trial would not be feasible today because it would be unethical to withhold the vaccine from some children. No one would risk the consequences of the disease to study whether the vaccine is necessary.

As noted previously, the design of a clinical trial is extremely important to ensure the generalizability and validity of the results. Well-designed clinical trials are very easy to analyze, whereas poorly designed trials are extremely difficult, sometimes impossible, to analyze. The issues that must be considered in designing clinical trials are outlined here. Some have been previously identified but are worth repeating.

The number of treatments involved. If there are two treatments involved, statistical analyses are straightforward because only one comparison is necessary. If more than two treatments are involved, then more complicated statistical analyses are required and the issue of multiple comparisons must be addressed (these issues are discussed in Chapter 7 and Chapter 9). The number of treatments involved in a clinical trial should always be based on clinical criteria and not be reduced to simplify statistical analysis.

The control treatment. In clinical trials, an experimental (or newly developed) treatment is compared against a control treatment. The control treatment may be a treatment that is currently in use and considered the standard of care, or the control treatment may be a placebo. If a standard treatment exists, it should be used as the control because it would be unethical to offer patients a placebo when a conventional treatment is available. (While clinical trials are considered the gold standard design to evaluate the effectiveness of an experimental treatment, there are instances where a control group is not available. Techniques to evaluate effectiveness in the absence of a control group are described in D'Agostino and Kwan.[4])

Outcome measures. The outcome or outcomes of interest must be clearly identified in the design phase of the clinical trial. The primary outcome is the one specified in the planned analysis and is used to determine the sample size required for the trial (this is discussed in detail in Chapter 8). The primary outcome is usually more objective than subjective in nature. It is appropriate to specify secondary outcomes, and results based on secondary outcomes should be reported as such. Analyses of secondary outcomes can provide important information and, in some cases, enough evidence for a follow-up trial in which the secondary outcomes become the primary outcomes.

Blinding. Blinding refers to the fact that patients are not aware of which treatment (experimental or control) they are receiving in the clinical trial. A *single blind* trial is one in which the investigator knows which treatment a patient is receiving but the patient does not. *Double blinding* refers to the situation in which both the patient and the investigator are not aware of which treatment is assigned. In many clinical trials, only the statistician knows which treatment is assigned to each patient.

Single-center versus multicenter trials. Some clinical trials are conducted at a single site or clinical center, whereas others are conducted—usually simultaneously—at several centers. There are advantages to including several centers, such as increased generalizability and an increased number of available patients. There are also disadvantages to including multiple centers, such as needing more resources to manage the trial and the introduction of center-specific characteristics (e.g., expertise of personnel, availability or condition of medical equipment, specific characteristics of participants) that could affect the observed outcomes.

Randomization. Randomization is a critical component of clinical trials. There are a number of randomization strategies that might be implemented in a given trial. The exact strategy depends on the specific details of the study protocol.

Sample size. The number of patients required in a clinical trial depends on the variation in the primary outcome and the expected difference in outcomes between the treated and control patients.

Population and sampling. The study population should be explicitly defined by the study investigators (patient inclusion and exclusion criteria). A strategy for patient recruitment must be carefully determined and a system for checking inclusion and exclusion criteria for each potential enrollee must be developed and followed.

Ethics. Ethical issues often drive the design and conduct of clinical trials. There are some ethical issues that are common to all clinical trials, such as the safety of the treatments involved. There are other issues that relate only to certain trials. Most institutions have institutional review boards (IRBs) that are responsible for approving research study protocols. Research protocols are evaluated on the basis of scientific accuracy and with respect to potential risks and benefits to participants. All participants in clinical trials must provide informed consent, usually on consent forms approved by the appropriate IRB.

Protocols. Each clinical trial should have a protocol, which is a manual of operations or procedures in which every aspect of the trial is clearly defined. The protocol details all aspects of subject enrollment, treatment assignment, data collection, monitoring, data management, and statistical analysis. The protocol ensures consistency in the conduct of the trial and is particularly important when a trial is conducted at several clinical centers (i.e., in a multicenter trial).

Monitoring. Monitoring is a critical aspect of all clinical trials. Specifically, participants are monitored with regard to their adherence to all aspects of the study protocol (e.g., attending all scheduled visits, completing study assessments, taking the prescribed medications or treatments). Participants are also carefully monitored for any side effects or adverse events. Protocol violations (e.g., missing scheduled visits) are summarized at the completion of a trial, as are the frequencies of adverse events and side effects.

Data management. Data management is a critical part of any study and is particularly important in clinical trials. Data management includes tracking subjects (ensuring that subjects complete each aspect of the trial on time), data entry, quality control (examining data for out-of-range values or inconsistencies), data cleaning, and constructing analytic databases. In most studies, a data manager is assigned to supervise all aspects of data management.

The statistical analysis in a well-designed clinical trial is straightforward. Assuming there are two treatments involved (an experimental treatment and a control), there are essentially three phases of analysis:

- Baseline comparisons, in which the participants assigned to the experimental treatment group are compared to the patients assigned to the control group with respect to relevant characteristics measured at baseline. These analyses are used to check that the randomization is successful in generating balanced groups.
- Crude analysis, in which outcomes are compared between patients assigned to the experimental and control treatments. In the case of a continuous outcome (e.g., weight), the difference in means is estimated; in the case of a dichotomous outcome (e.g., development of disease or not), relative risks are estimated; and in the case of time-to-event data (e.g., time to a heart attack), survival curves are estimated. (The specifics of these analyses are discussed in detail in Chapters 6, 7, 10, and 11.)
- Adjusted analyses are then performed, similar to the crude analysis, which incorporate important covariates (i.e., variables that are associated with the outcome) and confounding variables. (The specifics of statistical adjustment are discussed in detail in Chapters 9 and 11.)

There are several analytic samples considered in statistical analysis of clinical trials data. The first is the Intent to Treat (ITT) analysis sample. It includes all patients who were randomized. The second is the Per Protocol analysis sample, and it includes only patients who completed the treatment (i.e., followed the treatment protocol as designed). The third is the Safety analysis sample, and it includes all patients who took at least one dose of the assigned treatment even if they did not complete the treatment protocol. All aspects of the design, conduct, and analysis of a clinical trial should be carefully documented. Complete and accurate records of the clinical trial are essential for applications to the Food and Drug Administration (FDA).[5]

Clinical trials are focused on safety and efficacy. Safety is assessed by the nature and extent of adverse events and side effects. Adverse events may or may not be due to the drug being evaluated. In most clinical trials, clinicians indicate whether the adverse event is likely due to the drug or not. Efficacy is assessed by improvements in symptoms or other aspects of the indication or disease that the drug is designed to address.

There are several important stages in clinical trials. Preclinical studies are studies of safety and efficacy in animals. Clinical studies are studies of safety and efficacy in humans. There are three phases of clinical studies, described here.

Phase I: First Time in Humans Study. The main objectives in a Phase I study are to assess the toxicology and safety of the proposed treatment in humans and to assess the pharmacokinetics (how fast the drug is absorbed in, flows through, and is secreted from the body) of the proposed treatment. Phase I studies are not generally focused on efficacy (how well the treatment works); instead, safety is the focus. Phase I studies usually involve 10 to 15 patients, and many Phase I studies are performed in healthy, normal volunteers to assess side effects and adverse events. In Phase I studies, one goal is to determine the maximum tolerated dose (MTD) of the proposed drug in humans. Investigators start with very low doses and work up to higher doses. Investigations usually start with three patients, and three patients are added for each elevated dose. Data are collected at each stage to assess safety, and some Phase I studies are placebo-controlled. Usually, two or three separate Phase I studies are conducted.

Phase II: Feasibility or Dose-Finding Study. The focus of a Phase II study is still on safety, but of primary interest are side effects and adverse events (which may or may not be directly related to the drug). Another objective in the Phase II study is efficacy, but the efficacy of the drug is based on descriptive analyses in the Phase II study. In some cases, investigators do not know which specific aspects of the indication or disease the drug may affect or which outcome measure best captures this effect. Usually, investigators measure an array of outcomes to determine the best outcome for the next phase. In Phase II studies, investigators determine the optimal dosage of the drug with respect to efficacy (e.g., lower doses might be just as effective as the MTD). Phase II studies usually involve 50 to 100 patients who have the indication or disease of interest. Phase II studies are usually placebo-controlled or compared to a standard, currently available treatment. Subjects are randomized and studies are generally double blind. If a Phase II study indicates that the drug is safe but not effective, investigation cycles back to Phase I. Most Phase II studies proceed to Phase III based on observed safety and efficacy.

Phase III: Confirmatory Clinical Trial. The focus of the Phase III trial is efficacy, although data are also collected to monitor safety. Phase III trials are designed and executed to confirm the effect of the experimental treatment. Phase III trials usually involve two treatment groups, an experimental treatment at the determined optimal dose and a placebo or standard of care. Some Phase III trials involve three groups: placebo, standard of care, and experimental treatment. Sample sizes can range from 200 to 500 patients, depending on what is determined to be a clinically significant effect. (The exact number is determined by specific calculations that are described in Chapter 8.) At least two successful clinical trials performed by independent investigators at different clinical centers are required in Phase III studies to assess whether the effect of the treatment can be replicated by independent investigators in at least two different sets of participants. More details on the design and analysis of clinical trials can be found in Chow and Liu.[6]

Investigators need positive results (statistically proven efficacy) in at least two separate trials to submit an FDA application for drug approval. The FDA also requires clinical significance in two trials, with clinical significance specified by clinical investigators in the design phase when the number of subjects is determined (see Chapter 8).

The FDA New Drug Application (NDA) contains a summary of results of Phase I, Phase II, and Phase III studies. The FDA reviews an NDA within 6 months to 1 year after submission and grants approval or not. If a drug is approved, the sponsor may conduct Phase IV trials, also called post-marketing trials, that can be retrospective (e.g., based on medical record review) or prospective (e.g., a clinical trial involving many patients to study rare adverse events). These studies are often undertaken to understand the long-term effects (efficacy and safety) of the drug.

2.6 SAMPLE SIZE IMPLICATIONS

Biostatisticians have a critical role in designing studies, not only to work with investigators to select the most efficient design to address the study hypotheses but also to determine the appropriate number of participants to involve in the study. In Chapter 8, we provide formulas to compute the sample sizes needed to appropriately answer research questions. The sample size needed depends on the study design, the anticipated association between the risk factor and outcome or the effect of the drug (e.g., the difference between the experimental and control drugs) and also on the statistical analysis that will be used to answer the study questions. The sample size should not be too small such that an answer about the association or the effect of the drug under investigation is not possible, because in this instance, both participants and the investigators have wasted time and money. Alternatively, a sample size should not be too large because again time and money would be wasted but, in addition, participants may be placed at unnecessary risk. Both scenarios are unacceptable from an ethical standpoint, and therefore careful attention must be paid when determining the appropriate sample size for any study or trial.

2.7 SUMMARY

To determine which study design is most efficient for a specific application, investigators must have a specific, clearly defined research question. It is also important to understand current knowledge or research on the topic under investigation. The most efficient design depends on the expected association or effect, the prevalence or incidence of outcomes, the prevalence of risk factors or exposures, and the expected duration of the study. Also important are practical issues, costs, and—most importantly—ethical issues.

Choosing the appropriate study design to address a research question is critical. Whenever possible, prior to mounting a planned study, investigators should try to run a pilot or feasibility study, which is a smaller-scale version of the planned study, as a means to identify potential problems and issues. Whereas pilot studies can be time-consuming and costly, they are usually more than worthwhile.

2.8 PRACTICE PROBLEMS

1. An investigator wants to assess whether smoking is a risk factor for pancreatic cancer. Electronic medical records at a local hospital will be used to identify 50 patients with pancreatic cancer. One hundred patients who are similar but free of pancreatic cancer will also be selected. Each participant's medical record will be analyzed for smoking history. Identify the type of study proposed and indicate its specific strengths and weaknesses.

2. What is the most likely source of bias in the study described in Problem 1?

3. An investigator wants to assess whether the use of a specific medication given to infants born prematurely is associated with developmental delay. Fifty infants who were given the medication and 50 comparison infants who were also born prematurely but not given the medication will be selected for the analysis. Each infant will undergo extensive testing at age 2 for various aspects of development. Identify the type of study proposed and indicate its specific strengths and weaknesses.

4. Is bias or confounding more of an issue in the study described in Problem 3? Give an example of a potential source of bias and a potential confounding factor.

5. A study is planned to assess the effect of a new surgical intervention for gallbladder disease. One hundred patients with gallbladder disease will be randomly assigned to receive either the new surgical intervention or the standard surgical intervention. The efficacy of the new surgical intervention will be measured by the time a patient takes to return to normal activities, recorded in days. Identify the type of study proposed and indicate its specific strengths and weaknesses.

6. An investigator wants to assess the association between caffeine consumption and impaired glucose tolerance, a precursor to diabetes. A study is planned to include 70 participants. Each participant will be surveyed with regard to their daily caffeine consumption. In addition, each participant will submit a blood sample that will be used to measure his or her glucose level. Identify the type of study proposed and indicate its specific strengths and weaknesses.

7. Could the study described in Problem 6 be designed as a randomized clinical trial? If so, briefly outline the study design; if not, describe the barriers.

8. A study is planned to compare two weight-loss programs in patients who are obese. The first program is based on restricted caloric intake and the second is based on specific food combinations. The study will involve 20 participants and each participant will follow each program. The programs will be assigned in random order (i.e., some participants will first follow the restricted-calorie diet and then follow the food-combination diet, whereas others will first follow the food-combination diet and then follow the restricted-calorie diet). The number of pounds lost will be compared between diets. Identify the type of study proposed and indicate its specific strengths and weaknesses.

9. An orthopedic surgeon observes that many of his patients coming in for total knee replacement surgery played organized sports before the age of 10. He plans to collect more extensive data on participation in organized sports from four patients undergoing knee replacement surgery and to report the findings. Identify the type of study proposed and indicate its specific strengths and weaknesses.

10. Suggest an alternative design to address the hypothesis in Problem 9. What are the major issues in addressing this hypothesis?

11. In 1940, 2000 women working in a factory were recruited into a study. Half of the women worked in manufacturing and half in administrative offices. The incidence of bone cancer through 1970 among the 1000 women working in manufacturing was compared with that of the 1000 women working in administrative offices. Thirty of the women in manufacturing developed bone cancer as compared to 9 of the women in administrative offices. This study is an example of a
    a. randomized controlled trial
    b. case-control study
    c. cohort study
    d. crossover trial

12. An investigator reviewed the medical records of 200 children seen for care at Boston Medical Center in the past year who were between the ages of 8 and 12 years old, and identified 40 with asthma. He also identified 40 children of the same ages who were free of asthma. Each child and his or her family were interviewed to assess whether there might be an association between certain environmental factors, such as exposure to second-hand smoke, and asthma. This study is an example of a
    a. randomized controlled trial
    b. case-control study
    c. cohort study
    d. crossover trial

13. A study is designed to evaluate the impact of a daily multivitamin on students' academic performance. One hundred sixty students are randomly assigned to receive either the multivitamin or a placebo and are instructed to take the assigned drug daily for 20 days. On day 20, each student takes a standardized exam and the mean exam scores are compared between groups. This study is an example of a
    a. randomized controlled trial
    b. case-control study
    c. cohort study
    d. crossover trial

14. A study is performed to assess whether there is an association between exposure to second-hand cigarette smoke in infancy and delayed development. Fifty children with delayed development and 50 children with normal development are selected for investigation. Parents are asked whether their children were exposed to second-hand cigarette smoke in infancy or not. This study is an example of a
    a. prospective cohort study
    b. retrospective cohort study
    c. case-control study
    d. clinical trial

15. A study is planned to investigate risk factors for sudden cardiac death. A cohort of men and women between the ages of 35 and 70 is enrolled and followed for up to 20 years. As part of the study, participants provide data on demographic and behavioral characteristics; they also undergo testing for cardiac function and provide blood samples to assess lipid profiles and other biomarkers. A new measure of inflammation is hypothesized to be related to sudden cardiac death. What study design is most appropriate to assess the association between the new biomarker and sudden cardiac death? Describe its strengths and weaknesses.

REFERENCES

1. D'Agostino, R.B. and Kannel, W.B. "Epidemiological background and design: The Framingham Study." *Proceedings of the American Statistical Association, Sesquicentennial Invited Paper Sessions*, 1989: 707–719.
2. Gottlieb, M.S., Schroff, R., Scganker, H.M., Weisman, J.D., Fan, P.T., Wolf, R.A., and Saxon, A. "*Pneumocystis carinii* pneumonia and *mucosal candidiasis* in previously healthy homosexual men: Evidence of a new acquired cellular immunodeficiency." *New England Journal of Medicine* 1981; 305(24): 1425–1431.
3. Schelesselman, J.J. *Case-Control Studies: Design, Conduct, Analysis*. New York: Oxford University Press, 1982.
4. D'Agostino, R.B. and Kwan, H. "Measuring effectiveness: What to expect without a randomized control group." *Medical Care* 1995; 33(4 Suppl.): AS95–105.
5. United States Food and Drug Administration. Available at *http://www.fda.gov*.
6. Chow, S.C. and Liu, J.P. *Design and Analysis of Clinical Trials: Concepts and Methodologies*. New York: John Wiley & Sons, 1998.

CHAPTER 3

Quantifying the Extent of Disease

When and Why

Key Questions

- What is a disease outbreak?
- How do you know when a disease outbreak is occurring?
- How do you judge whether one group is more at risk for disease than another?

In the News

Zika is a growing public health issue, made even bigger with the 2016 Summer Olympics taking place in Brazil and the world trying to understand the risks associated with Zika virus infection.

Vacationers are rethinking their travel destinations and are advised to consult resources such as the CDC Traveler's Health website for updates on the spread of Zika around the world.[1]

The link between Zika virus and microcephaly is under intense study, and new findings are being reported regularly.[2]

Some cities and towns are taking preventive action against mosquitos. For example, in New York, city and state governments are reportedly investing more than $21 million over a 3-year period as part of their Zika response plan.[3]

Public health agencies around the globe regularly update their statistics on Zika infection. For example, the World Health Organization puts out weekly situation reports that detail the latest data and statistics from around the world.[4]

---

[1]Centers for Disease Control and Prevention. Travelers' health. Available at https://wwwnc.cdc.gov/travel/page/zika-travel-information.
[2]Johansson, M.A., Mier-y-Teran-Romero, L., Reefhuis, J., Gilboa, S.M., and Hills, S.L. *New England Journal of Medicine* 2016; 375: 1.
[3]*WNYC News*. Available at http://www.wnyc.org/story/nyc-invests-millions-mosquito-vigilance-zika/?hootPostID=0d3dda8cc9db99e9c80e4ae269e0f495. Accessed July 8, 2016.
[4]World Health Organization. Zika virus situation reports. Available at http://www.who.int/emergencies/zika-virus/situation-report/en/.

Dig In

- Which approach would you take to quantify the prevalence of Zika virus infection in your local community? How do you actually test for Zika virus infection?
- How might you investigate whether certain groups are at increased risk for Zika infection?
- What would you recommend to a family member or friend planning a trip to Central America? What if that person was thinking about starting a family? Do you have any recommendations for your local community to minimize the risk of Zika infection?

Learning Objectives

By the end of this chapter, the reader will be able to

- Define and differentiate prevalence and incidence
- Select, compute, and interpret the appropriate measure to compare the extent of disease between groups
- Compare and contrast relative risks, risk differences, and odds ratios
- Compute and interpret relative risks, risk differences, and odds ratios

In Chapter 2, we presented several different study designs that are popular in public health research. In subsequent chapters, we discuss statistical procedures to analyze data collected under different study designs. In statistical analyses, we first describe information we collect in our study sample and then estimate or make generalizations about the population based on data observed in the sample. The first step is called **descriptive statistics** and the second is called inferential statistics. Our goal is to present techniques to describe samples and procedures for generating inferences that appropriately account for uncertainty in our estimates. Remember that we analyze only a fraction or subset, called a sample, of the entire population, and based on that sample we make inferences about the larger population. Before we get to those procedures, we focus on some important measures for quantifying disease. Two quantities that are often used in epidemiological and biostatistical analysis are prevalence and incidence. We describe each in turn and then discuss measures that are used to compare groups in terms of prevalence and incidence of risk factors and disease.

3.1 PREVALENCE

Prevalence refers to the proportion of participants with a risk factor or disease at a particular point in time. Consider the prospective cohort study we described in Chapter 2, where a cohort of participants is enrolled at a specific time. We call the initial point or starting point of the study the baseline time point. Suppose in our cohort study each individual undergoes a complete physical examination at baseline. At the baseline examination, we determine—among other things—whether each participant has a history of (i.e., has been previously diagnosed with) cardiovascular disease (CVD). An estimate of the prevalence of CVD is computed by taking the ratio of the number of existing cases of CVD to the total number of participants examined. This is called the point prevalence (PP) of CVD as it refers to the extent of disease at a specific point in time (i.e., at baseline in our example).

$$\text{Point prevalence} = \frac{\text{Number of persons with disease}}{\text{Number of persons examined at baseline}}$$

Example 3.1. The fifth examination of the offspring in the Framingham Heart Study was conducted between 1991 and 1995. A total of $n = 3799$ participants participated in the fifth examination. **Table 3-1** shows the numbers of men and women with diagnosed CVD at the fifth examination. The point prevalence of CVD among all participants attending the fifth examination of the Framingham Offspring Study is 379 / 3799 = 0.0998, or 9.98%. The point prevalence of CVD among men is 244 / 1792 = 0.1362, or 13.62%, and the point prevalence of CVD among women is 135 / 2007 = 0.0673, or 6.73%.

TABLE 3-1 Men and Women with Diagnosed CVD

	Free of CVD	History of CVD	Total
Men	1548	244	1792
Women	1872	135	2007
Total	3420	379	3799

TABLE 3-2 Smoking and Diagnosed CVD

	Free of CVD	History of CVD	Total
Nonsmoker	2757	298	3055
Current smoker	663	81	744
Total	3420	379	3799

Table 3-2 contains data on prevalent CVD among participants who were and were not currently smoking cigarettes at the time of the fifth examination of the Framingham Offspring Study. Almost 20% (744 / 3799) of the participants attending the fifth examination of the Framingham Offspring Study reported that they were current smokers at the time of the exam. The point prevalence of CVD among nonsmokers is 298 / 3055 = 0.0975, or 9.75%, and the point prevalence of CVD among current smokers is 81 / 744 = 0.1089, or 10.89%.

3.2 INCIDENCE

In epidemiological studies, we are often more concerned with estimating the likelihood of developing disease rather than the proportion of people who have disease at a point in time. The latter reflects prevalence, whereas **incidence** reflects the likelihood of developing a disease among a group of participants free of the disease who are considered at risk of developing the disease over a specified observation period. Consider the study described previously, and suppose we remove participants with a history of CVD from our fixed cohort at baseline so that only participants free of CVD are included (i.e., those who are truly "at risk" of developing a disease). We follow these participants prospectively for 10 years and record, for each individual, whether or not they develop CVD during this follow-up period. If we are able to follow each individual for 10 years and can ascertain whether or not each develops CVD, then we can directly compute the likelihood or risk of developing CVD over 10 years. Specifically, we take the ratio of the number of new cases of CVD to the total number of participants free of disease at the outset. This is referred to as **cumulative incidence** (CI):

Cumulative incidence =

$$\frac{\text{Number of persons who develop a disease during a specified period}}{\text{Number of persons at risk (at baseline)}}$$

Cumulative incidence reflects the proportion of participants who become diseased during a specified observation period. The total number of persons at risk is the same as the total number of persons included at baseline who are disease-free. The computation of cumulative incidence assumes that all of these individuals are followed for the entire observation period. This may be possible in some applications—for example, during an acute disease outbreak with a short follow-up or observation period. However, in longer studies it can be difficult to follow every individual for the development of disease because some individuals may relocate, may not respond to investigators, or may die during the study follow-up period. In this example, the cohort is older (as is the case in most studies of cardiovascular disease, as well as in studies of any other diseases that occur more frequently in older persons) and the follow-up period is long (10 years), making it difficult to follow every individual. The issues that arise and the methods to handle incomplete follow-up are described later.

3.2.1 Problems Estimating the Cumulative Incidence

There are a number of problems that can arise that make estimating the cumulative incidence of disease difficult. Because studies of incidence are by definition longitudinal (e.g., 5 or 10 years of follow-up), some study participants may be lost over the course of the follow-up period. Some participants might choose to drop out of the study, others might relocate, and others may die during the follow-up period. Different study designs could also allow for participants to enter at different times (i.e., all participants are not enrolled at baseline, but instead there is a rolling or prolonged enrollment period). For these and other reasons, participants are often followed for different lengths of time. We could restrict attention to only those participants who complete the entire follow-up; however, this would result in ignoring valuable information. A better approach involves accounting for the varying follow-up times as described here.

3.2.2 Person-Time Data

Again, in epidemiological studies we are generally interested in estimating the probability of developing disease (incidence) over a particular time period (e.g., 10 years). The cumulative incidence assumes that the total population at risk is followed for the entire observation period and that the disease status is ascertained for each member of the population. For the reasons stated previously, it is not always possible to follow each individual for the entire observation period.

Making use of the varying amounts of time that different participants contribute to the study results in changing the unit of analysis from the person or study participant to one of person-time, which is explicitly defined. The time unit might be months or years (e.g., person-months or person-years). For example, suppose that an individual enters a study in 1990 and is followed until 2000, at which point they are determined to be disease-free. A second individual enters the same study in 1995 and develops the disease under study in 2000. The first individual contributes 10 years of disease-free follow-up time, whereas the second individual contributes five years of disease-free follow-up time and then contracts the disease. We would want to use all of this information to estimate the incidence of the disease. Together, these two participants contribute 15 years of disease-free time.

3.2.3 Incidence Rate

The **incidence rate** uses all available information and is computed by taking the ratio of the number of new cases to the total follow-up time (i.e., the sum of all disease-free person-time). Rates are estimates that attempt to deal with the problem of varying follow-up times and reflect the likelihood or risk of an individual changing disease status (e.g., developing disease) in a specified unit of time. The denominator is the sum of all of the disease-free follow-up time, specifically time during which participants are considered at risk for developing the disease. Rates are based on a specific time period (e.g., 5 years, 10 years) and are usually expressed as an integer value per a multiple of participants over a specified time (e.g., the incidence of disease is 12 per 1000 person-years).

The incidence rate (IR), also called the incidence density (ID), is computed by taking the ratio of the number of new cases of disease to the total number of person-time units available. These person-time units may be person-years (e.g., one individual may contribute 10 years of follow-up, whereas another may contribute 5 years of follow-up) or person-months (e.g., 360 months, 60 months). The denominator is the sum of all of the participants' time at risk (i.e., disease-free time). The IR or ID is reported as a rate relative to a specific time interval (e.g., 5 per 1000 person-years). The incidence rate is given as follows:

Incidence rate = IR =

$$\frac{\text{Number of persons who develop disease during a specified period}}{\text{Sum of the lengths of time during which persons are disease-free}}$$

For presentation purposes, the incidence rate is usually multiplied by some multiple of 10 (e.g., 100, 1000, 10,000) to produce an integer value (see Example 3.2).

Example 3.2. Consider again the fifth examination of the offspring in the Framingham Heart Study. As described in Example 3.1, a total of $n = 3799$ participants attended the fifth examination, and 379 had a history of CVD. This leaves a total of $n = 3420$ participants free of CVD at the fifth examination. Suppose we follow each participant for the development

CHAPTER 3 Quantifying the Extent of Disease

TABLE 3-3 Men and Women Who Develop CVD

	No CVD	Develop CVD	Total
Men	1358	190	1548
Women	1753	119	1872
Total	3111	309	3420

TABLE 3-4 Total Disease-Free Time in Men and Women

	Develop CVD	Total Follow-Up Time (years)	IR
Men	190	9984	0.01903
Women	119	12,153	0.00979
Total	309	22,137	0.01396

of CVD over the next 10 years. **Table 3-3** shows the numbers of men and women who develop CVD over a 10-year follow-up period.

Because each participant is not followed for the full 10-year period (the mean follow-up time is 7 years), we cannot correctly estimate cumulative incidence using the previous data. The estimate of the cumulative incidence assumes that each of the 3111 persons free of CVD is followed for 10 years. Because this is not the case, our estimate of cumulative incidence is incorrect; instead, we must sum all of the available follow-up time and estimate an incidence rate. **Table 3-4** displays the total disease-free follow-up times for men and women along with the incidence rates. The incidence rates can be reported as 190 per 10,000 person-years for men and 98 per 10,000 person-years for women; equivalent to this is 19 per 1000 men per 10 years and 9.8 per 1000 women per 10 years.

The denominator of the incidence rate accumulates disease-free time over the entire observation period, and the unit of analysis is person-time (e.g., person-years in Example 3.2). In comparison, the denominator of the cumulative incidence is measured at the beginning of the study (baseline) and the unit of analysis is the person. It is worth noting that rates have dimension (number of new cases per person-time units) and are often confused with proportions (or probabilities) or percentages, which are dimensionless and

range from 0 to 1, 0% to 100%. Example 3.3 illustrates the difference between the prevalence, cumulative incidence, and incidence rate. It is important to note that the time component is an integral part of the denominator of the incidence rate, whereas with the cumulative incidence the time component is only part of the interpretation.

Example 3.3. Consider a prospective cohort study including six participants. Each participant is enrolled at baseline, and the goal of the study is to follow each participant for 10 years. Over the course of the follow-up period, some participants develop CVD, some drop out of the study, and some die. **Figure 3-1** displays the follow-up experiences for each participant. In this example, Participant 1 develops CVD 6 years into the study, Participant 2 dies 9 years into the study but is free of CVD, Participant 3 survives the complete follow-up period disease-free, Participant 4 develops CVD 2 years into the study and dies after 8 years, Participant 5 drops out of the study after 7 disease-free years, and Participant 6 develops CVD 5 years into the study.

Using the data in Example 3.3, we now compute prevalence, cumulative incidence, and incidence rate.

Prevalence of CVD at baseline = 0 /6 = 0, or 0%

Prevalence of CVD at 5 years = 2 / 6 = 0.333, or 33%

Prevalence of CVD at 10 years = 2 / 3 = 0.666, or 67%
(Note that we can only assess disease status at 10 years in Participants 1, 3, and 6.)

Cumulative incidence of CVD at 5 years = 2 / 6 = 0.333, or 33%

The cumulative incidence of CVD at 10 years cannot be estimated because we do not have complete follow-up on Participants 2, 4, or 5. To make use of all available information, we compute the incidence rate.

The incidence rate of CVD =
3 / (6 + 9 + 10 + 2 + 7 + 5) = 3 / 39 = 0.0769

We can report this as an incidence rate of CVD of 7.7 per 100 person-years.

The incidence rate of death per person-year =
2 / (10 + 9 + 10 + 8 + 7 + 10) = 2 / 54 = 0.037

We can report this as an incidence rate of death of 3.7 per 100 person-years.

Notice that the prevalence and cumulative incidence are shown as percentages (these can also be shown as proportions or probabilities), whereas the incidence rates are reported as the number of events per person-time.

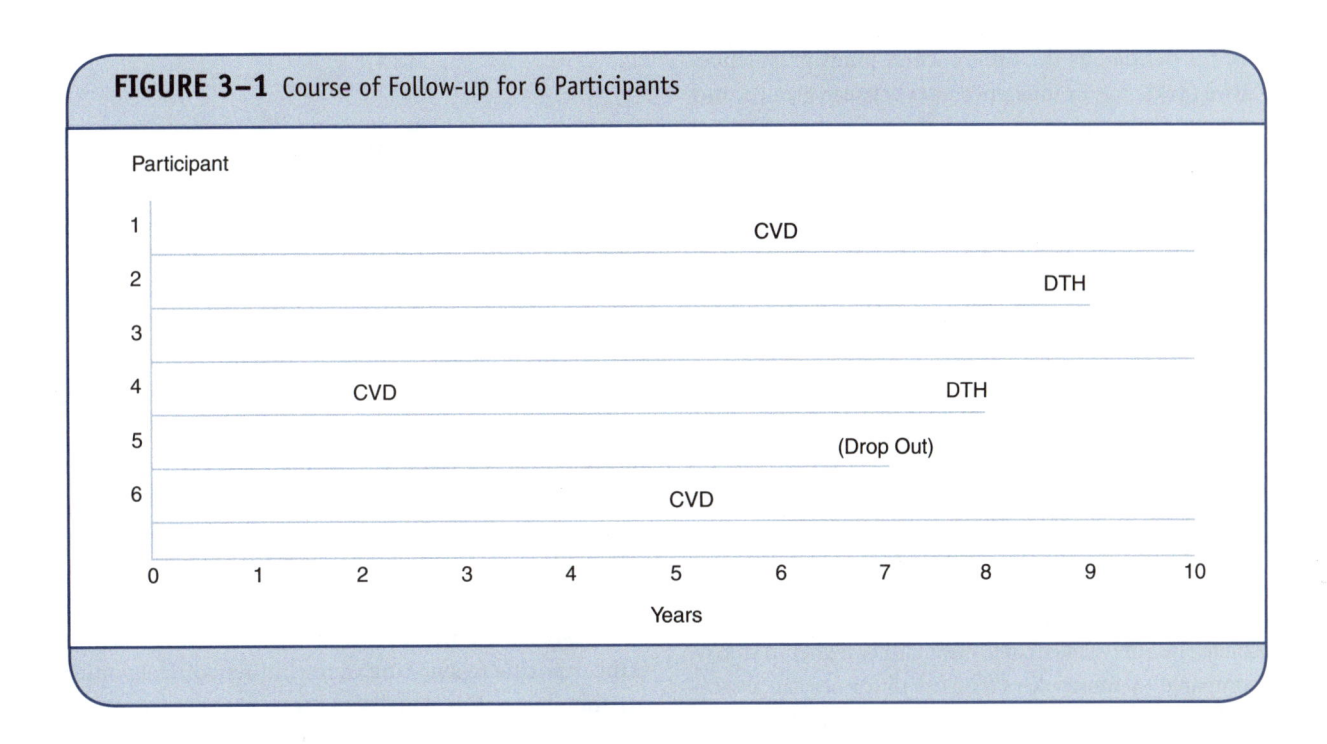

FIGURE 3–1 Course of Follow-up for 6 Participants

3.3 RELATIONSHIPS BETWEEN PREVALENCE AND INCIDENCE

The prevalence (proportion of the population with disease at a point in time) of a disease depends on the incidence (risk of developing disease within a specified time) of the disease as well as the duration of the disease. If the incidence is high but the duration is short, the prevalence (at a point in time) will be low by comparison. In contrast, if the incidence is low but the duration of the disease is long, the prevalence will be high. When the prevalence of disease is low (less than 1%), the prevalence is approximately equal to the product of the incidence and the mean duration of the disease (assuming that there have not been major changes in the course of the disease or its treatment).[1] Hypertension is an example of a disease with a relatively high prevalence and low incidence (due to its long duration). Influenza is an example of a condition with low prevalence and high incidence (due to its short duration).

The incidence rate is interpreted as the instantaneous potential to change the disease status (i.e., from non-diseased to diseased, also called the **hazard**) per unit time. An assumption that is implicit in the computation of the IR or ID is that the risk of disease is constant over time. This is not a valid assumption for many diseases because the risk of disease can vary over time and among certain subgroups (e.g., persons of different ages have different risks of disease). It is also very important that only persons who are at risk of developing the disease of interest are included in the denominator of the estimate of incidence. For example, if a disease is known to affect only people over 65 years of age, then including disease-free follow-up time measured on study participants less than 65 years of age will underestimate the true incidence. Person-times measured in these participants should not be included in the denominator as these participants are not truly at risk of developing the disease prior to age 65.

3.4 COMPARING THE EXTENT OF DISEASE BETWEEN GROUPS

It is often of interest to compare groups with respect to extent of disease or their likelihood of developing disease. These groups might be defined by an exposure to a potentially harmful agent (e.g., exposed or not), by a particular sociodemographic characteristic (e.g., men or women), or by a particular risk factor (e.g., current smoker or not). Popular comparative measures are generally categorized as difference measures or ratios. Difference measures are used to make absolute comparisons, whereas ratio measures are used to make relative comparisons. Differences or ratios can be constructed to compare prevalence measures or incidence measures between comparison groups.

3.4.1 Difference Measures: Risk Difference and Population Attributable Risk

The **risk difference** (RD), also called *excess risk*, measures the absolute effect of the exposure or the absolute effect of the risk factor of interest on prevalence or incidence. The risk difference is defined as the difference in point prevalence, cumulative incidence, or incidence rates between groups, and is given by the following:

$$\text{Risk difference} = RD = PP_{exposed} - PP_{unexposed}$$

$$RD = CI_{exposed} - CI_{unexposed}$$

$$RD = IR_{exposed} - IR_{unexposed}$$

In Example 3.1, we computed the point prevalence of CVD in smokers and nonsmokers. Exposed persons are those who reported smoking at the fifth examination of the Framingham Offspring Study. Using data from Example 3.1, the risk (prevalence) difference in CVD for smokers as compared to nonsmokers is computed by subtracting the point prevalence for nonsmokers from the point prevalence for smokers. The risk difference is 0.1089 − 0.0975 = 0.0114, and this indicates that the absolute risk (prevalence) of CVD is 0.0114 higher in smokers as compared to nonsmokers. The risk difference can also be computed by taking the difference in cumulative incidences or incidence rates between comparison groups, and the risk difference represents the excess risk associated with exposure to the risk factor. In Example 3.2 we estimated the incidence rates of CVD in men and women. Here the comparison groups are based on sex as opposed to exposure to a risk factor or not. Thus, we can compute the risk difference by either subtracting the incidence rate in men from the incidence rate in women or vice versa: the approach affects the interpretation. The incidence-rate difference between men and women, using data in Example 3.2, is 190/10,000 person-years in men − 98/10,000 person-years in women = 92/10,000 person-years. This indicates that there are 92 excess CVD events per 10,000 person-years in men as compared to women.

The range of possible values for the risk difference in point prevalence or cumulative incidence is −1 to 1. The range of possible values for the risk difference in incidence rates is − ∞ to ∞ events per person-time. The risk difference is positive when the risk for those exposed is greater than that for those unexposed. The risk difference is negative when the risk for those exposed is less than the risk for those unexposed. If exposure to the risk factor is unrelated to the risk of disease, then the risk difference is 0. A value of 0 is the null or no-difference value of the risk difference.

The **population attributable risk** (PAR) is another difference measure that quantifies the association between a risk factor and the prevalence or incidence of disease. The population attributable risk is computed as follows:

$$\text{Population attributable risk} = PAR = \frac{PP_{overall} - PP_{unexposed}}{PP_{overall}}$$

$$PAR = \frac{CI_{overall} - CI_{unexposed}}{CI_{overall}}$$

$$PAR = \frac{IR_{overall} - IR_{unexposed}}{IR_{overall}}$$

The population attributable risk is computed by first assessing the difference in overall risk (exposed and unexposed persons combined) and the risk of those unexposed. This difference is then divided by the overall risk and is usually presented as a percentage. Using data presented in Example 3.1, comparing prevalence of CVD in smokers and nonsmokers, the point prevalence of CVD for all participants attending the fifth examination of the Framingham Offspring Study is 379 / 3799 = 0.0998. The population attributable risk is computed as (0.0998 − 0.0975) / 0.0998 = 0.023 or 2.3% and suggests that 2.3% of the prevalent cases of CVD are attributable to smoking and could be eliminated if the exposure to smoking were eliminated. The population attributable risk is usually expressed as a percentage and ranges from 0% to 100%. The magnitude of the population attributable risk is interpreted as the percentage of risk (prevalence or incidence) associated with, or attributable to, the risk factor. If exposure to the risk factor is unrelated to the risk of disease, then the population attributable risk is 0% (i.e., none of the risk is associated with exposure to the risk factor). The population attributable risk assumes a causal relationship between the risk factor and disease and is also interpreted as the percentage of risk (prevalence or incidence) that could be eliminated if the exposure or risk factor were removed.

3.4.2 Ratios: Relative Risk, Odds Ratio, and Rate Ratio

The **relative risk** (RR), also called the risk ratio, is a useful measure to compare the prevalence or incidence of disease between two groups. It is computed by taking the ratio of the respective prevalences or cumulative incidences. Generally, the reference group (e.g., unexposed persons, persons without the risk factor, or persons assigned to the control group in a clinical trial setting) is considered in the denominator:

$$\text{Relative risk} = RR = \frac{PP_{exposed}}{PP_{unexposed}}$$

$$RR = \frac{CI_{exposed}}{CI_{unexposed}}$$

The ratio of incidence rates between two groups is called the **rate ratio** or the incidence density ratio.[2] Using data presented in Example 3.2, the rate ratio of incident CVD in men as compared to women is (190/10,000 person-years)/(98/10,000 person-years) = 1.94. Thus, the incidence of CVD is 1.94 times higher per person-year in men as compared to women.

The relative risk is often felt to be a better measure of the strength of the effect than the risk difference (or attributable risk) because it is relative to a baseline (or comparative) level. Using data presented in Example 3.1, the relative risk of CVD for smokers as compared to nonsmokers is 0.1089 / 0.0975 = 1.12; that is, the prevalence of CVD among smokers is 1.12 times that among nonsmokers. The range of the relative risk is 0 to ∞. If exposure to the risk factor is unrelated to the risk of disease, then the relative risk and the rate ratio will be 1. A value of 1 is considered the null or no-effect value of the relative risk or the rate ratio.

Under some study designs (e.g., the case-control study described in Chapter 2), it is not possible to compute a relative risk. Instead, an **odds ratio** is computed as a measure of effect. Suppose that in a case-control study we want to assess the relationship between exposure to a particular risk factor and disease status. Recall that in a case-control study, we select participants on the basis of their outcome—some have the condition of interest (cases) and some do not (controls).

Example 3.4. Table 3–5 shows the relationship between prevalent hypertension and prevalent cardiovascular disease at the fifth examination of the offspring in the Framingham Heart Study. The proportion of persons with hypertension who have CVD is 181 / 840 = 0.215. The proportion of persons free of hypertension but who have CVD is 188 / 2942 = 0.064. Odds are different from probabilities in that odds are computed as the ratio of the number of events to the number of nonevents, whereas a proportion is the ratio of the number of events to the total sample size. The odds that a person with hypertension has CVD are 181 / 659 = 0.275. The odds that a person free of hypertension has CVD are 188 / 2754 = 0.068. The relative risk of CVD for persons with as compared to without hypertension is 0.215 / 0.064 = 3.36,

or persons with hypertension are 3.36 times more likely to have prevalent CVD than persons without hypertension. The odds ratio is computed in a similar way but is based on the ratio of odds. The odds ratio is 0.275 / 0.068 = 4.04 and is interpreted as: People with hypertension have 4.04 times the odds of CVD compared to people free of hypertension.

Perhaps the most important characteristic of an odds ratio is its invariance property. Using the data in Table 3–5, the odds that a person with CVD has hypertension are 181 / 188 = 0.963. The odds that a person free of CVD has hypertension are 659 / 2754 = 0.239. The odds ratio for hypertension is therefore 0.963 / 0.239 = 4.04. People with CVD have 4.04 times the odds of hypertension compared to people free of CVD. This property does not hold for a relative risk. For example, the proportion of persons with CVD who have hypertension is 181 / 369 = 0.491. The proportion of persons free of CVD who have hypertension is 659 / 3413 = 0.193. The relative risk for hypertension is 0.491 / 0.193 = 2.54.

The invariance property of the odds ratio makes it an ideal measure of association for a case-control study. For example, suppose we conduct a case-control study to assess the association between cigarette smoking and a rare form of cancer (e.g., a cancer that is thought to occur in less than 1% of the general population). The cases are individuals with the rare form of cancer and the controls are similar to the cases but free of the rare cancer. Suppose we ask each participant whether he or she formerly or is currently smoking cigarettes or not. For this study, we consider former and current smokers as smokers. The data are shown in **Table 3–6**.

Using these data, we cannot calculate the incidence of cancer in the total sample or the incidence of cancer in smokers or in nonsmokers because of the way in which we collected the data. In this sample, 40 / 69 = 0.58 or 58% of the smokers have cancer and 10 / 31 = 0.32 or 32% of the nonsmokers have cancer—yet this is a rare cancer. These estimates do not reflect reality because the sample was

TABLE 3–5 Prevalent Hypertension and Prevalent CVD

	No CVD	CVD	Total
No hypertension	2754	188	2942
Hypertension	659	181	840
Total	3413	369	3782

TABLE 3–6 Smoking and Cancer

	Cancer (Case)	No Cancer (Control)	Total
Smoker	40	29	69
Nonsmoker	10	21	31
Total	50	50	100

specifically designed to include an equal number of cases and controls. Had we sampled individuals at random from the general population (using a cohort study design), we might have needed to sample more than 10,000 individuals to realize a sufficient number of cases for analysis. With this case-control study, we can estimate an association between smoking and cancer using the odds ratio. The odds of cancer in smokers are 40 / 29 = 1.379 and the odds of cancer in nonsmokers are 10 / 21 = 0.476. The odds ratio is 1.379 / 0.476 = 2.90, suggesting that smokers have 2.9 times the odds of cancer compared to nonsmokers. Note that this is equal to the odds ratio of smoking in cancer cases versus controls, i.e., (40 / 10) / (29 / 21) = 4 / 1.38 = 2.90.

The fact that we can estimate an odds ratio in a case-control study is a useful and important property. The odds ratio estimated in a study using a prospective sampling scheme (i.e., sampling representative groups of smokers and nonsmokers and monitoring for cancer incidence) is equivalent to the odds ratio based on a retrospective sampling scheme (i.e., sampling representative groups of cancer and patients free of cancer and recording smoking status).

The odds ratio can also be computed by taking the ratio of the point prevalence (PP) or cumulative incidence (CI) of disease to (1 − PP) or (1 − CI), respectively. The odds ratio is the ratio of the odds of developing disease for persons exposed as compared to those unexposed. Using cumulative incidences, the odds ratio is defined as:

$$\text{Odds ratio} = \frac{CI_{exposed} / (1 - CI_{exposed})}{CI_{unexposed} / (1 - CI_{unexposed})}$$

The odds ratio will approximate the relative risk when the disease under study is rare, usually defined as a prevalence or cumulative incidence less than 10%. For this reason, the interpretation of an odds ratio is often taken to be identical to that of a relative risk when the prevalence or cumulative incidence is low.

3.4.3 Issues with Person-Time Data

There are some special characteristics of person-time data that need attention, one of which is censoring. *Censoring* occurs when the event of interest (e.g., disease status) is not observed on every individual, usually due to time constraints (e.g., the study follow-up period ends, subjects are lost to follow-up, or they withdraw from the study). In epidemiological studies, the most common type of censoring that occurs is called **right censoring**. Suppose that we conduct a longitudinal study and monitor subjects prospectively over time for the development of CVD. For participants who develop CVD, their time to event is known; for the remaining subjects, all we know is that they did not develop the event during the study observation period. For these participants, their time-to-event (also called their survival time) is longer than the observation time. For analytic purposes, these times are censored, and are called Type I censored or right-censored times. Methods to handle survival time, also called time-to-event data, are discussed in detail in Allison[3] and in Chapter 11.

3.5 SUMMARY

Prevalence and incidence measures are important measures that quantify the extent of disease and the rate of development of disease in study populations. Understanding the difference between prevalence and incidence is critical. Prevalence refers to the extent of a disease at a point in time, whereas incidence refers to the development of disease over a specified time. Because it can be difficult to ascertain disease status in every participant in longitudinal studies—particularly when the follow-up period is long—measures that take into account all available data are needed. Incidence rates that account for varying follow-up times are useful measures in epidemiological analysis.

The formulas to estimate and compare prevalence and incidence are summarized in **Table 3–7**. In the next chapter, we present descriptive statistics. Specifically, we discuss how to estimate prevalence and incidence in study samples. We then move into statistical inference procedures, where we discuss estimating unknown population parameters based on sample statistics.

3.6 PRACTICE PROBLEMS

1. A cohort study is conducted to assess the association between clinical characteristics and the risk of stroke. The study involves $n = 1250$ participants who are free of stroke at the study start. Each participant is assessed at study start (baseline) and every year thereafter for 5 years. **Table 3–8** displays data on hypertension status measured at baseline and hypertension status measured 2 years later.
    a. Compute the prevalence of hypertension at baseline.
    b. Compute the prevalence of hypertension at 2 years.
    c. Compute the cumulative incidence of hypertension over 2 years.

TABLE 3–7 Summary of Key Formulas

Measure	Formula
Point prevalence (PP)*	$\dfrac{\text{Number of persons with disease}}{\text{Number of persons examined at baseline}}$
Cumulative incidence (CI)*	$\dfrac{\text{Number of persons who develop disease during a specified period}}{\text{Number of persons at risk (at baseline)}}$
Incidence rate (IR)	$\dfrac{\text{Number of persons who develop disease during a specified period}}{\text{Sum of the lengths of time during which persons are disease-free}}$
Risk difference (RD)	$PP_{exposed} - PP_{unexposed}$, $CI_{exposed} - CI_{unexposed}$, $IR_{exposed} - IR_{unexposed}$
Population attributable risk (PAR)	$\dfrac{PP_{overall} - PP_{unexposed}}{PP_{overall}}$, $\dfrac{CI_{overall} - CI_{unexposed}}{CI_{overall}}$, $\dfrac{IR_{overall} - IR_{unexposed}}{IR_{overall}}$
Relative risk (RR)	$\dfrac{PP_{exposed}}{PP_{unexposed}}$, $\dfrac{CI_{exposed}}{CI_{unexposed}}$
Odds ratio (OR)	$\dfrac{PP_{exposed}/(1-PP_{exposed})}{PP_{unexposed}/(1-PP_{unexposed})}$, $\dfrac{CI_{exposed}/(1-CI_{exposed})}{CI_{unexposed}/(1-CI_{unexposed})}$

* Can also be expressed as a percentage.

TABLE 3–8 Hypertension at Baseline and Two Years later

	Two Years Later: Not Hypertensive	Two Years Later: Hypertensive
Baseline: Not hypertensive	850	148
Baseline: hypertensive	45	207

TABLE 3–9 Hypertension at Baseline and Stroke Five Years Later

	Free of Stroke at Five Years	Stroke
Baseline: Not hypertensive	952	46
Baseline: hypertensive	234	18

2. The data shown in **Table 3-9** were collected in the study described in Problem 1 relating hypertensive status measured at baseline to incident stroke over 5 years.
    a. Compute the cumulative incidence of stroke in this study.
    b. Compute the cumulative incidence of stroke in patients classified as hypertensive at baseline.
    c. Compute the cumulative incidence of stroke in patients free of hypertension at baseline.
    d. Compute the risk difference of stroke in patients with hypertension as compared to patients free of hypertension.
    e. Compute the relative risk of stroke in patients with hypertension as compared to patients free of hypertension.
    f. Compute the population attributable risk of stroke due to hypertension.
3. A case-control study is conducted to assess the relationship between heavy alcohol use during

TABLE 3–10 Alcohol Use and Outcome of Pregnancy

	Miscarriage	Delivered Full Term
Heavy alcohol use	14	4
No heavy alcohol use	36	46

TABLE 3–11 Incident Coronary Artery Disease by Treatment

	Number of Participants	Number with Coronary Artery Disease
Cholesterol medication	400	28
Placebo	400	42

TABLE 3–12 Total Follow-Up Time by Treatment

	Number with Coronary Artery Disease	Total Follow-Up (years)
Cholesterol medication	28	3451
Placebo	42	2984

the first trimester of pregnancy and miscarriage. Fifty women who suffered miscarriage are enrolled, along with 50 who delivered full-term. Each participant's use of alcohol during pregnancy is ascertained. Heavy drinking is defined as four or more drinks on one occasion. The data are shown in **Table 3–10**.
   a. Compute the odds of miscarriage in women with heavy alcohol use during pregnancy.
   b. Compute the odds of miscarriage in women with no heavy alcohol use during pregnancy.
   c. Compute the odds ratio for miscarriage as a function of heavy alcohol use.

4. A randomized trial is conducted to evaluate the efficacy of a new cholesterol-lowering medication. The primary outcome is incident coronary artery disease. Participants are free of coronary artery disease at the start of the study and randomized to receive either the new medication or a placebo. Participants are followed for a maximum of 10 years for the development of coronary artery disease. The observed data are shown in **Table 3–11**.
   a. Compute the relative risk of coronary artery disease in patients receiving the new cholesterol medication as compared to those receiving a placebo.
   b. Compute the odds ratio of coronary artery disease in patients receiving the new cholesterol medication as compared to those receiving a placebo.
   c. Which measure is more appropriate in this design, the relative risk or odds ratio? Justify briefly.

5. In the study described in Problem 4, some patients were not followed for a total of 10 years. Some suffered events (i.e., developed coronary artery disease during the course of follow-up), whereas others dropped out of the study. **Table 3–12** displays the total number of person-years of follow-up in each group.
   a. Compute the incidence rate of coronary artery disease in patients receiving the new cholesterol medication.
   b. Compute the incidence rate of coronary artery disease in patients receiving a placebo.

6. A small cohort study is conducted in 13 patients with an aggressive cellular disorder linked to cancer. The clinical courses of the patients are depicted graphically in **Figure 3–2**.
   a. Compute the prevalence of cancer at 12 months.
   b. Compute the cumulative incidence of cancer at 12 months.
   c. Compute the incidence rate (per month) of cancer.
   d. Compute the incidence rate (per month) of death.

7. Five hundred people are enrolled in a 10-year cohort study. At the start of the study, 50 have diagnosed CVD. Over the course of the study, 40 people who were free of CVD at baseline develop CVD.
   a. What is the cumulative incidence of CVD over 10 years?
   b. What is the prevalence of CVD at baseline?
   c. What is the prevalence of CVD at 10 years?

8. A total of 150 participants are selected for a study of risk factors for cardiovascular disease. At baseline (study start), 24 are classified as hypertensive. At 1 year, an additional 12 have developed hypertension,

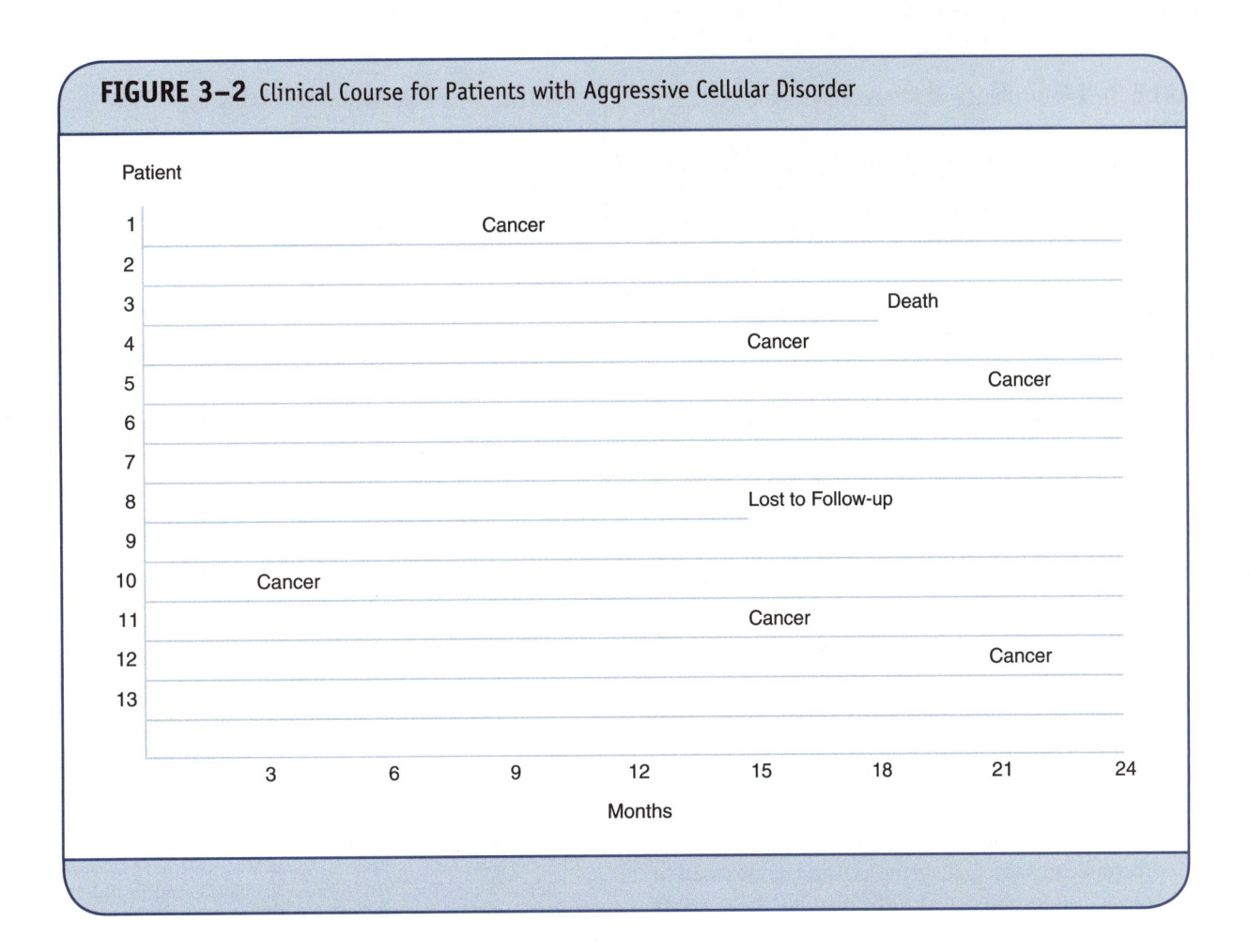

FIGURE 3–2 Clinical Course for Patients with Aggressive Cellular Disorder

```
Patient
 1                                  Cancer
 2
 3                                                      Death
 4                                          Cancer
 5                                                          Cancer
 6
 7
 8                                          Lost to Follow-up
 9
10          Cancer
11                                              Cancer
12                                                          Cancer
13
       3      6      9      12     15     18     21     24
                            Months
```

and at 2 years another 8 have developed hypertension. What is the prevalence of hypertension at 2 years in the study?

9. Consider the study described in Problem 8. What is the 2-year cumulative incidence of hypertension?

10. A national survey is conducted to assess the association between hypertension and stroke in persons over 75 years of age with a family history of stroke. Development of stroke is monitored over a 5-year follow-up period. The data are summarized in **Table 3–13** and the numbers are in millions.
    a. Compute the cumulative incidence of stroke in persons over 75 years of age.
    b. Compute the relative risk of stroke in hypertensive as compared to non-hypertensive persons.
    c. Compute the odds ratio of stroke in hypertensive as compared to non-hypertensive persons.

11. In a nursing home, a program is launched in 2005 to assess the extent to which its residents are affected by diabetes. Each resident has a blood test, and 48 of the 625 residents have diabetes in 2005. Residents who did not already have diabetes were again tested in 2010, and 57 residents had diabetes.
    a. What is the prevalence of diabetes in 2005?
    b. What is the cumulative incidence of diabetes over 5 years?
    c. What is the prevalence of diabetes in 2010 (assume that none of the residents in 2005 have died or left the nursing home)?

TABLE 3–13 Hypertension and Development of Stroke

	Developed Stroke	Did Not Develop Stroke
Hypertension	12	37
No hypertension	4	26

TABLE 3–14 Incidence of Stroke in Men and Women

	Number of Strokes	Number of Stroke-Free Person-Years
Men ($n = 125$)	9	478
Women ($n = 200$)	21	97

TABLE 3–15 Hypertension Status by Treatment

Group	Number Randomized	Number Free of Hypertension at 12 Weeks
Placebo	50	6
New drug	50	14

12. A prospective cohort study is run to estimate the incidence of stroke in persons 55 years of age and older. All participants are free of stroke at study start. Each participant is followed for a maximum of 5 years. The data are summarized in **Table 3–14**.
    a. What is the annual incidence rate of stroke in men?
    b. What is the annual incidence rate of stroke in women?
    c. What is the annual incidence rate of stroke (men and women combined)?

13. A clinical trial is run to assess the efficacy of a new drug to reduce high blood pressure. Patients with a diagnosis of hypertension (high blood pressure) are recruited to participate in the trial and randomized to receive either the new drug or placebo. Participants take the assigned drug for 12 weeks and their blood pressure status is recorded. At the end of the trial, participants are classified as still having hypertension or not. The data are shown in **Table 3–15**.
    a. What is the prevalence of hypertension at the start of the trial?
    b. What is the prevalence of hypertension at the end of the trial?
    c. Estimate the relative risk comparing the proportions of patients who are free of hypertension at 12 weeks between groups.

REFERENCES

1. Hennekens, C.H. and Buring, J.E. *Epidemiology in Medicine*. Philadelphia: Lippincott Williams & Wilkins, 1987.
2. Kleinbaum, D.G., Kupper, L.L., and Morgenstern, H. *Epidemiologic Research*. New York: Van Nostrand Reinhold Company Inc., 1982.
3. Allison, P. *Survival Analysis Using SAS: A Practical Guide*. Cary, NC: SAS Institute, 1995.

CHAPTER 4

Summarizing Data Collected in the Sample

When and Why

Key Questions

- What is the best way to make a case for action using data?
- Are investigators being deceptive or just confusing when they report relative differences instead of absolute differences?
- How can we be sure that we are comparing like statistics (apples to apples) when we attempt to synthesize data from various sources?

In the News

Summary statistics on key indicators in different groups and over time can make powerful statements. Simple tables or graphical displays of means, counts, or rates can shine a light on an issue that might be otherwise ignored. Some examples of current issues and a few key statistics are outlined here.

As of 2014, more than 21 million Americans 12 years and older had a substance use disorder. Of these disorders, nearly 2 million involved prescription painkillers and more than half a million involved heroin.[1]

The National Institute on Drug Abuse reports a 2.8-fold increase in overdose deaths in the United States from prescription drugs from 2001 to 2014, a 3.4-fold increase in deaths from opioid pain relievers, and a 6-fold increase in deaths due to heroin over the same period.[2]

Dig In

- How would you summarize the extent of prescription drug use in your community?
- What would you measure and how? What are the challenges in collecting these data?
- If you were to compare the extent of prescription drug use in your community with that in another community, how could you ensure that the data are comparable?

---

[1] American Society of Addiction Medicine. Opioid addiction 2016 facts and figures. Available at http://www.asam.org/docs/default-source/advocacy/opioid-addiction-disease-facts-figures.pdf. Accessed July 10, 2016.
[2] National Institute on Drug Abuse. Overdose death rates. Available at https://www.drugabuse.gov/related-topics/trends-statistics/overdose-death-rates.

Learning Objectives

By the end of this chapter, the reader will be able to

- Distinguish between dichotomous, ordinal, categorical, and continuous variables
- Identify appropriate numerical and graphical summaries for each variable type
- Compute a mean, median, standard deviation, quartiles, and range for a continuous variable
- Construct a frequency distribution table for dichotomous, categorical, and ordinal variables
- Provide an example of when the mean is a better measure of location than the median
- Interpret the standard deviation of a continuous variable
- Generate and interpret a box plot for a continuous variable
- Produce and interpret side-by-side box plots
- Differentiate between a histogram and a bar chart

Before any biostatistical analyses are performed, we must define the population of interest explicitly. The composition of the population depends on the investigator's research question. It is important to define the population explicitly as inferences based on the study sample will only be generalizable to the specified population. The population is the collection of all individuals about whom we wish to make generalizations. For example, if we wish to assess the prevalence of cardiovascular disease (CVD) among all adults 30 to 75 years of age living in the United States, then all adults in that age range living in the United States at the specified time of the study constitute the population of interest. If we wish to assess the prevalence of CVD among all adults 30 to 75 years of age living in the state of Massachusetts, then all adults in that age range living in Massachusetts at the specified time of the study constitute the population of interest. If we wish to assess the prevalence of CVD among all adults 30 to 75 years of age living in the city of Boston, then all adults in that age range living in Boston at the specified time of the study constitute the population of interest.

In most applications, the population is so large that it is impractical to study the entire population. Instead, we select a sample (a subset) from the population and make inferences about the population based on the results of an analysis on the sample. The **sample** is a subset of individuals from the population. Ideally, individuals are selected from the population into the sample at random. (We discuss this procedure and other concepts related to sampling in detail in Chapter 5.)

There are a number of techniques that can be used to select a sample. Regardless of the specific techniques used, the sample should be representative of the population (i.e., the characteristics of individuals in the sample should be similar to those in the population). By definition, the number of individuals in the sample is smaller than the number of individuals in the population. There are formulas to determine the appropriate number of individuals to include in the sample that depend on the characteristic being measured (i.e., exposure, risk factor, and outcome) and the desired level of precision in the estimate. We present details about sample size computations in Chapter 8.

Once a sample is selected, the characteristic of interest must be summarized in the sample using appropriate techniques. This is the first step in an analysis. Once the sample is appropriately summarized, statistical inference procedures are then used to generate inferences about the population based on the sample. We discuss statistical inference procedures in Chapters 6, 7, 9, 10, and 11.

In this chapter, we present techniques to summarize data collected in a sample. The appropriate numerical summaries and graphical displays depend on the type of characteristic under study. Characteristics—sometimes called variables, outcomes, or endpoints—are classified as one of the following types: dichotomous, ordinal, categorical, or continuous.

Dichotomous variables have only two possible responses. The response options are usually coded "yes" or "no." Exposure to a particular risk factor (e.g., smoking) is an example of a dichotomous variable. Prevalent disease status is another example of a dichotomous variable, where each individual in a sample is classified as having or not having the disease of interest at a point in time.

Ordinal and categorical variables have more than two possible responses but the response options are ordered and unordered, respectively. Symptom severity is an example of an **ordinal variable** with possible responses of minimal,

TABLE 4–1 Blood Pressure Categories

Classification of Blood Pressure	SBP and/or DBP
Normal	<120 and <80
Pre-hypertension	120–139 or 80–89
Stage I hypertension	140–159 or 90–99
Stage II hypertension	>160 and >100

moderate, and severe. The National Heart, Lung, and Blood Institute (NHLBI) issues guidelines to classify blood pressure as normal, pre-hypertension, Stage I hypertension, or Stage II hypertension.[1] The classification scheme is shown in **Table 4–1** and is based on specific levels of systolic blood pressure (SBP) and diastolic blood pressure (DBP). Participants are classified into the highest category, as defined by their SBP and DBP. Blood pressure category is an ordinal variable.

Categorical variables, sometimes called nominal variables, are similar to ordinal variables except that the responses are unordered. Race/ethnicity is an example of a categorical variable. It is often measured using the following response options: white, black, Hispanic, American Indian or Alaskan native, Asian or Pacific Islander, or other. Another example of a categorical variable is blood type, with response options A, B, AB, and O.

Continuous variables, sometimes called quantitative or measurement variables, in theory take on an unlimited number of responses between defined minimum and maximum values. Systolic blood pressure, diastolic blood pressure, total cholesterol level, CD4 cell count, platelet count, age, height, and weight are all examples of continuous variables. For example, systolic blood pressure is measured in millimeters of mercury (mmHg), and an individual in a study could have a systolic blood pressure of 120, 120.2, or 120.23, depending on the precision of the instrument used to measure systolic blood pressure. In Chapter 11 we present statistical techniques for a specific continuous variable that measures time to an event of interest, for example time to development of heart disease, cancer, or death.

Almost all numerical summary measures depend on the specific type of variable under consideration. One exception is the sample size, which is an important summary measure for any variable type (dichotomous, ordinal, categorical, or continuous). The sample size, denoted as n, reflects the number of independent or distinct units (participants) in the sample. For example, if a study is conducted to assess the total cholesterol in a population and a random sample of 100 individuals is selected for participation, then $n = 100$ (assuming all individuals selected agree to participate). In some applications, the unit of analysis is not an individual participant but might be a blood sample or specimen.

Suppose in the example study that each of the 100 participants provides blood samples for cholesterol testing at three distinct points in time (e.g., at the start of the study, and 6 and 12 months later). The unit of analysis could be the blood sample, in which case the sample size would be $n = 300$. It is important to note that these 300 blood samples are not 300 independent or unrelated observations because multiple blood samples are taken from each participant. Multiple measurements taken on the same individual are referred to as clustered or repeated measures data. Statistical methods that account for the clustering of measurements taken on the same individual must be used in analyzing the 300 total cholesterol measurements taken on participants over time. Details of these techniques can be found in Sullivan.[2] The sample size in most of the analyses discussed in this textbook refers to the number of individuals participating in the study. In the examples that follow, we indicate the sample size. It is always important to report the sample size when summarizing data as it gives the reader a sense of the precision of the analysis. The notion of precision is discussed in subsequent chapters in detail.

Numerical summary measures computed on samples are called **statistics**. Summary measures computed on populations are called parameters. The sample size is an example of an important statistic that should always be reported when summarizing data. In the following sections, we present sample statistics as well as graphical displays for each type of variable.

4.1 DICHOTOMOUS VARIABLES

Dichotomous variables take on one of only two possible responses. Sex is an example of a dichotomous variable, with response options of "male" or "female," as are current smoking status and diabetes status, with response options of "yes" or "no."

4.1.1 Descriptive Statistics for Dichotomous Variables

Dichotomous variables are often used to classify participants as possessing or not possessing a particular characteristic, having or not having a particular attribute. For example, in a study of cardiovascular risk factors we might collect information on participants such as whether or not they have diabetes, whether or not they smoke, and whether or not they are on treatment for high blood pressure or high cholesterol. The response options for each of these variables are "yes" or "no."

When analyzing dichotomous variables, responses are often classified as success or failure, with success denoting

TABLE 4–2 Frequency Distribution Table for Sex

	Frequency	Relative Frequency (%)
Male	1625	45.9
Female	1914	54.1
Total	3539	100.0

the response of interest. The success response is not necessarily the positive or healthy response but rather the response of interest. In fact, in many medical applications the focus is often on the unhealthy or "at-risk" response.

Example 4.1. The seventh examination of the offspring in the Framingham Heart Study was conducted between 1998 and 2001. A total of $n = 3539$ participants (1625 men and 1914 women) attended the seventh examination and completed an extensive physical examination. At that examination, numerous variables were measured including demographic characteristics, such as sex, educational level, income, and marital status; clinical characteristics, such as height, weight, systolic and diastolic blood pressure, and total cholesterol; and behavioral characteristics, such as smoking and exercise.

Dichotomous variables are often summarized in frequency distribution tables. **Table 4–2** displays a frequency distribution table for the variable sex measured in the seventh examination of the Framingham Offspring Study. The first column of the frequency distribution table indicates the specific response options of the dichotomous variable (in this example, male and female). The second column contains the frequencies (counts or numbers) of individuals in each response category (the numbers of men and women, respectively). The third column contains the relative frequencies, which are computed by dividing the frequency in each response category by the sample size (e.g., $1625 / 3539 = 0.459$). The relative frequencies are often expressed as percentages by multiplying by 100 and are most often used to summarize dichotomous variables. For example, in this sample 45.9% are men and 54.1% are women.

Another example of a frequency distribution table is presented in **Table 4–3**, showing the distribution of treatment with anti-hypertensive medication in persons attending the seventh examination of the Framingham Offspring Study. Notice that there are only $n = 3532$ valid responses, although the sample size is actually $n = 3539$. There are seven individuals with missing data on this particular question. Missing data occur in studies for a variety of reasons. When there is very little missing data (e.g., less than 5%) and there is no apparent pattern to the missingness (e.g., there is no systematic reason for missing data), then statistical analyses based on the available data are generally appropriate. However, if there is extensive missing data or if there is a pattern to the missingness, then caution must be exercised in performing statistical analyses. Techniques for handling missing data are beyond the scope of this book; more details can be found in Little and Rubin.[3] From Table 4–3, we can see that 34.5% of the participants are currently being treated for hypertension.

Sometimes it is of interest to compare two or more groups on the basis of a dichotomous outcome variable. For example, suppose we wish to compare the extent of treatment with anti-hypertensive medication in men and women. **Table 4–4** summarizes treatment with anti-hypertensive medication in men and women attending the seventh examination of the Framingham Offspring Study. The first column of the table indicates the sex of the participant. Sex is a dichotomous variable, and in this example it is used to distinguish the comparison groups (men and women). The outcome variable is also a dichotomous variable and represents treatment with anti-hypertensive medication or not. A total of $n = 611$ men and $n = 608$ women are on anti-hypertensive treatment. Because there are different numbers of men and women (1622 versus 1910) in the study sample, comparing frequencies (611 versus 608) is not the most appropriate comparison. The frequencies indicate that almost equal numbers of men and women are on treatment. A more appropriate comparison is based on relative frequencies, 37.7% versus 31.8%, which incorporate the different numbers of men and women in the sample. Notice that the sum of the rightmost column is not 100%, as it was in previous examples. In this example, the bottom row contains data on the total sample and 34.5% of all participants are being treated with anti-hypertensive medication. In Chapter 6 and Chapter 7, we will discuss formal methods to compare relative frequencies between groups.

4.1.2 Bar Charts for Dichotomous Variables

Graphical displays are very useful for summarizing data. There are many options for graphical displays, and many widely available software programs offer a variety of displays. However, it is important to choose the graphical display that accurately conveys information in the sample. We discuss data visualization in detail in Chapter 12. The appropriate graphical display depends on the type of variable being analyzed. Dichotomous variables are best summarized using **bar charts**. The response options (yes/no, present/absent) are shown on the horizontal axis, and either the frequencies or relative frequencies are plotted on the vertical axis, producing a frequency bar chart or relative frequency bar chart, respectively.

Figure 4–1 is a frequency bar chart depicting the distribution of men and women attending the seventh examination of the Framingham Offspring Study. The horizontal axis displays the two response options (male and female), and the vertical axis displays the frequencies (the numbers of men and women who attended the seventh examination).

Figure 4–2 is a relative frequency bar chart of the distribution of treatment with anti-hypertensive medication measured in the seventh examination of the Framingham Offspring Study. Notice that the vertical axis in Figure 4–2 displays relative frequencies and not frequencies, as was the case in Figure 4–1. In Figure 4–2, it is not necessary to show both responses as the relative frequencies, expressed as percentages, sum to 100%. If 65.5% of the sample is not being treated, then 34.5% must be on treatment. These types of bar charts are very useful for comparing relative frequencies between groups.

Figure 4–3 is a relative frequency bar chart describing treatment with anti-hypertensive medication in men versus women attending the seventh examination of the Framingham Offspring Study. Notice that the vertical axis displays relative frequencies and in this example, 37.7% of men were using anti-hypertensive medications as compared to 31.8% of women. **Figure 4–4** is an alternative display of the same data. Notice the scaling of the vertical axis. How do

TABLE 4–3 Frequency Distribution Table for Treatment with Anti-Hypertensive Medication

	Frequency	Relative Frequency (%)
No treatment	2313	65.5
Treatment	1219	34.5
Total	3532	100.0

TABLE 4–4 Treatment with Anti-Hypertensive Medication in Men and Women Attending the Seventh Examination of the Framingham Offspring Study

	n	Number on Treatment	Relative Frequency (%)
Male	1622	611	37.7
Female	1910	608	31.8
Total	3532	1219	34.5

Dichotomous Variables | 39

FIGURE 4–1 Frequency Bar Chart of Sex Distribution

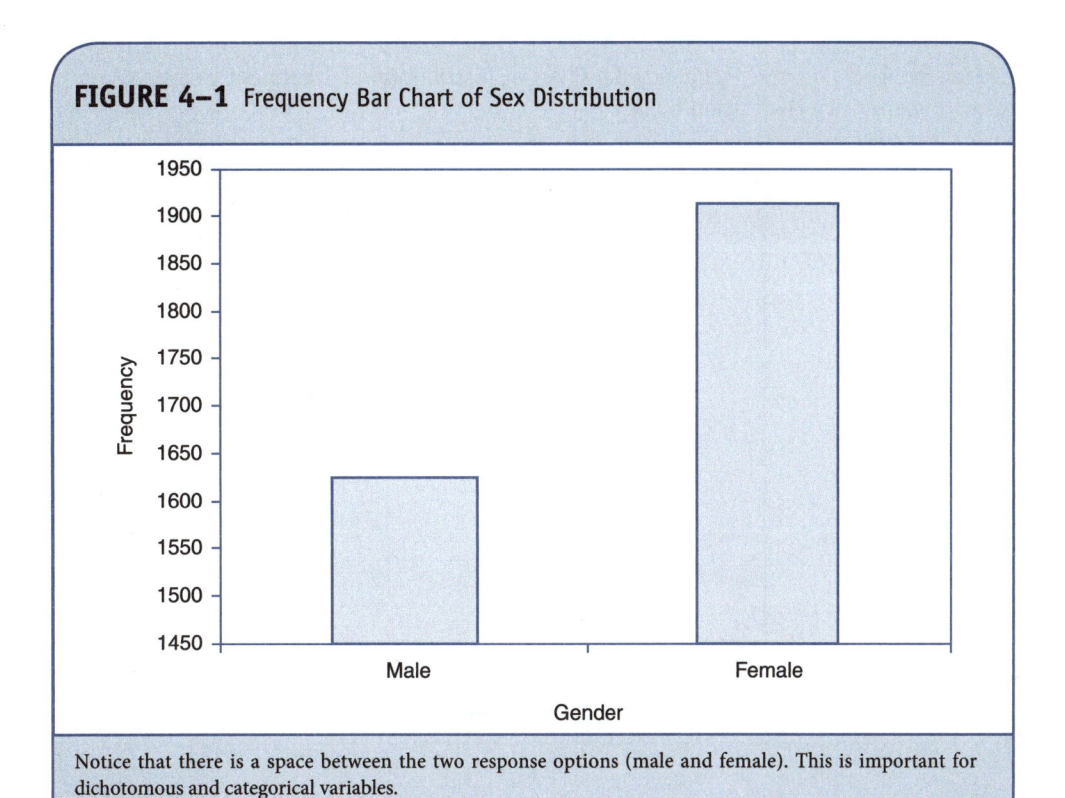

Notice that there is a space between the two response options (male and female). This is important for dichotomous and categorical variables.

FIGURE 4–2 Relative Frequency Bar Chart of Distribution of Treatment with Anti-Hypertensive Medication

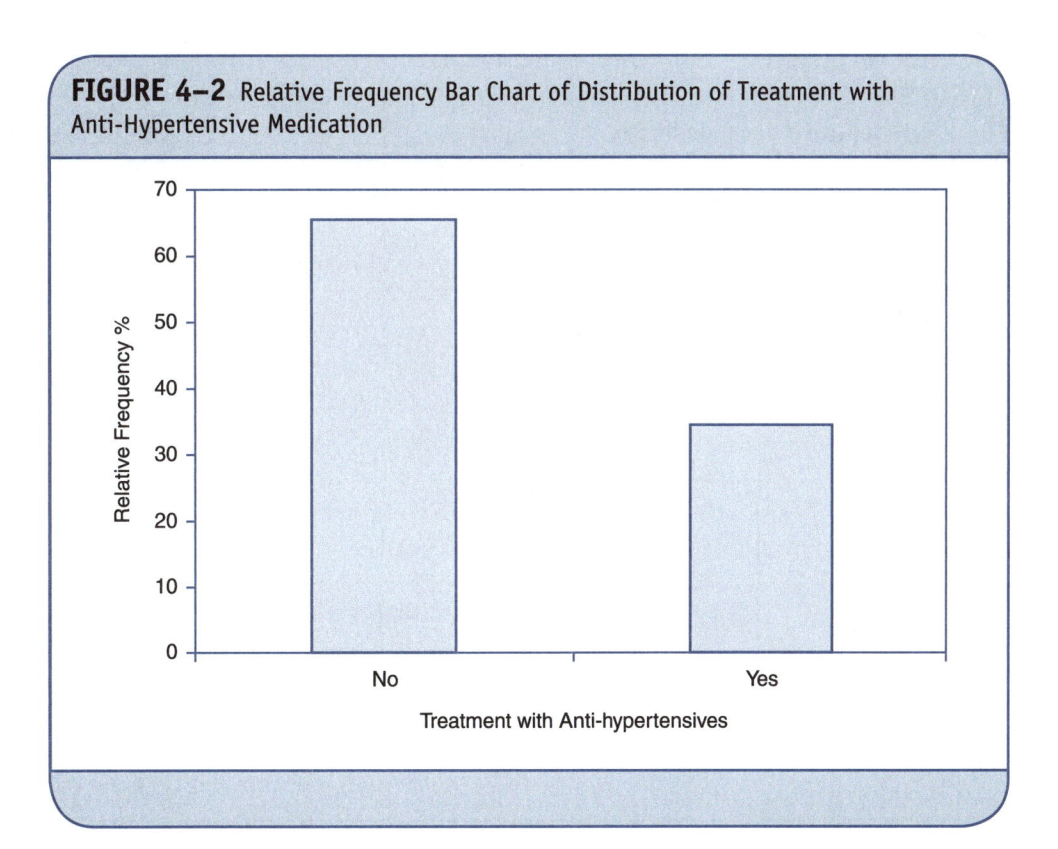

FIGURE 4–3 Relative Frequency Bar Chart of Distribution of Treatment with Anti-Hypertensive Medication by Sex

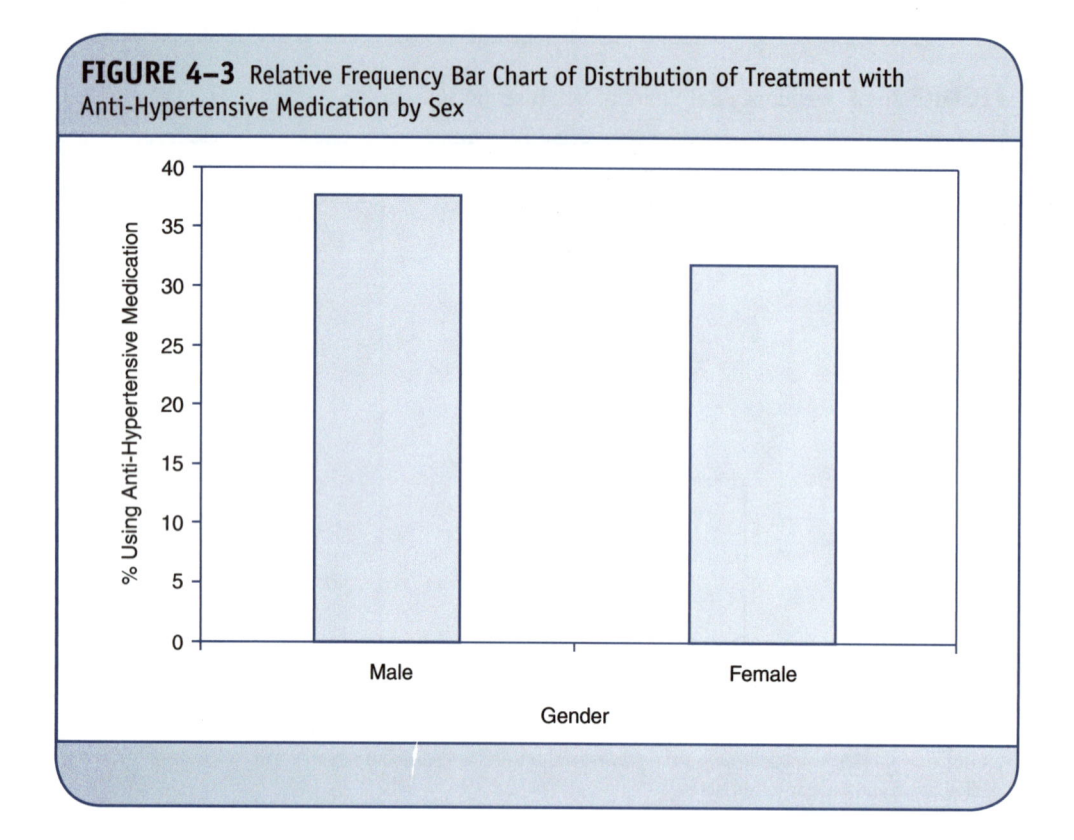

FIGURE 4–4 Relative Frequency Bar Chart of Distribution of Treatment with Anti-Hypertensive Medication by Sex

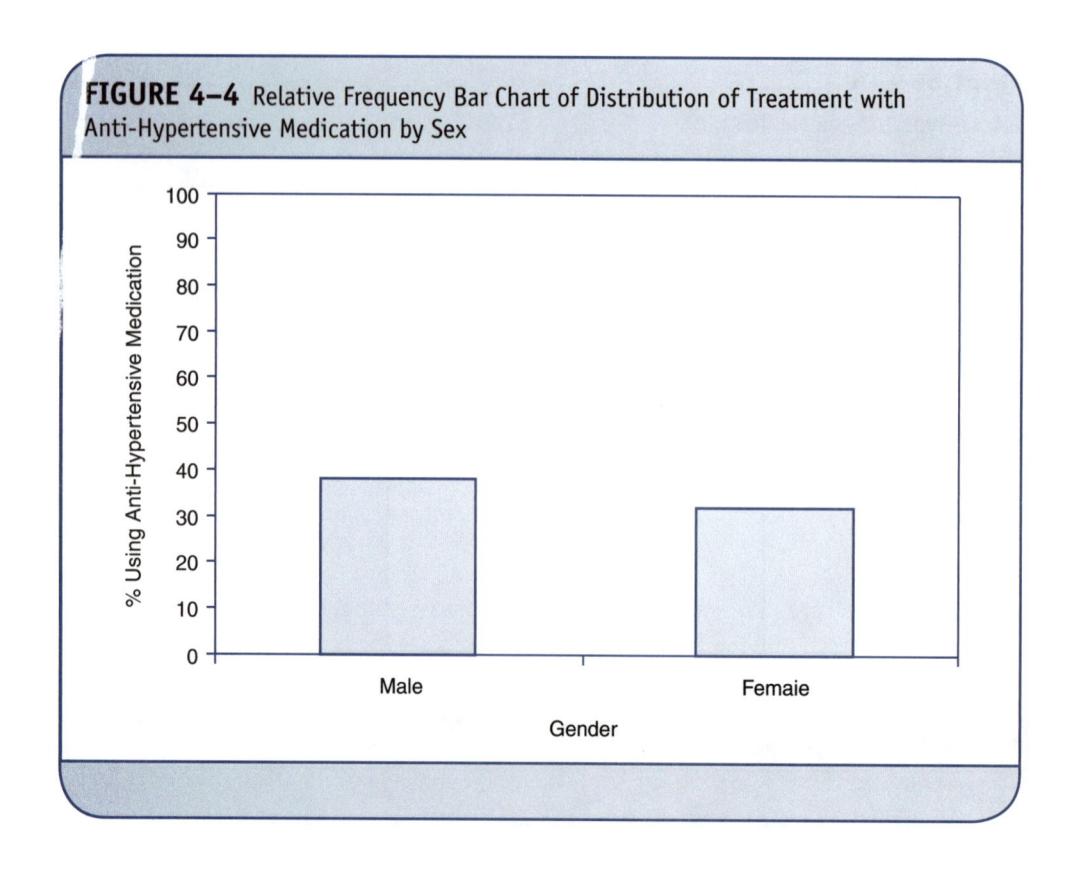

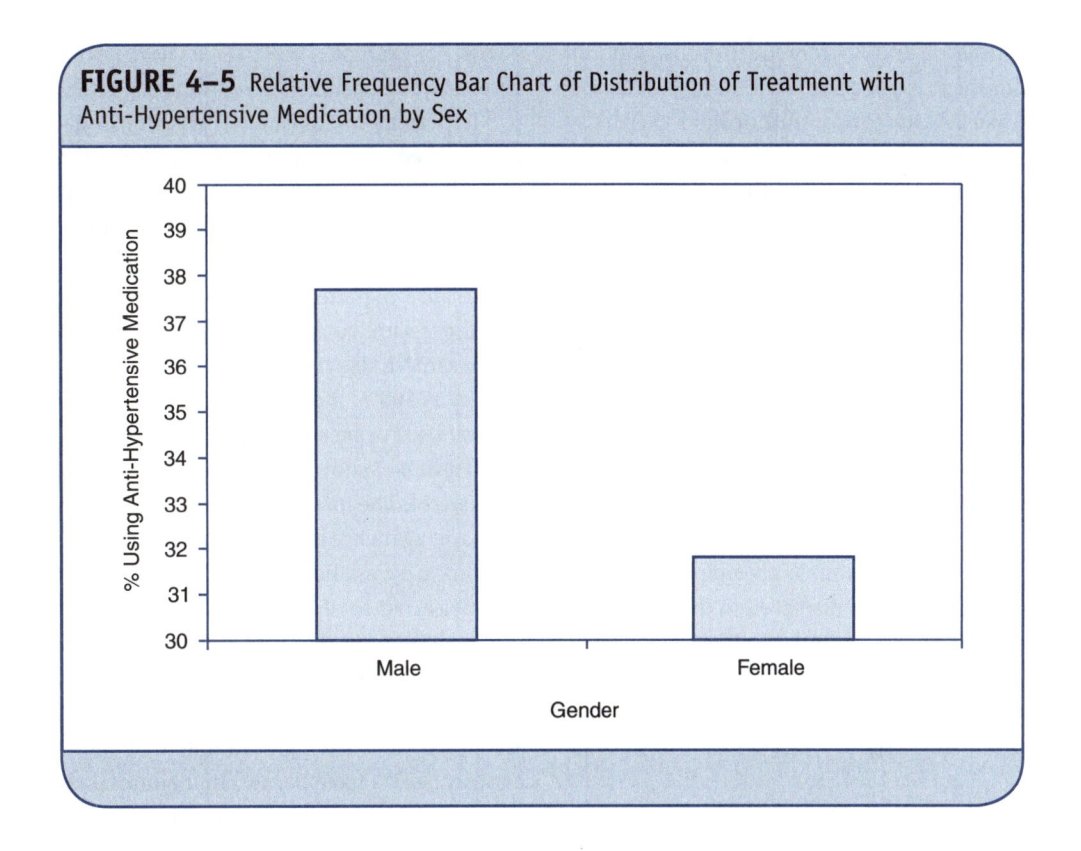

FIGURE 4–5 Relative Frequency Bar Chart of Distribution of Treatment with Anti-Hypertensive Medication by Sex

the relative frequencies compare visually? Finally, consider a third display of the same data, shown in **Figure 4–5**. How do the relative frequencies compare?

The axes in any graphical display should be scaled to accommodate the range of the data. While relative frequencies can, in theory, range from 0% to 100%, it is not necessary to always scale the axes from 0% to 100%. It is also potentially misleading to restrict the scaling of the vertical axis, as was done in Figure 4–3, to exaggerate the difference in the use of anti-hypertensive medication between men and women, at least from a visual standpoint. In this example, the relative frequencies are 31.8% and 37.7%, and thus scaling from 0% to 40% is appropriate to accommodate the data. It is always important to label axes clearly so that readers can appropriately interpret the data.

4.2 ORDINAL AND CATEGORICAL VARIABLES

Ordinal and categorical variables have a fixed number of response options that are ordered and unordered, respectively. Ordinal and categorical variables typically have more than two distinct response options, whereas dichotomous variables have exactly two response options. Summary statistics for ordinal and categorical variables again focus primarily on relative frequencies (or percentages) of responses in each response category.

4.2.1 Descriptive Statistics for Ordinal and Categorical Variables

Consider again a study of cardiovascular risk factors such as the Framingham Heart Study. In the study, we might collect information on participants such as their blood pressure, total cholesterol, and body mass index (BMI). Often, clinicians classify patients into categories. Each of these variables—blood pressure, total cholesterol, and BMI—are continuous variables. In this section, we organize continuous measurements into ordinal categories. For example, the NHLBI and the American Heart Association use the classification of blood pressure given in Table 4–1.[1,4,5] The American Heart Association uses the following classification for total cholesterol levels: desirable, less than 200 mg/dl; borderline high risk, 200–239 mg/dl; and high risk, 240 mg/dl or more.[4] Body mass index (BMI) is computed as the ratio of weight in kilograms to height in meters squared and the following categories are often used: underweight, less than 18.5; normal weight, 18.5–24.9; overweight, 25.0–29.9; and obese, 30.0 or greater. These are all examples of ordinal variables. In each case, it is healthier to be in a lower category.

Example 4.2. Using data from the seventh examination of the offspring in the Framingham Heart Study ($n = 3539$), we create the categories as defined previously for blood pressure, total cholesterol, and BMI. Frequency distribution tables, similar to those presented for dichotomous data, are also used to summarize categorical and ordinal variables. **Table 4–5** is a frequency distribution table for the ordinal blood pressure variable. The mutually exclusive (non-overlapping) and exhaustive (covering all possible options) categories are shown in the first column of the table. The frequencies, or numbers of participants in each response category, are shown in the middle column, and the relative frequencies, as percentages, are shown in the rightmost column.

Notice that there are only $n = 3533$ valid responses, whereas the sample size is $n = 3539$. There are six individuals with missing blood pressure data. More than a third of the sample (34.1%) has normal blood pressure, 41.1% are classified as pre-hypertension, 18.5% have Stage I hypertension, and 6.3% have Stage II hypertension. With ordinal variables, two additional columns are often displayed in the frequency distribution table, called the cumulative frequency and cumulative relative frequency, respectively (see **Table 4–6**).

The cumulative frequencies in this example reflect the number of patients at the particular blood pressure level or below. For example, 2658 patients have normal blood pressure or pre-hypertension. There are 3311 patients with normal, pre-hypertension, or Stage I hypertension. The cumulative relative frequencies are very useful for summarizing ordinal variables and indicate the percent of patients at a particular level or below. In this example, 75.2% of the patients are not classified as hypertensive (i.e., they have normal blood pressure or pre-hypertension). Notice that for the last (highest) blood pressure category, the cumulative frequency is equal to the sample size ($n = 3533$) and the cumulative relative frequency is 100%, indicating that all of the patients are at the highest level or below.

Table 4–7 shows the frequency distribution table for total cholesterol. The total cholesterol categories are defined as follows and again are based on the measured cholesterol values. Persons are classified as having desirable total cholesterol if the measured total cholesterol is less than 200 mg/dl; borderline high total cholesterol, 200–239 mg/dl; and high total cholesterol, 240 mg/dl or more. At the seventh examination of the Framingham Offspring Study, 51.6% of the patients are classified as having desirable total cholesterol levels and another 34.3% have borderline high total cholesterol. Using the cumulative relative frequencies, we can summarize the data as follows: 85.9% of patients have total cholesterol levels that are desirable or borderline high. The remaining 14.1% of patients are classified as having high cholesterol.

Table 4–8 is a frequency distribution table for the ordinal BMI variable. Both weight and height are measured directly and BMI is computed as described previously. In the seventh examination of the Framingham Offspring Study sample, 28% of the patients are classified as normal weight, 41.3% are classified as overweight, and 30.1% are classified as obese. Using the cumulative relative frequencies, almost 70% are underweight, normal weight, or overweight. This is equivalent to reporting that 70% of the sample is not obese.

TABLE 4–5 Frequency Distribution Table for Blood Pressure Categories

	Frequency	Relative Frequency (%)
Normal	1206	34.1
Pre-hypertension	1452	41.1
Stage I hypertension	653	18.5
Stage II hypertension	222	6.3
Total	3533	100.0

TABLE 4–6 Frequency Distribution Table for Blood Pressure Categories with Cumulative Frequency and Cumulative Relative Frequency

	Frequency	Relative Frequency (%)	Cumulative Frequency	Cumulative Relative Frequency (%)
Normal	1206	34.1	1206	34.1
Pre-hypertension	1452	41.1	2658	75.2
Stage I hypertension	653	18.5	3311	93.7
Stage II hypertension	222	6.3	3533	100.0
Total	3533	100.0		

TABLE 4–7 Frequency Distribution Table for Total Cholesterol Categories

	Frequency	Relative Frequency (%)	Cumulative Frequency	Cumulative Relative Frequency (%)
Desirable	1712	51.6	1712	51.6
Borderline high	1139	34.3	2851	85.9
High	469	14.1	3320	100.0
Total	3320	100.0		

TABLE 4–8 Frequency Distribution Table for Body Mass Index Categories

	Frequency	Relative Frequency (%)	Cumulative Frequency	Cumulative Relative Frequency (%)
Underweight	20	0.6	20	0.6
Normal weight	932	28.0	952	28.6
Overweight	1374	41.3	2326	69.9
Obese	1000	30.1	3326	100.0
Total	3326	100.0		

Table 4–5 through Table 4–8 contain summary statistics for ordinal variables. The key summary statistics for ordinal variables are relative frequencies and cumulative relative frequencies. Table 4–9 through Table 4–11 contain summary statistics for categorical variables. Categorical variables are variables with two or more distinct responses but the responses are unordered. Some examples of categorical variables measured in the Framingham Heart Study include marital status, handedness, and smoking status. For categorical variables, frequency distribution tables with frequencies and relative frequencies provide appropriate summaries. Cumulative frequencies and cumulative relative frequencies are generally not useful for categorical variables, as it is usually not of interest to combine categories as there is no inherent ordering.

Table 4–9 is a frequency distribution table for the categorical marital status variable. The mutually exclusive and exhaustive categories are shown in the first column of the table. The frequencies, or numbers of participants in each response category, are shown in the middle column, and the relative frequencies, as percents, are shown in the rightmost column. There are $n = 3530$ valid responses to the marital status question. A total of 9 participants did not

TABLE 4–9 Frequency Distribution Table for Marital Status

	Frequency	Relative Frequency (%)
Single	203	5.8
Married	2580	73.1
Widowed	334	9.5
Divorced	367	10.4
Separated	46	1.3
Total	3530	100.0

provide marital status data. The majority of the sample is married (73.1%) and approximately 10% of the sample is divorced, another 10% is widowed, 6% is single, and 1% is separated. The relative frequencies are the most relevant statistics used to describe a categorical variable. Cumulative frequencies and cumulative relative frequencies are not generally informative descriptive statistics for categorical variables.

TABLE 4–10 Frequency Distribution Table for Dominant Hand

	Frequency	Relative Frequency (%)
Right	3143	89.5
Left	370	10.5
Total	3513	100.0

TABLE 4–11 Frequency Distribution Table for Smoking Status

	Frequency	Relative Frequency (%)
Nonsmoker	1330	37.6
Former	1724	48.8
Current	482	13.6
Total	3536	100.0

TABLE 4–12 Current Smoking in the Framingham Offspring Study by Date of Exam

Exam Cycle	Dates	Current Smokers (%)
1	Aug 1971–Sept 1975	59.7
2	Oct 1979–Oct 1983	28.5
3	Dec 1983–Sept 1987	23.9
4	Apr 1987–Sept 1991	21.7
5	Jan 1991–June 1995	17.4
6	Jan 1995–Sept 1998	13.8
7	Sept 1998–Oct 2001	13.6

Marital status is a categorical variable, so there is no ordering to the responses and therefore the first column can be organized differently. For example, sometimes responses are presented from the most frequently to least frequently occurring in the sample, and sometimes responses are presented alphabetically. Any ordering is appropriate. In contrast, with ordinal variables there is an ordering to the responses and therefore response options can only be presented either from highest to lowest (healthiest to unhealthiest) or vice versa. The response options within an ordinal scale cannot be rearranged.

Table 4–10 is a frequency distribution table for a dichotomous categorical variable. Dichotomous variables are a special case of categorical variables with exactly two response options. Table 4–10 displays the distribution of the dominant hand of participants who attended the seventh examination of the Framingham Offspring Study. The response options are "right" or "left." There are $n = 3513$ valid responses to the dominant hand assessment. A total of 26 participants did not provide data on their dominant hand. The majority of the Framingham sample is right-handed (89.5%). **Table 4–11** is a frequency distribution table for a categorical variable reflecting smoking status. Smoking status here is measured as nonsmoker, former smoker, or current smoker. There are $n = 3536$ valid responses to the smoking status questions. Three participants did not provide adequate data to be classified. Almost half of the sample is former smokers (48.8%), over a third (37.6%) has never smoked, and approximately 14% are current smokers. The adverse health effects of smoking have been a major focus of public health messages in recent years, and the percentage of participants reporting as current smokers must be interpreted relative to the time of the study. **Table 4–12** shows the proportions of participants reporting currently smoking at the time of each examination of the Framingham offspring. The dates of each exam are also provided.

In the next two sections, we present graphical displays for ordinal and categorical variables, respectively. Whereas the numerical summaries for ordinal and categorical variables are identical (at least in terms of the frequencies and relative frequencies), graphical displays for ordinal and categorical variables are different in a very important way.

4.2.2 Histograms for Ordinal Variables

Histograms are appropriate graphical displays for ordinal variables. A histogram is different from a bar chart in one important feature. The horizontal axis of a histogram shows the distinct ordered response options of the ordinal variable. The vertical axis can show either frequencies or relative frequencies, producing a frequency histogram or relative frequency histogram, respectively. The bars are centered over each response option and scaled according to frequencies or relative frequencies as desired. The difference between a histogram and a bar chart is that the bars in a histogram run together; there is no space between adjacent responses. This reinforces the idea that the response categories are ordered and based on an underlying continuum. This underlying continuum may or may not be measurable.

Figure 4–6 is a frequency histogram for the blood pressure data displayed in Table 4–5. The horizontal axis displays the ordered blood pressure categories and the vertical axis displays the frequencies or numbers of participants classified in each category. The histogram immediately conveys the message that the majority of participants are in the lower (healthier) two categories of the distribution. A small number of participants are in the Stage II hypertension category. The histogram in **Figure 4–7** is a relative frequency histogram for the same data. Notice that the figure is the same except for the

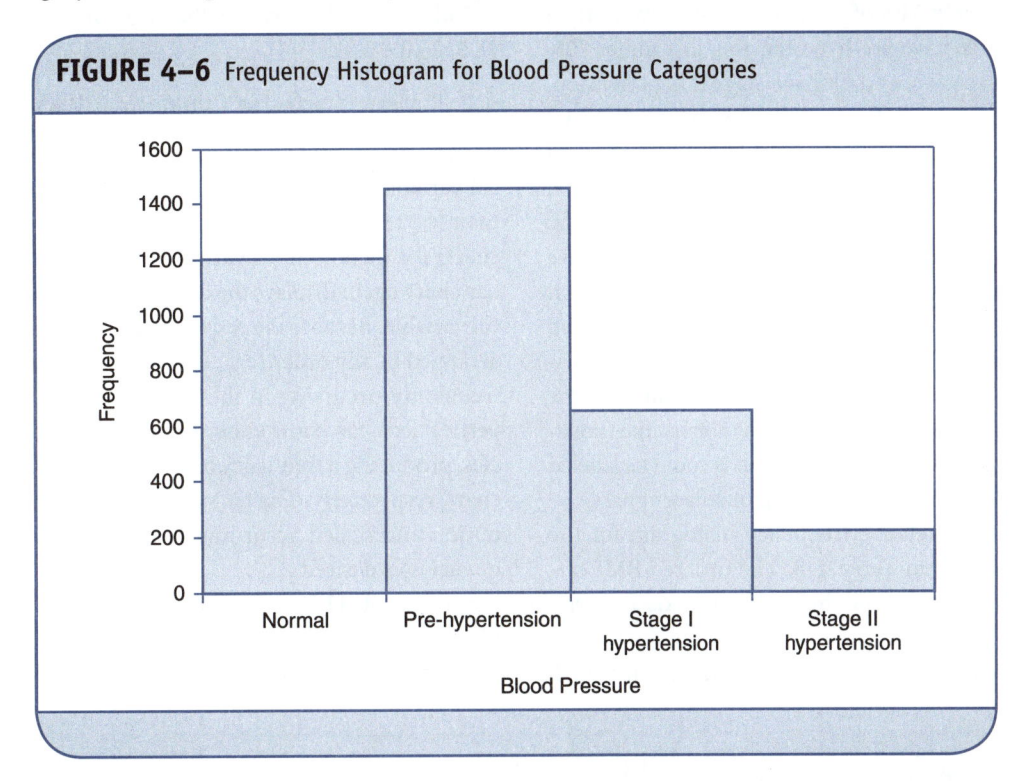

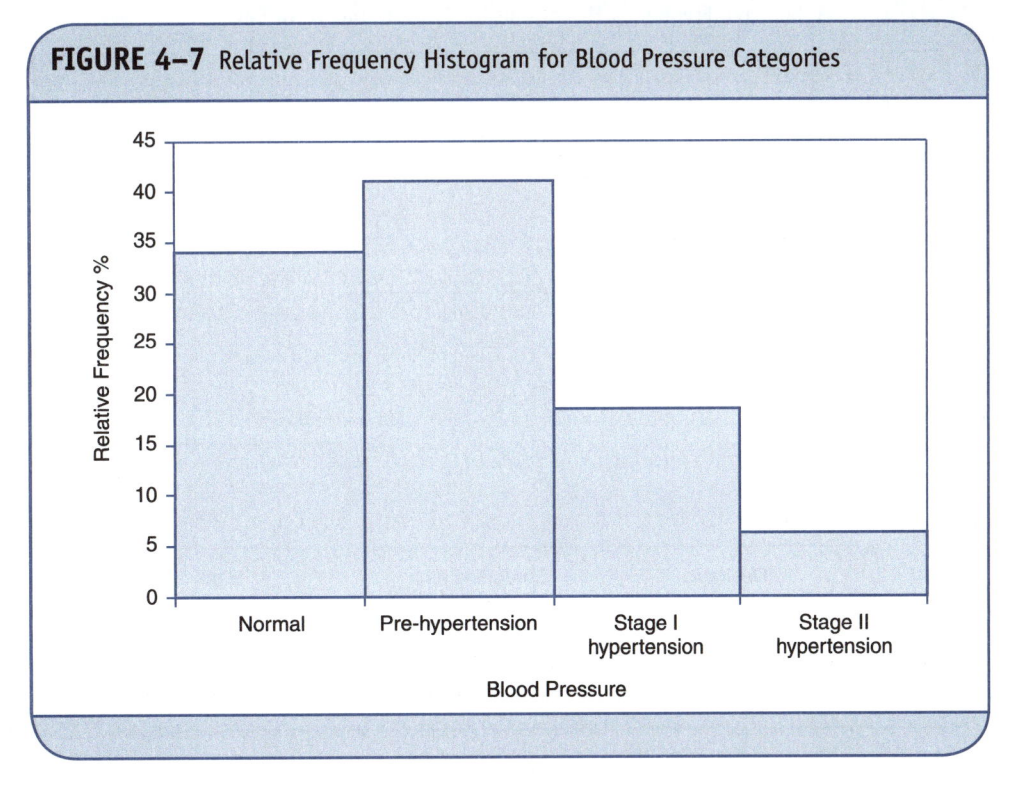

FIGURE 4–6 Frequency Histogram for Blood Pressure Categories

FIGURE 4–7 Relative Frequency Histogram for Blood Pressure Categories

vertical axis, which is scaled to accommodate relative frequencies instead of frequencies.

Usually, relative frequency histograms are preferred over frequency histograms, as the relative frequencies are most appropriate for summarizing the data. From Figure 4–7, we can see that approximately 34% of the participants have normal blood pressure, 41% have pre-hypertension, just under 20% have Stage I hypertension, and 6% have Stage II hypertension.

Figure 4–8 is a relative frequency histogram for the total cholesterol variable summarized in Table 4–7. The bars of the histogram run together to reflect the fact that there is an underlying continuum of total cholesterol measurements. From Figure 4–8, we see that over 50% of the participants have desirable total cholesterol levels and just under 15% have high total cholesterol levels. The horizontal axis could be scaled differently. **Figure 4–9** makes the continuum of total cholesterol underlying the categories used here to summarize the data more obvious. Another alternative is to mark the transition points. In Figure 4–9, the horizontal axis could be labeled 200 and 240 at the points of intersection of adjacent bars.

Figure 4–10 is a relative frequency histogram for the BMI data summarized in Table 4–8. The ordered BMI categories are shown in text along the horizontal axis and relative frequencies, as percents, are displayed along the vertical axis. From Figure 4–10, it is immediately evident that a small percentage of the participants are underweight and that the majority of the participants are overweight or obese, with the former more likely than the latter. The horizontal axis of Figure 4–10 could be scaled differently to show the numerical values of BMI that define the ordinal categories or with labels to indicate the BMI values that separate adjacent bars (e.g., 18.5, 25.0, 30.0).

4.2.3 Bar Charts for Categorical Variables

Bar charts are appropriate graphical displays for categorical variables. Bar charts for categorical variables with more than two responses are constructed in the same fashion as bar charts for dichotomous variables. The horizontal axis of the bar chart again displays the distinct responses of the categorical variable. Because the responses are unordered, they can be arranged in any order (e.g., from the most frequently to least frequently occurring in the sample, or alphabetically). The vertical axis can show either frequencies or relative frequencies, producing a frequency bar chart or relative frequency bar chart, respectively. The bars are centered over each response option and scaled according to frequencies or relative frequencies as desired.

Figure 4–11 is an example of a frequency bar chart displaying the distribution of marital status in the participants who attended the seventh examination of the Framingham

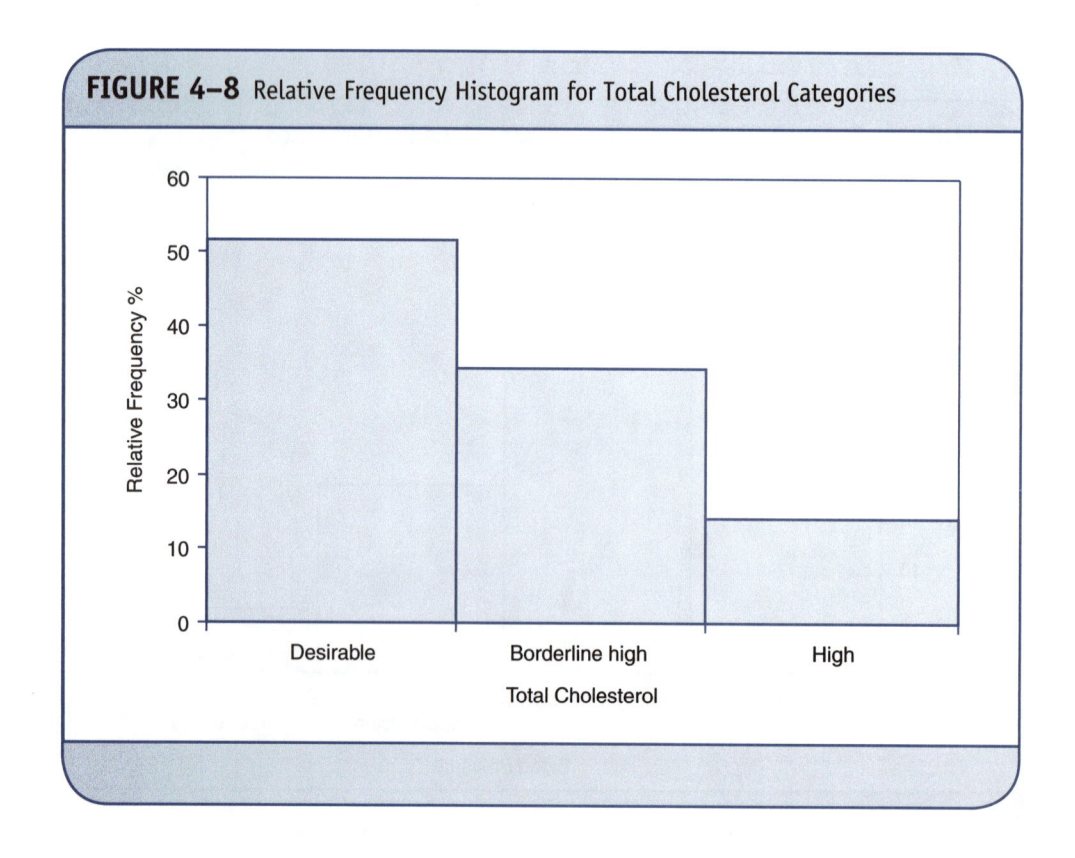

FIGURE 4–8 Relative Frequency Histogram for Total Cholesterol Categories

FIGURE 4–9 Relative Frequency Histogram for Total Cholesterol Categories

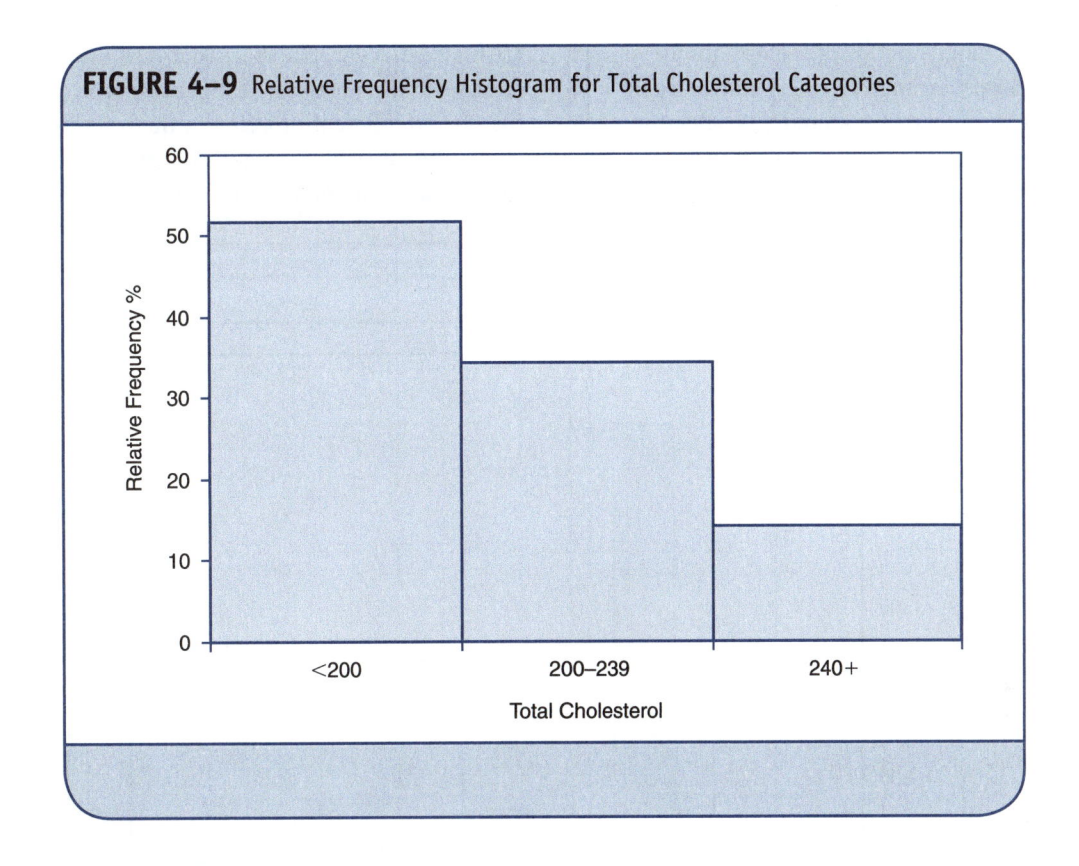

FIGURE 4–10 Relative Frequency Histogram for BMI Categories

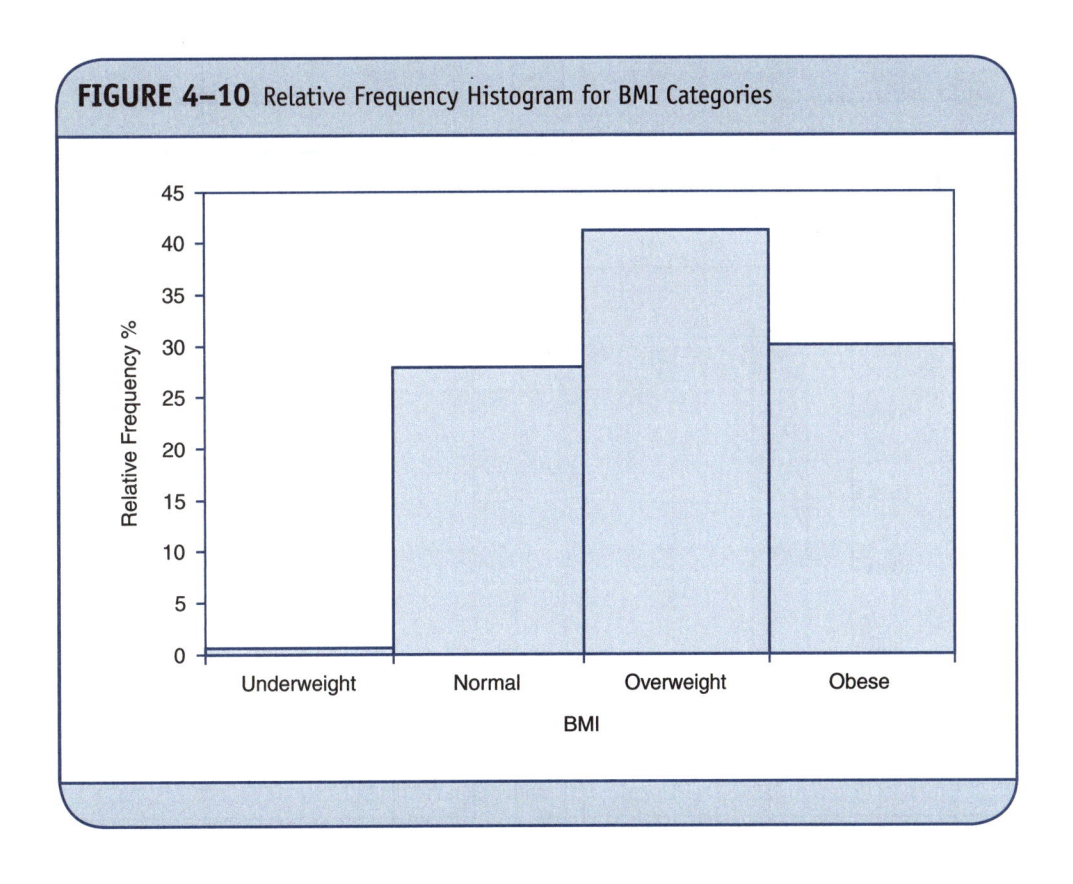

Offspring Study. The distinct response options are shown on the horizontal axis, arranged as they were in Table 4–9. Frequencies, or numbers of respondents selecting each response, are shown on the vertical axis. It is immediately evident from the bar chart that the majority of respondents are married. Notice the spaces between the bars, indicating a clear separation between distinct response options. Because there is no inherent ordering to the response options, as is always the case with a categorical variable, the horizontal axis can be scaled differently. **Figure 4–12** and **Figure 4–13** show

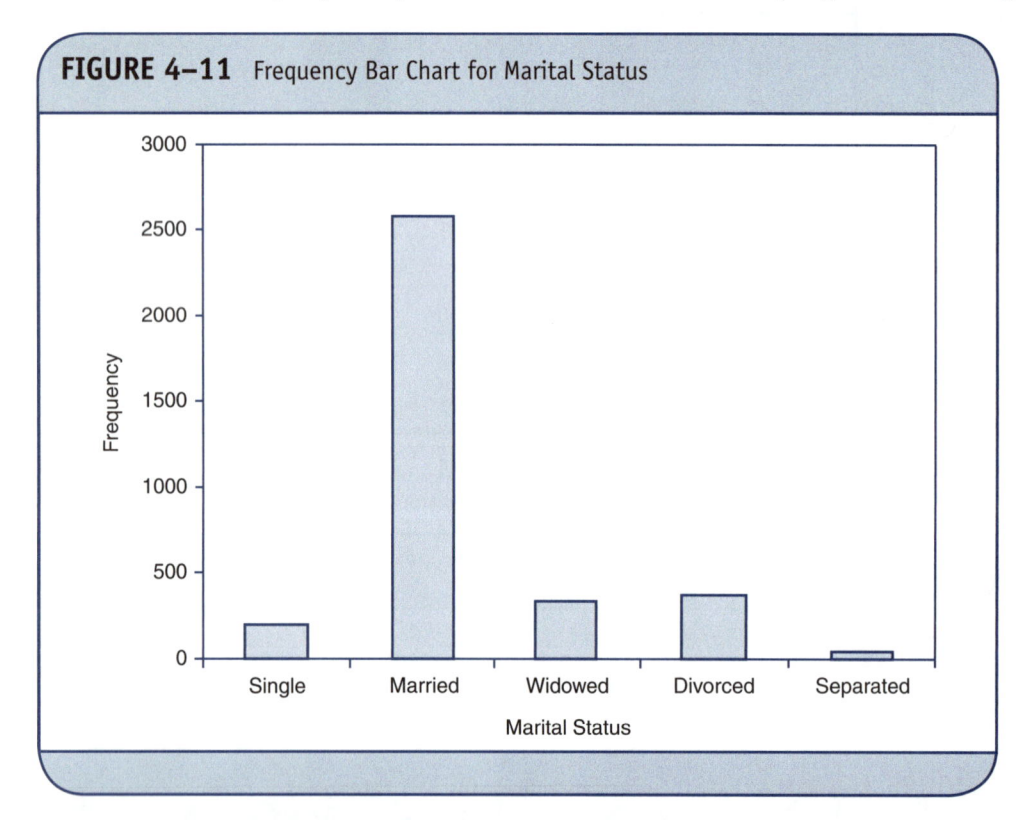

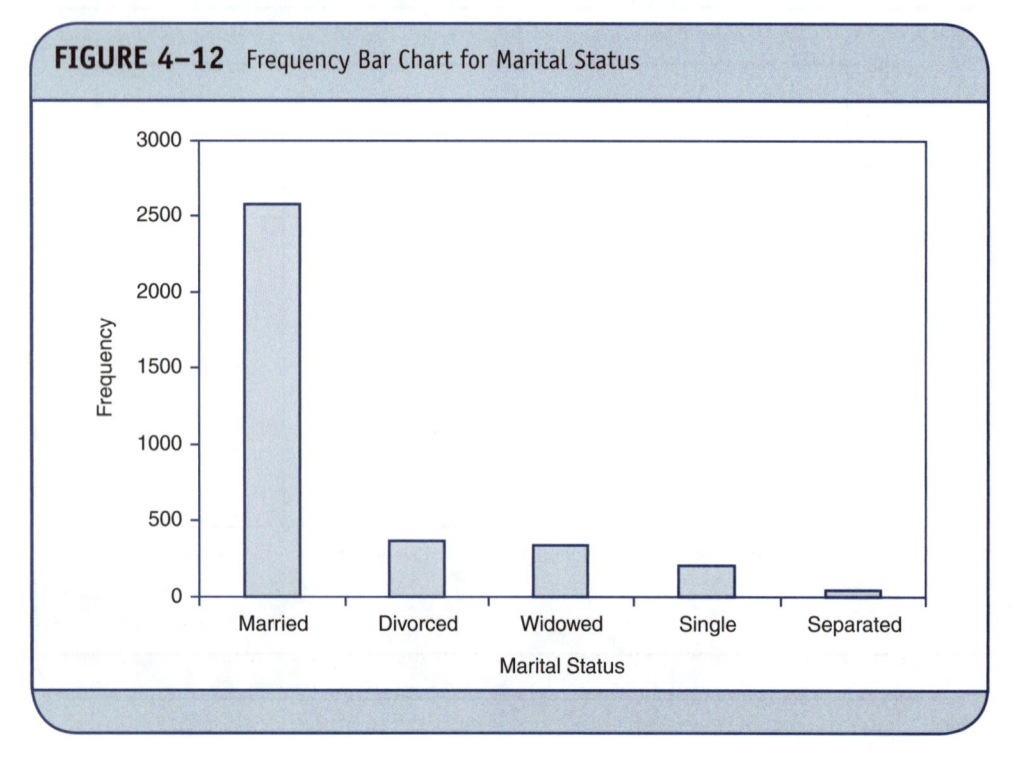

FIGURE 4–11 Frequency Bar Chart for Marital Status

FIGURE 4–12 Frequency Bar Chart for Marital Status

FIGURE 4–13 Frequency Bar Chart for Marital Status

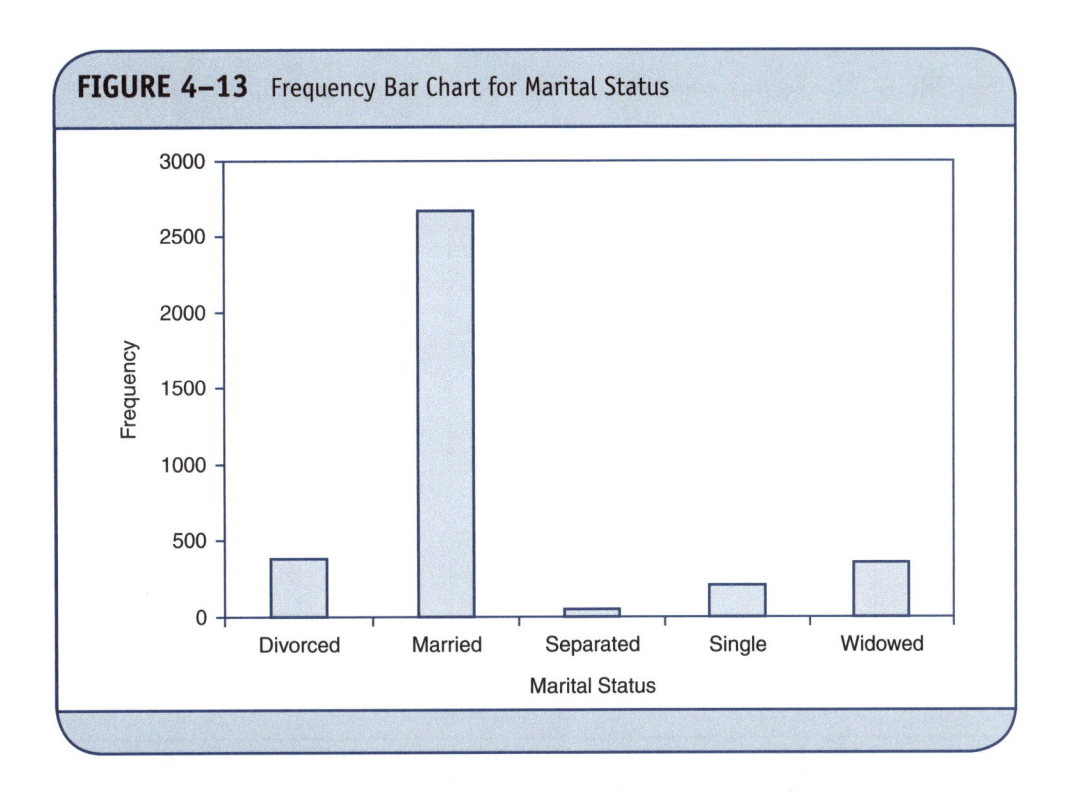

FIGURE 4–14 Relative Frequency Bar Chart for Dominant Hand

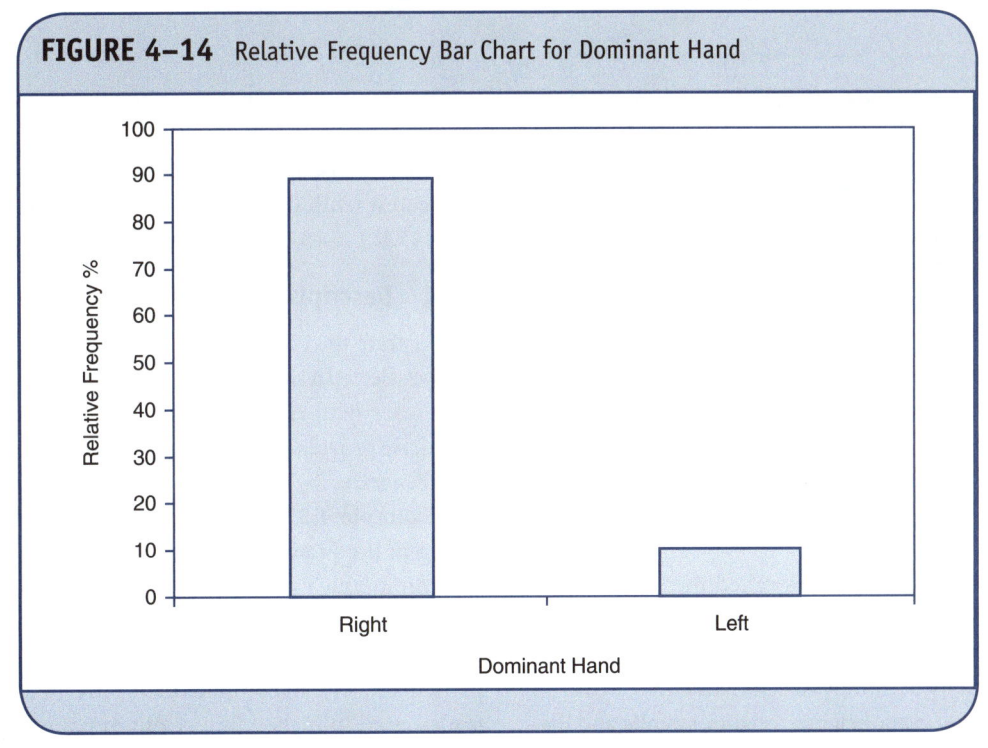

alternative arrangements of the response options. Each of the bar charts displays identical data. All three presentations are appropriate as the categorical responses can be ordered in any way. Figure 4–12 arranges the responses from most frequently to least frequently occurring, and Figure 4–13 arranges the responses alphabetically.

It is more typical to present relative frequency bar charts with relative frequencies displayed as proportions or percentages on the vertical axis. **Figure 4–14** is a relative frequency bar chart displaying the distribution of the dominant hand and is summarized in Table 4–10. Figure 4–14 clearly displays that the vast majority of participants in the seventh examination of

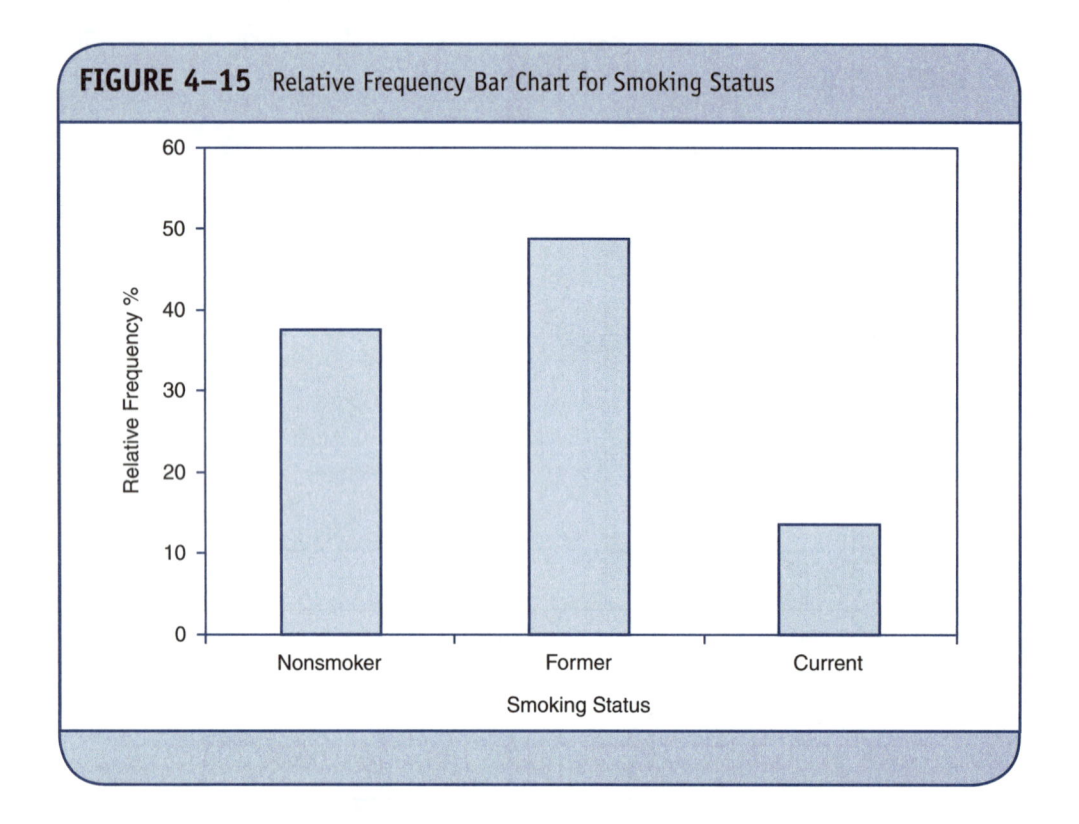

FIGURE 4–15 Relative Frequency Bar Chart for Smoking Status

the Framingham Heart Study are right-handed. Because the dominant hand is a dichotomous variable, we only need to display the relative frequency for one bar, as the second is determined from the first.

Figure 4–15 is a relative frequency bar chart displaying the distribution of smoking status measured at the seventh examination of the Framingham Offspring Study and summarized in Table 4–11. Figure 4–15 gives the clear impression that a small percentage of the participants are current smokers. Almost 50% are former smokers. Again, the response options in Figure 4–14 and Figure 4–15 may be rearranged, as there is no inherent ordering to these ordinal scales.

4.3 CONTINUOUS VARIABLES

Continuous variables, sometimes called measurement or quantitative variables, take on an unlimited number of distinct responses between a theoretical minimum value and maximum value. In a study of cardiovascular risk factors, we might measure participants' ages, heights, weights, systolic and diastolic blood pressures, total serum cholesterol levels, and so on. The measured values for each of these continuous variables depend on the scale of measurement. For example, in adult studies such as the Framingham Heart Study, age is usually measured in years. Studies of infants might measure age in days or even hours, whichever is more appropriate. Heights can be measured in inches or centimeters, weights can be measured in pounds or kilograms. Assuming weight is measured in pounds, measurements might be to the nearest pound, the nearest tenth of a pound, or the nearest hundredth (e.g., 145, 145.1, 145.13), depending on the precision of the scale.

4.3.1 Descriptive Statistics for Continuous Variables

To illustrate the computations of descriptive statistics in detail, we selected a small subset of the Framingham Heart Study data. After performing computations by hand on the small subset, we provide descriptive statistics for the full sample that were generated by computer.

Example 4.3. At the seventh examination of the offspring in the Framingham Heart Study ($n = 3539$), numerous continuous variables were measured, including systolic and diastolic blood pressure, total serum cholesterol, height, and weight. Using the heights and weights measured on each participant, we can compute their BMI. In this study, height is measured in inches and weight in pounds. The following formula is used to compute BMI using these metrics:

$$\text{BMI} = 703.03 \times \frac{\text{Weight in pounds}}{(\text{Height in inches})^2}$$

To illustrate the computation of descriptive statistics for continuous variables, we randomly selected a subset of

TABLE 4–13 Subsample of $n = 10$ Participants Attending the Seventh Examination of the Framingham Offspring Study

Participant ID	Systolic Blood Pressure	Diastolic Blood Pressure	Total Serum Cholesterol	Weight (lbs)	Height (in.)	BMI
1	141	76	199	138	63.00	24.4
2	119	64	150	183	69.75	26.4
3	122	62	227	153	65.75	24.9
4	127	81	227	178	70.00	25.5
5	125	70	163	161	70.50	22.8
6	123	72	210	206	70.00	29.6
7	105	81	205	235	72.00	31.9
8	113	63	275	151	60.75	28.8
9	106	67	208	213	69.00	31.5
10	131	77	159	142	61.00	26.8

10 participants who completed the seventh examination of the Framingham Offspring Study. The data values are shown in **Table 4–13**. The first column contains a unique identification number for each participant, the second through sixth columns contain the actual measurements on participants, and the rightmost column contains the BMI computed using the formula shown. Descriptive statistics for each continuous variable are now computed. Formulas for the computations are presented within examples and then summarized at the end of this chapter.

The first summary statistic for a continuous variable (as well as for dichotomous, categorical, and ordinal variables) is the sample size. The sample size here is $n = 10$. It is always important to report the sample size to convey the size of the study. Larger studies are generally viewed more favorably, as larger sample sizes generally produce more precise results. However, there is a point at which increasing the sample size does not materially increase the precision of the analysis. (Sample size computations are discussed in detail in Chapter 8.)

Because this sample is small ($n = 10$), it is relatively easy to summarize the sample by inspecting the observed values. Suppose we first consider the diastolic blood pressures. To facilitate interpretation, we order the diastolic blood pressures from smallest to largest:

62 63 64 67 70
72 76 77 81 81

Diastolic blood pressures less than 80 are considered normal (see Table 4–1); thus the participants in this sample can be summarized as generally having normal diastolic pressures. There are two participants with diastolic blood pressures of 81, but they hardly exceed the upper limit of the "normal" classification. The diastolic blood pressures in this sample are not all identical (with the exception of the two measured values of 81) but they are relatively similar. In general, the participants in this sample can be described as having healthy diastolic blood pressures from a clinical standpoint.

For larger samples, such as the full seventh examination of the Framingham Offspring Study with $n = 3539$, it is impossible to inspect individual values to generate a summary, so summary statistics are necessary. There are two general aspects of a useful summary for a continuous variable. The first is a description of the center or average of the data (i.e., what is a typical value), and the second addresses variability in the data.

Using the diastolic blood pressures, we now illustrate the computation of several statistics that describe the average value and the variability of the data. In biostatistics, the term "average" is a very general term. There are several statistics that describe the average value of a continuous variable. The first is probably the most familiar—the *sample mean*. The sample mean is computed by summing all of the values and dividing by the sample size. For the sample of diastolic blood pressures, the sample mean is computed as follows:

$$\text{Sample Mean} = \frac{62+63+64+67+70+72+76+77+81+81}{10}$$

$$= \frac{713}{10} = 71.3$$

To simplify the formulas for sample statistics (and for population parameters), we usually denote the variable of interest as *X*. *X* is simply a placeholder for the variable being analyzed. Here *X* = diastolic blood pressure. The sample mean is denoted $\bar{X}$ (read "X bar"), and the formula for the sample mean is

$$\bar{X} = \frac{\Sigma X}{n}$$

where Σ indicates summation (i.e., the sum of the diastolic blood pressures in this sample). The mean diastolic blood pressure is $\bar{X} = 71.3$.

When reporting summary statistics for a continuous variable, the convention is to report one more decimal place than the number of decimal places measured. Here, systolic and diastolic blood pressures, total serum cholesterol, and weight are measured to the nearest integer, therefore the summary statistics are reported to the nearest tenths place. Height is measured to the nearest quarter inch (hundredths); therefore the summary statistics are reported to the nearest thousandths place. BMI is computed to the nearest tenth, so summary statistics are reported to the nearest hundredths place.

The sample mean is one measure of the average diastolic blood pressure. A second measure of the average value is the **sample median**. The sample median is the middle value in the ordered dataset, or the value that separates the top 50% of the values from the bottom 50%. When there is an odd number of observations in the sample, the median is the value that holds as many values above it as below it in the ordered dataset. When there is an even number of observations in the sample, the median is defined as the mean of the two middle values in the ordered dataset. In the sample of *n* = 10 diastolic blood pressures, the two middle values are 70 and 72, and thus the median is (70 + 72)/2 = 71. Half of the diastolic blood pressures are above 71 and half are below.

The mean and median provide different information about the average value of a continuous variable. Suppose the sample of 10 diastolic blood pressures looked like this:

62	63	64	67	70
72	76	77	81	140

The sample mean for this sample is $\bar{X}$ = 772/10 = 77.2. This does not represent a typical value, as the majority of diastolic blood pressures in this sample are below 77.2. The extreme value of 140 is affecting the computation of the mean. For this same sample, the median is 71. The median is unaffected by extreme or outlying values. For this reason, the median

TABLE 4–14 Means and Medians of Variables in Subsample of Size *n* = 10

	Mean	Median
Diastolic blood pressure	71.3	71.0
Systolic blood pressure	121.2	122.5
Total serum cholesterol	202.3	206.5
Weight (lbs)	176.0	169.5
Height (in.)	67.175	69.375
Body mass index (BMI)	27.26	26.60

is preferred over the mean when there are extreme values (values either very small or very large relative to the others). When there are no extreme values, the mean is the preferred measure of a typical value, in part because each observation is considered in the computation of the mean. When there are no extreme values in a sample, the mean and median of the sample will be close in value.

Table 4–14 displays the sample means and medians for each of the continuous measures in the sample of *n* = 10. For each continuous variable measured in this subsample of participants, the means and medians are not identical but are relatively close in value, suggesting that the mean is the most appropriate summary of a typical value for each of these variables. (If the mean and median are very different, it suggests that there are outliers affecting the mean.)

A third measure of a typical value for a continuous variable is the mode. The **mode** is defined as the most frequent value. The mode of the diastolic blood pressures is 81, the mode of the total cholesterol levels is 227, and the mode of the heights is 70 because these values each appear twice, whereas the other values only appear once. For each of the other continuous variables, there are 10 distinct values and thus there is no mode (because no value appears more frequently than any other). Suppose the diastolic blood pressures looked like:

62	63	64	64	70
72	76	77	81	81

In this sample, there are two modes, 64 and 81. The mode is a useful summary statistic for a continuous variable. It is presented not instead of either the mean or the median but rather in addition to the mean or median.

The second aspect of a continuous variable that must be summarized is the variability in the sample. A relatively crude

TABLE 4–15 Ranges of Variables in Subsample of Size n = 10

	Minimum	Maximum	Range
Diastolic blood pressure	62	81	19
Systolic blood pressure	105	141	36
Total serum cholesterol	150	275	125
Weight (lbs)	138	235	97
Height (in.)	60.75	72.00	11.25
Body mass index (BMI)	22.8	31.9	9.1

TABLE 4–16 Deviations from the Mean

Diastolic Blood Pressure (X)	Deviation from the Mean ($X - \overline{X}$)
76	4.7
64	−7.3
62	−9.3
81	9.7
70	−1.3
72	0.7
81	9.7
63	−8.3
67	−4.3
77	5.7
$\Sigma X = 713$	$\Sigma(X - \overline{X}) = 0$

yet important measure of variability in a sample is the **sample range**. The sample range is computed as follows:

Sample range = Maximum value − Minimum value

Table 4–15 displays the sample ranges for each of the continuous measures in the subsample of $n = 10$ observations. The range of a variable depends on the scale of measurement. The blood pressures are measured in millimeters of mercury, total cholesterol is measured in milligrams per deciliter, weight in pounds, and so on. The range of total serum cholesterol is large, with the minimum and maximum in the sample of size $n = 10$ differing by 125 units. In contrast, the heights of participants are more homogeneous, with a range of 11.25 inches. The range is an important descriptive statistic for a continuous variable, but it is based on only two values in the dataset. Like the mean, the sample range can be affected by extreme values and thus it must be interpreted with caution. The most widely used measure of variability for a continuous variable is called the **standard deviation**, which is described now.

Assuming that there are no extreme or outlying values of the variable, the mean is the most appropriate summary of a typical value. To summarize variability in the data, we specifically estimate the variability in the sample around the sample mean. If all of the observed values in a sample are close to the sample mean, the standard deviation is small (i.e., close to zero), and if the observed values vary widely around the sample mean, the standard deviation is large. If all of the values in the sample are identical, the sample standard deviation is zero.

In the sample of $n = 10$ diastolic blood pressures, we found $\overline{X} = 71.3$. **Table 4–16** displays each of the observed values along with the respective deviations from the sample mean. The deviations from the mean reflect how far each individual's diastolic blood pressure is from the mean diastolic blood pressure. The first participant's diastolic blood pressure is 4.7 units above the mean, whereas the second participant's diastolic blood pressure is 7.3 units below the mean. What we need is a summary of these deviations from the mean, in particular a measure of how far (on average) each participant is from the mean diastolic blood pressure. If we compute the mean of the deviations by summing the deviations and dividing by the sample size, we run into a problem: the sum of the deviations from the mean is zero. This will always be the case as it is a property of the sample mean—the sum of the deviations below the mean will always equal the sum of the deviations above the mean.

The goal is to capture the magnitude of these deviations in a summary measure. To address this problem of the deviations summing to zero, we could take the absolute values or the squares of each deviation from the mean. Both methods address the problem. The more popular method to summarize the deviations from the mean involves squaring the deviations. (Absolute values are difficult in mathematical proofs, which are beyond the scope of this book.) **Table 4–17** displays each of the observed values, the respective deviations from the sample mean, and the squared deviations from the mean.

The squared deviations are interpreted as follows. The first participant's squared deviation is 22.09, meaning that his or her diastolic blood pressure is 22.09 units squared from the mean diastolic blood pressure. The second participant's diastolic blood pressure is 53.29 units squared from the mean diastolic blood pressure. A quantity that is often used to measure variability in a sample is called the **sample variance**, and it is essentially the mean of the squared deviations. The sample variance is denoted s^2 and is computed as follows:

$$s^2 = \frac{\sum (X - \bar{X})^2}{n - 1}$$

The sample variance is not actually the mean of the squared deviations because we divide by $(n - 1)$ instead of n. In statistical inference (which is described in detail in Chapters 6, 7, 9, 10, and 11), we make generalizations or estimates of population parameters based on sample statistics. If we were to compute the sample variance by taking the mean of the squared deviations and divide by n, we would consistently underestimate the true population variance. Dividing by $(n - 1)$ produces a better estimate of the population variance. The sample variance is nonetheless usually interpreted as the average squared deviation from the mean. In this sample of $n = 10$ diastolic blood pressures, the sample variance is $s^2 = 472.10 / 9 = 52.46$. Thus, on average, diastolic blood pressures are 52.46 units squared from the mean diastolic blood pressure.

Because of the squaring, the variance is not particularly interpretable. The more common measure of variability in a sample is the **sample standard deviation**, defined as the square root of the sample variance:

$$s = \sqrt{s^2} = \sqrt{\frac{\sum (X - \bar{X})^2}{n - 1}}$$

The sample standard deviation of the diastolic blood pressures is $s = \sqrt{52.46} = 7.2$. On average, diastolic blood pressures are 7.2 units from (above or below) the mean diastolic blood pressure.

When a dataset has outliers, or extreme values, we summarize a typical value using the median as opposed to the mean. When a dataset has outliers, variability is often summarized by a statistic called the interquartile range (IQR). The **interquartile range** is the difference between the first and third quartiles. The first quartile, denoted Q_1, is the value in the dataset that holds 25% of the values below it. The third quartile, denoted Q_3, is the value in the dataset that holds 25% of the values above it. The IQR is defined as

$$IQR = Q_3 - Q_1$$

For the sample of $n = 10$ diastolic blood pressures, the median is 71 (50% of the values are above 71 and 50% are below). The quartiles can be computed in the same way we computed the median, but we consider each half of the dataset separately (see **Figure 4–16**).

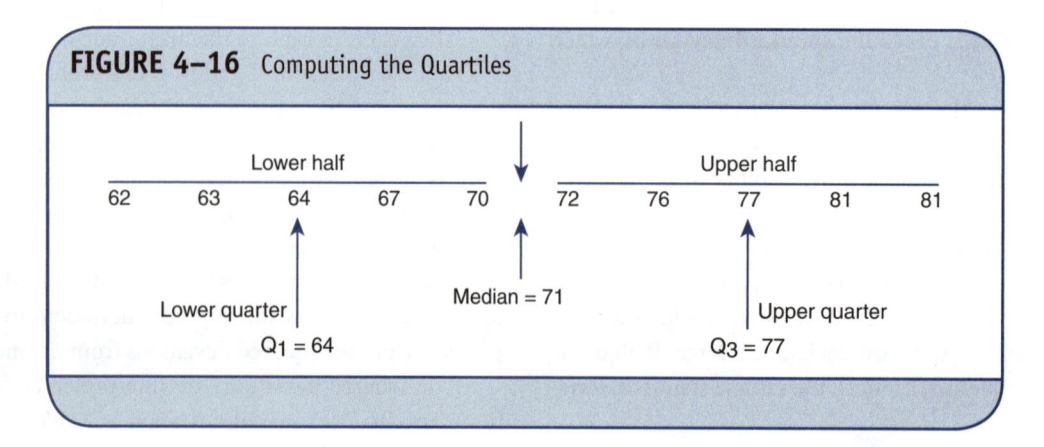

TABLE 4–17 Squared Deviations from the Mean

Diastolic Blood Pressure (X)	Deviation from the Mean $(X-\bar{X})$	Squared Deviation from the Mean $(X-\bar{X})^2$
76	4.7	22.09
64	−7.3	53.29
62	−9.3	86.49
81	9.7	94.09
70	−1.3	1.69
72	0.7	0.49
81	9.7	94.09
63	−8.3	68.89
67	−4.3	18.49
77	5.7	32.49
$\Sigma X = 713$	$\Sigma(X-\bar{X}) = 0$	$\Sigma(X-\bar{X})^2 = 472.10$

FIGURE 4–16 Computing the Quartiles

Lower half: 62 63 64 67 70
Upper half: 72 76 77 81 81
Median = 71
Lower quarter: $Q_1 = 64$
Upper quarter: $Q_3 = 77$

There are five values below the median (lower half) and the middle value is 64, which is the first quartile. There are five values above the median (upper half) and the middle value is 77, which is the third quartile. The IQR is 77 − 64 = 13; the IQR is the range of the middle 50% of the data. When the sample size is odd, the median and quartiles are determined in the same way. Suppose in the previous example that the lowest value (62) was excluded and the sample size was $n = 9$. The median and quartiles are indicated graphically in **Figure 4–17**. When the sample size is 9, the median is the middle number, 72. The quartiles are determined in the same way, looking at the lower and upper halves, respectively. There are four values in the lower half, so the first quartile is the mean of the two middle values in the lower half, (64 + 67) / 2 = 65.5. The same approach is used in the upper half to determine the third quartile, (77 + 81) / 2 = 79. Some statistical computing packages use slightly different algorithms to compute the quartiles. Results can be different, especially for small samples.

When there are no outliers in a sample, the mean and standard deviation are used to summarize a typical value and the variability in the sample, respectively. When there are outliers in a sample, the median and IQR are used to summarize a typical value and the variability in the sample, respectively.

An important issue is determining whether a sample has outliers or not. There are several methods to determine outliers in a sample. A very popular method is based on the following:

Outliers are values below $Q_1 − 1.5 \times (Q_3 − Q_1)$
or above $Q_3 + 1.5 \times (Q_3 − Q_1)$,
or equivalently, values below $Q_1 − 1.5 \times IQR$
or above $Q_3 + 1.5 \times IQR$

These are referred to as *Tukey fences*.[6] For the diastolic blood pressures, the lower limit is $64 − 1.5 \times (77 − 64) = 44.5$ and the upper limit is $77 + 1.5 \times (77 − 64) = 96.5$. The diastolic blood pressures range from 62 to 81; therefore there are no outliers. The best summary of a typical diastolic blood pressure is the mean ($\bar{X} = 71.3$) and the best summary of variability is given by the standard deviation ($s = 7.2$).

Table 4–18 displays the means, standard deviations, medians, quartiles, and IQRs for each of the continuous

FIGURE 4–17 Median and Quartiles for n = 9

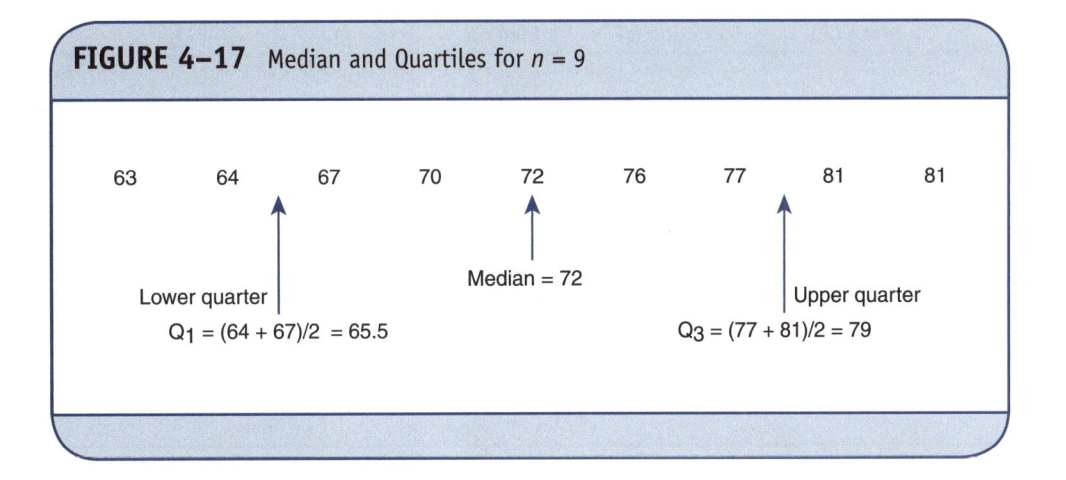

63   64   67   70   72   76   77   81   81

Lower quarter
$Q_1 = (64 + 67)/2 = 65.5$

Median = 72

Upper quarter
$Q_3 = (77 + 81)/2 = 79$

TABLE 4–18 Summary Statistics on n = 10 Participants Attending the Seventh Examination of the Framingham Offspring Study

	Mean ($\bar{X}$)	Standard Deviation (s)	Median	Q_1	Q_3	IQR
Systolic blood pressure	121.2	11.1	122.5	113.0	127.0	14.0
Diastolic blood pressure	71.3	7.2	71.0	64.0	77.0	13.0
Total serum cholesterol	202.3	37.7	206.5	163.0	227.0	64.0
Weight (lbs)	176.0	33.0	169.5	151.0	206.0	55.0
Height (in.)	67.175	4.205	69.375	63.0	70.0	7.0
Body mass index (BMI)	27.26	3.10	26.60	24.9	29.6	4.7

TABLE 4–19 Limits for Assessing Outliers in Characteristics Measured in $n = 10$ Participants Attending the Seventh Examination of the Framingham Offspring Study

	Minimum	Maximum	Lower Limit[a]	Upper Limit[b]
Systolic blood pressure	105	141	92	148
Diastolic blood pressure	62	81	44.5	96.5
Total serum cholesterol	150	275	67	323
Weight (lbs)	138	235	68.5	288.5
Height (in.)	60.75	72.00	52.5	80.5
Body mass index (BMI)	22.8	31.9	17.85	36.65

[a] Determined by $Q_1 - 1.5 \times (Q_3 - Q_1)$.
[b] Determined by $Q_3 + 1.5 \times (Q_3 - Q_1)$.

TABLE 4–20 Summary Statistics on Sample of Participants Attending the Seventh Examination of the Framingham Offspring Study ($n = 3539$)

	Mean ($\overline{X}$)	Standard Deviation (s)	Median	Q_1	Q_3	IQR
Systolic blood pressure	127.3	19.0	125.0	114.0	138.0	24.0
Diastolic blood pressure	74.0	9.9	74.0	67.0	80.0	13.0
Total serum cholesterol	200.3	36.8	198.0	175.0	223.0	48.0
Weight (lbs)	174.4	38.7	170.0	146.0	198.0	52.0
Height (in.)	65.957	3.749	65.750	63.000	68.750	5.75
Body mass index (BMI)	28.15	5.32	27.40	24.5	30.8	6.3

variables displayed in Table 4–13 in the subsample of $n = 10$ participants who attended the seventh examination of the Framingham Offspring Study. **Table 4–19** displays the observed minimum and maximum values along with the limits to determine outliers using the quartile rule for each of the variables in the subsample of $n = 10$ participants. Are there outliers in any of the variables? Which statistics are most appropriate to summarize the average or typical value and the dispersion or variability? Because there are no suspected outliers in the subsample of $n = 10$ participants, the mean and standard deviation are the most appropriate statistics to summarize average values and dispersion, respectively, of each of these characteristics.

Table 4–20 displays the means, standard deviations, medians, quartiles, and IQRs for each of the continuous variables displayed in Table 4–13 in the full sample ($n = 3539$) of participants who attended the seventh examination of the Framingham Offspring Study. Looking just at the means and medians, does it appear that any of the characteristics are subject to outliers in the full sample?

Table 4–21 displays the observed minimum and maximum values along with the limits to determine outliers using the quartile rule for each of the variables in the full sample ($n = 3539$) of participants who attended the seventh examination of the Framingham Offspring Study. Are there outliers in any of the variables? Which statistics are most

TABLE 4–21 Limits for Assessing Outliers in Characteristics Measured in Participants Attending the Seventh Examination of the Framingham Offspring Study

	Minimum	Maximum	Lower Limit[a]	Upper Limit[b]
Systolic blood pressure	81.0	216.0	78	174
Diastolic blood pressure	41.0	114.0	47.5	99.5
Total serum cholesterol	83.0	357.0	103	295
Weight (lbs)	90.0	375.0	68.0	276.0
Height (in.)	55.00	78.75	54.4	77.4
Body mass index (BMI)	15.8	64.0	15.05	40.25

[a] Determined by $Q_1 - 1.5 \times (Q_3 - Q_1)$.
[b] Determined by $Q_3 + 1.5 \times (Q_3 - Q_1)$.

appropriate to summarize the average or typical values and the dispersion or variability for each variable?

In the full sample, each of the characteristics has outliers on the upper end of the distribution, as the maximum values exceed the upper limits in each case. There are also outliers on the low end for diastolic blood pressure and total cholesterol, as the minimums are below the lower limits. For some of these characteristics, the difference between the upper limit and the maximum (or the lower limit and the minimum) is small (e.g., height, systolic and diastolic blood pressures), whereas for others (e.g., total cholesterol, weight, and BMI) the difference is much larger. This method for determining outliers is a popular one but is not generally applied as a hard and fast rule. In this application, it would be reasonable to present means and standard deviations for height and systolic and diastolic blood pressures, and medians and IQRs for total cholesterol, weight, and BMI. Another method for assessing whether a distribution is subject to outliers or extreme values is through graphical displays.

4.3.2 Box-Whisker Plots for Continuous Variables

Box-whisker plots are very useful plots for displaying the distribution of a continuous variable. In Example 4.3, we considered a subsample of $n = 10$ participants who attended the seventh examination of the Framingham Offspring Study. We computed the following summary statistics on diastolic blood pressures. These statistics are sometimes referred to as quantiles, or **percentiles**, of the distribution. A specific quantile or percentile is a value in the dataset that holds a specific percentage of the values at or below it. For example, the first quartile is the 25th percentile, meaning that it holds 25% of the values at or below it. The median is the 50th percentile, the third quartile is the 75th percentile, and the maximum is the 100th percentile (i.e., 100% of the values are at or below it).

Minimum	62
Q_1	64
Median	71
Q_3	77
Maximum	81

A **box-whisker** plot is a graphical display of these percentiles. **Figure 4–18** is a box-whisker plot of the diastolic blood pressures measured in the subsample of $n = 10$ participants described in Example 4.3. The horizontal lines represent (from the top) the maximum, the third quartile, the median (also indicated by the dot), the first quartile, and the minimum. The shaded box represents the middle 50% of the distribution (between the first and third quartiles). A box-whisker plot is meant to convey the distribution of a variable at a quick glance.

Figure 4–19 is a box-whisker plot of the diastolic blood pressures measured in the full sample of participants who attended the seventh examination of the Framingham Offspring Study. In the full sample, we determined that there were outliers both at the low and the high end (see Table 4–21). In Figure 4–19, the outliers are displayed as horizontal lines at the top and bottom of the distribution. At the low end of the distribution, there are five values that are considered outliers (i.e., values below 47.5, which was the lower limit for determining outliers). At the high end of the distribution, there are 12 values that are considered

FIGURE 4–18 Box-Whisker Plot of Diastolic Blood Pressures in Subsample of $n = 10$

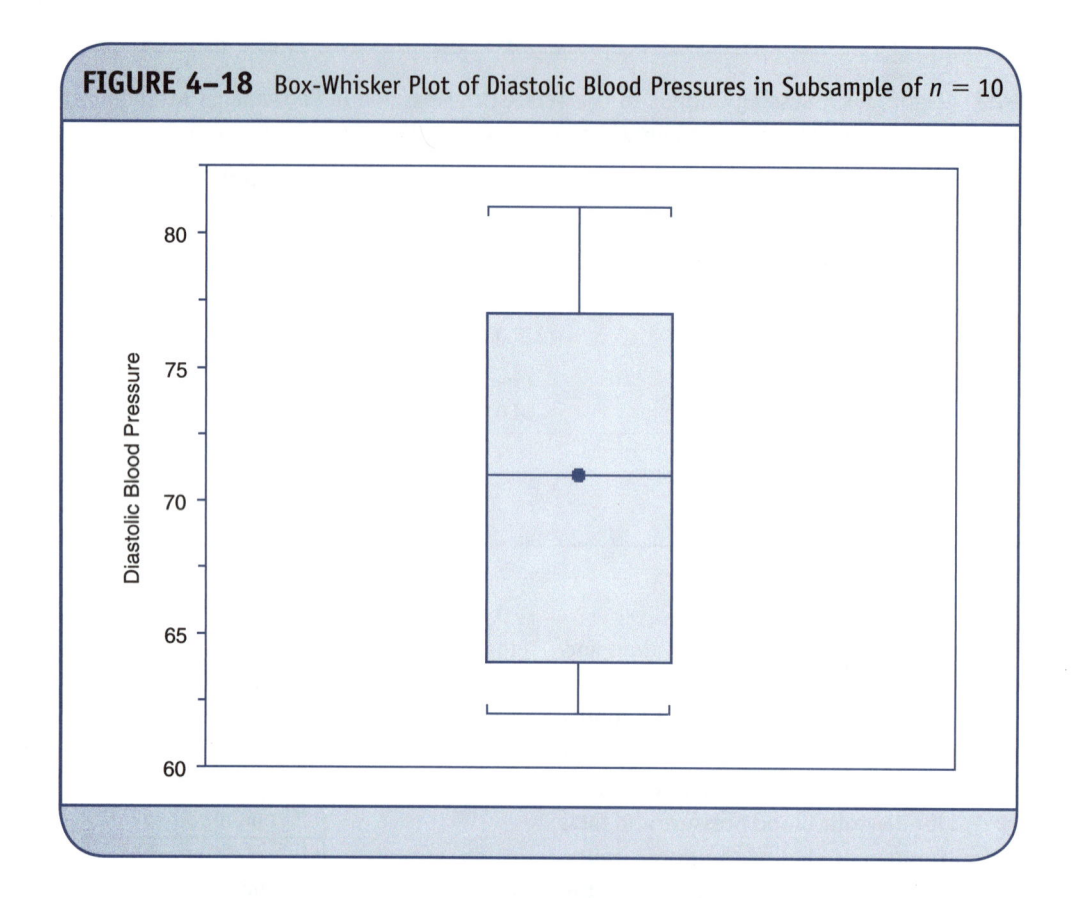

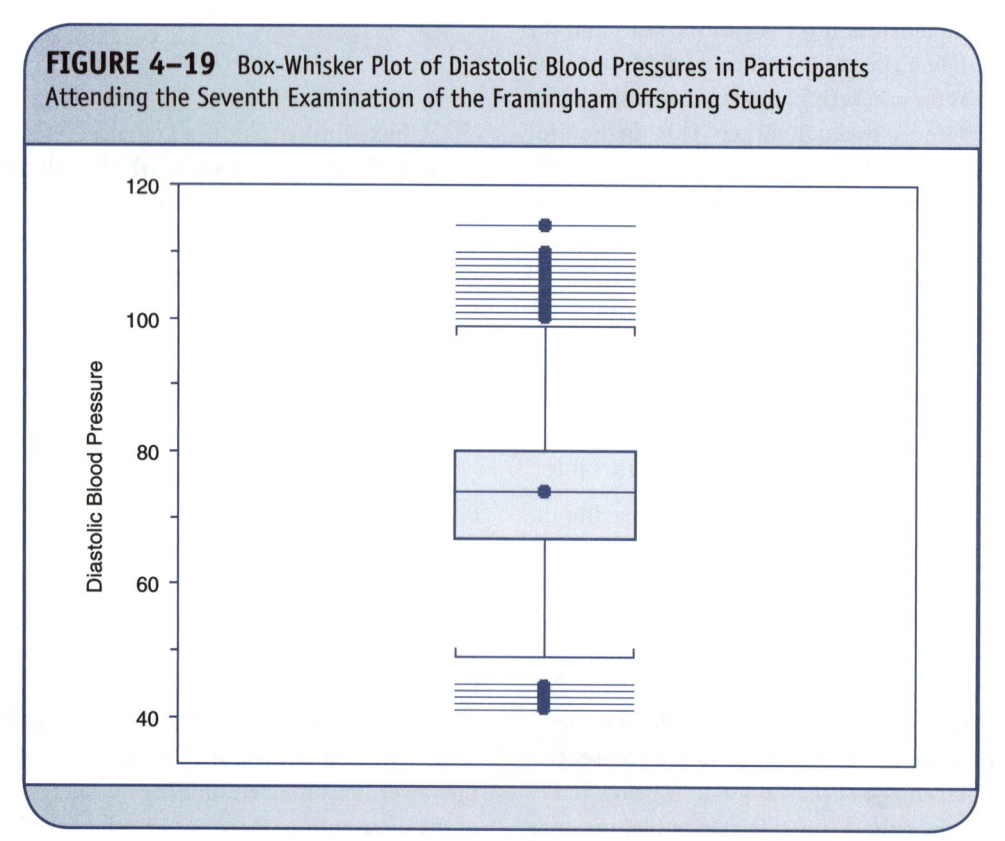

FIGURE 4–19 Box-Whisker Plot of Diastolic Blood Pressures in Participants Attending the Seventh Examination of the Framingham Offspring Study

outliers (i.e., values above 99.5, which was the upper limit for determining outliers). The "whiskers" of the plot (the notched horizontal lines) are the limits we determined for detecting outliers (47.5 and 99.5).

Figure 4–20 is a box-whisker plot of the total serum cholesterol levels measured in the full sample of participants who attended the seventh examination of the Framingham Offspring Study. In the full sample, we determined that there were outliers both at the low and the high end (see Table 4–21). Again in Figure 4–20, the outliers are displayed as horizontal lines at the top and bottom of the distribution. The outliers of total cholesterol are more numerous than those we observed for diastolic blood pressure, particularly on the high end of the distribution.

Box-whisker plots are very useful for comparing distributions. **Figure 4–21** shows side-by-side box-whisker plots of the distributions of weight (in pounds) for men and women attending the seventh examination of the Framingham Offspring Study. The figure clearly shows a shift in the distributions, with men having much higher weights. In fact, the 25th percentile of weight in men is approximately 180 lbs, equal to the 75th percentile in women. Specifically, 25% of the men weigh 180 lbs or less as compared to 75% of the women. There are a substantial number of outliers at the high end of the distribution among both men and women. There are two outlying low values among men.

Because men are generally taller than women (see **Figure 4–22**), it is not surprising that men have greater weights than women. A more appropriate comparison is of BMI (see **Figure 4–23**). The distributions of BMI are similar for men and women. There are again a substantial number of outliers in the distributions in both men and women. However, when taking height into account (by comparing BMI instead of comparing weight), we see that the most extreme outliers are among the women. What statistics would be most appropriate to summarize typical BMI for men and women?

In the box-whisker plots, outliers are values that either exceed $Q_3 + 1.5 \times IQR$ or fall below $Q_1 - 1.5 \times IQR$. Some statistical computing packages use the following to determine outliers: values which either exceed $Q_3 + 3 \times IQR$

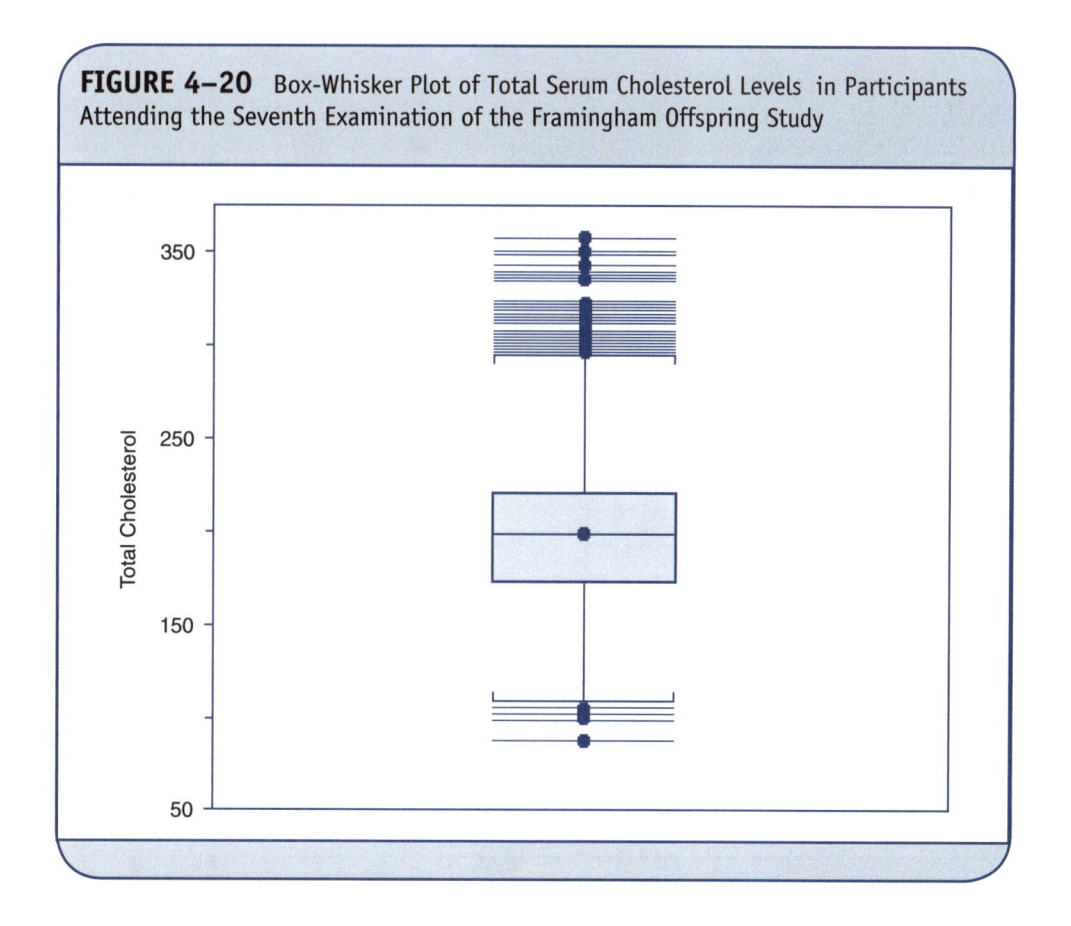

FIGURE 4–20 Box-Whisker Plot of Total Serum Cholesterol Levels in Participants Attending the Seventh Examination of the Framingham Offspring Study

FIGURE 4–21 Side-by-Side Box-Whisker Plots of Weight in Men and Women Attending the Seventh Examination of the Framingham Offspring Study

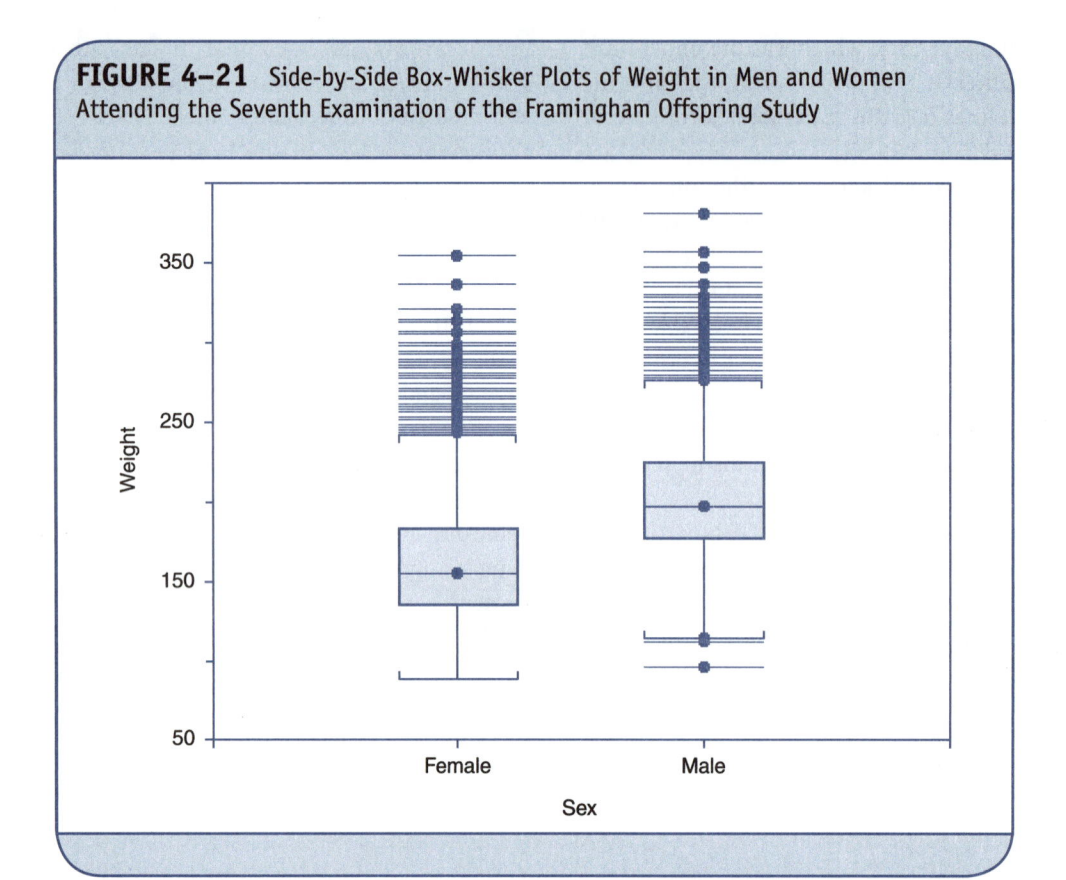

FIGURE 4–22 Side-by-Side Box-Whisker Plots of Height in Men and Women Attending the Seventh Examination of the Framingham Offspring Study

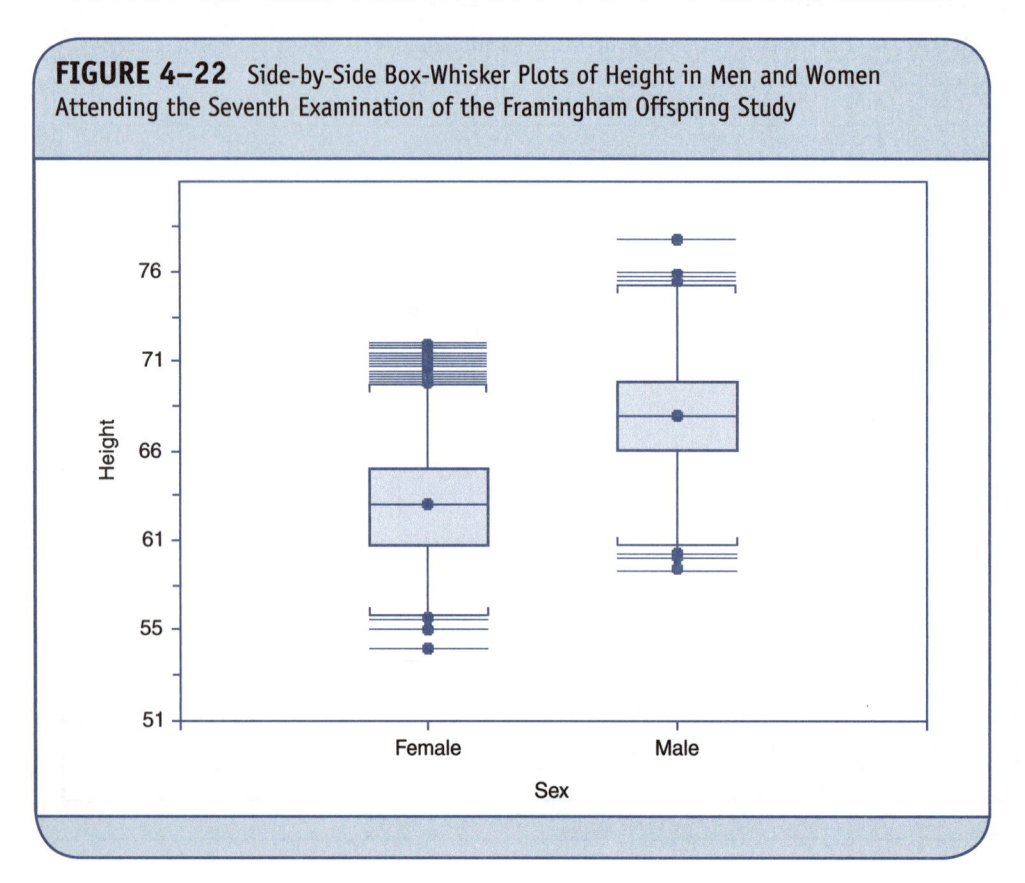

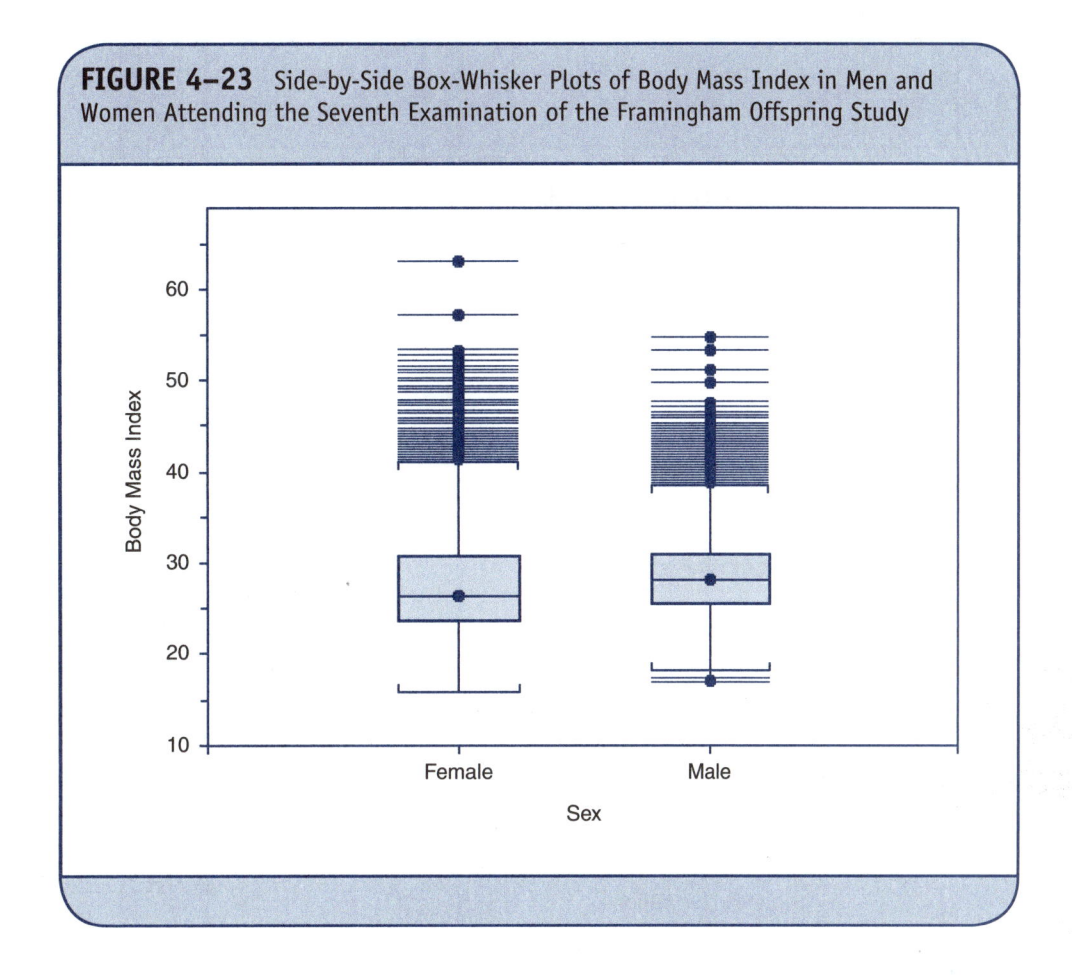

FIGURE 4–23 Side-by-Side Box-Whisker Plots of Body Mass Index in Men and Women Attending the Seventh Examination of the Framingham Offspring Study

or fall below $Q_1 - 3 \times IQR$, which would result in fewer observations being classified as outliers.[7,8] The rule using $1.5 \times IQR$ is the more commonly applied rule to determine outliers.

4.4 SUMMARY

The first important aspect of any statistical analysis is an appropriate summary of the key variables. This involves first identifying the type of variable being analyzed. This step is extremely important, as the appropriate numerical and graphical summaries depend on the type of variable being analyzed. Variables are dichotomous, ordinal, categorical, or continuous. The best numerical summaries for dichotomous, ordinal, and categorical variables are relative frequencies. The best numerical summaries for continuous variables include the mean and standard deviation or the median and interquartile range, depending on whether or not there are outliers in the sample. The mean and standard deviation, or the median and interquartile range, summarize location and dispersion, respectively. The best graphical summary for dichotomous and categorical variables is a bar chart, and the best graphical summary for an ordinal variable is a histogram. Both bar charts and histograms can be designed to display frequencies or relative frequencies, with the latter being the more popular display. Box-whisker plots provide a very useful and informative summary for continuous variables. Box-whisker plots are also useful for comparing the distributions of a continuous variable among mutually exclusive (i.e., non-overlapping) comparison groups. **Figure 4–24** summarizes key statistics and graphical displays organized by variable type.

FIGURE 4-24 Key Statistics and Graphical Displays

Variable Type	Statistic/Graphical Display	Definition
Dichotomous, Ordinal, or Categorical	Relative Frequency	Frequency/n
Dichotomous or Categorical	Frequency or Relative Frequency Bar Chart	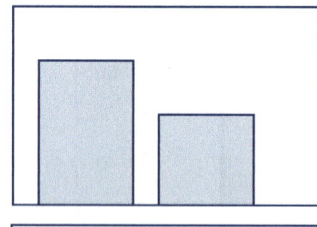
Ordinal	Frequency or Relative Frequency Histogram	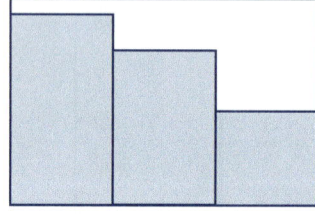
Continuous	Mean	$\overline{X} = \dfrac{\sum X}{n}$
	Standard Deviation	$s = \sqrt{\dfrac{\sum(X - \overline{X})^2}{n-1}}$
	Median	Middle value in ordered dataset
	First Quartile	Q_1 = Value holding 25% below it
	Third Quartile	Q_3 = Value holding 25% above it
	Interquartile Range	IQR = $Q_3 - Q_1$
	Criteria for Outliers	Values below $Q_1 - 1.5 \times (Q_3 - Q_1)$ or above $Q_3 + 1.5 \times (Q_3 - Q_1)$
	Box-Whisker Plot	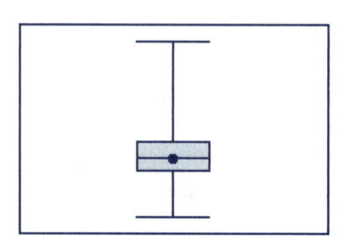

4.5 PRACTICE PROBLEMS

1. A study is run to estimate the mean total cholesterol level in children 2 to 6 years of age. A sample of 9 participants is selected and their total cholesterol levels are measured as follows.

185	225	240	196	175
180	194	147	223	

   a. Compute the sample mean.
   b. Compute the sample standard deviation.
   c. Compute the median.
   d. Compute the first and third quartiles.
   e. Which measure, the mean or median, is a better measure of a typical value? Justify.
   f. Which measure, the standard deviation or the interquartile range, is a better measure of dispersion? Justify.

2. Generate a box-whisker plot for the data in Problem 1.

3. The box-whisker plots in **Figure 4–25** show the distributions of total cholesterol levels in boys and girls 10 to 15 years of age.
   a. What is the median total cholesterol level in boys?
   b. Are there any outliers in total cholesterol in boys? Justify briefly.
   c. What proportion of the boys has total cholesterol less than 205?
   d. What proportion of the girls has total cholesterol less than 205?

4. The following data were collected as part of a study of coffee consumption among graduate students. The following reflect cups per day consumed:

3	4	6	8	2	1	0	2

   a. Compute the sample mean.
   b. Compute the sample standard deviation.
   c. Compute the median.
   d. Compute the first and third quartiles.
   e. Which measure, the mean or median, is a better measure of a typical value? Justify.
   f. Which measure, the standard deviation or the interquartile range, is a better measure of dispersion? Justify.

5. In a study of a new anti-hypertensive medication, systolic blood pressures are measured at baseline. The data are as follows:

120	112	138	145	135
150	145	163	148	128
143	156	160	142	150

   a. Compute the sample mean.
   b. Compute the sample median.
   c. Compute the sample standard deviation.

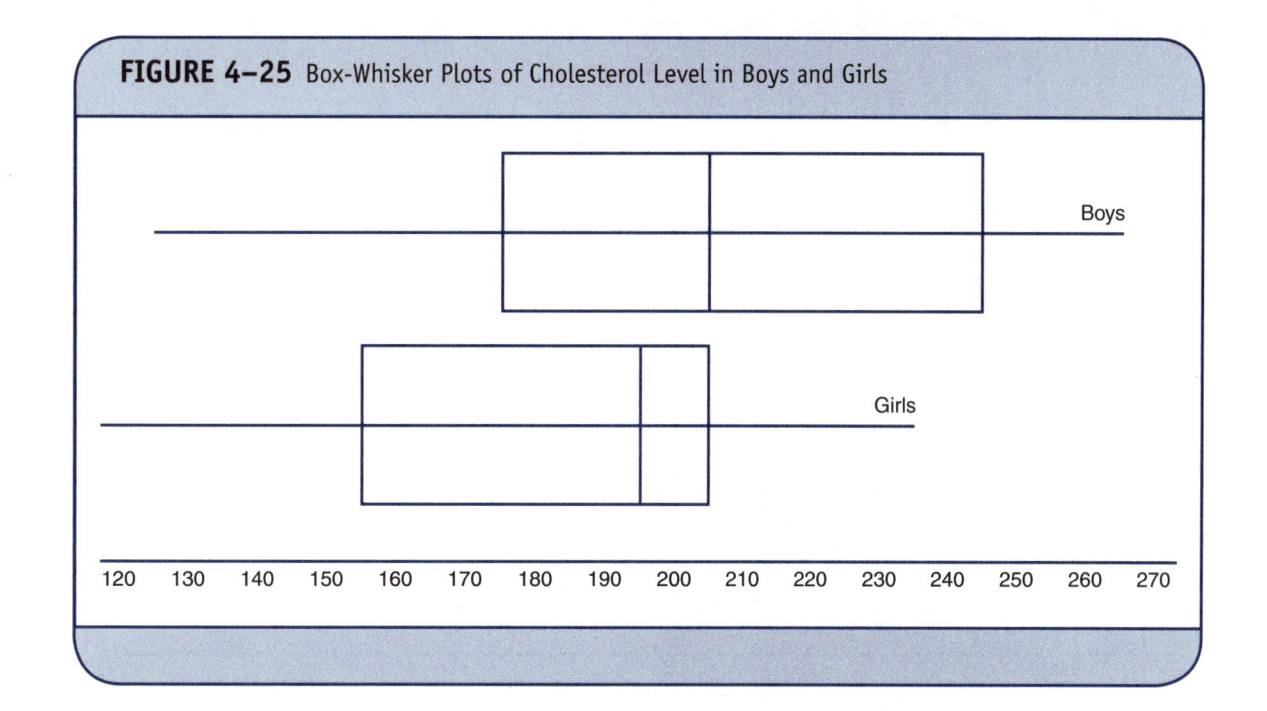

FIGURE 4–25 Box-Whisker Plots of Cholesterol Level in Boys and Girls

d. Compute the sample range.
e. Are there any outliers? Justify.
6. Organize the systolic blood pressures in Problem 5 into categories and develop a frequency distribution table.
7. Generate a relative frequency histogram using the data in Problem 6.
8. **Figure 4–26** shows birth weight (in grams) of infants who were delivered full-term, classified by their mother's weight gain during pregnancy:
   a. What is the median birth weight for mothers who gained 50 to 74 pounds?
   b. What is the interquartile range for mothers who gained 75 pounds or more?
   c. Are there any outliers in birth weight among mothers who gained less than 50 pounds?
   d. What are the best measures of a typical birth weight and the variability in birth weight among mothers who gained less than 50 pounds?
9. The following are baby height measurements (in centimeters) for a sample of infants participating in a study of infant health:

   28 30 41 48 29
   48 62 49 51 39

   a. Compute the sample mean.
   b. Compute the sample standard deviation.
   c. Compute the median.
   d. Compute the first and third quartiles.
   e. Which measure, the mean or median, is a better measure of a typical value? Justify.
   f. Which measure, the standard deviation or the interquartile range, is a better measure of dispersion? Justify.
10. Construct a frequency distribution table using the data in Problem 9 and the following categories: less than 35, 35–44, 45–54, 55 or more.
11. Generate a relative frequency histogram using the data in Problem 10.

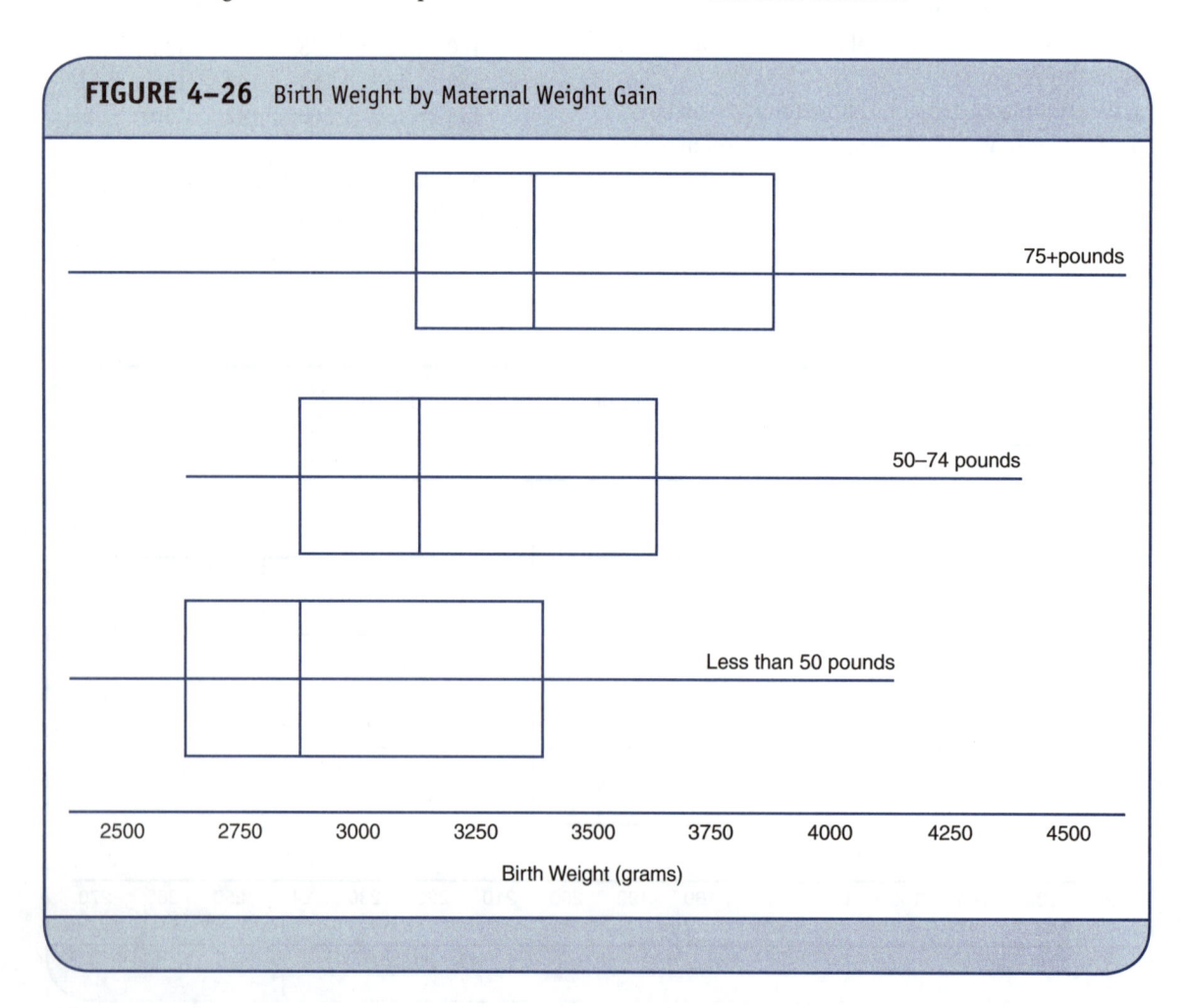

FIGURE 4–26 Birth Weight by Maternal Weight Gain

12. The following data were collected from 10 randomly selected patients undergoing physical therapy following knee surgery. The data represent the percent gain in range of motion after 3 weeks of physical therapy.

    24% 32% 50% 62% 21%
    45% 80% 24% 30% 10%

    a. Compute the sample mean percent gain in range of motion.
    b. Compute the sample standard deviation.
    c. Compute the median.
    d. Compute the first and third quartiles.

13. The following are HDL levels measured in healthy females.

    65 60 58 52 70 54 72 80 38

    a. Compute the sample mean HDL in healthy females.
    b. Compute the sample standard deviation of HDL in healthy females.
    c. Compute the median HDL in healthy females.

14. In a clinical trial for a new cancer treatment, participants are asked to rate their quality of life at baseline. Data for all participants are summarized (in **Table 4–22**). Generate a graphical display of the quality of life at baseline.

15. The following are summary statistics on triglyceride levels measured in a sample of $n = 200$ participants. Are there outliers? Justify your answer.

    $n = 200$ $\bar{X} = 203$ $s = 64$ Min = 38 $Q_1 = 59$
    Median = 180 $Q_3 = 243$ Max = 394

16. **Figure 4–27** is a graphical display of the ages of participants in a study of risk factors for dementia.

TABLE 4–22 Quality of Life

Quality of Life	Number of Participants
Excellent	25
Very good	29
Good	36
Fair	53
Poor	57

Identify the median age, and the first and third quartiles of age.

17. A clinical trial is conducted to evaluate the efficacy of a new drug to increase HDL cholesterol ("good" cholesterol) in men at high risk for cardiovascular disease. Participants are randomized to receive the new drug or placebo and followed for 12 weeks, at which time HDL cholesterol is measured. The data are summarized in **Figure 4–28**.
    a. What is the 25th percentile of HDL in patients receiving the placebo?
    b. What is median HDL in patients receiving the new drug?
    c. Are there any outliers in HDL in patients receiving the new drug? Justify.

18. A pilot study is run to investigate the feasibility of recruiting pregnant women into a study of risk factors for preterm delivery. Women are invited to participate at their first clinical visit for prenatal care. The following represent the gestational ages in weeks of women who consent to participate in the study.

    11 14 21 22 9 10 13 18

    a. Compute the sample mean gestational age.
    b. Compute the sample standard deviation of gestational age.
    c. Compute the median gestational age.
    d. What proportion of the sample enroll in the first trimester of pregnancy (i.e., between 1 week and 13 weeks, inclusive, of pregnancy)?

19. If there are outliers in a sample, which of the following is always true?
    a. Mean > Median
    b. Standard deviation is smaller than expected (smaller than if there were no outliers)
    c. Mean < Median
    d. Standard deviation is larger than expected (larger than if there were no outliers)

20. The following data are a sample of white blood counts in thousands of cells per cubic millimeter for nine patients entering a hospital in Boston, MA, on a given day.

    7 35 5 9 8 3 10 12 8

    a. Are there any outliers in this data? Justify.
    b. Appropriately summarize the data.

FIGURE 4–27 Age of Participants in a Study of Risk Factors for Dementia

FIGURE 4–28 HDL Cholesterol

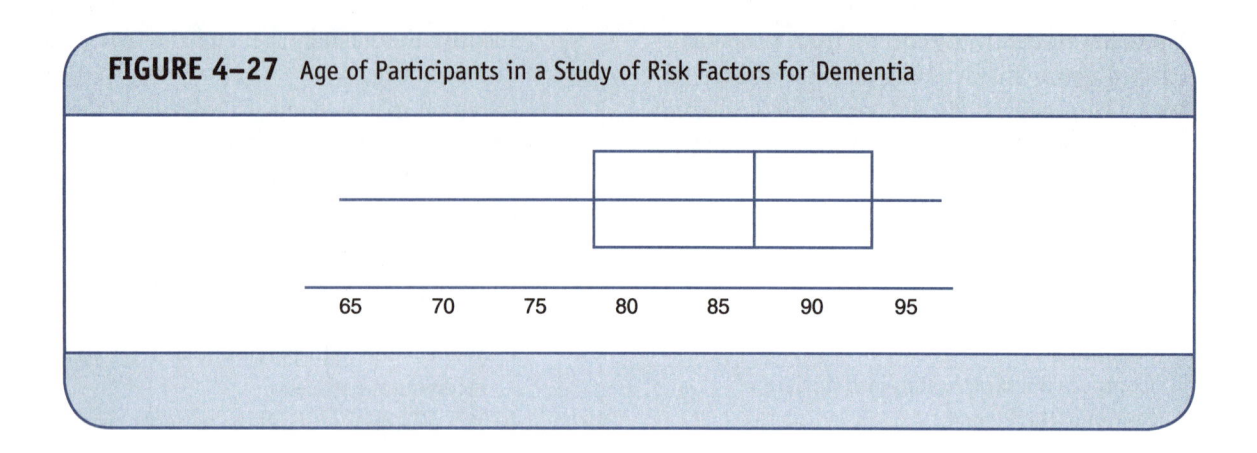

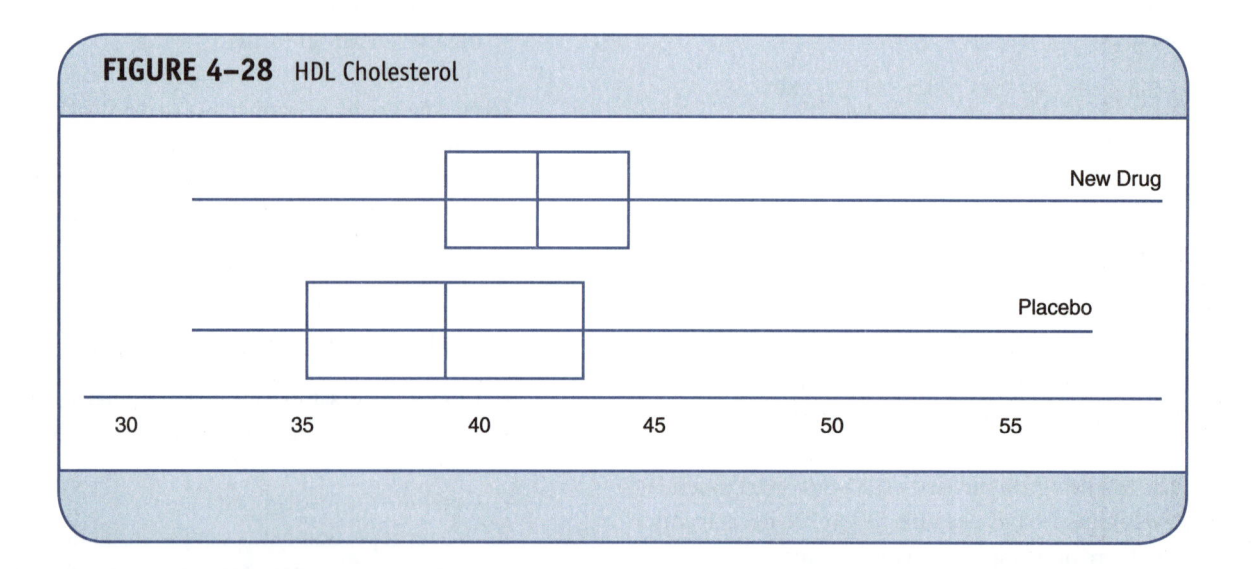

REFERENCES

1. National Heart, Lung, and Blood Institute. Available at *http://www.nhlbi.nih.gov*.
2. Sullivan, L.M. "Repeated measures." *Circulation* 2008; 117(9): 1238–1243.
3. Little, R.J. and Rubin, D.B. *Statistical Analysis with Missing Data*. New York: John Wiley & Sons, 1987.
4. American Heart Association. Available at *http://www.americanheart.org*.
5. The Expert Panel. "Expert panel on detection, evaluation, and treatment of high blood cholesterol in adults: summary of the second report of the NCEP expert panel (Adult Treatment Panel II)." *Journal of the American Medical Association* 1993; 269: 3015–3023.
6. Hoaglin, D.C. "John W. Tukey and data analysis." *Statistical Science* 2003; 18(3): 311–318.
7. SAS version 9.1. © 2002–2003 by SAS Institute, Cary, NC.
8. S-PLUS version 7.0. © 1999–2006 by Insightful Corp., Seattle, WA.

CHAPTER 5

The Role of Probability

When and Why

Key Questions
- How much variability in a measure is reasonable?
- What does it mean when you have a medical test and your result is "within normal limits"?
- Should you be concerned if you take a medical test and your result is not "normal"?

In the News

Hemoglobin is a protein found in red blood cells that carries oxygen. People with low levels might have anemia, whereas people with high levels might have heart or lung problems. In men, the "normal" range is 13.8–17.2 g/dL; in women, the "normal" range is 12.1–15.1 g/dL.[1]

The American Heart Association categorizes systolic (when the heart muscle contracts) and diastolic (when the heart is at rest and refilling) blood pressure levels as normal, pre-hypertensive, stage 1 hypertension, stage 2 hypertension, or hypertensive crisis.[2] Normal systolic blood pressure is less than 120 mm Hg and normal diastolic blood pressure is less than 80 mmHg.

Cholesterol levels—for example, total serum cholesterol, high-density lipoprotein (HDL) cholesterol, and low-density lipoprotein (LDL) cholesterol—are not categorized as normal or not normal, but rather as desirable, borderline high, and high. For some of the cholesterol subfractions, different values define desirable versus borderline high categories in men versus women. Desirable total serum cholesterol for both men and women is less than 200 mg/dL.[3]

Normal ranges of characteristics are usually determined by measurements in a reasonably large sample of healthy volunteers and reflect the range of values that extend two standard deviations from the mean in either direction. Approximately 5% of values outside of that range are also observed in healthy volunteers.

Dig In
- Are there any psychological risks in defining "normal ranges" for tests that, by design, include healthy values that are outside of those ranges?
- Can you think of a currently available test that is particularly problematic in terms of psychological risk? How might we minimize the risks for patients interested in taking that test?
- Can you think of a better approach to defining "normal ranges"?

Learning Objectives

By the end of this chapter, the reader will be able to

- Define the terms "equally likely" and "at random"
- Compute and interpret unconditional and conditional probabilities
- Evaluate and interpret independence of events
- Explain the key features of the binomial distribution model
- Calculate probabilities using the binomial formula
- Explain the key features of the normal distribution model
- Calculate probabilities using the standard normal distribution table
- Compute and interpret percentiles of the normal distribution
- Define and interpret the standard error
- Explain sampling variability
- Apply and interpret the results of the Central Limit Theorem

Probabilities are numbers that reflect the likelihood that a particular event occurs. We hear about probabilities in many everyday situations, ranging from weather forecasts

---

[1] Available at https://www.nlm.nih.gov/medlineplus/ency/article/003645.htm.
[2] Available at http://www.heart.org/HEARTORG/Conditions/HighBloodPressure/AboutHighBloodPressure/Understanding-Blood-Pressure-Readings_UCM_301764_Article.jsp#.V4pYkrgrIps.
[3] Available at https://medlineplus.gov/magazine/issues/summer12/articles/summer12pg6-7.html.

(probability of rain or snow) to the lottery (probability of hitting the big jackpot). In biostatistical applications, it is probability theory that underlies statistical inference. **Statistical inference** involves making generalizations or inferences about unknown population parameters based on sample statistics. A population parameter is any summary measure computed on a population (e.g., the population mean, which is denoted μ; the population variance, which is denoted σ^2; or the population standard deviation, which is denoted σ). In Chapter 4, we presented a number of different statistics to summarize sample data. In Chapters 6, 7, 9, 10, and 11, we discuss statistical inference in detail. Specifically, we present formulas and procedures to make inferences about a population based on a single sample. There are many details to consider but, in general, we select a sample from the population of interest, measure the characteristic under study, summarize this characteristic in our sample, and then make inferences about the population based on what we observe in the sample.

This process can be very powerful. In many statistical applications, we make inferences about a large number of individuals in a population based on a study of only a small fraction of the population (i.e., the sample). This is the power of biostatistics. If a study is replicated or repeated on another sample from the population, it is possible that we might observe slightly different results. This can be due to the fact that the second study involves a slightly different sample, the sample again being only a subset of the population. A third replication might produce yet different results. From any given population, there are many different samples that can be selected. The results based on each sample can vary, and this variability is called sampling variability.

In practice, we select only one sample. Understanding how that study sample is produced is important for statistical inference. In our discussion of probability, we focus on probability as it applies to selecting individuals from a population into a sample. We also focus on quantifying sampling variability. Whenever we make statistical inferences, we must recognize that we are making these inferences based on incomplete information—specifically, we are making inferences about a population based only on a single sample.

5.1 SAMPLING

Probability is important in selecting individuals from a population into a sample, and again in statistical inference when we make generalizations about the population based on that sample. When we select a sample from a population, we want that sample to be representative of the population.

Specifically, we want the sample to be similar to the population in terms of key characteristics. For example, suppose we are interested in studying obesity in residents of the state of California. The Centers for Disease Control and Prevention (CDC) reports data on demographic information on the residents of each state, and some of these demographic characteristics are summarized for the state of California in **Table 5–1**.[1]

In California, the residents are almost equally split by sex, their median age is 33.3 years, and 72.7% are 18 years of age and older. The state has 63.4% whites, 7.4% African Americans, and the remainder are other races. Almost 77% of the residents have a high school diploma, and over 26% have a bachelors degree or higher. If we select a sample of the residents of California for a study, we would not want the sample to be 75% men with a median age of 50. That sample would not be representative of the population. **Table 5–2** contains the same demographic information for California, Massachusetts, Florida, and Alabama.[1]

Notice the differences in demographic characteristics among these states. Previous studies have shown that obesity is related to sex, race, and educational level; therefore, when we select a sample to study obesity, we must be careful to accurately reflect these characteristics.[2-4] Studies have shown that prevalent obesity is inversely related to educational level (i.e., persons with higher levels of education are less likely to be obese). If we select a sample to assess obesity in a state, we would want to take care not to over-represent (or under-represent) persons with lower educational levels as this could inflate (or diminish) our estimate of obesity in that state.

Sampling individuals from a population into a sample is a critically important step in any biostatistical analysis. There are two popular types of sampling, probability sampling and non-probability sampling. In **probability sampling**, each member of the population has a known probability of being

TABLE 5–1 Demographic Characteristics for California

Men (%)	49.8
Median age (years)	33.3
18 years of age and older (%)	72.7
White (%)	63.4
African-American (%)	7.4
High school graduate or higher (%)	76.8
Bachelors degree or higher (%)	26.6

TABLE 5–2 Demographic Characteristics for California, Massachusetts, Florida, and Alabama

	California	Massachusetts	Florida	Alabama
Men (%)	49.8	48.2	48.8	48.3
Median age (years)	33.3	36.5	38.7	35.8
18 years of age and older (%)	72.7	76.4	77.2	74.7
White (%)	63.4	86.2	79.7	72.0
African-American (%)	7.4	6.3	15.5	26.3
High school graduate or higher (%)	76.8	84.8	79.9	75.3
Bachelors degree or higher (%)	26.6	33.2	22.3	19.0

selected. In **non-probability sampling**, each member of the population is selected without the use of probability. Popular methods of each type of sampling are described here. More details can be found in Cochran and Kish.[5,6]

5.1.1 Probability Sampling: Simple Random Sampling

In **simple random sampling**, we start with what is called the **sampling frame**, a complete list or enumeration of the entire population. Each member of the population is assigned a unique identification number, and then a set of numbers is selected at random to determine the individuals to be included in the sample. Many introductory statistical textbooks contain tables of random numbers that can be used, and there are also statistical computing packages that can be used to generate random numbers.[7,8] Simple random sampling is a technique against which many other sampling techniques are compared. It is most useful when the population is relatively small because it requires a complete enumeration of the population as a starting point. In this sampling scheme, each individual in the population has the same chance of being selected. We use N to represent the number of individuals in the population, or the population size. Using simple random sampling, the probability that any individual is selected into the sample is $1 / N$.

5.1.2 Probability Sampling: Systematic Sampling

In **systematic sampling**, we again start with the complete sampling frame and members of the population are assigned unique identification numbers. However, in systematic sampling every third or every fifth person is selected. The spacing or interval between selections is determined by the ratio of the population size to the sample size (N / n). For example, if the population size is $N = 1000$ and a sample size of $n = 100$ is desired, then the sampling interval is $1000 / 100 = 10$; so every tenth person is selected into the sample. The first person is selected at random from among the first ten in the list, and the first selection is made at random using a random numbers table or a computer-generated random number. If the desired sample size is $n = 175$, then the sampling fraction is $1000 / 175 = 5.7$. Clearly, we cannot take every 5.7th person, so we round this down to 5 and take every fifth person. Once the first person is selected at random from the first five in the list, every fifth person is selected from that point on through the end of the list.

There is a possibility that a systematic sample might not be representative of a population. This can occur if there is a systematic arrangement of individuals in the population. Suppose that the population consists of married couples and that the sampling frame is set up to list each husband and then his wife. Selecting every tenth person (or any even-numbered multiple) would result in selecting all men or women depending on the starting point. This is a very extreme example, but as a general principle, all potential sources of systematic bias should be considered in the sampling process.

5.1.3 Probability Sampling: Stratified Sampling

In **stratified sampling**, we split the population into non-overlapping groups or strata (e.g., men and women; people under 30 years of age and people 30 years of age and older) and then sample within each strata. Sampling within each strata can be by simple random sampling or systematic sampling. The idea behind stratified sampling is to ensure adequate representation of individuals within each strata. For example, if a population contains 70% men and 30% women and we want to ensure the same representation in the sample, we can stratify and sample the requisite numbers of men and women to ensure the same representation. For example, if the desired sample size is $n = 200$, then $n = 140$ men and $n = 60$

women could be sampled either by simple random sampling or by systematic sampling.

5.1.4 Non-Probability Sampling: Convenience Sampling

Non-probability samples are often used in practice because in many applications, it is not possible to generate a sampling frame. In non-probability samples, the probability that any individual is selected into the sample is unknown. Whereas it is informative to know the likelihood that any individual is selected from the population into the sample (which is only possible with a complete sampling frame), it is not possible here. However, what is most important is selecting a sample that is representative of the population.

In **convenience sampling**, we select individuals into our sample by any convenient contact. For example, we might approach patients seeking medical care at a particular hospital in a waiting or reception area. Convenience samples are useful for collecting preliminary data. They should not be used for statistical inference as they are generally not constructed to be representative of any specific population.

5.1.5 Non-Probability Sampling: Quota Sampling

In **quota sampling**, we determine a specific number of individuals to select into our sample in each of several non-overlapping groups. The idea is similar to stratified sampling in that we develop non-overlapping groups and sample a predetermined number of individuals within each group. For example, suppose we wish to ensure that the distribution of participants' ages in the sample is similar to that in the population. Suppose our desired sample size is $n = 300$ and we know from census data that in the population, approximately 30% are under age 20, 40% are between 20 and 49, and 30% are 50 years of age and older. We then sample $n = 90$ persons under age 20, $n = 120$ between the ages of 20 and 49, and $n = 90$ who are 50 years of age and older. Sampling proceeds until these totals, or quotas, are reached in each group. Quota sampling is different from stratified sampling because in a stratified sample, individuals within each stratum are selected at random. Here we enroll participants until the quota is reached.

5.2 BASIC CONCEPTS

We now discuss how probabilities are determined and we consider sampling by simple random sampling. A **probability** is a number that reflects the likelihood that a particular event, such as sampling a particular individual from a population into a sample, will occur. Probabilities range from 0 to 1. Sometimes probabilities are converted to percentages, in which case the range is 0% to 100%. A probability of 0 indicates that there is no chance that a particular event will occur, whereas a probability of 1 indicates that an event is certain to occur. In most applications, we are dealing with probabilities between 0 and 1, not inclusive.

Example 5.1. Suppose we wish to conduct a study of obesity in children 5 to 10 years of age who are seeking medical care at a particular pediatric practice. The population includes all children who were seen in the practice in the past 12 months and is summarized in **Table 5–3**.

If we select a child at random (by simple random sampling), then each child has the same probability of being selected. In statistics, the phrase "at random" is synonymous with "equally likely." In this application, each child has the same probability of being selected and that probability is determined by $1 / N$, where N is the population size. Thus, the probability that any child is selected is $1 / 5290 = 0.0002$. In most sampling situations, we are generally not concerned with sampling a specific individual but instead concern ourselves with the probability of sampling certain "types" of individuals. For example, what is the probability of selecting a boy or a child 7 years of age? The following formula can be used to compute probabilities of selecting individuals with specific attributes or characteristics.

$$P(\text{characteristic}) = \frac{\text{Number of persons with characteristic}}{\text{Total number of persons in the population } (N)}$$

If we select a child at random, the probability that we select a boy is computed as follows: $P(\text{boy}) = 2560 / 5290 = 0.484$. This probability can be interpreted in two ways. The probability that we select a boy at random from the population is 0.484, or 48.4%. A second interpretation is that the proportion of boys in the population is 48.4%. The probability that we select a child who is 7 years of age is $P(7 \text{ years of age}) = 913 / 5290 = 0.173$.

This formula can also be used to compute probabilities of characteristics defined more narrowly or more broadly. For example, what is the probability of selecting a boy who is

TABLE 5–3 Children Seen over the Past 12 Months

	Age (years)						Total
	5	6	7	8	9	10	
Boys	432	379	501	410	420	418	2560
Girls	408	513	412	436	461	500	2730
Total	840	892	913	846	881	918	5290

10 years of age? P(boy who is 10 years of age) = 418 / 5290 = 0.079. What is the probability of selecting a child (boy or girl) who is at least 8 years of age? P(at least 8 years of age) = (846 + 881 + 918) / 5290 = 2645 / 5290 = 0.500.

To use this formula to compute probabilities, we must enumerate the population as in Table 5–3. Once the population is enumerated, we can compute probabilities relatively easily by counting the numbers of individuals who possess the characteristic of interest and dividing by the population size.

5.3 CONDITIONAL PROBABILITY

All of the probabilities computed in the previous section are unconditional probabilities. In each case, the denominator is the total population size, $N = 5290$, reflecting the fact that everyone in the entire population is eligible to be selected. Sometimes it is of interest to focus on a particular subset of the population (e.g., a subpopulation). For example, suppose we are interested just in the girls and ask the question, what is the probability of selecting a 9-year-old from the subpopulation of girls? There are a total of $N_G = 2730$ girls (here N_G refers to the population size of girls) and the probability of selecting a 9-year-old from the subpopulation of girls is written as follows: P(9-year-old | girls), where "|" refers to the fact that we are conditioning on or referring to a specific subgroup, and that subgroup is specified to the right of "|".

The conditional probability is computed using the same approach we used to compute unconditional probabilities. P(9-year-old | girls) = 461 / 2730 = 0.169. This means that 16.9% of the girls are 9 years of age. Note that this is not the same as the probability of selecting a 9-year-old girl, which is P(girl who is 9 years of age) = 461 / 5290 = 0.087. What is the probability of selecting a boy from among the 6-year-olds? P(boy | 6 years of age) = 379 / 892 = 0.425. Thus, 42.5% of the 6-year-olds are boys (and 57.5% of the 6-year-olds are girls). Some popular applications of conditional probability in public health and medicine are described here.

5.3.1 Evaluating Screening Tests

Screening tests are often used in clinical practice to assess the likelihood that a person has a particular medical condition. Many screening tests are based on laboratory tests that detect particular markers that are related to a specific disease. For example, a popular screening test for prostate cancer is called the prostate-specific antigen (PSA) test and is recommended for men over 50 years of age. The test measures PSA, which is a protein produced by the prostate gland. The PSA screening test measures the amount of PSA in the blood. Low levels of PSA are normal but elevated levels have been shown to be associated with prostate cancer. The PSA test alone does not provide a diagnosis of prostate cancer. A more invasive test, such as a biopsy—in which tissue is extracted and examined under a microscope—is needed to make a diagnosis. Data such as that shown in Example 5.2 are often compiled to evaluate the usefulness of a screening test.

Example 5.2. Suppose that we wish to investigate whether the PSA test is a useful screening test for prostate cancer. To be useful clinically, we want the test to distinguish between men with and without prostate cancer. Suppose that a population of $N = 120$ men over 50 years of age who are considered high-risk for prostate cancer have both the PSA screening test and a biopsy. The PSA results are reported as low, slightly to moderately elevated, or highly elevated. The classifications are based on the following levels of measured protein, respectively: 0–2.5, 2.6–19.9, and 20 or more nanograms per milliliter (ng/ml).[9] The biopsy provides the diagnosis of prostate cancer. The results of the study are shown in **Table 5–4**.

The probability that a man has prostate cancer given that he has a low level of PSA is P(prostate cancer | low PSA) = 3 / 64 = 0.047. The probability that a man has prostate cancer given that he has a slightly to moderately elevated level of PSA is P(prostate cancer | slightly to moderately elevated PSA) = 13 / 41 = 0.317. The probability that a man has prostate cancer given that he has a highly elevated level of PSA is

TABLE 5–4 Evaluation of PSA Test

PSA Level	Biopsy Results		Total
	Prostate Cancer	No Prostate Cancer	
Low	3	61	64
Slightly to moderately elevated	13	28	41
Highly elevated	12	3	15
Total	28	92	120

P(prostate cancer | highly elevated PSA) = 12 / 15 = 0.80. The probability or likelihood that a man has prostate cancer is highly dependent on his PSA level. Based on these data, is the PSA test a clinically useful screening test?

Some screening tests are based on laboratory tests, whereas others are based on a set of self-administered questions that address a specific state, such as depression, anxiety, or stress. Patients who screen positive (e.g., score high or low on a set of questions) might be referred to undergo subsequent screening or more involved testing to make a diagnosis.

Example 5.3. In pregnancy, women often undergo screening to assess whether their fetus is likely to have Down Syndrome. The screening test evaluates levels of specific hormones in the blood. The screening tests are reported as positive or negative, indicating that a woman is more or less likely to be carrying an affected fetus. Suppose that a population of $N = 4810$ pregnant women undergoes the screening test and the women scored as either positive or negative depending on the levels of hormones in the blood. In addition, suppose that each woman in the study has an amniocentesis. Amniocentesis is an invasive procedure that provides a more definitive assessment as to whether a fetus is affected with Down Syndrome. Amniocentesis is called a diagnostic test, or the gold standard (i.e., the true state). The results of the screening tests and the amniocentesis (diagnostic test) are summarized in **Table 5–5**.

Using the data from Table 5–5, the probability that a woman with a positive test has an affected fetus is P(affected fetus | screen positive) = 9 / 360 = 0.025, and the probability that a woman with a negative test has an affected fetus is P(affected fetus | negative screen positive) = 1 / 4450 = 0.0002. Is the screening test a clinically useful test?

5.3.2 Sensitivity and Specificity

Screening tests are not used to make medical diagnoses but instead to identify individuals most likely to have a certain condition. Ideally, screening tests are not excessively costly and pose little risk to the patients undergoing the test. There are a number of widely used screening tests available today, including the PSA test for prostate cancer, mammograms for breast cancer, and serum and ultrasound assessments for prenatal diagnosis.

When a screening test is proposed, there are two measures that are often used to evaluate its performance, the sensitivity and specificity of the test. Suppose that the results of the screening test are dichotomous—specifically, each person is classified as either positive or negative for having the condition of interest. **Sensitivity** is also called the true positive fraction and is defined as the probability that a diseased person screens positive. **Specificity** is also called the true negative fraction and is defined as the probability that a disease-free person screens negative. To evaluate the screening test, each participant undergoes the test and is classified as positive or negative based on criteria that are specific to the test (e.g., high levels of a marker in a serum test, or the presence of a mass on a mammogram). Each participant also undergoes a diagnostic test, which provides a definitive diagnosis (e.g., an amniocentesis, or a biopsy). A total of N patients complete both the screening test and the diagnostic test. The data are often organized as shown in **Table 5–6**. The results of the screening test are shown in the rows of the table, and the results of the diagnostic test are shown in the columns.

The definitions of sensitivity and specificity are given as:

Sensitivity = True Positive Fraction =
P(screen positive | disease) = $a / (a + c)$.

Specificity = True Negative Fraction =
P(screen negative | disease free) = $d / (b + d)$.

False Positive Fraction =
P(screen positive | disease free) = $b / (b + d)$.

False Negative Fraction =
P(screen negative | disease) = $c / (a + c)$.

TABLE 5–5 Evaluation of Prenatal Screening Test

	Affected Fetus	Unaffected Fetus	Total
Positive	9	351	360
Negative	1	4449	4450
Total	10	4800	4810

TABLE 5–6 Data Layout for Evaluating Screening Tests

	Diseased	Disease Free	Total
Screen positive	a	b	a+b
Screen negative	c	d	c+d
Total	a+c	b+d	N

The false positive fraction is 1−specificity and the false negative fraction is 1−sensitivity. Therefore, knowing sensitivity and specificity captures the information in the false positive and false negative fractions. These are simply alternate ways of expressing the same information. Often, sensitivity and the false positive fraction are reported for a test.

Example 5.4. Consider again the study presented in Example 5.3 in which a population of $N = 4810$ pregnant women has a screening test and an amniocentesis to assess the likelihood of carrying a fetus with Down Syndrome. The results are summarized in Table 5–5.

The performance characteristics of the test are:

$$\text{Sensitivity} = P(\text{screen positive} \mid \text{affected fetus}) = 9/10 = 0.900.$$

$$\text{Specificity} = P(\text{screen negative} \mid \text{unaffected fetus}) = 4449/4800 = 0.927.$$

$$\text{False Positive Fraction} = P(\text{screen positive} \mid \text{unaffected fetus}) = 351/4800 = 0.073.$$

$$\text{False Negative Fraction} = P(\text{screen negative} \mid \text{affected fetus}) = 1/10 = 0.100.$$

These results are interpreted as follows. If a woman is carrying an affected fetus, there is a 90% chance that the screening test will be positive. If the woman is carrying an unaffected fetus (defined here as a fetus free of Down Syndrome, because this test does not assess other abnormalities), there is a 92.7% chance that the screening test will be negative. If a woman is carrying an unaffected fetus, there is a 7.3% chance that the screening test will be positive, and if the woman is carrying an affected fetus, there is a 10% chance that the test will be negative. The false positive and false negative fractions quantify errors in the test. The errors are often of greatest concern. For example, if a woman is carrying an unaffected fetus, there is a 7.3% chance that the test will incorrectly come back positive. This is potentially a serious problem, as a positive test result would likely produce great anxiety for the woman and her family. A false negative result is also problematic. If a woman is carrying an affected fetus, there is a 10% chance that the test will come back negative. The woman and her family might feel a false sense of assurance that the fetus is not affected when, in fact, the screening test missed the abnormality.

The sensitivity and false positive fractions are often reported for screening tests. However, for some tests the specificity and false negative fractions might be the most important. The most important characteristics of any screening test depend on the implications of an error. In all cases, it is important to understand the performance characteristics of any screening test to appropriately interpret results and their implications.

5.3.3 Positive and Negative Predictive Value

Patients undergoing medical testing often ask the following questions: What is the probability that I have the disease if my screening test comes back positive? What is the probability that I do not have the disease if my test comes back negative? These questions again can be answered with conditional probabilities. These quantities are called the **positive predictive value** and **negative predictive value** of a test and are defined in this section. Consider again the data layout of Table 5–6 summarizing the results of screening and diagnostic tests applied to a population of N individuals. The definitions of positive and negative predictive values are given as follows: Positive predictive value = $P(\text{disease} \mid \text{screen positive}) = a/(a + b)$. Negative predictive value = $P(\text{disease free} \mid \text{screen negative}) = d/(c + d)$. Notice the difference in the definitions of the positive and negative predictive values as compared to the sensitivity and specificity provided previously.

Example 5.5. Consider again the study presented in Example 5.3 and Example 5.4, in which a population of $N = 4810$ pregnant women has a screening test and an amniocentesis to assess the likelihood of carrying a fetus with Down Syndrome. The results are summarized in Table 5–5. The positive predictive value $P(\text{affected fetus} \mid \text{screen positive}) = 9/360 = 0.025$ and negative predictive value $P(\text{unaffected} \mid \text{screen negative}) = 4449/4450 = 0.999$. These results are interpreted as follows. If a woman screens positive, there is a 2.5% chance that she is carrying an affected fetus. If a woman screens negative, there is a 99.9% chance that she is carrying an unaffected fetus.

Patients often want to know the probability of having the disease if they screen positive (positive predictive value). As with this example, the positive predictive value can be very low (2.5%) because it depends on the prevalence of the disease in the population. Thankfully, the prevalence of Down Syndrome in the population is low. Using the data in Example 5.5, the prevalence of Down Syndrome in the population of $N = 4810$ women is $10/4810 = 0.002$ (i.e., Down Syndrome affects 2 fetuses per 1000). Whereas this screening test has good performance characteristics (sensitivity of 90% and specificity of 92.7%) because the prevalence of the condition is so low, even a test with a high probability of detecting an affected fetus does not translate into a high positive predictive value. Because positive and negative predictive values depend on the prevalence of the disease, they cannot be estimated in case control designs (see Section 2.2.4 for more details).

5.4 INDEPENDENCE

In probability, two events are said to be **independent** if the probability of one is not affected by the occurrence or non-occurrence of the other. The definition alone is difficult to digest.

Example 5.6. In Example 5.2, we analyzed data from a population of $N = 120$ men who had both a PSA test and a biopsy for prostate cancer. Suppose we have a different test for prostate cancer. This prostate test produces a numerical risk that classifies a man as at low, moderate, or high risk for prostate cancer. A sample of 120 men undergo the new test and also have a biopsy. The data are summarized in **Table 5–7**.

The probability that a man has prostate cancer given he has a low risk is P(prostate cancer | low risk) = 10 / 60 = 0.167. The probability that a man has prostate cancer given he has a moderate risk is P(prostate cancer | moderate risk) = 6 / 36 = 0.167. The probability that a man has prostate cancer given he has a high risk is P(prostate cancer | high risk) = 4 / 24 = 0.167. The probability or likelihood that a man has prostate cancer here is unrelated to or independent of his risk based on the prostate test. Knowing a man's prostate test result does not affect the likelihood that he has prostate cancer in this example. Thus, the likelihood that a man has prostate cancer is independent of his prostate test result.

Independence can be demonstrated in several ways. Consider two events—call them A and B (A could be a low risk based on the prostate test and B a diagnosis of prostate cancer). Two events are independent if

$$P(A \mid B) = P(A) \text{ or if } P(B \mid A) = P(B).$$

For example, define A = *low risk* and B = *prostate cancer*. To check independence, we must compare a conditional and an unconditional probability: P(A | B) = P(low risk | prostate cancer) = 10 / 20 = 0.50, and P(A) = P(low risk) = 60 / 120 = 0.50. The equality of the conditional and unconditional probabilities here indicates independence. Independence can also be checked using P(B | A) = P(prostate cancer | low risk) = 10 / 60 = 0.167, and P(B) = P(prostate cancer) = 20 / 120 = 0.167. Both versions of the definition of independence will give the same result.

Example 5.7. Table 5–8 contains information on a population of $N = 6732$ individuals who are classified as having or not having prevalent cardiovascular disease (CVD). Each individual is also classified in terms of having a family history of cardiovascular disease or not. In this analysis, family history is defined as a first-degree relative (parent or sibling) with diagnosed CVD before age 60.

Are family history and prevalent CVD independent? Another way to ask the question is as follows: Is there a relationship between family history and prevalent CVD? This is a question of independence of events. Let A = *prevalent CVD* and B = *family history of CVD*. (Note that it does not matter how we define A and B; the result will be identical.) We now must check whether P(A | B) = P(A) or if P(B | A) = P(B). Again, it makes no difference which definition is used, the results will be identical. We compute P(A | B) = P(prevalent CVD | family history of CVD) = 491 / 859 = 0.572 and compare it to P(A) = P(prevalent CVD) = 643 / 6732 = 0.096. These probabilities are not equal; therefore family history and prevalent CVD are not independent. In the population, the probability of prevalent CVD is 0.096%, or 9.6% of the population has prevalent CVD. Individuals with a family history of CVD are much more likely to have prevalent CVD. The chance of prevalent CVD given a family history is 57.2%, as compared to 2.6% (152/5873 = 0.026) among patients with no family history.

5.5 BAYES' THEOREM

Bayes' Theorem is a probability rule that can be used to compute a conditional probability based on specific available information.

TABLE 5–7 Prostate Test and Biopsy Results

Prostate Test Risk	Biopsy Results		Total
	Prostate Cancer	No Prostate Cancer	
Low	10	50	60
Moderate	6	30	36
High	4	20	24
Total	20	100	120

TABLE 5–8 Family History and Prevalent CVD

	Prevalent CVD	Free of CVD	Total
Family history of CVD	491	368	859
No family history of CVD	152	5721	5873
Total	643	6089	6732

There are several versions of the theorem, ranging from simple to more involved. The following is the simple statement of the rule:

$$P(A|B) = \frac{P(B|A)\,P(A)}{P(B)}$$

Example 5.8. Suppose a patient exhibits symptoms raising concern with his physician that he may have a particular disease. Suppose the disease is relatively rare with a prevalence of 0.2% (meaning it affects 2 out of every 1000 persons). The physician recommends testing, starting with a screening test. The screening test is noninvasive, based on a blood sample and costs $250. Before agreeing to the screening test, the patient wants to know what will be learned from the test—specifically, he wants to know his chances of having the disease if the test comes back positive. The physician reports that the screening test is widely used and has a reported sensitivity of 85%. In addition, the test comes back positive 8% of the time and negative 92% of the time.

The patient wants to know the positive predictive value or P(disease | screen positive). Using Bayes' Theorem, we can compute this as follows:

P(disease | screen positive) = P(screen positive | disease) P(disease)/P(screen positive).

We know that P(disease) = 0.002, P(screen positive | disease) = 0.85 and P(screen positive) = 0.08. We can now substitute the values into the above equation to compute the desired probability:

P(disease | screen positive) = (0.85)(0.002) / (0.08) = 0.021.

If the patient undergoes the test and it comes back positive, there is a 2.1% chance that he has the disease. Without the test, there is a 0.2% chance that he has the disease (the prevalence in the general population). Should the patient have the screening test?

Another important question that the patient might ask is, What is the chance of a false positive result? Specifically, what is P(screen positive | no disease)? We can compute this conditional probability with the available information using Bayes' Theorem:

P(screen positive | no disease) =

$$\frac{P(\text{no disease} \mid \text{screen positive})P(\text{screen positive})}{P(\text{no disease})}$$

Note that if P(disease) = 0.002, then P(no disease) = 1 − 0.002. The events *disease* and *no disease* are called complementary events. The *no disease* group includes all members of the population not in the *disease* group. The probabilities of complementary events must sum to 1, i.e., P(disease) + P(no disease) = 1. Similarly, P(no disease | screen positive) + P(disease | screen positive) = 1. Hence P(screen positive | no disease) = (1 − 0.021)(0.08) / (1 − 0.002) = 0.078. Using Bayes' Theorem, there is a 7.8% chance that the screening test will be positive in patients free of disease, which is the false positive fraction of the test.

5.6 PROBABILITY MODELS

To compute the probabilities in the previous sections, we counted the number of participants that had the characteristic of interest and divided by the population size. For conditional probabilities, the population size (denominator) was modified to reflect the subpopulation of interest. In each of the examples in the previous sections, we had a tabulation of the population (the sampling frame) that allowed us to compute the desired probabilities. There are instances where a complete enumeration or tabulation is not available. In some of these instances, probability models or mathematical equations can be used to generate probabilities.

There are many probability models, and the model appropriate for a specific application depends on the specific attributes of the application. If an application satisfies specific attributes of a probability model, the model can be used to generate probabilities. In the following two sections, we describe two very popular probability models. These probability models are extremely important in statistical inference, which is discussed in detail in Chapters 6, 7, 9, 10, and 11.

5.6.1 A Probability Model for a Discrete Outcome: The Binomial Distribution

The **binomial distribution** model is an important probability distribution model that is appropriate when a particular experiment or process has two possible outcomes. The name itself, binomial, reflects the dichotomy of responses.

There are many different probability models and if a particular process results in more than two distinct outcomes, a multinomial probability model might be appropriate. Here we focus on the situation in which the outcome is dichotomous. For example, adults with allergies might report relief with medication or not, children with a bacterial infection might respond to antibiotic therapy or not, adults who suffer a myocardial infarction might survive or not, or a medical device such as a coronary stent might be successfully implanted or not. These are just a few examples of applications or processes where the outcome of interest has two possible values.

The two outcomes are often labeled "success" and "failure," with "success" denoting the outcome of interest. In some medical and public health applications, we are interested in quantifying the extent of disease (the unhealthy response). Clearly, the unhealthy response is not a "success"; this nomenclature is simply used for the application of the binomial distribution model. In any application of the binomial distribution, we must clearly specify which outcome is the success and which is the failure.

The binomial distribution model allows us to compute the probability of observing a specified number of successes when the process is repeated a specific number of times (e.g., in a set of patients) and the outcome for a given patient is either a success or a failure. We must first introduce some notation that is necessary for the binomial distribution model. First, we let n denote the number of times the application or process is repeated and x denote the number of successes, out of n, of interest. We let p denote the probability of success for any individual. The binomial distribution model is defined as

$$P(x \text{ successes}) = \frac{n!}{x!(n-x)!} p^x (1-p)^{n-x},$$

where ! denotes factorial, defined as $k! = k(k-1)(k-2)\ldots 1$. For example, $4! = 4(3)(2)1 = 24$, $2! = 2(1) = 2$ and $1! = 1$. There is one special case where $0! = 1$.

Appropriate use of the binomial distribution model depends on satisfying the following three assumptions about the application or process under investigation: each replication of the process results in one of two possible outcomes (success or failure); the probability of success is the same for each replication; and the replications are independent, meaning here that a success in one patient does not influence the probability of success in another.

Example 5.9. Consider the example where adults with allergies report relief from allergic symptoms with a specific medication. Suppose we know that the medication is effective in 80% of patients with allergies who take it as prescribed. If we provide the medication to 10 patients with allergies, what is the probability that it is effective in exactly 7 patients?

Do we satisfy the three assumptions of the binomial distribution model? Each replication of this application involves providing medication to a patient and assessing whether he or she reports relief from symptoms. The outcome is relief from symptoms ("yes" or "no"), and here we will call a reported relief from symptoms a success. The probability of success for each person is 0.8. The final assumption is that the replications are independent. In this setting, this essentially means that the probability of success for any given patient does not depend on the success or failure of any other patient (For more details, see D'Agostino, Sullivan and Beiser.[10]) This is a reasonable assumption in this case. We now need to set the notation: $n = 10$, $x = 7$, and $p = 0.80$. The desired probability is

$$P(7 \text{ successes}) = \frac{10!}{7!(10-7)!} 0.80^7 (1-0.80)^{10-7}.$$

The first (factorial) portion of the formula is computed as follows:

$$\frac{10!}{7!(10-7)!} = \frac{10(9)(8)(7)(6)(5)(4)(3)(2)1}{\{7(6)(5)(4)(3)(2)(1)\}\{(3)(2)(1)\}} =$$

$$\frac{10(9)(8)}{3(2)} = 120.$$

Substituting,

$$P(7 \text{ successes}) = (120)(0.2097)(0.008) = 0.2013.$$

There is a 20.13% chance that exactly 7 of 10 patients will report relief from symptoms when the probability that anyone reports relief is 0.80.

What is the probability that none report relief? We can again use the binomial distribution model with $n = 10$, $x = 0$, and $p = 0.80$:

$$P(0 \text{ successes}) = \frac{10!}{0!(10-0)!} 0.80^0 (1-0.80)^{10-0}.$$

The first (factorial) portion of the formula is computed as follows:

$$\frac{10!}{0!(10-0)!} = \frac{10!}{\{1\}\{10!\}} = 1.$$

Substituting,

$$P(0 \text{ successes}) = (1)(1)(0.0000001024) = 0.0000001024.$$

There is practically no chance that none of the 10 report relief from symptoms when the probability of reporting relief for any individual patient is 0.80.

What is the most likely number of patients who will report relief when the medication is given to 10 patients? If 80% report relief and we consider 10 patients, we expect 8 to report relief. What is the probability that 8 of 10 report relief?

$$P(8 \text{ successes}) = \frac{10!}{8!(10-8)!} 0.80^8 (1-0.80)^{10-8}.$$

The first (factorial) portion of the formula is computed as follows:

$$\frac{10!}{8!(10-8)!} = \frac{10(9)(8)(7)(6)(5)(4)(3)(2)1}{\{8(7)(6)(5)(4)(3)(2)(1)\}\{(2)(1)\}} =$$

$$\frac{10(9)}{2} = 45.$$

Substituting,

$$P(8 \text{ successes}) = (45)(0.1678)(0.04) = 0.3020.$$

There is a 30.20% chance that exactly 8 of 10 patients will report relief from symptoms when the probability that anyone reports relief is 0.80. The probability that exactly 8 report relief will be the highest probability of all possible outcomes (0 through 10).

Many statistical textbooks provide tables of binomial probabilities for specific combinations of n, x, and p.[10] These tables can be used to determine the probability of observing x successes out of n replications when the probability of success is p. The tables can be of limited use as they often contain only some combinations of n, x, and p. The binomial formula can always be used as long as assumptions are appropriately satisfied.

Example 5.10. The likelihood that a patient with a heart attack dies of the attack is 4% (i.e., 4 of 100 die of the attack). Suppose we have 5 patients who suffer a heart attack. What is the probability that all will survive?

Here we are provided with the probability that an attack is fatal. For this example, we call a success a fatal attack, and thus $p = 0.04$. We have $n = 5$ patients and want to know the probability that all survive or, in other words, that none are fatal (0 successes). We again need to assess the assumptions. Each attack is fatal or non-fatal, the likelihood of a fatal attack is 4% for all patients, and the outcome of individual patients is independent. It should be noted that the assumption that the probability of success is constant must be evaluated carefully. The probability that a patient dies from a heart attack depends on many factors including age, the severity of the attack, and other co-morbid conditions.

To apply the binomial formula, we must be convinced that all patients are at the same risk of a fatal attack. The assumption of the independence of events must be evaluated carefully. As long as the patients are unrelated, the assumption is usually appropriate. Prognosis of disease could be higher in members of the same family or in individuals who are co-habitating. In this example, suppose that the 5 patients being analyzed are unrelated, of similar age, and free of co-morbid conditions:

$$P(0 \text{ successes}) = \frac{5!}{0!(5-0)!} 0.04^0 (1-0.04)^{5-0}.$$

The first (factorial) portion of the formula is computed as follows:

$$\frac{5!}{0!(5-0)!} = \frac{5!}{\{1\}\{5\}} = 1.$$

Substituting,

$$P(0 \text{ successes}) = (1)(1)(0.8154) = 0.8154.$$

There is an 81.54% chance that all patients will survive the attack when the chance that any one dies is 0.04. In this example, the possible outcomes are 0, 1, 2, 3, 4, or 5 successes (fatalities). Because the probability of fatality is so low, the most likely response is 0 (all patients survive).

The binomial formula generates the probability of observing exactly x successes out of n. If we want to compute the probability of a range of outcomes, we need to apply the formula more than once. Suppose in this example we wanted to compute the probability that no more than 1 person dies of the attack. Specifically, we want P(no more than 1 success) = P(0 or 1 successes) = P(0 successes) + P(1 success). To compute this probability, we apply the binomial formula twice. We already computed P(0 successes), so we now compute P(1 success):

$$P(1 \text{ success}) = \frac{5!}{1!(5-1)!} 0.04^1 (1-0.04)^{5-1}.$$

The first (factorial) portion of the formula is computed as follows:

$$\frac{5!}{1!(5-1)!} = \frac{5(4)(3)(2)1}{\{1\}\{4(3)(2)(1)\}} = 5.$$

Substituting,

$$P(1 \text{ success}) = (5)(0.04)(0.8493) = 0.1697.$$

P(no more than 1 success) = P(0 or 1 successes) = P(0 successes) + P(1 success) = 0.8154 + 0.1697 = 0.9851.

The chance that no more than 1 of 5 (or equivalently, that at most 1 of 5) die from the attack is 98.51%.

What is the probability that 2 or more of 5 die from the attack? Here we want to compute P(2 or more successes). The possible outcomes are 0, 1, 2, 3, 4, or 5, and the sum of the probabilities of each of these outcomes is 1 (i.e., we are certain to observe either 0, 1, 2, 3, 4, or 5 successes). We just computed P(0 or 1 successes) = 0.9851, so P(2, 3, 4, or 5 successes) = 1 − P(0 or 1 successes) = 1 − 0.9851 = 0.0149. There is a 1.49% chance that 2 or more of 5 will die from the attack.

The mean or expected number of successes of a binomial population is defined as $\mu = np$, and the standard deviation is $\sigma = \sqrt{n(p)(1-p)}$. In Example 5.9, we considered a binomial distribution with $n = 10$ and $p = 0.80$. The mean, or expected, number of successes is $\mu = np = 10(0.80) = 8$, and the standard deviation is $\sigma = \sqrt{n(p)(1-p)} = \sqrt{10(0.8)(0.2)} = 1.3$. In Example 5.10, we considered a binomial distribution with $n = 5$ and $p = 0.04$. The mean, or expected, number of successes is $\mu = np = 5(0.04) = 0.2$ (i.e., the most likely number of successes is 0 out of 5).

5.6.2 A Probability Model for a Continuous Outcome: The Normal Distribution

The normal distribution model is an important probability distribution model that is appropriate when a particular experiment or process results in a continuous outcome. There are many different probability models for continuous outcomes and the appropriate model depends on the distribution of the outcome of interest. The normal probability model applies when the distribution of the continuous outcome follows what is called the Gaussian distribution, or is well described by a bell-shaped curve (**Figure 5–1**).

The horizontal or *x*-axis is used to display the scale of the characteristic being analyzed (e.g., height, weight, systolic blood pressure). The vertical axis reflects the probability of observing each value. Notice that the curve is highest in the middle, suggesting that the middle values have higher probabilities or are more likely to occur. The curve tails off above and below the middle, suggesting that values at either extreme are much less likely to occur. Similar to the binomial distribution model, there are some assumptions for appropriate use of the normal distribution model. The normal distribution model is appropriate for a continuous outcome if the following conditions are true. First, in a normal distribution the mean is equal to the median and also equal to the mode, which is defined as the most frequently observed value. As we discussed in Chapter 4, it is not always the case that the mean and median are equal. For example, if a particular characteristic is subject to outliers, then the mean will not be equal to the median and therefore the characteristic will not follow a normal distribution. An example might be the length of stay in the hospital (measured in days) following a specific procedure. Length of stay often follows a skewed distribution, as illustrated in **Figure 5–2**.

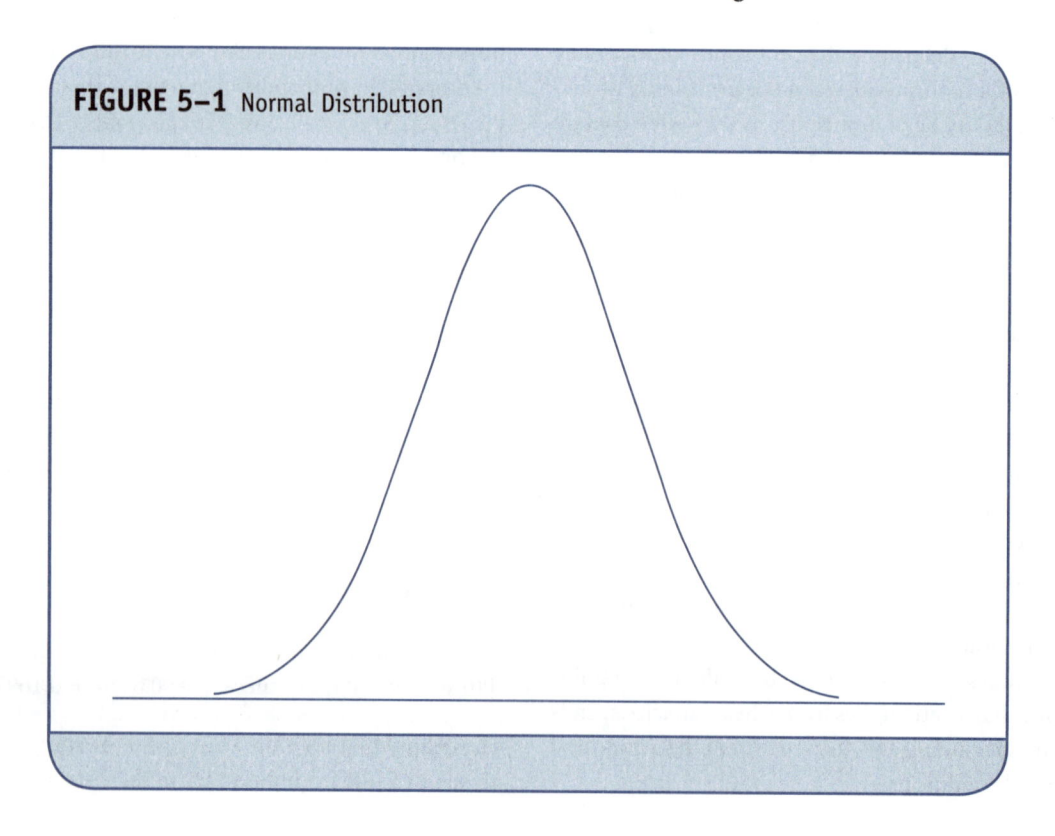

FIGURE 5–1 Normal Distribution

FIGURE 5–2 Non-Normal Distribution

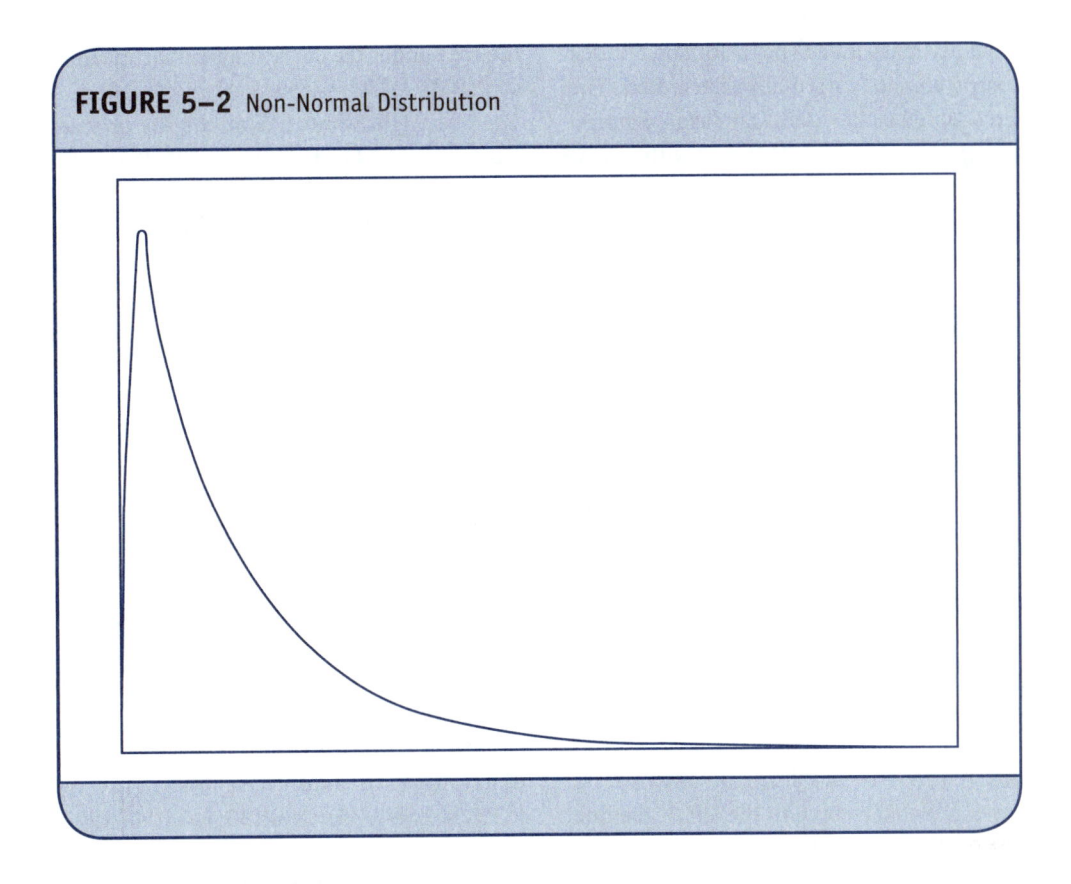

Many characteristics are approximately normally distributed, such as height and weight for specific age and sex groups, as are many laboratory and clinical measures such as cholesterol level and systolic blood pressure. The first property of the normal distribution implies the following: $P(x > \mu) = P(x < \mu) = 0.5$, where x denotes the continuous variable of interest and μ is the population mean. The probability that a value exceeds the mean is 0.5 and equivalent to the probability that a value is below the mean, which is the definition of the median. In a normal distribution, the mean, median, and mode (the most frequent value) are equal. A continuous variable that follows a normal distribution is one for which the following three statements are also true.

i. Approximately 68% of the values fall between the mean and one standard deviation (in either direction), i.e., $P(\mu - \sigma < x < \mu + \sigma) = 0.68$, where μ is the population mean and σ is the population standard deviation.
ii. Approximately 95% of the values fall between the mean and two standard deviations (in either direction), i.e., $P(\mu - 2\sigma < x < \mu + 2\sigma) = 0.95$.
iii. Approximately 99.9% of the values fall between the mean and three standard deviations (in either direction), i.e., $P(\mu - 3\sigma < x < \mu + 3\sigma) = 0.999$.

Part (iii) of the preceding indicates that for a continuous variable with a normal distribution, almost all of the observations fall between $\mu - 3\sigma$ and $\mu + 3\sigma$; thus the minimum value is approximately $\mu - 3\sigma$ and the maximum is approximately $\mu + 3\sigma$. In the following examples, we will illustrate how these probabilities 0.68, 0.95, and 0.999 were derived.

Another attribute of a normal distribution is that it is symmetric about the mean. The curve to the right of the mean is a mirror image of that to the left. A continuous variable with a distribution like that displayed in Figure 5–1—whose mean, median, and mode are equal—is symmetric and satisfies the preceding conditions (i) through (iii) follows a normal distribution. Similar to the binomial case, there is a normal distribution model that can be used to compute probabilities. The normal probability model is shown below and computing probabilities with the normal distribution model requires calculus,

$$P(x) = \frac{1}{\sigma\sqrt{2\pi}} e^{-(x-\mu)^2/(2\sigma^2)},$$

where μ is the population mean and σ is the population standard deviation. There is an alternative to using calculus to compute probabilities for normal variables, and it involves the use of probability tables. This is the approach we use.

Example 5.11. Body mass index (BMI) for specific sex and age groups is approximately normally distributed. The mean BMI for men aged 60 is 29 with a standard deviation of 6, and for women aged 60 the mean BMI is 28 with a standard deviation of 7. Suppose we consider the distribution of BMI among men aged 60. Knowing that the distribution is normal and having the mean and standard deviation allow us to completely generate the distribution. The distribution of BMI among men aged 60 is shown in **Figure 5–3**.

Notice that the mean ($\mu = 29$) is in the center of the distribution, the horizontal axis is scaled in units of the standard deviation ($\sigma = 6$), and the distribution essentially ranges from $\mu - 3\sigma$ to $\mu + 3\sigma$. This is not to say that there are not BMI values below 11 or above 47; there are such values, but they occur very infrequently. To compute probabilities about normal distributions, we compute areas under the curve. For example, suppose a man aged 60 is selected at random—what is the probability his BMI is less than 29? The probability is displayed graphically and represented by the area under the curve to the left of the value 29 in **Figure 5–4**.

The probability that a male has a BMI less than 29 is equivalent to the area under the curve to the left of the line drawn at 29. For any probability distribution, the total area under the curve is 1. For the normal distribution, we know that the mean is equal to the median, and thus half (50%) of the area under the curve is above the mean and half is below, so P(BMI < 29) = 0.50.

We might want to compute the probability that a male has a BMI of 29 or less, P($x \leq 29$). This can be thought of as P($x \leq 29$) = P($x < 29$) + P($x = 29$). We know that P($x < 29$) = 0.50. With the normal distribution, P($x = 29$) is defined as 0. In fact, the probability of being exactly equal to any value is always defined as 0. That is not to say that there are no men with a BMI of 29. There is no area in a single line, and thus P(X = 29) is defined as 0. Thus, P($x \leq 29$) = P($x < 29$). This concept will be illustrated further.

Suppose we want to know the probability that a male has a BMI less than 35. The probability is displayed graphically and represented by the area under the curve to the left of the value 35 in **Figure 5–5**. The probability that a male has a BMI less than 35 is equivalent to the area under the curve to the left of the line drawn at 35. For the normal distribution, we know that approximately 68% of the area under the curve lies between the mean plus or minus one standard deviation. For men aged 60, 68% of the area under the curve lies between 23 and 35. We also know that the normal distribution is symmetric about the mean; therefore P($29 < x < 35$) = P($23 < x < 29$) = 0.34. Thus, P($x < 35$) = 0.5 + 0.34 = 0.84.

What is the probability that a male aged 60 has a BMI less than 41? Using similar logic and the fact that approximately

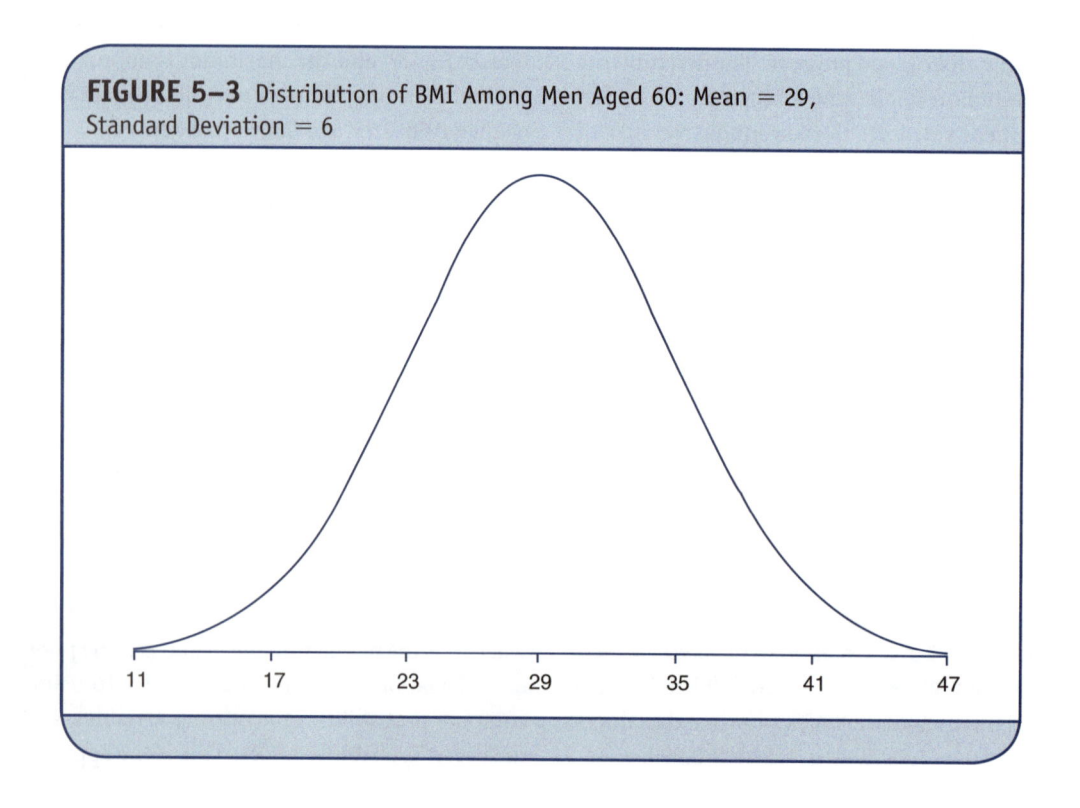

FIGURE 5–3 Distribution of BMI Among Men Aged 60: Mean = 29, Standard Deviation = 6

FIGURE 5–4 P(BMI < 29)

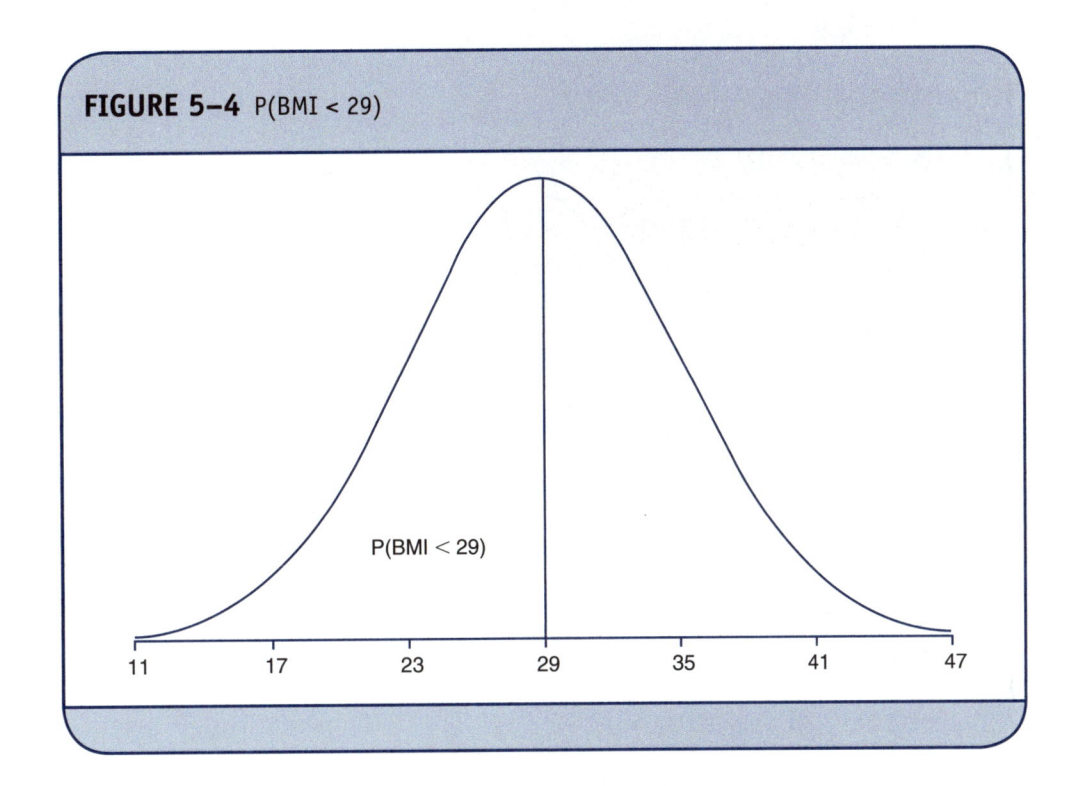

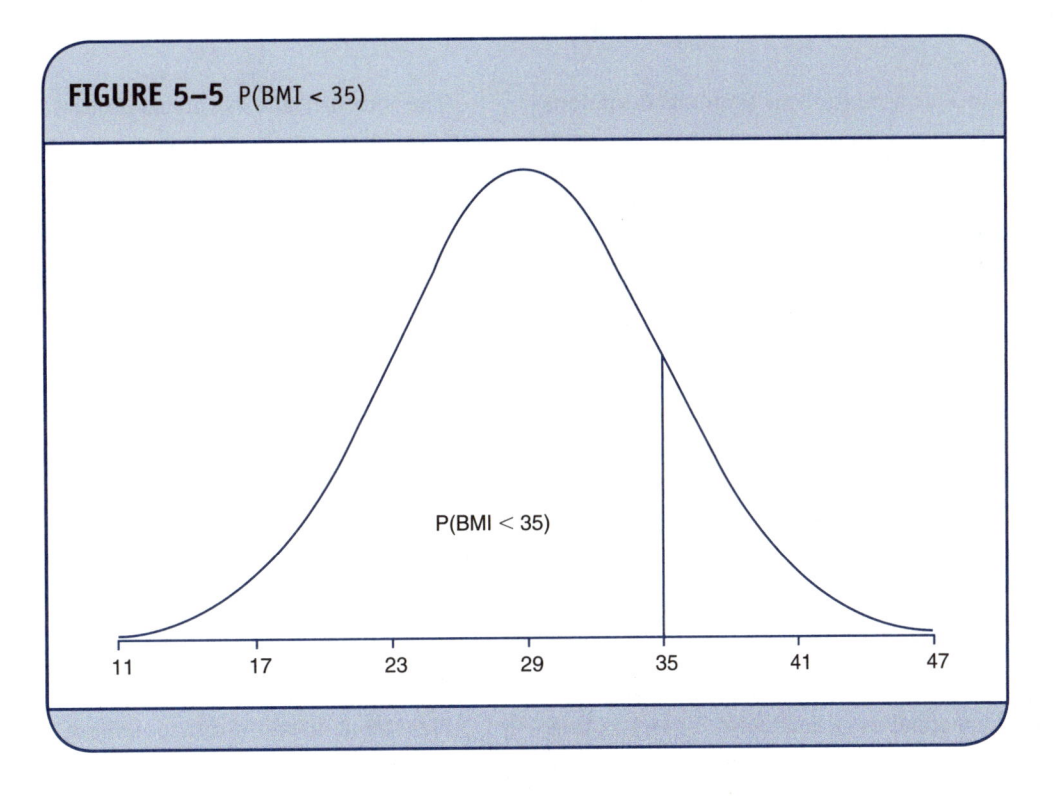

P(BMI < 29)

11   17   23   29   35   41   47

FIGURE 5–5 P(BMI < 35)

P(BMI < 35)

11   17   23   29   35   41   47

95% of the area under the curve lies between the mean plus or minus two standard deviations—i.e., P(29 < x < 41) = P(17 < x < 29) = 0.475—we can compute P(x < 41) = 0.5 + 0.475 = 0.975.

Suppose we now want to compute the probability that a male aged 60 has a BMI less than 30 (the threshold for classifying someone as obese). The area of interest, reflecting the probability, is displayed graphically in **Figure 5–6**. Because

FIGURE 5-6 P(BMI < 30)

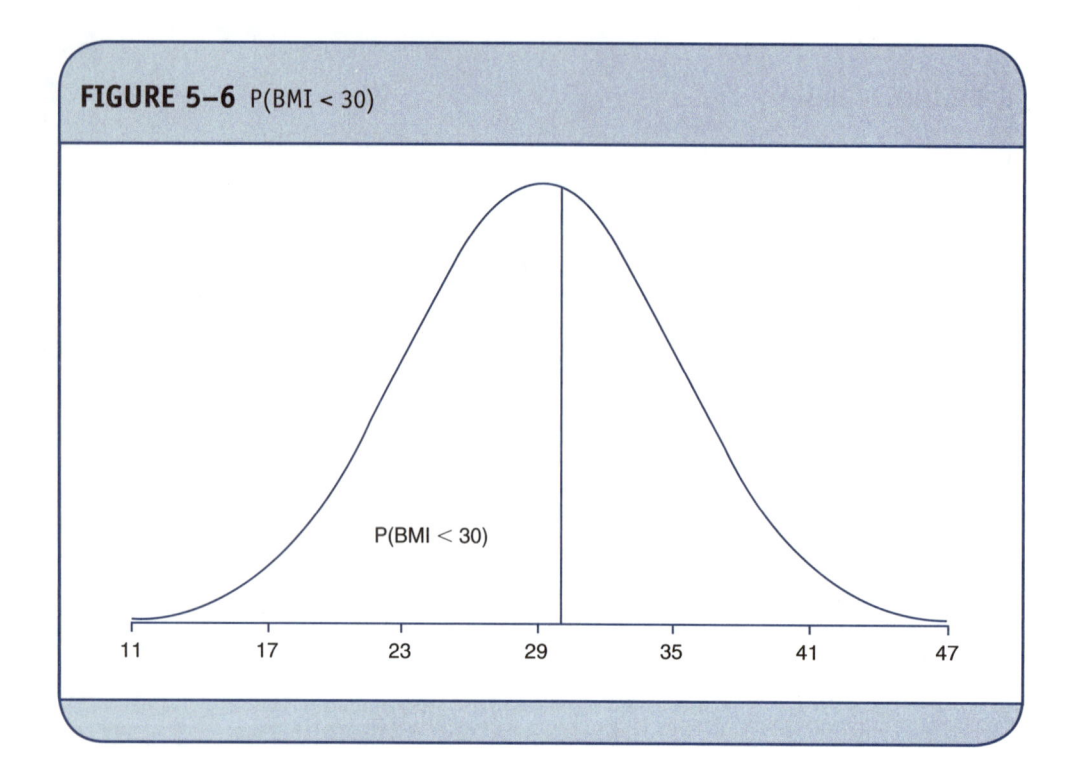

30 is not the mean nor a multiple of standard deviations above or below the mean, we cannot use the properties of a normal distribution to determine P(x < 30). The P(x < 30) is certainly between 0.5 and 0.84, based on the previous computations, but we can determine a more exact value. To do so, we need a table of probabilities for the normal distribution.

Because every application we face could potentially involve a normal distribution with a different mean and standard deviation, we will use a table of probabilities for the standard normal distribution. The standard normal distribution is a normal distribution with a mean of 0 and standard deviation of 1. We will always use z to refer to a standard normal variable. Up to this point, we have been using x to denote the variable of interest (e.g., x = BMI, x = height, x = weight). z will be reserved to refer to the standard normal distribution. The standard normal distribution is displayed in **Figure 5-7**.

The mean of the standard normal distribution is 0; thus the distribution is centered at 0. Multiples of the standard deviation above and below the mean are by units of the standard deviation ($\sigma = 1$). The range of the standard normal distribution is approximately −3 to 3. Table 1 in the Appendix contains probabilities for the standard normal distribution.

The body of Table 1 contains probabilities for the standard normal distribution, which correspond to areas under the standard normal curve. Specifically, Table 1 is organized to provide the area under the curve to the left of or less than the specified z value. Table 1 can accommodate two decimal places of z. The units place and the first decimal place are shown in the left column and the second decimal place is displayed across the top row. For example, suppose we want to compute P(z < 0). Because Table 1 contains z to two decimal places, this is equivalent to P(z < 0.00). We locate 0.0 in the left column (units and tenths place) and 0.00 across the top row (hundredths place). P(z < 0.00) = 0.5000. Similarly, P(z < 0.52) = 0.6985.

The question of interest is P(x < 30). We now have Table 1, which contains all of the probabilities for the standard normal distribution. BMI follows a normal distribution with a mean of 29 and a standard deviation of 6. We can use the standard normal distribution to solve this problem.

Figure 5-8 shows the distributions of BMI for men aged 60 and the standard normal distribution side-by-side. The areas under the curve are identical; only the scaling of the x-axis is different. BMI ranges from 11 to 47 while the standard normal variable, z, ranges from −3 to 3. We want to compute P(x < 30). We determine the z value that corresponds to x = 30 and then use Table 1 to find the probability or area under

FIGURE 5–7 The Standard Normal Distribution: $\mu = 0$, $\sigma = 1$

FIGURE 5–8 Distribution of BMI and Standard Normal Distribution

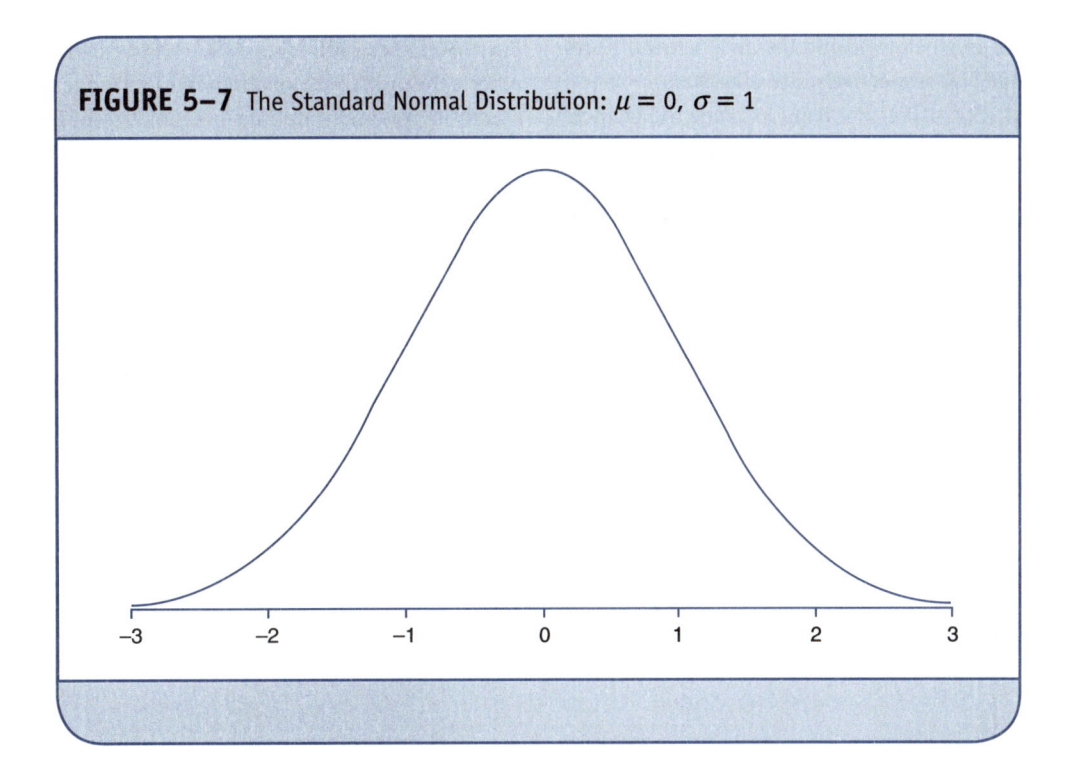

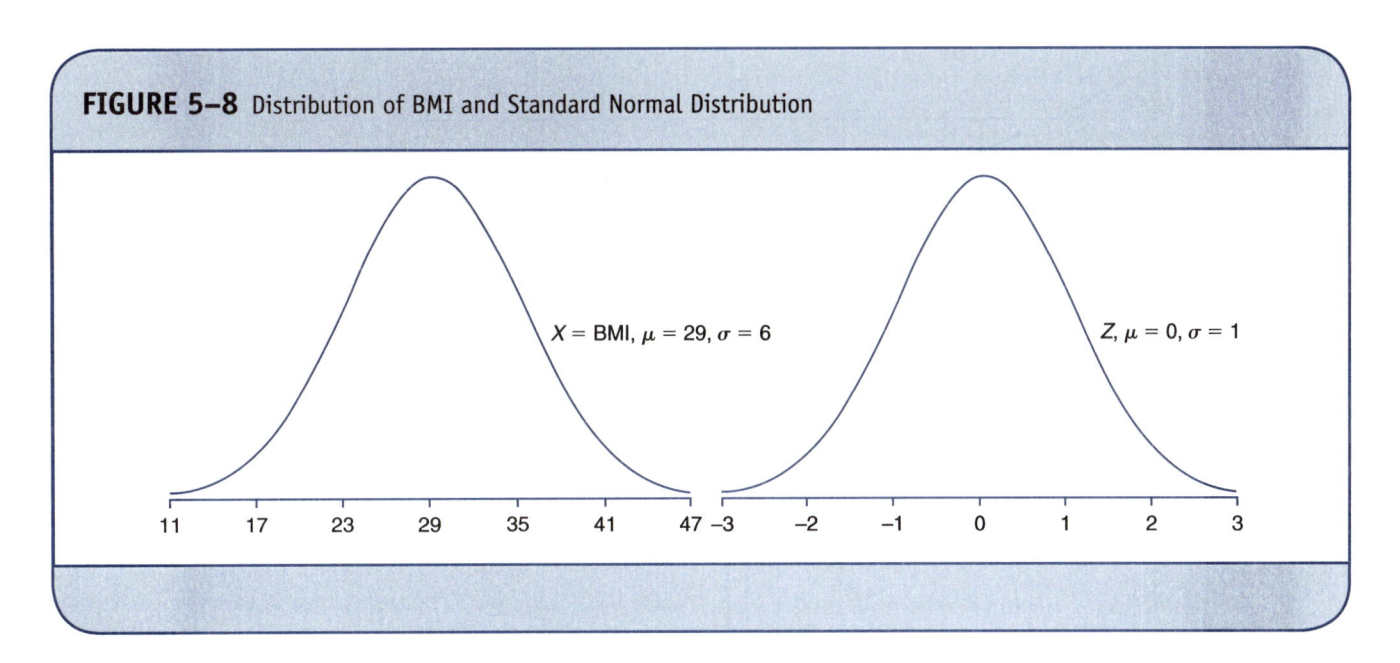

the curve. The following formula converts an x value into a z score, also called a standardized score:

$$Z = \frac{x - \mu}{\sigma},$$

where μ is the mean and σ is the standard deviation of the variable x. We want to compute $P(x < 30)$. Using the preceding formula, we convert ($x = 30$) to its corresponding z score (this is called *standardizing*):

$$Z = \frac{30 - 29}{6} = \frac{1}{6} = 0.17.$$

Thus, $P(x < 30) = P(z < 0.17)$. We can solve the latter using Table 1: $P(x < 30) = P(z < 0.17) = 0.5675$. Notice in

Figure 5–9 that the area below 30 and the area below 0.17 in the *x* and *z* distributions, respectively, are identical.

Using Table 1, P(z < 0.17) = 0.5675. Thus, the chance that a male aged 60 has a BMI less than 30 is 56.75%.

Suppose we now want to compute the probability that a male aged 60 has a BMI of 30 or less. Specifically, we want P(x ≤ 30). P(x ≤ 30) = P(x < 30) + P(x = 30). The second term reflects the probability of observing a male age 60 with a BMI of exactly 30. We are computing probabilities for the normal distribution as areas under the curve. There is no area in a single line and thus P(x = 30) is defined as 0. For the normal distribution, and for other probability distributions for continuous variables,

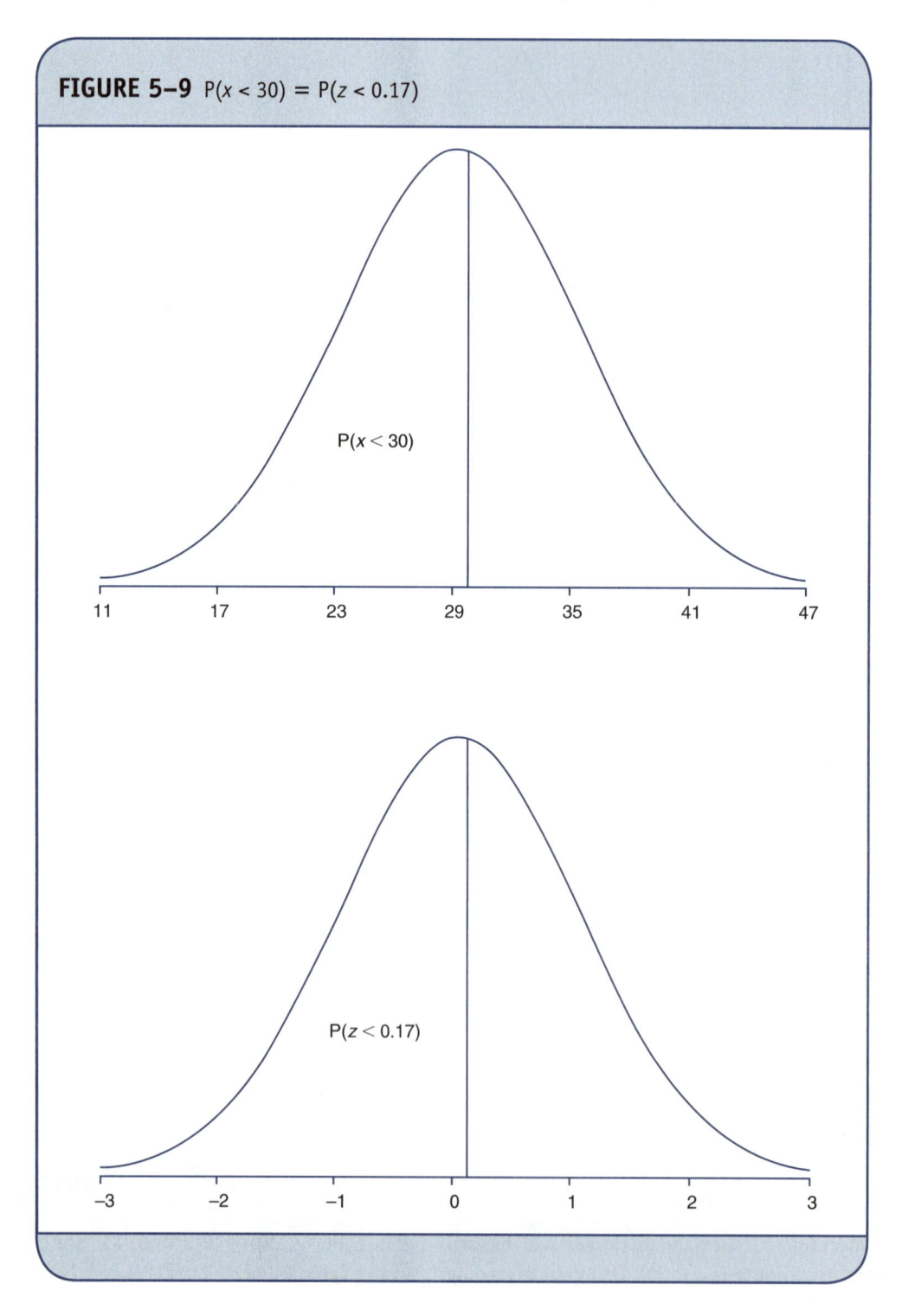

FIGURE 5–9 P(x < 30) = P(z < 0.17)

this will be the case. Therefore, $P(x \leq 30) = P(x < 30) = 0.5675$. Note that for the binomial distribution and for other probability distributions for discrete variables, the probability of taking on a specific value is not defined as 0 (see Section 5.6.1).

Consider again Example 5.11. What is the probability that a male aged 60 has a BMI exceeding 35? Specifically, what is $P(x > 35)$? Again we standardize:

$$P(x > 35) = P\left(z > \frac{35 - 29}{6} = \frac{6}{6} = 1\right).$$

We now need to compute $P(z > 1)$. If we look up $z = 1.00$ in Table 1, we find that $P(z < 1.00) = 0.8413$. Table 1 always gives the probability that z is less than the specified value. We want $P(z > 1)$ (see **Figure 5-10**). Table 1 gives $P(z < 1) = 0.8413$; thus $P(z > 1) = 1 - 0.8413 = 0.1587$. Almost 16% of men aged 60 have BMI over 35.

What is the probability that a male aged 60 has a BMI between 30 and 35? Note that this is the same as asking what proportion of men aged 60 has a BMI between 30 and 35. Specifically, we want $P(30 < x < 35)$. To solve this, we standardize and use Table 1. From the preceding examples, $P(30 < x < 35) = P(0.17 < z < 1)$. This can be computed as $P(0.17 < z < 1) = 0.8413 - 0.5675 = 0.2738$. This probability can be thought of as $P(0.17 < z < 1) = P(z < 1) - P(z < 0.17)$.

Now consider BMI in women. What is the probability that a female aged 60 has a BMI less than 30? We use the same approach, but recall that for women aged 60, the mean is 28 and the standard deviation is 7,

$$P(x < 30) = P\left(z < \frac{30 - 28}{7} = \frac{2}{7} = 0.29\right).$$

Using Table 1, $P(z < 0.29) = 0.6141$. Therefore, 61.41% of women aged 60 have a BMI less than 30, and 38.59% of women have a BMI of 30 or more.

What is the probability that a female aged 60 has a BMI exceeding 40? Specifically, what is $P(x > 40)$? Again, we standardize:

$$P(x > 40) = P\left(z > \frac{40 - 28}{7} = \frac{12}{7} = 1.71\right).$$

We need now to compute $P(z > 1.71)$. If we look up $z = 1.71$ on Table 1, we find that $P(z < 1.71) = 0.9564$. $P(z > 1.71) = 1 - 0.9564 = 0.0436$. Less than 5% of the women aged 60 have a BMI exceeding 40.

Table 1 is very useful for computing probabilities about normal distributions. To do this, we first standardize or convert a problem about a normal distribution (x) into a problem about the standard normal distribution (z). Once we have the problem in terms of z, we use Table 1 in the Appendix to compute the desired probability.

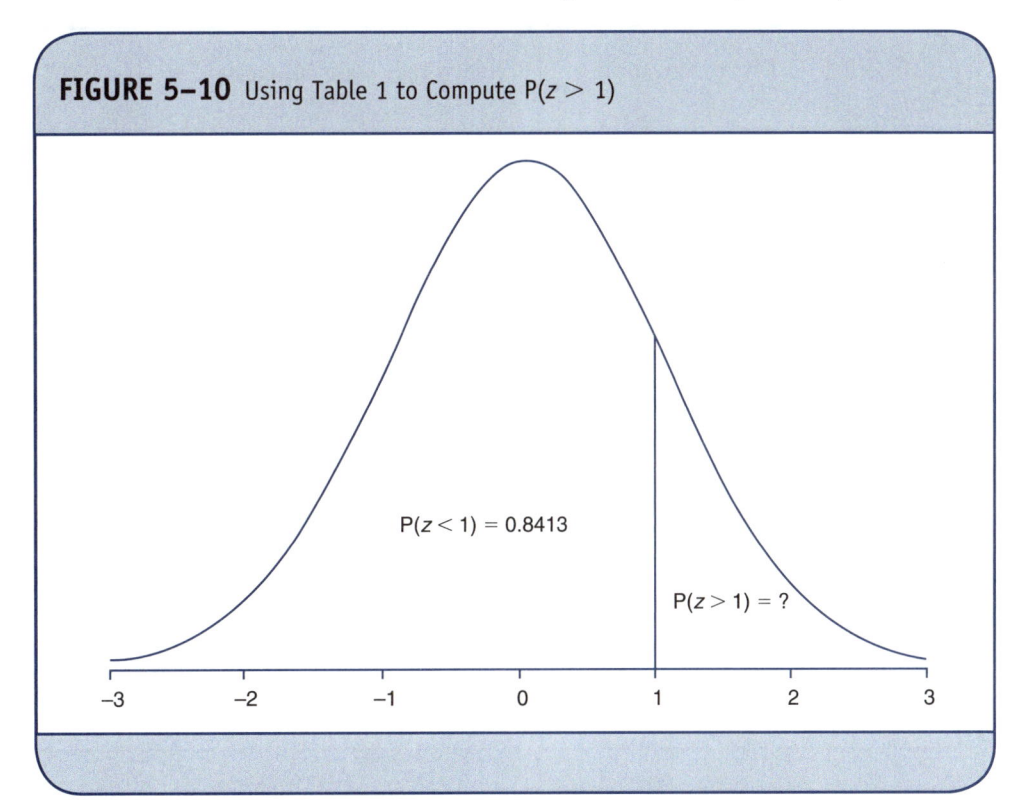

FIGURE 5-10 Using Table 1 to Compute $P(z > 1)$

$P(z < 1) = 0.8413$

$P(z > 1) = ?$

The standard normal distribution can also be useful for computing percentiles. A **percentile** is a value in the distribution that holds a specified percentage of the population below it. The pth percentile is the value that holds $p\%$ of the values below it. For example, the median is the 50th percentile, the first quartile is the 25th percentile, and the third quartile is the 75th percentile. In some instances, it may be of interest to compute other percentiles, for example, the 5th or 95th. The following formula is used to compute percentiles of a normal distribution.

$$x = \mu + z\sigma,$$

where μ is the mean, σ is the standard deviation of the variable x, and z is the value from the standard normal distribution for the desired percentile.

Example 5.12. Consider again Example 5–11, where we analyzed BMI in men and women aged 60. The mean BMI for men aged 60 is 29 with a standard deviation of 6, and for women aged 60 the mean BMI is 28 with a standard deviation of 7. What is the 90th percentile of BMI for men?

Figure 5–11 shows the distribution of BMI in men aged 60. The 90th percentile is the BMI that holds 90% of the BMI values below it. It therefore must be a value in the high (right) end of the distribution if 90% of the values are below it, and therefore only 10% are above it. The vertical line in Figure 5–11 is an estimate of the value of the 90th percentile.

To compute the 90th percentile, we use the formula $x = \mu + z\sigma$. The mean and standard deviation are 29 and 6, respectively; what is needed is the z value reflecting the 90th percentile of the standard normal distribution. To compute this value, we use Table 1—however, we use Table 1 almost in "reverse." When computing percentiles, we know the area under the curve (or probability) and want to compute the z score. In Example 5.11, we computed z scores and used Table 1 to determine areas under the curve, or probabilities. Here we know that the area under the curve below the desired z value is 0.90 (or 90%). What z score holds 0.90 below it? The interior of Table 1 contains areas under the curve below z. If the area under the curve below z is 0.90, we find 0.90 in the body (center) of Table 1. The value 0.90 is not there exactly; however, the values 0.8997 and 0.9015 are contained in Table 1. These correspond to z values of 1.28 and 1.29, respectively (i.e., 89.97% of the area under the standard normal curve is below 1.28). The exact z value holding 90% of the values below it is 1.282. This value is determined by a statistical computing package (e.g., Microsoft Excel®) with more precision than that shown in Table 1. Using $z = 1.282$, we can now compute the 90th percentile of BMI for men: $x = 29 + 1.282(6) = 36.69$. Ninety percent of the BMIs in men aged 60 are below 36.69, and 10% of the BMIs in men aged 60 are above 36.69. What is the 90th percentile of BMI among women aged 60? $x = 28 + 1.282(7) = 36.97$. Ninety percent of

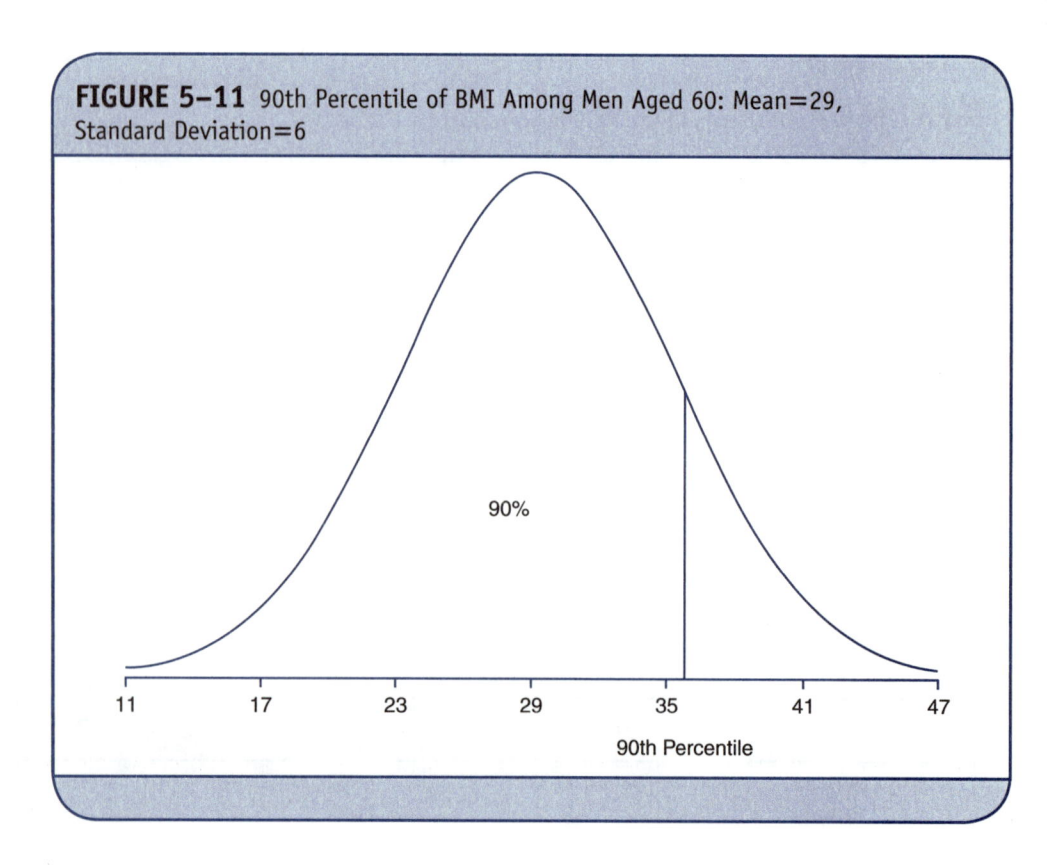

FIGURE 5–11 90th Percentile of BMI Among Men Aged 60: Mean=29, Standard Deviation=6

the BMIs in women aged 60 are below 36.97, and 10% of the BMIs in women aged 60 are above 36.97.

Table 1A in the Appendix was developed using Table 1 and contains the z values for popular percentiles. It can be used to compute percentiles for normal distributions. A popular application of percentiles is in anthropometrics, which is the study of human measurements. These measures, such as weight and height, are used to study patterns in body size. For example, pediatricians often measure a child's weight, length, or height, and head circumference. The observed values are often converted into percentiles to assess where a particular child falls relative to his or her peers (i.e., children of the same sex and age). For example, if a child's weight for his or her age is extremely low, it might be an indication of malnutrition. Growth charts for boys and girls including length-for-age, weight-for-age, BMI-for-age, head circumference-for-age, and other anthropometric charts are available for infants from birth to 36 months and children 2 to 20 years old on the Centers for Disease Control (CDC) website at *http://www.cdc.gov/nchs/about/major/nhanes/growthcharts*.

Example 5.13. For infant girls, the mean length at 10 months is 72 cm with a standard deviation of 3 cm. Suppose a girl of 10 months has a measured length of 67 cm. How does her length compare to other girls of 10 months?

We can compute her percentile by determining the proportion of girls with lengths below 67. Specifically,

$$P(x < 67) = P\left(z < \frac{67 - 72}{3} = \frac{-5}{3} = -1.67\right).$$

Using Table 1, $P(z < -1.67) = 0.0475$. This girl is in the 4.75th percentile. Fewer than 5% of girls of 10 months are below 67 cm. This may be an instance where some intervention is needed.

5.6.3 Sampling Distributions

In Chapters 6, 7, 9, 10, and 11, we focus on statistical inference, where we make inferences or generalizations about population parameters based on observed sample statistics. In Chapter 4, we presented techniques and a number of statistics to summarize sample data. Suppose we want to generate an estimate of a continuous variable in a population (e.g., weight, HDL cholesterol level). It is very typical to estimate the mean of a continuous variable in a population. The mean of a representative sample is a very good estimate of the unknown population mean. If a second sample is selected, that sample might produce a slightly different estimate (i.e., the mean of the second sample might be slightly different than the mean of the first). Whenever we perform statistical inference, we must recognize that we are essentially working with incomplete information—specifically, only a fraction of the population. When we make estimates about population parameters based on sample statistics, it is extremely important to quantify the precision in our estimates. This is done using probability and, in particular, the probability models we have just discussed.

Example 5.14. Consider the following small population consisting of $N = 6$ patients who recently underwent total hip replacement. We are interested in a patient's self-reported pain-free function, rated on a scale of 0 to 100, with higher scores representing better function (e.g., 0 = severely limited and painful functioning to 100 = completely pain-free functioning), measured 3 months post-procedure. The data are shown below and are ordered from smallest to largest:

25   50   80   85   90   100

The population mean is $\mu = \frac{\Sigma X}{N} = 71.7$ and the standard deviation is $\sigma = \sqrt{\frac{\Sigma(X - \mu)^2}{N}} = 25.9$. A box-whisker plot of the population data is shown in **Figure 5–12**.

The distribution of pain-free function scores is slightly skewed, with the majority of patients reporting high scores. Suppose we did not have the population data and instead were interested in estimating the mean pain-free function score based on a sample. Suppose we planned to take a sample of size $n = 4$. Table 5–9 shows all possible samples of size $n = 4$ from the population of $N = 6$ when sampling without replacement. (Sampling without replacement means that we select an individual and with that person aside, we select a second from those remaining, and so on. In contrast, when sampling with replacement, we make a selection, record that selection, and place that person back before making a second selection. When sampling with replacement, the same individual can be selected into the sample multiple times.) The right column shows the sample mean based on the four observations contained in that sample.

The probability of selecting any particular sample in **Table 5–9** is $1 / 15 = 0.07$. Suppose by chance we select Sample 1. The mean of Sample 1 is 60. If we based our estimate of the unknown population mean on Sample 1, and particularly on the sample mean of Sample 1, we would underestimate the true population mean ($\mu = 71.7$). If we selected Sample 15, we would overestimate the population mean because the mean of Sample 15 is $\overline{X} = 88.8$.

The collection of all possible sample means (in this example, there are 15 distinct samples that are produced by sampling four individuals at random without replacement) is called the **sampling distribution** of the sample means. We

FIGURE 5–12 Box-Whisker Plot of Population Data

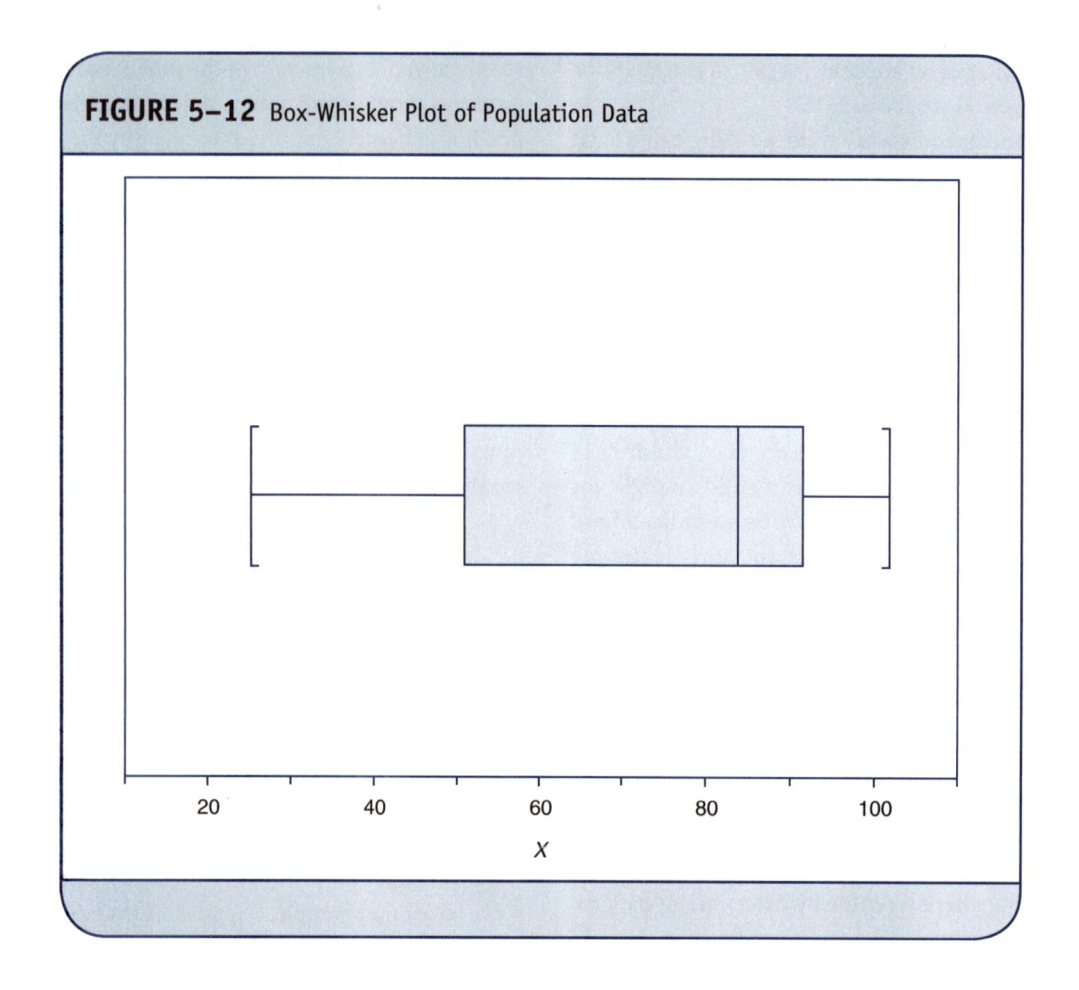

TABLE 5–9 All Samples of Size $n = 4$

Sample	Observations in the Sample				Sample Mean ($\bar{X}$)
1	25	50	80	85	60.0
2	25	50	80	90	61.3
3	25	50	80	100	63.8
4	25	50	85	90	62.5
5	25	50	85	100	65.0
6	25	50	90	100	66.3
7	25	80	85	90	70.0
8	25	80	85	100	72.5
9	25	80	90	100	73.8
10	25	85	90	100	75.0
11	50	80	85	90	76.3
12	50	80	85	100	78.8
13	50	80	90	100	80.0
14	50	85	90	100	81.3
15	80	85	90	100	88.8

consider it a population because it includes all possible values produced in this case by a specific sampling scheme. If we compute the mean and standard deviation of this population of sample means, we get the following: $\mu_{\bar{X}} = 71.7$ and a standard deviation of $\sigma_{\bar{X}} = 8.5$. The subscripts here are to distinguish these parameters from those based on the population data (x). To be consistent, the parameters based on the population data could include a subscript x.

Notice that the mean of the sample means is $\mu_{\bar{X}} = 71.7$, which is precisely the value of the population mean (μ). This will always be the case. Specifically, the mean of the sampling distribution of the sample means will always be equivalent to the population mean. This is important as it indicates that, on average, the sample mean is equal to the population mean. This is the definition of an unbiased estimator. Unbiasedness is a desirable property in an estimator. Notice also that the variability in the sample means is much smaller than the variability in the population; this will also always be the case. A box-whisker plot of the population of sample means is shown in **Figure 5–13**.

FIGURE 5–13 Box-Whisker Plot of Population of Sample Means

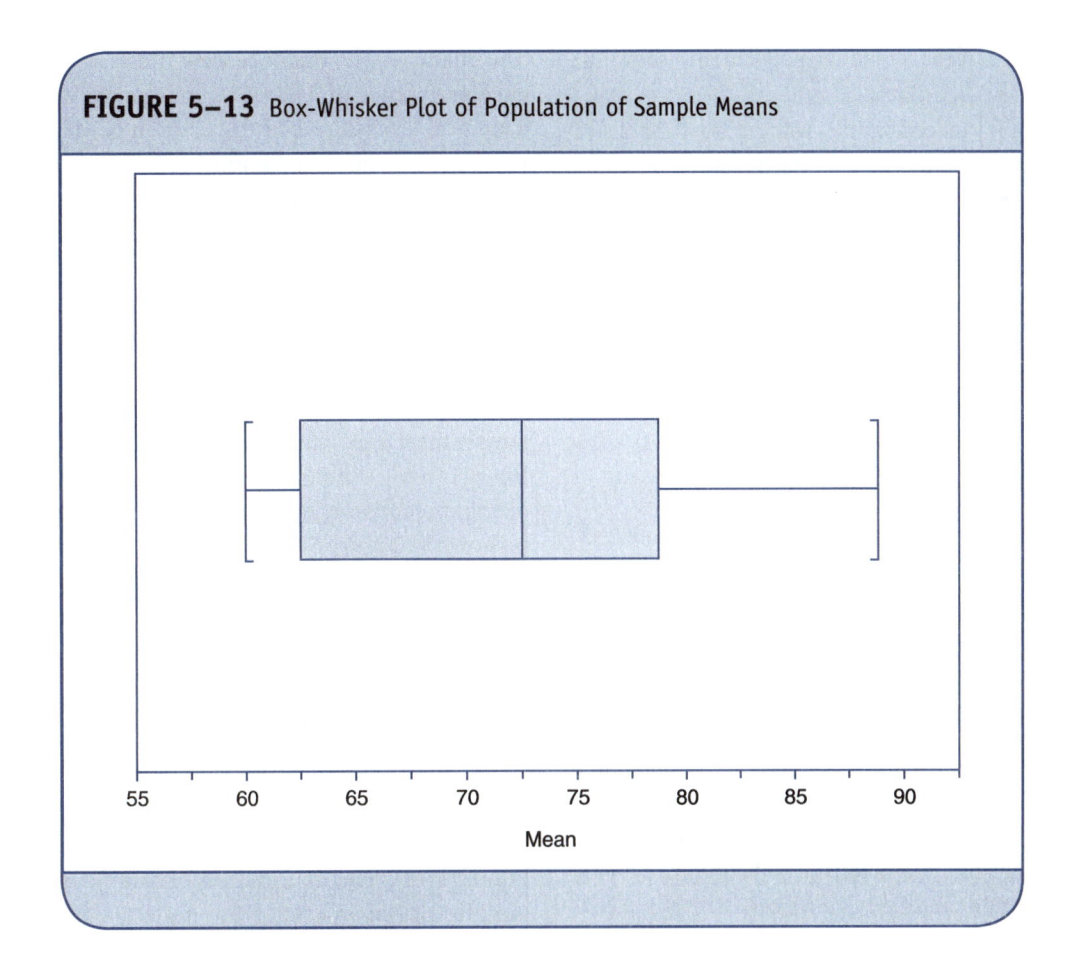

Notice that the distribution of the sample means is more symmetric and has a much more restricted range (60 to 88.8) than the distribution of the population data (25 to 100) shown in Figure 5–12. The importance of these observations is stated formally in the Central Limit Theorem.

Central Limit Theorem. Suppose we have a population with known mean and standard deviation, μ and σ, respectively. The distribution of the population can be normal or it can be non-normal (e.g., skewed toward the high or low end, or flat). If we take simple random samples of size n from the population with replacement, then for large samples (usually defined as samples with $n > 30$), the sampling distribution of the sample means is approximately normally distributed with a mean of $\mu_{\bar{X}} = \mu$ and a standard deviation $\sigma_{\bar{X}} = \frac{\sigma}{\sqrt{n}}$.

The importance of this theorem for applications is as follows. The theorem states that, regardless of the distribution of the population (normal or not), as long as the sample is sufficiently large (usually $n > 30$), then the distribution of the sample means is approximately normal. In Section 5.6.2, we learned that it is relatively straightforward to compute probabilities about a normal distribution. Therefore, when we make inferences about a population mean based on the sample mean, we can use the normal probability model to quantify uncertainty. This will become explicit in Chapters 6, 7, 9, 10, and 11.

Before illustrating the use of the Central Limit Theorem, we first illustrate the result. For the result of the Central Limit Theorem to hold, the sample must be sufficiently large ($n > 30$). There are two exceptions to this: If the outcome in the population is normal, then the result holds for samples of any size (i.e., the sampling distribution of the sample means is approximately normal even for samples of size less than 30). If the outcome in the population is dichotomous, then the result holds for samples that meet the following criterion: $\min[np, n(1 - p)] > 5$, where n is the sample size and p is the probability of success in the population. We next illustrate the results of the Central Limit Theorem for populations with different distributions.

Example 5.15. Suppose we measure a characteristic in a population and that this characteristic follows a normal distribution with a mean of 75 and standard deviation of 8. The distribution of the characteristic in the population is shown in **Figure 5–14**.

If we take simple random samples with replacement of size $n = 10$ from the population and for each sample we compute the sample mean, the Central Limit Theorem states that the distribution of sample means is approximately normal. Note that the sample size $n = 10$ does not meet the criterion of $n > 30$, but in this case, our original population is normal and therefore the result holds. The distribution of sample means based on samples of size $n = 10$ is shown in **Figure 5–15**. The mean of the sample means is 75 and the standard deviation of the sample means is 2.5 (i.e., $\mu_{\bar{X}} = \mu = 75$ and a standard deviation of $\sigma_{\bar{X}} = \frac{\sigma}{\sqrt{n}} = \frac{8}{\sqrt{10}} = 2.5$).

If we take simple random samples with replacement of size $n = 5$, we get a similar distribution. The distribution of sample means based on samples of size $n = 5$ is shown in **Figure 5–16**. The mean of the sample means is again 75 and the standard deviation of the sample means is 3.6 (i.e., $\mu_{\bar{X}} = \mu = 75$ and a standard deviation of $\sigma_{\bar{X}} = \frac{\sigma}{\sqrt{n}} = \frac{8}{\sqrt{5}} = 3.6$). Notice that the variability in sample means is larger for samples of size 5 as compared to samples of size 10.

Example 5.16. Suppose we measure a characteristic in a population and that this characteristic is dichotomous with 30% of the population classified as a success (i.e., $p = 0.30$).

The characteristic might represent disease status, the presence or absence of a genetic abnormality, or the success of a medical procedure. The distribution of the outcome in the population is shown in **Figure 5–17**.

This population is clearly not normal. The results of the Central Limit Theorem are said to apply to binomial populations as long as the minimum of np and $n(1 - p)$ is at least 5, where n refers to the sample size. If we take simple random samples with replacement of size $n = 20$ from this binomial population and for each sample we compute the sample mean, the Central Limit Theorem states that the distribution of sample means should be approximately normal because min$[np, n(1 - p)] = \min[20(0.3), 20(0.7)] = \min(6, 14) = 6$. The distribution of sample means based on samples of size $n = 20$ is shown in **Figure 5–18**. The mean of the sample means is 6 and the standard deviation of the sample means is 0.4 (i.e., $\mu_{\bar{X}} = \mu = 6$ and a standard deviation of $\sigma_{\bar{X}} = \frac{\sigma}{\sqrt{n}} = \frac{2}{\sqrt{20}} = 0.4$).

Suppose we take simple random samples with replacement of size $n = 10$. In this scenario, we do not meet the sample size requirement for the results of the Central Limit Theorem to hold—i.e., min$[np, n(1 - p)] = \min[10(0.3), 10(0.7)] = \min(3, 7) = 3$. The distribution of sample means based on samples of size $n = 10$ is shown in **Figure 5–19**. The distribution of sample means based on samples of size $n = 10$ is not quite normally distributed.

FIGURE 5–14 Normal Population

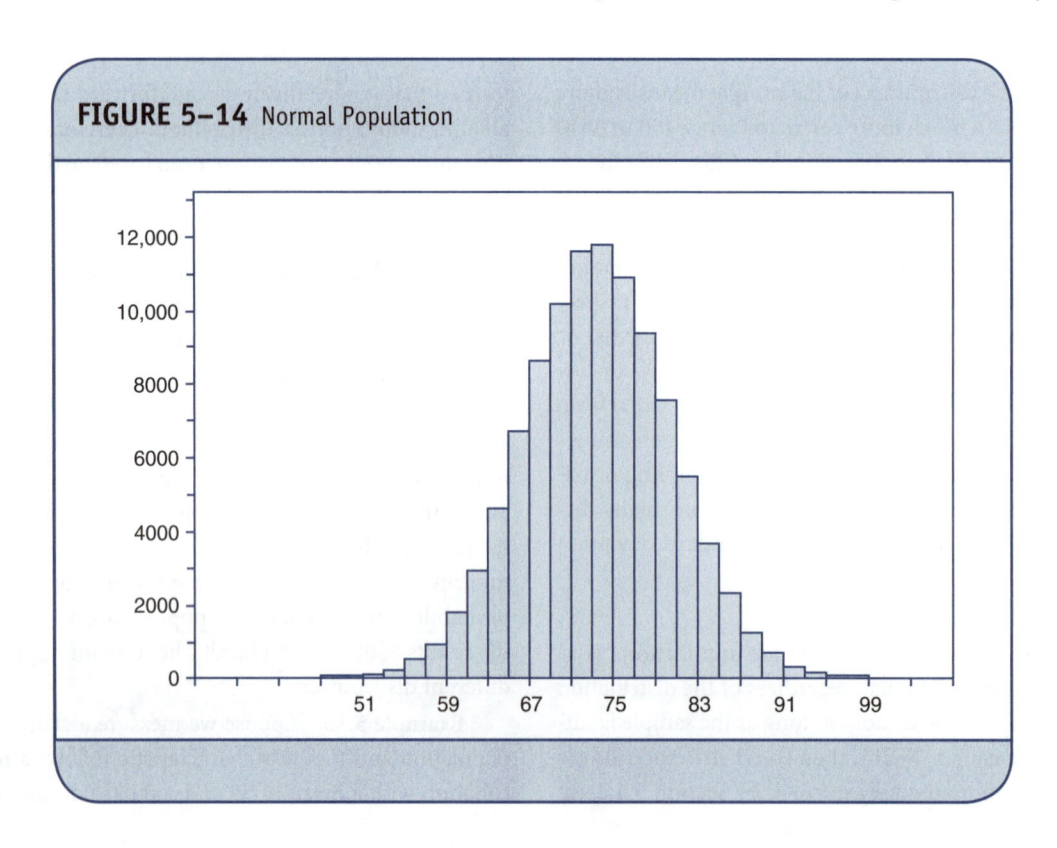

FIGURE 5–15 Distribution of Sample Means Based on Samples of Size $n = 10$

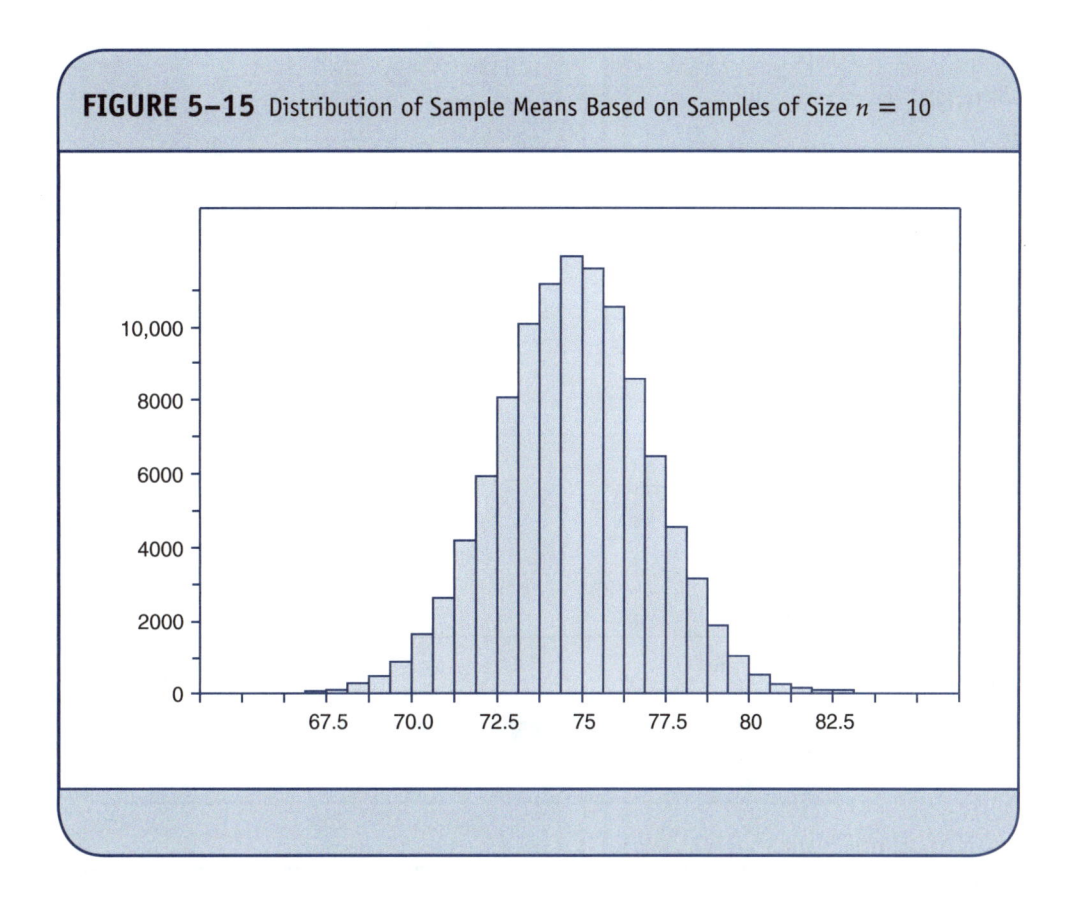

FIGURE 5–16 Distribution of Sample Means Based on Samples of Size $n = 5$

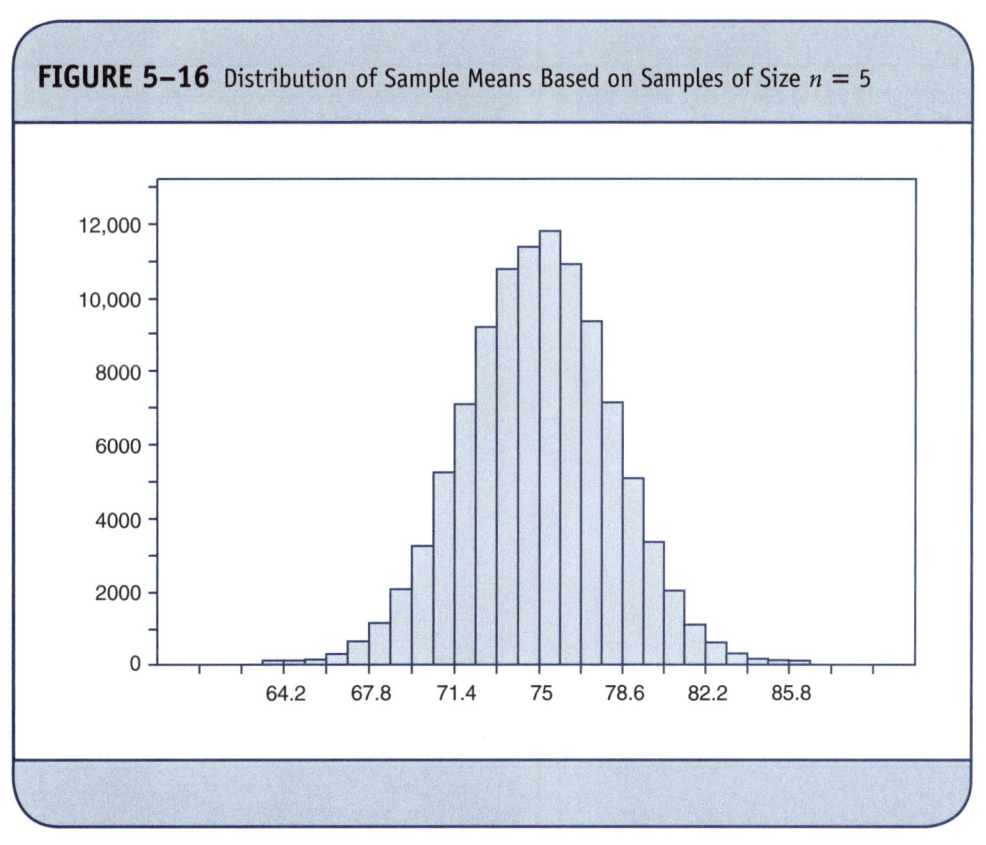

CHAPTER 5 The Role of Probability

FIGURE 5–17 Population with a Dichotomous Outcome

FIGURE 5–18 Distribution of Sample Means Based on Samples of Size $n = 20$

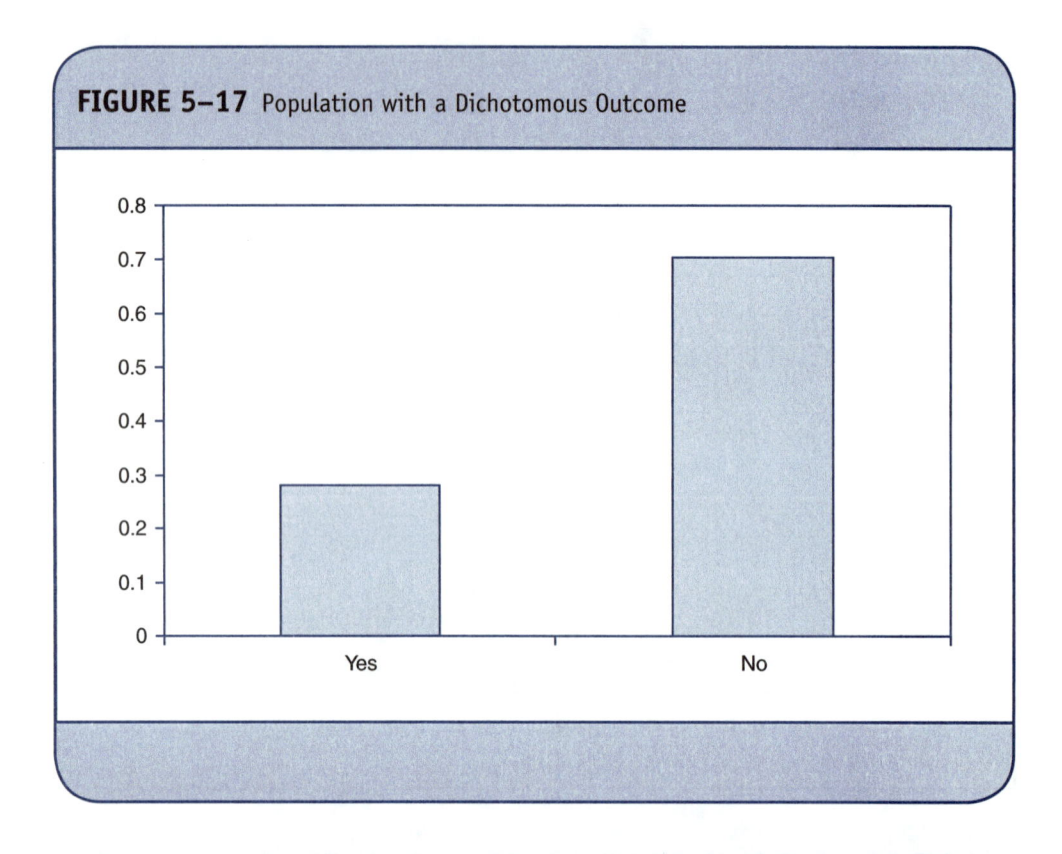

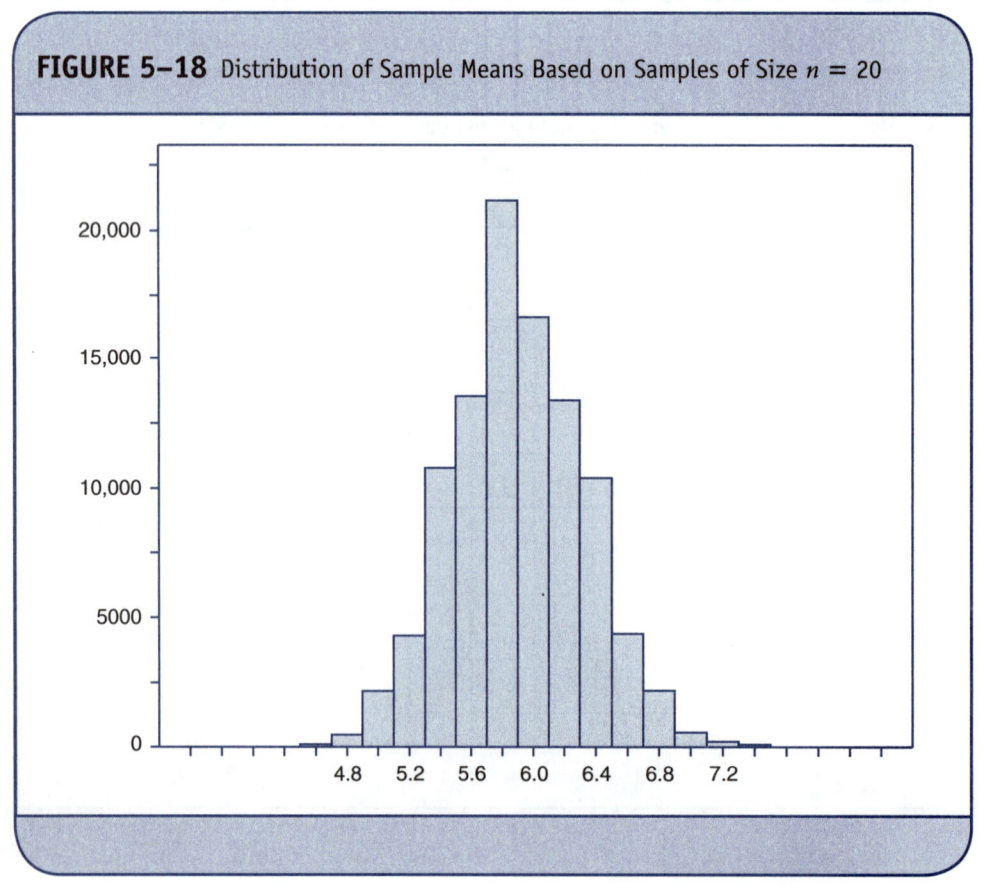

FIGURE 5–19 Distribution of Sample Means Based on Samples of Size $n = 10$

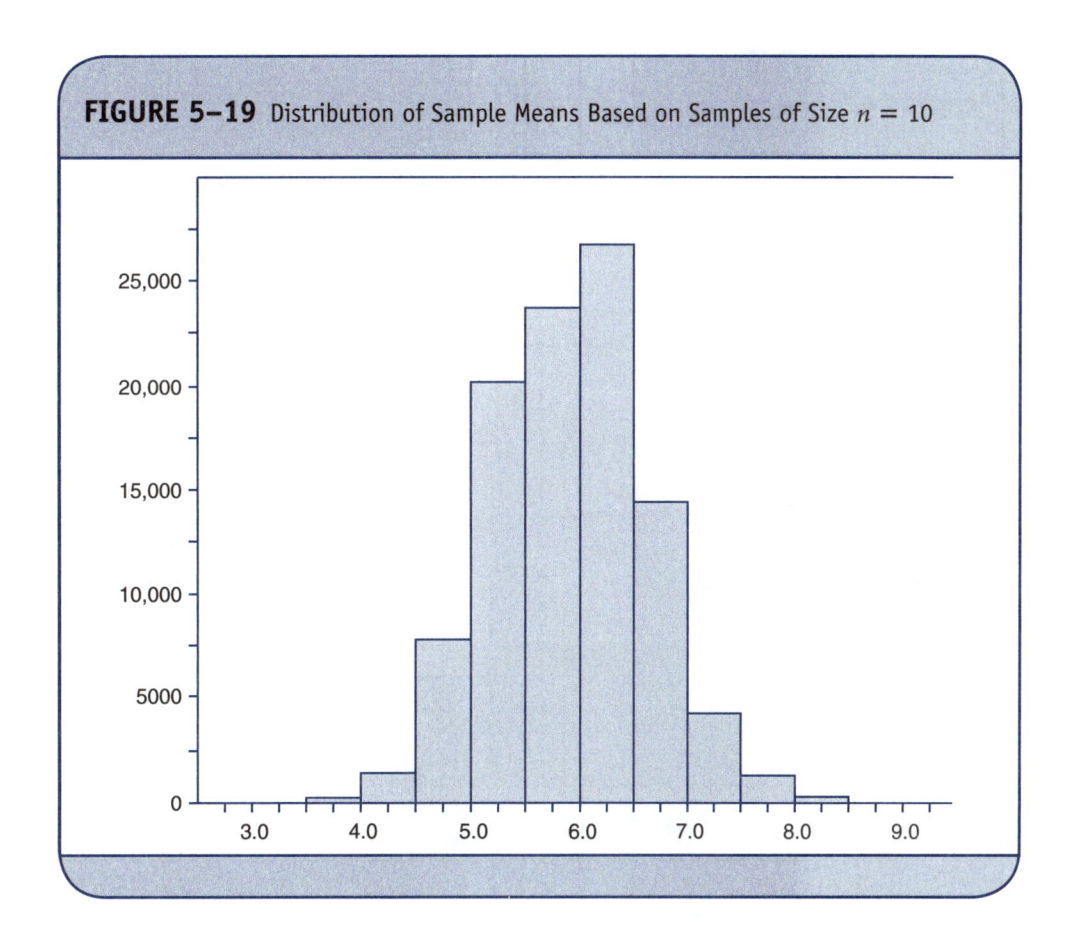

The sample size must be larger for the distribution to approach normality.

Example 5.17. Suppose we measure a characteristic in a population and that this characteristic follows a Poisson distribution with $\mu = 3$ and $\sigma = 1.7$. The Poisson is another probability model for a discrete variable and its distribution is shown in **Figure 5–20**. This population is not normal. The results of the Central Limit Theorem are said to apply when $n > 30$. The distribution of sample means based on samples of size $n = 30$ is shown in **Figure 5–21**.

The mean of the sample means is 3 and the standard deviation of the sample means is 0.3 (i.e., $\mu_{\bar{X}} = \mu = 3$ and a standard deviation of $\sigma_{\bar{X}} = \frac{\sigma}{\sqrt{n}} = \frac{1.7}{\sqrt{30}} = 0.3$).

Samples of smaller size from this population do not meet the requirements of the Central Limit Theorem, and thus the result would not hold. For example, suppose we take simple random samples with replacement of size $n = 10$. The distribution of sample means based on samples of size $n = 10$ is shown in **Figure 5–22** and is not quite normally distributed. Samples of size 30 or greater will be approximately normally distributed. The distribution of sample means based on samples of size $n = 50$ is shown in **Figure 5–23** and is normally distributed.

The mean of the sample means will always be equal to the population mean ($\mu_{\bar{X}} = \mu$). The standard deviation of the sample means, defined as $\sigma_{\bar{X}}$, equals $\frac{\sigma}{\sqrt{n}}$ (i.e., $\sigma_{\bar{X}} = \frac{\sigma}{\sqrt{n}}$) and is also called the **standard error**. The standard error decreases as the sample size increases. Specifically, the variability in the sample means is smaller for larger sample sizes. This is intuitively sensible as extreme values will have less impact in samples of larger size. Notice in the previous examples how the standard errors decrease as the sample sizes increase.

Example 5.18. High density lipoprotein (HDL) cholesterol—the "good" cholesterol—has a mean of 54 and a standard deviation of 17 in patients over age 50. Suppose a physician has 40 patients over age 50 and wants to determine the probability that their mean HDL cholesterol is 60 or more. Specifically, we want to know $P(\bar{X} > 60)$.

Probability questions about a sample mean can be addressed with the Central Limit Theorem as long as the sample size is sufficiently large. In this example, we have

CHAPTER 5 The Role of Probability

FIGURE 5–20 Poisson Population

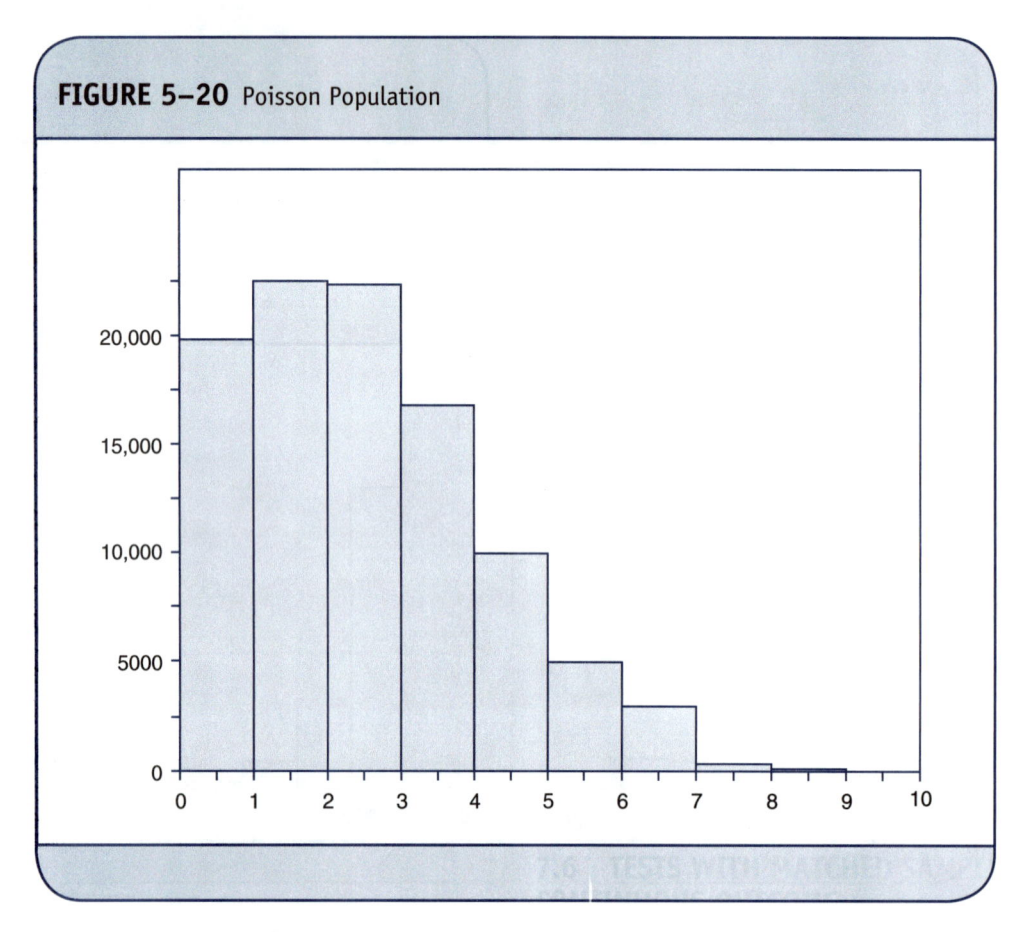

FIGURE 5–21 Distribution of Sample Means Based on Samples of Size $n = 30$

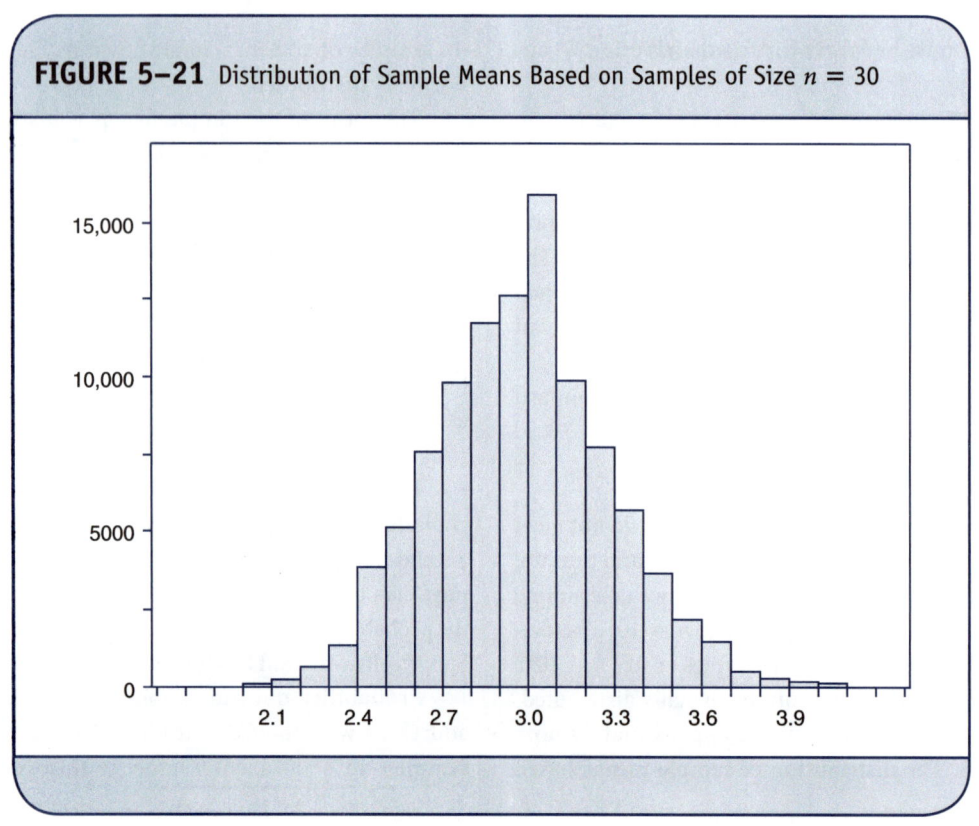

FIGURE 5–22 Distribution of Sample Means Based on Samples of Size $n = 10$

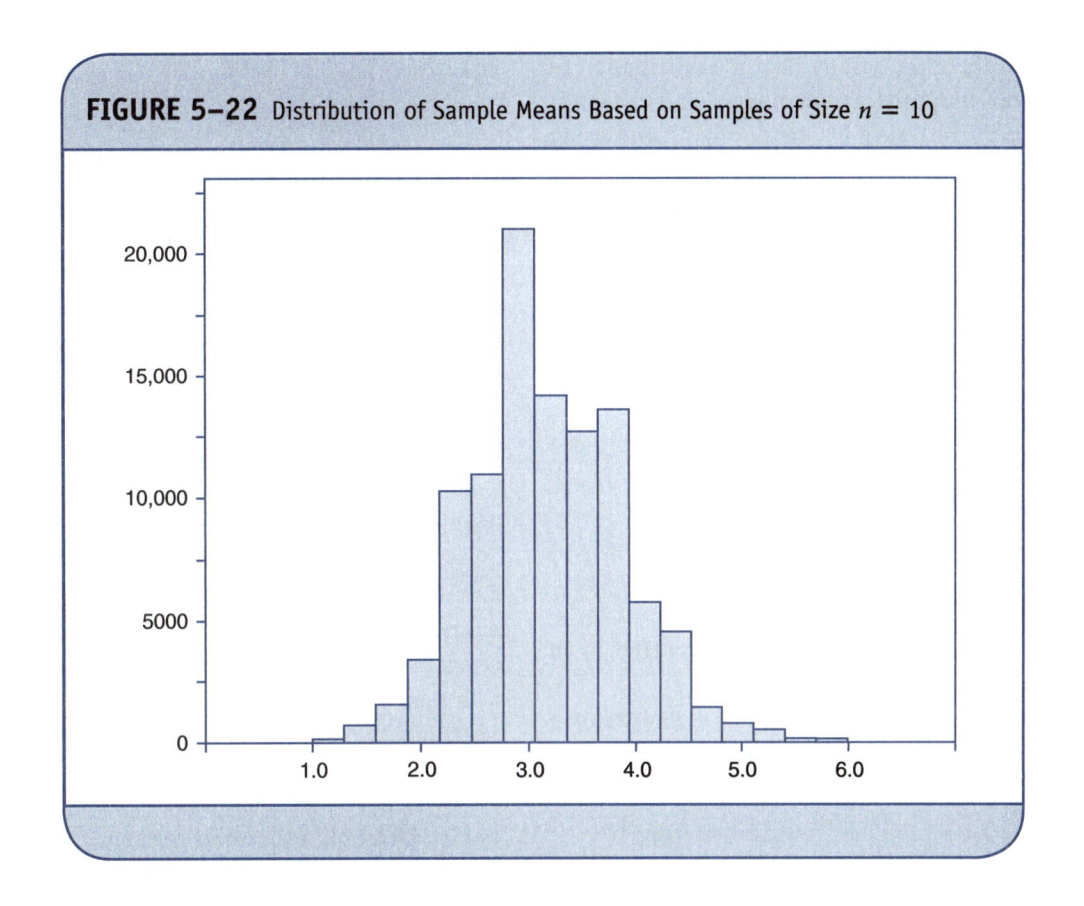

FIGURE 5–23 Distribution of Sample Means Based on Samples of Size $n = 50$

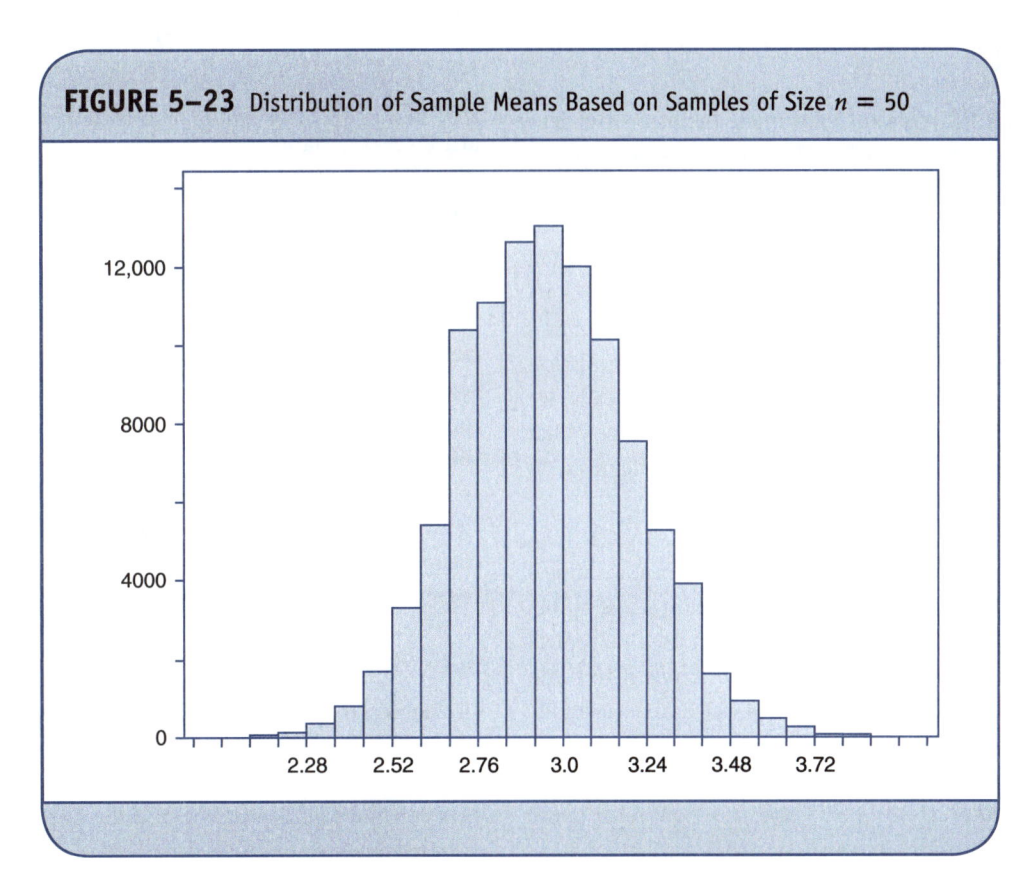

$n = 40$. We can therefore assume that the distribution of the sample mean is approximately normal and can compute the desired probability by standardizing and using Table 1.

Standardizing involves subtracting the mean and dividing by the standard deviation. The mean of $\overline{X}$ is $\mu_{\overline{X}} = \mu$ and the standard deviation of $\overline{X}$ is $\sigma_{\overline{X}} = \dfrac{\sigma}{\sqrt{n}}$. Therefore, the formula to standardize a sample mean is

$$Z = \dfrac{\overline{X} - \mu_{\overline{X}}}{\sigma_{\overline{X}}} = \dfrac{\overline{X} - \mu}{\sigma/\sqrt{n}}.$$

We want to compute

$$P(\overline{X} > 60) = P\left(z > \dfrac{60 - 54}{17/\sqrt{40}} = \dfrac{6}{2.7} = 2.22\right).$$

$P(z > 2.22)$ can be solved with Table 1 with $P(z > 2.22) = 1 - 0.9868 = 0.0132$. The chance that the mean HDL in 40 patients exceeds 60 is 1.32%.

What is the probability that the mean HDL cholesterol in 40 patients is less than 50? We now want

$$P(\overline{X} < 50) = P\left(z > \dfrac{50 - 54}{17/\sqrt{40}} = \dfrac{-4}{2.7} = -1.48\right).$$

$P(z < -1.48)$ can be solved with Table 1, $P(z < -1.48) = 0.0694$. The chance that the mean HDL in 40 patients is less than 50 is 6.94%.

Example 5.19. We want to estimate the mean low density lipoprotein (LDL)—the "bad" cholesterol—in the population of adults 65 years of age and older. Suppose that we know from studies of adults under age 65 that the standard deviation is 13 and we assume that the variability in LDL in adults 65 years of age and older is the same. We select a sample of $n = 100$ participants 65 years of age and older and use the mean of the sample as an estimate of the population mean. We want our estimate to be precise—specifically, we want it to be within 3 units of the true mean LDL value. What is the probability that our estimate (i.e., the sample mean) is within 3 units of the true mean?

We can represent this question as $P(\mu - 3 < \overline{X} < \mu + 3)$. Because this is a probability about a sample mean, we appeal to the Central Limit Theorem. With a sample of size $n = 100$, we satisfy the sample size criterion and can use the Central Limit Theorem to solve the problem (i.e., convert to z and use Table 1). In the previous example, we asked questions around specific values of the sample mean (e.g., 50, 60) and converted those to z scores and worked with Table 1. Here the values of interest are $\mu - 3$ and $\mu + 3$. We use these values below:

$P(\mu - 3 < \overline{X} < \mu + 3) =$

$$P\left(\dfrac{(\mu - 3) - \mu}{13/\sqrt{100}} < z < \dfrac{(\mu + 3) - \mu}{13/\sqrt{100}}\right) =$$

$$P\left(\dfrac{-3}{1.3} < z < \dfrac{3}{1.3}\right) = P(-2.31 < z < 2.31).$$

This we can solve with Table 1, $P(-2.31 < z < 2.31) = 0.9896 - 0.0104 = 0.9792$. There is a 97.92% chance that the sample mean, based on a sample of size $n = 100$, will be within 3 units of the true population mean. This is a very powerful statement. When looking only at 100 individuals aged 65 and older, there is almost a 98% chance that the sample mean is within 3 units of the population mean.

Example 5.20. Alpha fetoprotein (AFP), a substance produced by a fetus, is often measured in pregnant women as a means of assessing whether there might be problems with fetal development. High levels of AFP have been seen in babies with neural-tube defects. When measured at 15–20 weeks gestation, AFP is normally distributed with a mean of 58 and a standard deviation of 18.

What is the probability that AFP exceeds 75 in a pregnant woman measured at 18 weeks gestation? Specifically, what is $P(x > 75)$? Because AFP is normally distributed, we standardize:

$P(x > 75) = P\left(z > \dfrac{75 - 58}{18} = \dfrac{17}{18} = 0.94\right).$ Using Table 1, $P(x > 75) = P(z > 0.94) = 1 - 0.8264 = 0.1736$. There is a 17% chance that AFP exceeds 75 in a pregnant woman measured at 18 weeks gestation.

In a sample of 50 women, what is the probability that their mean AFP exceeds 75? Specifically, what is $P(\overline{X} > 75)$? Using the Central Limit Theorem, we standardize:

$P(\overline{X} > 75) = P\left(z > \dfrac{75 - 58}{18/\sqrt{50}} = \dfrac{17}{2.55} = 6.67\right).$ It is extremely unlikely (probability very close to 0) to observe a z score exceeding 6.67. There is virtually no chance that in a sample of 50 women their mean AFP exceeds 75. Notice that the first part of the question addresses the probability of observing a single woman with an AFP exceeding 75, whereas the second part of the question addresses the probability that the mean AFP in a sample of 50 women exceeds 75. The latter requires the application of the Central Limit Theorem.

5.7 SUMMARY

In Chapters 6, 7, 9, 10, and 11, we discuss statistical inference in detail. We present formulas and procedures to make inferences about populations based on a single sample. The relationship between the sample statistic and the population parameter is based on the sampling distribution of that statistic and probability theory. Here we discussed probability as it applies to selecting individuals from a population into a sample. There are some basic concepts of probability that can be applied when the entire population can be enumerated.

When the population enumeration is not available, probability models can be used to determine probabilities as long as specific attributes are satisfied. The binomial and normal distribution models are popular models for discrete and continuous outcomes, respectively. A key theorem is the Central Limit Theorem, which brings together the concepts of probability and inference. We rely heavily on the Central Limit Theorem in the next chapters, where we discuss statistical inference in detail. **Table 5–10** summarizes key formulas and concepts in probability.

5.8 PRACTICE PROBLEMS

1. A recent study reported that the prevalence of hyperlipidemia (defined as total cholesterol over 200) is 30% in children 2 to 6 years of age. If 12 children are analyzed:
   a. What is the probability that at least 3 are hyperlipidemic?
   b. What is the probability that exactly 3 are hyperlipidemic?
   c. How many would be expected to meet the criteria for hyperlipidemia?
2. Hyperlipidemia in children has been hypothesized to be related to high cholesterol in their parents. The data in **Table 5–11** were collected on parents and children.
   a. What is the probability that one or both parents are hyperlipidemic?
   b. What is the probability that a child and both parents are hyperlipidemic?
   c. What is the probability that a child is hyperlipidemic if neither parent is hyperlipidemic?

TABLE 5–10 Summary of Key Formulas

Concept	Formula
Basic probability	$P(\text{Characteristic}) = \dfrac{\text{Number of persons with characteristic}}{N}$
Conditional probability rule	$P(A|B) = \dfrac{P(A \text{ and } B)}{P(B)}$
Sensitivity	$P(\text{Screen positive} \mid \text{Disease})$
Specificity	$P(\text{Screen negative} \mid \text{Disease free})$
False positive fraction	$P(\text{Screen positive} \mid \text{Disease free})$
False negative fraction	$P(\text{Screen negative} \mid \text{Disease})$
Positive predictive value	$P(\text{Disease} \mid \text{Screen positive})$
Negative predictive value	$P(\text{Disease free} \mid \text{Screen negative})$
Independent events	$P(A|B) = P(A)$ or $P(B|A) = P(B)$
Bayes Theorem	$P(A|B) = \dfrac{P(B|A)P(A)}{P(B)}$
Binomial distribution	$P(x \text{ successes}) = \dfrac{n!}{x!(n-x)!} p^x (1-p)^{n-x}$
Standard normal distribution	$z = \dfrac{X - \mu}{\sigma}$ (Table 1)
Percentiles of the normal distribution	$X = \mu + z\sigma$ (Table 1A)
Application of Central Limit Theorem	$z = \dfrac{\overline{X} - \mu}{\sigma/\sqrt{n}}$ (Table 1)

TABLE 5–11 Hyperlipidemia in Parents and Children

Child	Both Parents Hyperlipidemic	One Parent Hyperlipidemic	Neither Parent Hyperlipidemic
Not hyperlipidemic	13	34	83
Hyperlipidemic	45	42	6

    d. What is the probability that a child is hyperlipidemic if both parents are hyperlipidemic?
3. Total cholesterol in children aged 10 to 15 is assumed to follow a normal distribution with a mean of 191 and a standard deviation of 22.4.
    a. What proportion of children 10 to 15 years of age has total cholesterol between 180 and 190?
    b. What proportion of children 10 to 15 years of age would be classified as hyperlipidemic? (Assume that hyperlipidemia is defined as a total cholesterol level over 200.)
    c. If a sample of 20 children is selected, what is the probability that the mean cholesterol level in the sample will exceed 200?
4. A national survey of graduate students is conducted to assess their consumption of coffee. **Table 5–12** summarizes the data.
    a. What proportion of students drinks decaffeinated coffee only?
    b. What proportion of coffee drinkers (caffeinated and decaffeinated) is female?
    c. What proportion of the females does not drink coffee?
5. Among coffee drinkers, men drink a mean of 3.2 cups per day with a standard deviation of 0.8 cup. Assume the number of cups per day follows a normal distribution.
    a. What proportion drinks 2 cups per day or more?
    b. What proportion drinks no more than 4 cups per day?
    c. If the top 5% of coffee drinkers are considered "heavy" coffee drinkers, what is the minimum number of cups consumed by a heavy coffee drinker?
    d. If a sample of 20 men is selected, what is the probability that the mean number of cups per day is greater than 3?
6. A study is conducted to assess the impact of caffeine consumption, smoking, alcohol consumption, and physical activity on cardiovascular disease. Suppose that 40% of participants consume caffeine and smoke. If 8 participants are evaluated, what is the probability that:
    a. Exactly half of them consume caffeine and smoke?
    b. More than 6 consume caffeine and smoke?
    c. Exactly 4 do not consume caffeine or smoke?
7. As part of the study described in Problem 6, investigators wanted to assess the accuracy of self-reported smoking status. Participants are asked whether they currently smoke or not. In addition, laboratory tests are performed on hair samples to determine the presence or absence of nicotine. The laboratory assessment is considered the gold standard, or the truth about nicotine consumption. The data are shown in **Table 5–13**.
    a. What is the sensitivity of the self-reported smoking status?
    b. What is the specificity of the self-reported smoking status?

TABLE 5–12 Coffee Consumption in Graduate Students

	Do Not Drink Coffee	Drink Decaffeinated Only	Drink Caffeinated Coffee
Male	145	94	365
Female	80	121	430

TABLE 5–13 Self-Reported Smoking Status

	Nicotine Absent	Nicotine Present
Self-reported nonsmoker	82	14
Self-reported smoker	12	52

8. A recent study of cardiovascular risk factors reported that 30% of adults meet criteria for hypertension. If 15 adults are assessed, what is the probability that:
   a. Exactly 5 are the criteria for hypertension?
   b. None meet the criteria for hypertension?
   c. How many would you expect to meet the criteria for hypertension?

9. Table 5–14 displays blood pressure status by gender.
   a. What proportion of the participants has optimal blood pressure?
   b. What proportion of men has optimal blood pressure?
   c. What proportion of participants with hypertension is men?
   d. Are hypertensive status and male gender independent?

10. Diastolic blood pressures are assumed to follow a normal distribution with a mean of 85 and a standard deviation of 12.
    a. What proportion of people has diastolic blood pressures less than 90?
    b. What proportion has diastolic blood pressures between 80 and 90?
    c. If someone has a diastolic blood pressure of 100, what percentile does this represent?

11. Consider the data described in Problem 10. If 15 participants are sampled, what is the probability that their mean diastolic blood pressure exceeds 90?

12. A large national study finds that 10% of pregnant women deliver prematurely. A local obstetrician is seeing 16 pregnant women in his next clinic session.
    a. What is the probability that none will deliver prematurely?
    b. What is the probability that fewer than 3 will deliver prematurely?
    c. What is the probability that none will deliver prematurely if, in fact, the true percentage who deliver prematurely is 5.5%?
    d. If the true percentage is 10% and this obstetrician has 146 pregnant women under his care, how many would be expected to deliver prematurely?

13. Table 5–15 cross-classifies pregnant women in a study by their body mass index (BMI) at 16 weeks gestation and whether they had a pre-term delivery.
    a. What is the probability that a woman delivers pre-term?
    b. What is the probability a woman has a BMI less than 30 and delivers pre-term?
    c. What proportion of women with a BMI greater than 35 delivers pre-term?
    d. Are BMI and pre-term delivery independent? Justify.

14. In the study described in Problem 13, suppose the mean BMI at 16 weeks gestation is 28.5 with a standard deviation of 3.6, and that BMI is assumed to follow a normal distribution. Find the following:
    a. The proportion of women with a BMI greater than 30.
    b. The proportion of women with a BMI greater than 40.
    c. The BMI that separates the top 10% from the rest.

15. Suppose we want to estimate the mean BMI for women in pregnancy at 20 weeks gestation. If we have a sample of 100 women and measure their BMI at 20 weeks gestation, what is the probability that the sample mean is within 1 unit of the true BMI if the standard deviation in the BMI is taken to be 3.6?

TABLE 5–14 Blood Pressure by Gender

	Optimal	Normal	Hypertension	Total
Male	22	73	55	150
Female	43	132	65	240
Total	65	205	120	390

TABLE 5–15 Body Mass Index and Preterm Delivery

	BMI < 30	BMI 30–34.9	BMI 35+
Pre-term	320	80	120
Full term	4700	480	300

16. Diastolic blood pressures are approximately normally distributed with a mean of 75 and a standard deviation of 10.
    a. What is the 90th percentile of diastolic blood pressure?
    b. If we consider samples of 20 patients, what is the 90th percentile of the mean diastolic blood pressure?
17. **Table 5–16** displays the number of children in a local town classified as normal weight, overweight, and obese by year in school.
    a. What proportion of the children is obese?
    b. What proportion of the elementary school children is overweight or obese?
    c. What proportion of the normal weight children is in high school?
18. In a primary care clinic, 25% of all patients who have appointments fail to show up. Each clinic session has 10 scheduled appointments.
    a. What is the probability that half of the patients fail to show up?
    b. What is the probability that all patients show up?
    c. In a week, there are 10 clinic sessions. How many patients would you expect to fail to show up?
19. High density lipoprotein (HDL) in healthy males follows a normal distribution with a mean of 50 and a standard deviation of 8.
    a. What proportion of healthy males has HDL exceeding 60?
    b. What proportion of healthy males has HDL lower than 40?
    c. What is the 90th percentile of HDL in healthy males?
20. **Table 5–17** summarizes data collected in a study to evaluate a new screening test for ovarian cancer.

TABLE 5–16 Obesity Status in Children

	Normal Weight	Overweight	Obese	Total
Elementary	50	120	80	250
Middle	30	45	45	120
Junior high	50	50	40	140
High school	30	85	85	200
Total	160	300	250	710

TABLE 5–17 New Screening Test for Ovarian Cancer

Screening Test	Ovarian Cancer	Free of Ovarian Cancer
Positive	28	23
Negative	22	127
Total	50	150

There are 200 women are involved in the study. Fifty have ovarian cancer and 150 do not. The results are tabulated below.
   a. Find the sensitivity of the screening test.
   b. Find the false positive fraction of the screening test.
   c. What proportion of women who screen positive actually has ovarian cancer?

21. An experimental drug has been shown to be 75% effective in eliminating symptoms of allergies in animal studies. A small human study involving six participants is conducted. What is the probability that the drug is effective on more than half of the participants?

REFERENCES

1. National Center for Health Statistics. *Health, United States, 2003, with Chartbook on Trends in the Health of Americans.* Hyattsville, MD: US Government Printing Office, 2003.
2. Hedley, A.A., Ogden, C.L., Johnson, C.L., Carroll, M.D., Curtin, L.R., and Flegal, K.M. "Prevalence of overweight and obesity among U.S. children, adolescents, and adults, 1999–2002." *Journal of the American Medical Association* 2004; 291: 2847–2850.
3. Cope, M.B. and Allison, D.B. "Obesity: Person and population." *Obesity* 2006; 14: S156–S159.
4. Kim, J., Peterson, K.F., Scanlon, K.S., Fitzmaurice, G.M., Must, A., Oken, E., Rifas-Shiman, S.L., Rich-Edwards, J.W., and Gillman, M.W. "Trends in overweight from 1980 through 2001 among preschool-aged children enrolled in a health maintenance organization." *Obesity* 2006; 14: 1107–1112.
5. Cochran, W.G. *Sampling Techniques* (3rd ed.). New York: John Wiley & Sons, 1977.
6. Kish, L. *Survey Sampling (Wiley Classics Library).* New York: John Wiley & Sons, 1995.
7. Rosner, B. *Fundamentals of Biostatistics.* Belmont, CA: Duxbury-Brooks/Cole, 2006.
8. SAS Version 9.1. © 2002–2003 by SAS Institute Inc., Cary, NC.
9. Thompson, I.M., Pauler, D.K., Goodman P.J., et al. "Prevalence of prostate cancer among men with a prostate-specific antigen level ≤ 4.0 ng per milliliter." *The New England Journal of Medicine* 2004; 350(22): 2239–2246.
10. D'Agostino, R.B., Sullivan, L.M., and Beiser, A. *Introductory Applied Biostatistics.* Belmont, CA: Duxbury–Brooks/Cole, 2004.

CHAPTER 6

Confidence Interval Estimates

When and Why

Key Questions

- What is the difference between a 95% confidence interval and the mid-range of a sample defined by the 2.5th and 97.5th percentiles?
- What do pollsters mean when they show different percentages of voters in favor of specific candidates and call the results "a statistical dead heat"?
- Why not generate a 99% (even 100%) confidence interval instead of 95% to be sure about results?

In the News

Traumatic brain injury (TBI) and concussion, a special type of TBI, are the source of extensive research and debate. The Centers for Disease Control and Prevention (CDC) collects data on TBIs and produces fact sheets on this topic regularly.[1] There are nearly 2 million reported TBIs in the United States each year, and approximately 300,000 of these injuries are sports related. Among men, such an injury is most likely to be sustained while playing football. Among women, TBI is most likely to be sustained while playing soccer.

A study of almost 9 million insurance records reported that concussions are on the rise, particularly for adolescents, and that prior estimates of concussion incidence were actually underestimates.[2]

There are a number of challenges in collecting accurate data on concussions, as approximately 47% of people who suffer concussions do not experience symptoms immediately after the injury occurs.[3]

---

[1] Available at http://www.cdc.gov/traumaticbraininjury/get_the_facts.html.
[2] "Concussions on the rise for adolescents." *ScienceDaily*. Available at https://www.sciencedaily.com/releases/2016/07/160710094045.htm. Accessed July 10, 2016.
[3] Sports Concussion Institute. Available at http://www.concussiontreatment.com/concussionfacts.html.

Dig In

- How would you go about estimating the proportion of athletes in your school who suffered a concussion in the past year?
- How would you be sure that you captured all of the concussion events?
- Suppose that you collect data and estimate that 7% of students engaged in organized sports at your school experienced a concussion during the academic year, with a 95% confidence interval of (1%–13%). What is the appropriate interpretation of this confidence interval?

Learning Objectives

By the end of this chapter, the reader will be able to

- Define point estimate, standard error, confidence level, and margin of error
- Compare and contrast standard error and margin of error
- Compute and interpret confidence intervals for means and proportions
- Differentiate independent and matched or paired samples
- Compute confidence intervals for the difference in means and proportions in independent samples and for the mean difference in paired samples
- Identify the appropriate confidence interval formula based on type of outcome variable and number of samples

---

We now begin statistical inference. In Chapter 4, we presented descriptive statistics used to summarize sample data. In Chapter 5, we presented key concepts in probability and the Central Limit Theorem. In statistical inference, we use all of these concepts to make inferences about unknown population parameters based on sample statistics. There are two broad areas of statistical inference: estimation and hypothesis testing. In estimation, sample statistics are used to generate estimates about unknown population parameters. In hypothesis testing, a specific statement or hypothesis is generated about a population parameter and sample statistics are used to assess the likelihood that the hypothesis is true. We discuss estimation here and introduce hypothesis testing in Chapter 7.

Estimation is the process of determining a likely value for a population parameter (e.g., the true population mean or population proportion) based on a random sample. In practice, we select a sample from the population and use sample statistics (e.g., the sample mean or the sample proportion) to estimate the unknown parameter. The sample should be representative of the population, with participants selected at random from the population. Because different samples can produce different results, it is necessary to quantify the precision—or lack thereof—that might exist among estimates from different samples.

The techniques for estimation as well as for other procedures in statistical inference depend on the appropriate classification of the key study variable (which we also call the **outcome** or endpoint) as continuous or dichotomous. (There are other types of variables, which are discussed in Chapters 7, 10, and 11; here we focus on continuous and dichotomous outcomes.) Another key issue is the number of comparison groups in the investigation. For example, in the two-comparison-group case it is important to determine whether the samples from the groups are independent (i.e., physically separate, such as men versus women) or dependent (also called **matching** or pairing). These issues dictate the appropriate estimation technique. **Table 6–1** outlines these issues and identifies the estimation techniques that we discuss here. Each is discussed in detail.

6.1 INTRODUCTION TO ESTIMATION

There are two types of estimates that can be produced for any population parameter, a point estimate and a confidence interval estimate. A **point estimate** for a population parameter is a single-valued estimate of that parameter. A **confidence interval** (CI) estimate is a range of values for a population parameter with a level of confidence attached (e.g., 95% confidence that the interval contains the unknown parameter). The level of confidence is similar to a probability. The CI starts with the point estimate and builds in what is called a margin of error. The margin of error incorporates the confidence level (e.g., 90% or 95%, which is chosen by the investigator) and the sampling variability or the standard error of the point estimate.

A CI is a range of values that is likely to cover the true population parameter, and its general form is point estimate ± **margin of error**. The point estimate is determined first. The point estimates for the population mean and proportion are the sample mean and sample proportion, respectively. These are our best single-valued estimates of the unknown population parameters. Recall from Chapter 5 that the sample mean is an unbiased estimator of the population mean. The same holds true for the sample proportion with regard to estimating the population proportion. Thus the starting place, or point estimate, for the CI for the population mean is the sample mean, and the point estimate for the population proportion is the sample proportion.

Next, a level of confidence is selected that reflects the likelihood that the CI contains the true, unknown parameter. Usually, confidence levels of 90%, 95%, and 99% are chosen, although theoretically any confidence level between 0% and 100% can be selected.

Suppose we want to generate a CI estimate for an unknown population mean. Again, the form of the CI is point estimate ± margin of error, or $\bar{X} \pm$ margin of error. Suppose we select a 95%

TABLE 6–1 Estimation Techniques

Number of Samples	Outcome Variable	Parameter to be Estimated
One sample	Continuous	Mean
Two independent samples	Continuous	Difference in means
Two dependent, matched samples	Continuous	Mean difference
One sample	Dichotomous	Proportion (e.g., prevalence, cumulative incidence)
Two independent samples	Dichotomous	Difference or ratio of proportions (e.g., attributable risk, relative risk, odds ratio)

confidence level. This means that there is a 95% probability that a CI will contain the true population mean. Thus,

$$P(\overline{X} - \text{margin of error} < \mu < \overline{X} + \text{margin of error}) = 0.95.$$

In Chapter 5, we presented the Central Limit Theorem, which stated that for large samples, the distribution of the sample means is approximately normal with a mean $\mu_{\overline{X}} = \mu$ and standard deviation $\sigma_{\overline{X}} = \dfrac{\sigma}{\sqrt{n}}$. We use the Central Limit Theorem to develop the margin of error.

For the standard normal distribution, the following is a true statement: $P(-1.96 < z < 1.96) = 0.95$, i.e., there is a 95% chance that a standard normal variable (z) will fall between -1.96 and 1.96. The Central Limit Theorem states that for large samples, $z = \dfrac{\overline{X} - \mu}{\sigma/\sqrt{n}}$. If we make this substitution, the following statement is true: $P(-1.96 < \dfrac{\overline{X} - \mu}{\sigma/\sqrt{n}} < 1.96) = 0.95$. Using algebra, we can rework this inequality such that the mean (μ) is the middle term. The steps are outlined below:

$$P(-1.96 < \dfrac{\overline{X} - \mu}{\sigma/\sqrt{n}} < 1.96) = 0.95$$

$$P(-1.96 \dfrac{\sigma}{\sqrt{n}} < \overline{X} - \mu < 1.96 \dfrac{\sigma}{\sqrt{n}}) = 0.95$$

$$P(-\overline{X} - 1.96 \dfrac{\sigma}{\sqrt{n}} < -\mu < -\overline{X} + 1.96 \dfrac{\sigma}{\sqrt{n}}) = 0.95$$

$$P(\overline{X} - 1.96 \dfrac{\sigma}{\sqrt{n}} < \mu < \overline{X} + 1.96 \dfrac{\sigma}{\sqrt{n}}) = 0.95$$

The 95% CI for the population mean is the interval in the last probability statement and is given by: $\overline{X} \pm 1.96\,\sigma/\sqrt{n}$. The margin of error is $1.96\,\sigma/\sqrt{n}$, where 1.96 reflects the fact that a 95% confidence level is selected and $\sigma/\sqrt{n}$ is the standard error (or the standard deviation of the point estimate, $\overline{X}$). The general form of a CI can be rewritten as follows:

point estimate $\pm z$ SE (*point estimate*),

where z is the value from the standard normal distribution reflecting the selected confidence level (e.g., for a 95% confidence level, $z = 1.96$). Table 1B in the Appendix contains the z values for popular confidence levels such as 90%, 95%, and 99%. In Table 1B, we find for 90%, $z = 1.645$; for 95%, $z = 1.96$; and for 99%, $z = 2.576$. Higher confidence levels have larger z values, which translate to larger margins of error and wider CIs. For example, to be 99% confident that a CI contains the true unknown parameter, we need a wider interval. In many applications, a confidence level of 95% is used. This is a generally accepted, but not prescribed, value.

In practice, we often do not know the value of the population standard deviation (σ). If the sample size is large ($n > 30$), then the sample standard deviation (s) can be used to estimate the population standard deviation. Note that the prior derivation is based on the Central Limit Theorem, which requires a large sample size. There are instances in which the sample size is not sufficiently large (e.g., $n < 30$), and therefore the general result of the Central Limit Theorem does not apply. In this case, we cannot use the standard normal distribution (z) in the confidence interval. Instead we use another probability distribution, called the t distribution, which is appropriate for small samples.

The t distribution is another probability model for a continuous variable. The t distribution is similar to the standard normal distribution but takes a slightly different shape depending on the exact sample size. Specifically, the t values for CIs are larger for smaller samples, resulting in larger margins of error (i.e., there is more imprecision with small samples). t values for CIs are contained in Table 2 in the Appendix. t values are indexed by degrees of freedom (df) in Table 2, which is defined as $n - 1$. **Table 6-2** is an excerpt of Table 2 showing the t values for small samples ranging in size from 5 to 10 (thus, the degrees of freedom range from 4 to 9, as $df = n - 1$).

Specific guidelines for using the standard normal (z) or t distributions are provided in subsequent sections as we discuss the CI formulas for specific applications. It is important to note that appropriate use of the t distribution assumes that the outcome of interest is approximately normally distributed.

Before providing specific formulas, we first discuss the interpretation of CIs in general. Suppose we want to estimate a population mean using a 95% confidence level. If we take 100 different samples (in practice, we take only one) and for each sample we compute a 95% CI, in theory 95 out of the 100 CIs will contain the true mean value (μ). This leaves 5 of 100 CIs that will not include the true mean value. In practice, we select one random sample and generate one CI. This interval may or may

TABLE 6–2 *t* Values for Confidence Intervals

		Confidence Level			
df	80%	90%	95%	98%	99%
4	1.533	2.132	2.776	3.747	4.604
5	1.476	2.015	2.571	3.365	4.032
6	1.440	1.943	2.447	3.143	3.707
7	1.415	1.895	2.365	2.998	3.499
8	1.397	1.860	2.306	2.896	3.355
9	1.383	1.833	2.262	2.821	3.250

not contain the true mean; the observed interval may overestimate μ or underestimate μ. The 95% CI is the likely range of the true, unknown parameter. It is important to note that a CI does not reflect the variability in the unknown parameter but instead provides a range of values that are likely to include the unknown parameter.

6.2 CONFIDENCE INTERVALS FOR ONE SAMPLE, CONTINUOUS OUTCOME

We wish to estimate the mean of a continuous outcome variable in a single population. For example, we wish to estimate the mean systolic blood pressure, body mass index (BMI), total cholesterol level, or white blood cell count in a single population. We select a sample and compute descriptive statistics on the sample data using the techniques described in Chapter 4. Specifically, we compute the sample size (n), the sample mean ($\overline{X}$), and the sample standard deviation (s). The formulas for CIs for the population mean depend on the sample size and are given in **Table 6–3**.

Example 6.1. In Chapter 4, we presented data on $n = 3539$ participants who attended the seventh examination of the offspring in the Framingham Heart Study. Descriptive statistics on variables measured in the sample are shown in **Table 6–4** (these and other statistics were presented in Table 4–20). The numbers of participants (n) who provided information on each characteristic are shown in the second column of Table 6–4.

We wish to generate a 95% CI for systolic blood pressure using data collected in the Framingham Offspring Study. Because the sample size is large, we use the following formula,

$$\overline{X} \pm z\frac{s}{\sqrt{n}}.$$

The z value for 95% confidence is $z = 1.96$. Substituting the sample statistics and the z value for 95% confidence, we have

$$127.3 \pm 1.96\frac{19.0}{\sqrt{3534}}.$$

TABLE 6–3 Confidence Intervals for μ

$n \geq 30$	$\overline{X} \pm z\dfrac{s}{\sqrt{n}}$	(Find z in Table 1B)
$n < 30$	$\overline{X} \pm t\dfrac{s}{\sqrt{n}}$	(Find t in Table 2, $df = n - 1$)

TABLE 6–4 Summary Statistics on Participants Attending the Seventh Examination of the Framingham Offspring Study ($n = 3539$)

	n	Mean ($\overline{X}$)	Standard Deviation (s)
Systolic blood pressure	3534	127.3	19.0
Diastolic blood pressure	3532	74.0	9.9
Total serum cholesterol	3310	200.3	36.8
Weight (lbs)	3506	174.4	38.7
Height (in.)	3326	65.957	3.749
Body mass index (BMI)	3326	28.15	5.32

Working through the computations, we have

$$127.3 \pm 0.63.$$

Adding and subtracting the margin of error, we get (126.7, 127.9). A point estimate for the true mean systolic blood pressure in the population is 127.3, and we are 95% confident that the true mean is between 126.7 and 127.9. The margin of error is very small here because of the large sample size.

A 90% CI for BMI is given below. Notice that $z = 1.645$ to reflect the 90% confidence level:

$$28.15 \pm 1.645\frac{5.32}{\sqrt{3326}},$$

$$28.15 \pm 0.152,$$

$$(28.00, 28.30).$$

We are 90% confident that the true mean BMI in the population is between 28.00 and 28.30. Again, the CI is very precise or narrow due to the large sample size.

Example 6.2. In Chapter 4, we also presented data on a subsample of $n = 10$ participants who attended the seventh examination of the Framingham Offspring Study. Descriptive statistics on variables measured in the subsample are shown in **Table 6–5** (these and other statistics were presented in Table 4–18).

Suppose we compute a 95% CI for the true systolic blood pressure using data in the subsample. Because the sample size is small, we must now use the CI formula that involves t rather than z,

$$\overline{X} \pm t\frac{s}{\sqrt{n}}.$$

6.3 CONFIDENCE INTERVALS FOR ONE SAMPLE, DICHOTOMOUS OUTCOME

There are many applications where the outcome of interest is dichotomous. The parameter of interest is the unknown population proportion, denoted p. For example, suppose we wish to estimate the proportion of people with diabetes in a population, or the proportion of people with hypertension or obesity. The latter are defined by specific levels of blood pressure and BMI, respectively. Again, we select a sample and compute descriptive statistics on the sample data using the techniques described in Chapter 4. When the outcome of interest is dichotomous, we record on each member of the sample whether they have the characteristic of interest or not. The sample size is denoted by n and we let x denote the number of successes in the sample. Recall that for dichotomous outcomes, we define one of the outcomes a success and the other a failure. The specific response that is considered a success is defined by the investigator. For example, if we wish to estimate the proportion of people with diabetes in a population, we consider a diagnosis of diabetes (the outcome of interest) a success and lack of diagnosis a failure. In this example, x represents the number of people with a diagnosis of diabetes in the sample. The sample proportion is denoted $\hat{p}$, and is computed by taking the ratio of the number of successes in the sample to the sample size, $\hat{p} = \frac{x}{n}$. The formula for the CI for the population proportion is given in **Table 6–6**.

The CI for the population proportion takes the same form as the CI for the population mean (i.e., point estimate ± margin of error). The point estimate for the population proportion is the sample proportion. The margin of error is the product of the z value for the desired confidence level (e.g., $z = 1.96$ for 95% confidence) and the standard error of the point estimate,

$$SE(\hat{p}) = \sqrt{\frac{\hat{p}(1-\hat{p})}{n}}.$$

The preceding formula is appropriate for large samples, defined as at least five successes ($n\hat{p}$) and at least five failures

TABLE 6–5 Summary Statistics on $n = 10$ Participants Attending the Seventh Examination of the Framingham Offspring Study

	n	Mean ($\bar{X}$)	Standard Deviation (s)
Systolic blood pressure	10	121.2	11.1
Diastolic blood pressure	10	71.3	7.2
Total serum cholesterol	10	202.3	37.7
Weight (lbs)	10	176.0	33.0
Height (in.)	10	67.175	4.205
Body mass index (BMI)	10	27.26	3.10

We first need to determine the appropriate t value from Table 2. To do this, we need $df = n - 1 = 10 - 1 = 9$. The t value for 95% confidence with $df = 9$ is $t = 2.262$. Substituting the sample statistics and the t value for 95% confidence, we have

$$121.2 \pm 2.262 \frac{11.1}{\sqrt{10}}.$$

Working through the computations, we have

$$121.2 \pm 7.94.$$

Adding and subtracting the margin of error, we get (113.3, 129.1). Based on this sample of size $n = 10$, our best estimate of the true mean systolic blood pressure in the population is 121.2. Based on this sample, we are 95% confident that the true mean systolic blood pressure in the population is between 113.3 and 129.1. Notice that the margin of error is larger here primarily due to the smaller sample size.

Using the subsample, we now compute a 90% CI for the mean BMI. Because the sample size is small, we again need to determine an appropriate value from the t distribution. For 90% confidence and $df = 9$, $t = 1.833$.

$$27.26 \pm 1.833 \frac{3.10}{\sqrt{10}},$$

$$27.26 \pm 1.80,$$

$$(25.46, 29.06).$$

We are 90% confident that the true mean BMI in the population is between 25.46 and 29.06. Again, because of the small sample size, the CI is less precise.

TABLE 6–6 Confidence Interval for p

$\min[n\hat{p}, n(1-\hat{p})] \geq 5$	$\hat{p} \pm z\sqrt{\dfrac{\hat{p}(1-\hat{p})}{n}}$	(Find z in Table 1B)

$[n(1-\hat{p})]$ in the sample. If there are fewer than five successes or failures, then alternative procedures—called exact methods—must be used to estimate the population proportion.[1,2]

Example 6.3. In Chapter 4, we presented data on $n = 3539$ participants who attended the seventh examination of the Offspring in the Framingham Heart Study. One particular characteristic measured was treatment with antihypertensive medication. There were a total of 1219 participants on treatment and 2313 participants not on treatment (Table 4–3). If we call treatment a success, then $x = 1219$ and $n = 3532$. The sample proportion is

$$\hat{p} = \frac{x}{n} = \frac{1219}{3532} = 0.345.$$

Thus, a point estimate for the population proportion is 0.345, or 34.5%. Our best estimate of the proportion of participants in the population on treatment for hypertension is 0.345. Suppose we now wish to generate a 95% CI. To use the preceding formula, we need to satisfy the sample size criterion—specifically, we need at least five successes and five failures. Here we more than satisfy that requirement, so the CI formula in Table 6–6 can be used.

$$\hat{p} \pm z\sqrt{\frac{\hat{p}(1-\hat{p})}{n}},$$

$$0.345 \pm 1.96\sqrt{\frac{0.345(1-0.345)}{3532}},$$

$$0.345 \pm 0.016,$$

$$(0.329, 0.361).$$

Thus, we are 95% confident that the true proportion of persons on antihypertensive medication is between 0.329 and 0.361, or between 32.9% and 36.1%.

Specific applications of estimation for a single population with a dichotomous outcome involve estimating prevalence and cumulative incidence. In Chapter 3, we generated point estimates for prevalence and incidence data; in the following examples, we add CI estimates.

Example 6.4. In Example 3.1, we presented data on $n = 3799$ participants who attended the fifth examination of the offspring in the Framingham Heart Study. **Table 6–7** shows the numbers of men and women with diagnosed, or prevalent, cardiovascular disease (CVD) at the fifth examination.

The prevalence of CVD for all participants attending the fifth examination of the Framingham Offspring Study is $379 / 3799 = 0.0998$. The prevalence of CVD among men is $244 / 1792 = 0.1362$, and the prevalence of CVD among women is $135 / 2007 = 0.0673$. These are point estimates. Following, we generate 95% CI estimates for prevalence in the total population and in the populations of men and women. To use the preceding formula, we need to satisfy the sample size criterion—specifically, we need at least five successes and five failures in each sample. In this example, a success is prevalent CVD and a failure is freedom from CVD. Here we more than satisfy that requirement for men, women, and the pooled or total sample. For the total sample,

TABLE 6–7 Prevalent CVD in Men and Women

	Free of CVD	Prevalent CVD	Total
Men	1548	244	1792
Women	1872	135	2007
Total	3420	379	3799

$$0.0998 \pm 1.96\sqrt{\frac{0.0998(1-0.0998)}{3799}},$$

$$0.0998 \pm 0.0095,$$

$$(0.090, 0.109).$$

We are 95% confident that the true prevalence of CVD in the population is between 9% and 10.9%.

The CI estimates for men and women are as follows. For men:

$$0.1362 \pm 1.96\sqrt{\frac{0.1362(1-0.1362)}{1792}},$$

$$0.1362 \pm 0.016,$$

$$(0.120, 0.152).$$

For women:

$$0.0673 \pm 1.96\sqrt{\frac{0.0673(1-0.0673)}{2007}},$$

$$0.0673 \pm 0.011,$$

$$(0.056, 0.078).$$

We are 95% confident that the true prevalence of CVD in men is between 12% and 15.2%, and in women, between 5.6% and 7.8%. Each of these CI estimates is very precise (small margins of error) because of the large sample sizes.

6.4 CONFIDENCE INTERVALS FOR TWO INDEPENDENT SAMPLES, CONTINUOUS OUTCOME

There are many applications where it is of interest to compare two groups with respect to their mean scores on a continuous outcome. For example, the comparison groups might be men versus women, patients assigned to an experimental treatment versus a placebo in a clinical trial, or patients with a history of cardiovascular disease versus patients free of cardiovascular disease. We can compare mean systolic blood pressures in men versus women, or mean BMI or total cholesterol levels in patients assigned to experimental treatment versus placebo.

A key feature here is that the two comparison groups are independent, or physically separate. The two groups might be determined by a particular attribute or characteristic (e.g., sex, diagnosis of cardiovascular disease) or might be set up by the investigator (e.g., participants assigned to receive an experimental drug or placebo).

Similar to the approach we used to estimate the mean of a continuous variable in a single population (Section 6.2), we first compute descriptive statistics on each of the two samples using the techniques described in Chapter 4. Specifically, we compute the sample size, mean, and standard deviation in each sample. We denote these summary statistics as $n_1, \overline{X}_1$, and s_1 for Sample 1 and $n_2, \overline{X}_2$, and s_2 for Sample 2. (The designation of Sample 1 and Sample 2 is essentially arbitrary. In a clinical trial setting, the convention is to call the experimental treatment Group 1 and the control treatment Group 2. However, when comparing groups such as men and women, either group can be 1 or 2.) The interpretation of the CI estimate depends on how the samples are assigned, but this is just for interpretation and does not affect the computations. This is illustrated in the following examples.

In the two-independent-samples application with a continuous outcome, the parameter of interest is the difference in population means, $\mu_1 - \mu_2$. The point estimate for the difference in population means is the difference in sample means, $\overline{X}_1 - \overline{X}_2$. The form of the CI is again point estimate $\pm$ margin of error, and the margin of error incorporates a value from either the z or t distribution reflecting the selected confidence level and the standard error of the point estimate. The use of z or t depends on whether the sample sizes are large or small. (The specific guidelines follow.) The standard error of the point estimate incorporates the variability in the outcome of interest in each of the comparison groups. **Table 6–8** contains the formulas for CIs for the difference in population means.

The formulas in Table 6–8 assume equal variability in the two populations (i.e., that the population variances are equal, or $\sigma_1^2 = \sigma_2^2$). This means that the outcome is equally variable in each of the comparison populations. For analysis, we have samples from each of the comparison populations. If the sample variances are similar, then the assumption about variability in the populations is reasonable. As a guideline, if the ratio of the sample variances, s_1^2/s_2^2, is between 0.5 and 2 (i.e., if one variance is no more than double the other), then the formulas in Table 6–8 are appropriate. If the ratio of the sample variances is greater than 2 or less than 0.5, then alternative formulas must be used to account for the heterogeneity in variances. Specific details can be found in D'Agostino, Sullivan, and Beiser and in Rosner.[3,4]

In the formulas in Table 6–8, $\overline{X}_1$ and $\overline{X}_2$ are the means of the outcome in the independent samples, z or t are values from the z or t distributions reflecting the desired confidence level, and $S_p\sqrt{\dfrac{1}{n_1} + \dfrac{1}{n_2}}$ is the standard error of the point estimate, $SE(\overline{X}_1 - \overline{X}_2)$.

Here, S_p is the pooled estimate of the common standard deviation (again, assuming that the variances in the populations are similar) computed as the weighted average of the standard deviations in the samples:

$$S_p = \sqrt{\frac{(n_1 - 1)s_1^2 + (n_2 - 1)s_2^2}{n_1 + n_2 - 2}}.$$

TABLE 6–8 Confidence Intervals for $(\mu_1 - \mu_2)$

$n_1 \geq 30$ and $n_2 \geq 30$	$(\overline{X}_1 - \overline{X}_2) \pm zS_p\sqrt{\dfrac{1}{n_1} + \dfrac{1}{n_2}}$	(Find z in Table 1B)
$n_1 < 30$ or $n_2 < 30$	$(\overline{X}_1 - \overline{X}_2) \pm tS_p\sqrt{\dfrac{1}{n_1} + \dfrac{1}{n_2}}$	(Find t in Table 2, $df = n_1 + n_2 - 2$)

Because we are assuming equal variances in the groups, we pool the information on variability (sample variances) to generate an estimate of the population variability.

Example 6.5. In Example 6.1, we recapped the data presented in Chapter 4 on $n = 3539$ participants who attended the seventh examination of the offspring in the Framingham Heart Study. **Table 6-9** contains descriptive statistics on the same continuous characteristics, stratified (or generated separately) by sex.

Suppose we want to compare mean systolic blood pressures in men versus women using a 95% CI. The summary statistics shown in Table 6-9 indicate that we have large samples (more than 30) of both men and women, and therefore we use the CI formula from Table 6-8 with z as opposed to t. However, before implementing the formula, we first check whether the assumption of equality of the population variances is reasonable. The guideline suggests investigating the ratio of the sample variances, s_1^2/s_2^2. Suppose we call men Group 1 and women Group 2. (Again, this is arbitrary. It only needs to be noted when interpreting the results.) The ratio of the sample variances is $17.5^2 / 20.1^2 = 0.76$, which falls between 0.5 and 2, suggesting that the assumption of equality of the population variances is reasonable. The appropriate CI formula for the difference in mean systolic blood pressures between men and women is:

$$(\overline{X}_1 - \overline{X}_2) \pm zS_p\sqrt{\frac{1}{n_1} + \frac{1}{n_2}}$$

Before substituting, we first compute S_p, the pooled estimate of the common standard deviation:

$$S_p = \sqrt{\frac{(n_1-1)s_1^2 + (n_2-1)s_2^2}{n_1 + n_2 - 2}},$$

$$S_p = \sqrt{\frac{(1623-1)(17.5)^2 + (1911-1)(20.1)^2}{1623 + 1911 - 2}},$$

$$S_p = \sqrt{359.12} = 19.0.$$

Notice that the pooled estimate of the common standard deviation, S_p, falls between the standard deviations in the comparison groups (i.e., 17.5 and 20.1). S_p is slightly closer in value to the standard deviation in the women (20.1) as there are slightly more women in the sample. Recall that S_p is a weighted average of the standard deviations in the comparison groups, weighted by the respective sample sizes.

The 95% CI for the difference in mean systolic blood pressures is:

$$(128.2 - 126.5) \pm 1.96(19.0)\sqrt{\frac{1}{1623} + \frac{1}{1911}},$$

$$1.7 \pm 1.26,$$

$$(0.44, 2.96).$$

The CI is interpreted as follows: We are 95% confident that the difference in mean systolic blood pressures between men and women is between 0.44 and 2.96 units. Our best estimate of the difference, the point estimate, is 1.7 units. The standard error of the difference is 0.641, and the margin of error is 1.26 units.

Note that when we generate estimates for a population parameter in a single sample (e.g., the mean μ or population proportion p), the resulting CI provides a range of likely values for that parameter. In contrast, when there are two independent samples and the goal is to compare means (or proportions, discussed in Section 6.6), the resultant CI does not provide a range of values for the parameters in the

TABLE 6-9 Summary Statistics on Men and Women Attending the Seventh Examination of the Framingham Offspring Study

	Men			Women		
	n	$\overline{X}$	s	n	$\overline{X}$	s
Systolic blood pressure	1623	128.2	17.5	1911	126.5	20.1
Diastolic blood pressure	1622	75.6	9.8	1910	72.6	9.7
Total serum cholesterol	1544	192.4	35.2	1766	207.1	36.7
Weight (lbs)	1612	194.0	33.8	1894	157.7	34.6
Height (in.)	1545	68.9	2.7	1781	63.4	2.5
Body mass index (BMI)	1545	28.8	4.6	1781	27.6	5.9

comparison populations; instead, the CI provides a range of values for the difference. In this example, we estimate that the difference in mean systolic blood pressures is between 0.44 and 2.96 units, with men having the higher values. The last aspect of the interpretation is based on the fact that the CI is positive and that we called men Group 1 and women Group 2. Had we designated the groups the other way (i.e., women as Group 1 and men as Group 2), the CI would have been −2.96 to −0.44, suggesting that women have lower systolic blood pressures (anywhere from 0.44 to 2.96 units lower than men).

Table 6–10 includes 95% CIs for the difference in means in each characteristic, computed using the same formula we used for the CI for the difference in mean systolic blood pressures. Notice that the 95% CI for the difference in mean total cholesterol levels between men and women is −17.16 to −12.24. Men have lower mean total cholesterol levels than women, anywhere from 12.24 to 17.16 units lower. The men have higher mean values on each of the other characteristics considered (indicated by the positive CIs).

Again, the CI for the difference in means provides an estimate of the absolute difference in means of the outcome variable of interest between the comparison groups. It is often of interest to make a judgment as to whether there is a statistically meaningful difference between comparison groups. This judgment is based on whether the observed difference is beyond that expected by chance.

The CI for the difference in means provides a range of likely values for $(\mu_1 - \mu_2)$. It is important to note that all values in the CI are equally likely estimates of the true value of $(\mu_1 - \mu_2)$. If there is no difference between the population means, then the difference is zero (i.e., $\mu_1 - \mu_2 = 0$). Zero is the no difference or null value of the parameter (in this case, no difference in means). If a 95% CI for the difference in means includes 0 (the null value), then we say that there is no statistically meaningful or statistically significant difference in the means. If the CI does not include 0, then we conclude that there is a statistically significant difference in means. (We discuss statistical significance in much greater detail in Chapter 7.) For each of the characteristics considered in Example 6.5, there is a statistically significant difference in means between men and women because none of the CIs include the null value, 0. However, notice that some of the means are not very different between men and women (e.g., systolic blood pressure and BMI). Because the 95% CIs do not include 0, we conclude that there are statistically meaningful differences between means. This is primarily due to the large sample sizes. We discuss this in further detail in Chapter 7.

Example 6.6. In Example 6.2, we recapped the data presented in Chapter 4 on the subsample of $n = 10$ participants who attended the seventh examination of the offspring in the Framingham Heart Study. **Table 6–11** contains descriptive statistics on the same continuous characteristics in the subsample, stratified (or generated separately) by sex.

Using these data, suppose we wish to construct a 95% CI for the difference in mean systolic blood pressures between men and women. Again, we call men Group 1 and women Group 2. Because the sample sizes are small (i.e., $n_1 < 30$ and $n_2 < 30$), the CI formula with t is appropriate. However, before implementing the formula we first check whether the assumption of equality of the population variances is reasonable. The ratio of the sample variances is $9.7^2 / 12.0^2 = 0.65$, which falls between 0.5 and 2, suggesting that the assumption of equality of the population variances is reasonable. The appropriate CI formula for comparing mean systolic blood pressures between men and women is:

$$(\overline{X}_1 - \overline{X}_2) \pm tS_p\sqrt{\frac{1}{n_1} + \frac{1}{n_2}}.$$

TABLE 6–10 Confidence Intervals for Differences in Means

	Men		Women		95% CI
	$\overline{X}$	s	$\overline{X}$	s	
Systolic blood pressure	128.2	17.5	126.5	20.1	(0.44, 2.96)
Diastolic blood pressure	75.6	9.8	72.6	9.7	(2.38, 3.67)
Total serum cholesterol	192.4	35.2	207.1	36.7	(−17.16, −12.24)
Weight (lbs)	194.0	33.8	157.7	34.6	(33.98, 38.53)
Height (in.)	68.9	2.7	63.4	2.5	(5.31, 5.66)
Body mass index (BMI)	28.8	4.6	27.6	5.9	(0.82, 1.56)

TABLE 6–11 Summary Statistics on the Subsample of $n = 10$ Men and Women Attending the Seventh Examination of the Framingham Offspring Study

	Men			Women		
	n	$\bar{X}$	s	n	$\bar{X}$	s
Systolic blood pressure	6	117.5	9.7	4	126.8	12.0
Diastolic blood pressure	6	72.5	7.1	4	69.5	7.9
Total serum cholesterol	6	193.8	34.2	4	215.0	42.8
Weight (lbs)	6	196.9	26.9	4	146.0	27.2
Height (in.)	6	70.2	2.7	4	62.6	2.3
Body mass index (BMI)	6	28.0	3.6	4	26.2	2.9

Before substituting, we first compute S_p, the pooled estimate of the common standard deviation:

$$S_p = \sqrt{\frac{(6-1)(9.7)^2 + (4-1)(12.0)^2}{6+4-2}}$$

$$= \sqrt{112.81} = 10.6.$$

Notice that again the pooled estimate of the common standard deviation, S_p, falls between the standard deviations in the comparison groups (i.e., 9.7 and 12). We next need to find the appropriate t value for 95% confidence. The degrees of freedom for the two sample procedures are defined as $df = n_1 + n_2 - 2 = 6 + 4 - 2 = 8$. From Table 2, $t = 2.306$. The 95% CI for the difference in mean systolic blood pressures is:

$$(117.5 - 126.8) \pm 2.306(10.6)\sqrt{\frac{1}{6} + \frac{1}{4}},$$

$$-9.3 \pm 15.78,$$

$$(-25.08, 6.48).$$

We are 95% confident that the difference in mean systolic blood pressures between men and women is between −25.08 and 6.48 units. Our best estimate of the difference, the point estimate, is −9.3 units. The standard error of the difference is 6.84 units, and the margin of error is 15.78 units. In this sample, the men have lower mean systolic blood pressures than women by 9.3 units. However, based on this interval, we cannot conclude that there is a statistically significant difference in mean systolic blood pressures between men and women because the 95% CI includes the null value, 0. Again, the CI is a range of likely values for the difference in means. Because the interval contains some positive values (suggesting men have higher mean blood pressures) and some negative values (suggesting that women have higher values) and therefore also 0 (no difference), we cannot conclude that there is a statistically significant difference. This 95% CI for the difference in mean blood pressures is much wider than the one based on the full sample derived in Example 6.5. The small sample size produces a very imprecise estimate of the difference in mean systolic blood pressures.

6.5 CONFIDENCE INTERVALS FOR MATCHED SAMPLES, CONTINUOUS OUTCOME

There is a study design alternative to that described in the previous section, again where it is of interest to compare two groups with respect to their mean scores on a continuous outcome. Here the two comparison groups are dependent (or matched, or paired). One such possible scenario involves a single sample of participants and each participant is measured twice, possibly before and after an intervention or under two experimental conditions (e.g., in a crossover trial). The goal of the analysis is to compare the mean score measured before the intervention with the mean score measured afterward. Another scenario is one in which matched samples are analyzed. For example, we might be interested in the difference in an outcome between twins or siblings. Again, we have two samples and the goal is to compare the two means; however, the samples are related or dependent. In the first scenario, before and after measurements are taken in the same individual. In the second scenario, measures are taken in pairs of individuals from the same family. When the samples are dependent, we cannot use the techniques described in Section 6.4 to compare means. Because the samples are dependent, statistical techniques that account for the dependency must be used. The technique here focuses on difference scores (e.g., the difference between measures taken before versus after the intervention, or the difference between measures taken in twins or siblings).

In estimation and other statistical inference applications, it is critically important to appropriately identify the unit of analysis. Units of analysis are independent entities. In one-sample and two-independent-samples applications, participants are the units of analysis. In the two-dependent-samples application, the pair is the unit and not the number of measurements, which is twice the number of pairs or units.

The parameter of interest is the mean difference, μ_d. Again, the first step is to compute descriptive statistics on the sample data. Specifically, descriptive statistics are computed on difference scores. We compute the sample size (which, in this case, is the number of distinct participants or distinct pairs) and the mean and standard deviation of the difference scores. We denote these summary statistics as n, $\overline{X}_d$, and s_d, respectively. The appropriate formula for the CI for the mean difference depends on the sample size. The formulas are shown in **Table 6–12** and are identical to those we presented for estimating the mean of a single sample (see Table 6–3), except here we focus on difference scores.

Here n is the number of participants or pairs, $\overline{X}_d$ and s_d are the mean and standard deviation of the difference scores (where differences are computed on each participant or between members of a matched pair), and z or t are the values from the z or t distributions reflecting the desired confidence level. $s_d/\sqrt{n}$ is the standard error of the point estimate, $SE(\overline{X}_d)$.

Example 6.7. In the Framingham Offspring Study, participants attend clinical examinations approximately every four years. The data in **Table 6–13** are systolic blood pressures measured at the sixth and seventh examinations in a subsample of $n = 15$ randomly selected participants. The first column contains unique identification numbers, assigned only to distinguish individual participants.

Suppose we want to compare systolic blood pressures between examinations (i.e., changes over four years). Because the data in the two samples (Examination 6 and

TABLE 6–13 Systolic Blood Pressures Measured at the Sixth and Seventh Examinations of the Framingham Offspring Study

Subject Identification Number	Examination 6	Examination 7
1	168	141
2	111	119
3	139	122
4	127	127
5	155	125
6	115	123
7	125	113
8	123	106
9	130	131
10	137	142
11	130	131
12	129	135
13	112	119
14	141	130
15	122	121

Examination 7) are matched, we compute difference scores. The difference scores can be computed by subtracting the blood pressure measured at Examination 7 from that measured at Examination 6, or vice versa. If we subtract the blood pressure measured at Examination 6 from that measured at Examination 7, then positive differences represent increases over time and negative differences represent decreases over time. **Table 6–14** contains the difference scores for each participant.

Notice that several participants' systolic blood pressures decreased over four years (e.g., Participant 1's blood pressure decreased by 27 units from 168 to 141), whereas others increased (e.g., Participant 2's blood pressure increased by 8 units from 111 to 119). We now estimate the mean difference in blood pressures over four years. This is similar to a one-sample problem with a continuous outcome (Section 6.2), except here we focus on the difference scores. In this sample, we have $n = 15$, $\overline{X}_d = -5.3$, and $s_d = 12.8$, respectively. The data for the calculations are shown in **Table 6–15**, and $\overline{X}_d = \dfrac{-79.0}{15} = -5.3$ and $s_d = \sqrt{\dfrac{2296.95}{(15-1)}} = \sqrt{164.07} = 12.8$.

We now use these descriptive statistics to compute a 95% CI for the mean difference in systolic blood pressures in the

TABLE 6–12 Confidence Intervals for μ_d

$n \geq 30$ $\quad \overline{X}_d \pm z \dfrac{s_d}{\sqrt{n}} \quad$ (Find z in Table 1B)

$n < 30$ $\quad \overline{X}_d \pm t \dfrac{s_d}{\sqrt{n}} \quad$ (Find t in Table 2, $df = n-1$)

TABLE 6–14 Difference Scores

Subject Identification Number	Examination 6	Examination 7	Difference
1	168	141	−27
2	111	119	8
3	139	122	−17
4	127	127	0
5	155	125	−30
6	115	123	8
7	125	113	−12
8	123	106	−17
9	130	131	1
10	137	142	5
11	130	131	1
12	129	135	6
13	112	119	7
14	141	130	−11
15	122	121	−1

TABLE 6–15 Summary Statistics on Difference Scores

Subject Identification Number	Difference	Difference − $\overline{X}_d$	(Difference − $\overline{X}_d$)^2
1	−27	−21.7	470.89
2	8	13.3	176.89
3	−17	−11.7	136.89
4	0	5.3	28.09
5	−30	−24.7	610.09
6	8	13.3	176.89
7	−12	−6.7	44.89
8	−17	−11.7	136.89
9	1	6.3	39.69
10	5	10.3	106.09
11	1	6.3	39.69
12	6	11.3	127.69
13	7	12.3	151.29
14	−11	−5.7	32.49
15	−1	4.3	18.49
	−79.0	0.5*	2296.95

*The sum of the differences from the mean should sum to 0 but here sum to 0.5 due to rounding.

population. Because the sample size is small ($n = 15$), we use the following formula:

$$\overline{X}_d \pm t \frac{s_d}{\sqrt{n}}.$$

We first need the t value for 95% confidence. The degrees of freedom are $df = n - 1 = 14$. From Table 2, $t = 2.145$. We now substitute the descriptive statistics on the difference scores and the t value for 95% confidence:

$$-5.3 \pm 2.145 \frac{12.8}{\sqrt{15}},$$

$$-5.3 \pm 7.1,$$

$$(-12.4, 1.8).$$

We are 95% confident that the mean difference in systolic blood pressures between Examination 6 and Examination 7 (approximately four years apart) is between -12.4 and 1.8. The null (or no effect) value of the CI for the mean difference is 0. Therefore, based on the 95% CI we cannot conclude that there is a statistically significant difference in blood pressures over time because the CI for the mean difference includes 0.

In Chapter 2, we discussed various study designs, particularly a special case of the randomized trial called the crossover trial (Section 2.3.2). In a crossover trial with two treatments (e.g., an experimental treatment and a control), each participant receives both treatments. Outcomes are measured after each treatment in each participant (see Figure 2.6). A major advantage to the crossover trial is that each participant acts as his or her own control, and therefore fewer participants are required to demonstrate an effect. When the outcome is continuous, the assessment of a treatment effect in a crossover trial is performed using the techniques described here.

Example 6.8. A crossover trial is conducted to evaluate the effectiveness of a new drug designed to reduce the symptoms of depression in adults over 65 years of age following a stroke. Symptoms of depression are measured on a scale of 0 to 100, with higher scores indicative of more frequent and severe symptoms of depression. Patients who suffered a stroke are eligible for the trial. The trial is run as a crossover trial where each patient receives both the new drug and a placebo. Patients are blind to the treatment assignment, and the order of treatments (e.g., placebo and then new drug or new drug and then placebo) are randomly assigned. After each treatment, depressive symptoms are measured in each patient. The difference in depressive symptoms is measured in each patient by subtracting the depressive symptom score after taking the placebo from the depressive symptom score after taking the new drug. A total of 100 participants completed the trial and the data are summarized in **Table 6–16**.

The mean difference in the sample is $\overline{X}_d = -12.7$, meaning on average patients scored 12.7 points lower on the depressive symptoms scale after taking the new drug as compared to placebo (i.e., improved by 12.7 points on average). We now construct a 95% CI for the mean difference in the population. Because the sample size is large, we use the following formula:

$$\overline{X}_d \pm z \frac{s_d}{\sqrt{n}},$$

$$-12.7 \pm 1.96 \frac{8.9}{\sqrt{100}},$$

$$-12.7 \pm 1.74,$$

$$(-14.4, -10.7).$$

We are 95% confident that the mean improvement in depressive symptoms after taking the new drug as compared to placebo is between 10.7 and 14.4 units (or alternatively, the depressive symptoms scores are 10.7 to 14.4 units lower after taking the new drug as compared to the placebo). Because we computed the differences by subtracting the scores after taking the placebo from the scores after taking the new drug, and because higher scores are indicative of worse or more severe depressive symptoms, negative differences reflect improvement (i.e., lower depressive symptoms scores after taking the new drug as compared to the placebo). Because the 95% CI

TABLE 6–16 Summary Statistics on Differences in Depressive Symptoms

	n	Mean Difference	Std Dev Difference
Depressive symptoms after taking the new drug – Depressive symptoms after taking placebo	100	−12.7	8.9

for the mean difference does not include 0, we can conclude that there is a statistically significant difference (in this case, a significant improvement) in depressive symptom scores after taking the new drug as compared to the placebo.

6.6 CONFIDENCE INTERVALS FOR TWO INDEPENDENT SAMPLES, DICHOTOMOUS OUTCOME

It is very common to compare two groups in terms of the presence or absence of a particular characteristic or attribute. There are many instances in which the outcome variable is dichotomous (e.g., prevalent cardiovascular disease or diabetes, current smoking status, incident coronary heart diesease, cancer remission, successful device implant). Similar to the applications described in Section 6.4 for continuous outcomes, we focus here on the case where there are two comparison groups that are independent or physically separate and the outcome is dichotomous. The two groups might be determined by a particular attribute or characteristic of the participant (e.g., sex, age less than 65 versus age 65 and older) or might be set up by the investigator (e.g., participants assigned to receive an experimental drug or a placebo, a pharmacological versus a surgical treatment). When the outcome is dichotomous, the analysis involves comparing the proportions of successes between the two groups.

We discussed several methods that are used to compare proportions in two independent groups in Chapter 3, including the **risk difference**, which is computed by taking the difference in proportions between comparison groups and is similar to the estimate of the difference in means for a continuous outcome described in Section 6.4. The **relative risk**, also called the risk ratio, is another useful measure to compare proportions between two independent populations, and it is computed by taking the ratio of proportions. Generally, the reference group (e.g., unexposed persons, persons without a risk factor, or persons assigned to the control group in a clinical trial setting) is considered in the denominator of the ratio. The relative risk is often felt to be a better measure of the strength of an effect than the risk difference because it is a relative as opposed to an absolute measure. A third popular measure is the odds ratio. The **odds ratio** is computed by taking the ratio of the odds of success in the comparison groups. The odds is defined as the ratio of the number of successes to the number of failures (see Section 3.4.2 for more details). CI estimates for the risk difference, the relative risk, and the odds ratio are described in the following.

6.6.1 Confidence Intervals for the Risk Difference

The risk difference (RD) is similar to the difference in means when the outcome is continuous. The parameter of interest is the difference in proportions in the population, $RD = p_1 - p_2$. The point estimate is the difference in sample proportions, $\hat{RD} = \hat{p}_1 - \hat{p}_2$. The sample proportions are computed by taking the ratio of the number of successes (x) to the sample size (n) in each group, $\hat{p}_1 = x_1/n_1$ and $\hat{p}_2 = x_2/n_2$, respectively. The formula for the CI for the difference in proportions, or the RD, is given in **Table 6–17**.

The formula in Table 6–17 is appropriate for large samples, defined as at least five successes ($n\hat{p}$) and at least five failures [$n(1 - \hat{p})$] in each sample. If there are fewer than five successes or five failures in either comparison group, then alternative procedures called exact methods must be used to estimate the difference in population proportions.[5]

Example 6.9. In Example 3.1, we presented data measured in participants who attended the fifth examination of the offspring in the Framingham Heart Study. A total of $n = 3799$ participants attended the fifth examination, and **Table 6–18** contains data on prevalent CVD among participants who were and were not currently smoking cigarettes at the time of the fifth examination in the Framingham Offspring Study.

The outcome is prevalent CVD and the two comparison groups are defined by current smoking status. The point estimate of prevalent CVD among nonsmokers is 298 / 3055 = 0.0975, and the point estimate of prevalent CVD among current smokers is 81 / 744 = 0.1089. When constructing CIs for the RD, the convention is to call the exposed or treated Group 1 and the unexposed or untreated Group 2. Here smoking status

TABLE 6–17 Confidence Interval for $(p_1 - p_2)$

$\min[n_1\hat{p}_1,\ n_1(1 - \hat{p}_1), n_2\hat{p}_2, n_2(1 - \hat{p}_2)] \geq 5$ $\qquad \hat{p}_1 - \hat{p}_2 \pm z\sqrt{\dfrac{\hat{p}_1(1-\hat{p}_1)}{n_1} + \dfrac{\hat{p}_2(1-\hat{p}_2)}{n_2}}$ $\qquad$ (Find z in Table 1B)

TABLE 6–18 Prevalent CVD in Smokers and Nonsmokers

	Free of CVD	History of CVD	Total
Nonsmoker	2757	298	3055
Current smoker	663	81	744
Total	3420	379	3799

Table 6–19 Pain Reduction by Treatment

Treatment Group	n	Reduction of 3+ Points Number	Proportion
New pain reliever	50	23	0.46
Standard pain reliever	50	11	0.22

defines the comparison groups, and we will call the current smokers Group 1 and the nonsmokers Group 2. A CI for the difference in prevalent CVD (or RD) between smokers and nonsmokers is given below. In this example, we have more than enough successes (cases of prevalent CVD) and failures (persons free of CVD) in each comparison group.

$$\hat{p}_1 - \hat{p}_2 \pm z \sqrt{\frac{\hat{p}_1(1-\hat{p}_1)}{n_1} + \frac{\hat{p}_2(1-\hat{p}_2)}{n_2}},$$

$$(0.1089 - 0.0975) \pm 1.96\sqrt{\frac{0.1089(1-0.1089)}{744} + \frac{0.0975(1-0.0975)}{3055}},$$

$$0.0114 \pm 0.0247,$$

$$(-0.0133, 0.0361).$$

We are 95% confident that the difference in proportions of smokers as compared to nonsmokers with prevalent CVD is between −0.0133 and 0.0361. The null, or no difference, value for the RD is 0. Because the 95% CI includes 0, we cannot conclude that there is a statistically significant difference in prevalent CVD between smokers and nonsmokers.

Example 6.10. A randomized trial is conducted to evaluate the effectiveness of a newly developed pain reliever designed to reduce pain in patients following joint replacement surgery. The trial compares the new pain reliever to the pain reliever currently used (called the standard of care). A total of 100 patients undergoing joint replacement surgery agree to participate in the trial. Patients are randomly assigned to receive either the new pain reliever or the standard pain reliever following surgery. The patients are blind to the treatment assignment. Before receiving the assigned treatment, patients are asked to rate their pain on a scale of 0 to 10, with higher scores indicative of more pain. Each patient is then given the assigned treatment and after 30 minutes is again asked to rate his or her pain on the same scale. The primary outcome is a reduction in pain of 3 or more scale points (defined by clinicians as a clinically meaningful reduction). The data shown in **Table 6–19** are observed in the trial.

A point estimate for the difference in proportions of patients reporting a clinically meaningful reduction in pain between treatment groups is 0.46 − 0.22 = 0.24. (Notice that we call the experimental or new treatment Group 1 and the standard Group 2.) There is a 24 percentage point increase in patients reporting a meaningful reduction in pain with the new pain reliever as compared to the standard pain reliever. We now construct a 95% CI for the difference in proportions. The sample sizes in each comparison group are adequate—i.e., each treatment group has at least five successes (patients reporting reduction in pain) and at least five failures—therefore the formula for the CI is

$$(0.46 - 0.22) \pm 1.96\sqrt{\frac{0.46(1-0.46)}{50} + \frac{0.22(1-0.22)}{50}},$$

$$0.24 \pm 0.180,$$

$$(0.06, 0.42).$$

We are 95% confident that the difference in proportions of patients reporting a meaningful reduction in pain is between 0.06 and 0.42 comparing the new and standard pain relievers. Our best estimate is an increase of 24 percentage points with the new pain reliever. Because the 95% CI does not contain 0 (the null value), we can conclude that there is a statistically significant difference between pain relievers—in this case, in favor of the new pain reliever.

6.6.2 Confidence Intervals for the Relative Risk

The RD quantifies the absolute difference in risk (or proportions), whereas the relative risk is, as the name indicates, a relative measure. Both the RD and the RR are useful but they convey different information. A relative risk (RR) is computed by taking the ratio of proportions, p_1/p_2. Again, the convention is to place the proportion of successes or the risk for the control or unexposed group in the denominator. The parameter of interest is the relative risk in the population, $RR = p_1/p_2$, and the point estimate is the relative risk in the sample, $\hat{RR} = \hat{p}_1/\hat{p}_2$. The RR is a ratio and does not follow a normal distribution, regardless of the sample sizes in the comparison groups. However, the natural log (ln) of the RR is approximately normally distributed and is used to produce a CI for the RR. First, a CI is generated for ln(RR), and then the antilog of the upper and lower limits of the CI for ln(RR) are taken to give the upper and lower limits of the CI for the RR. The two steps are detailed in **Table 6–20**.

The null or no difference value of the CI for the RR is 1. If a 95% CI for the RR does not include 1, then the risks (or proportions) are said to be statistically significantly different.

Example 6.11. Consider again the data shown in Table 6–18, reflecting the prevalence of CVD among participants who were and were not currently smoking cigarettes at the time of the fifth examination in the Framingham Offspring Study.

The outcome is prevalent CVD and the two comparison groups are defined by current smoking status. In Example 6.9, we generated a point estimate for the difference in prevalence of CVD between smokers and nonsmokers of 0.0114. Smokers had a higher prevalence by 1.14%. We then generated a 95% CI for the difference in prevalent CVD between smokers and nonsmokers of $(-0.0133, 0.0361)$. Because the 95% CI included 0, we could not conclude that there was a statistically significant difference in prevalent CVD between smokers and nonsmokers. We now use these data to generate a point estimate and 95% CI estimate for the RR.

The point estimate for the relative risk is

$$\hat{RR} = \frac{\hat{p}_1}{\hat{p}_2} = \frac{0.1089}{0.0975} = 1.12$$

and the interpretation is as follows: Smokers have 1.12 times the prevalence of CVD compared to nonsmokers. The 95% CI estimate for the RR is computed using the two-step procedure outlined in Table 6–20:

$$\ln(\hat{RR}) \pm z\sqrt{\frac{(n_1 - x_1)/x_1}{n_1} + \frac{(n_2 - x_2)/x_2}{n_2}},$$

$$\ln(1.12) \pm 1.96\sqrt{\frac{663/81}{744} + \frac{2757/298}{3055}},$$

$$0.113 \pm 0.232,$$

$$(-0.119, 0.345).$$

A 95% CI for ln(RR) is $(-0.119, 0.345)$. To generate the CI for the RR, we take the antilog (exp) of the lower and upper limits:

$$(\exp(-0.119), \exp(0.345)),$$
$$(0.89, 1.41).$$

Thus, we are 95% confident that the RR of prevalent CVD in smokers as compared to nonsmokers is between 0.89 and 1.41. The null, or no difference, value for a relative risk is 1. Because this CI includes the null value, we cannot conclude that there is a statistically significant difference in prevalent CVD between smokers and nonsmokers.

Example 6.12. Consider again the data shown in Table 6–19 from the randomized trial assessing the effectiveness of a newly developed pain reliever as compared to the standard pain reliever. The primary outcome is reduction in pain of 3 or more scale points (defined by clinicians as a clinically meaningful reduction).

We generated a point estimate for the difference in proportions of patients reporting a clinically meaningful reduction in pain between pain relievers as $0.46 - 0.22 = 0.24$, and the 95% CI was $(0.06, 0.42)$. Because the 95% CI for the RD did not contain 0 (the null value), we concluded that there was a statistically significant difference between pain relievers.

TABLE 6–20 Confidence Interval for $RR = p_1/p_2$

CI for ln(RR)	$\ln(\hat{RR}) \pm z\sqrt{\frac{(n_1 - x_1)/x_1}{n_1} + \frac{(n_2 - x_2)/x_2}{n_2}}$	(Find z in Table 1B)
CI for RR	exp(Lower limit), exp (Upper limit)	

We now use these data to generate a point estimate and 95% CI estimate for the RR.

The point estimate for the relative risk is

$$\hat{RR} = \frac{\hat{p}_1}{\hat{p}_2} = \frac{0.46}{0.22} = 2.09.$$

Patients receiving the new pain reliever have 2.09 times the likelihood of reporting a meaningful reduction in pain as compared to patients receiving the standard pain reliever. The 95% CI estimate for the RR is computed using the two-step procedure outlined in Table 6–20.

$$\ln(2.09) \pm 1.96\sqrt{\frac{27/23}{50} + \frac{39/11}{50}},$$

$$0.737 \pm 0.602,$$

$$(0.135, 1.339).$$

A 95% CI for ln(RR) is (0.135, 1.339). To generate the CI for the RR, we take the antilog (exp) of the lower and upper limits:

$$(\exp(0.135), \exp(1.339)),$$

$$(1.14, 3.82).$$

Thus, we are 95% confident that patients receiving the new pain reliever have between 1.14 and 3.82 times the likelihood of reporting a meaningful reduction in pain as compared to patients receiving the standard pain reliever. Because this CI does not include 1 (the null value), we can conclude that this difference is statistically significant.

6.6.3 Confidence Intervals for the Odds Ratio

There are some study designs in which it is not possible to estimate an RR (e.g., a case-control study), and therefore an odds ratio is computed as a relative measure of effect.[6] An odds ratio is similar to the RR in that it is a ratio, as opposed to a difference. The odds ratio (OR) is computed by taking the ratio of odds, where the odds in each group is computed by dividing the number of successes (x) by the number of failures ($n - x$) in that group. Similar to the RR, the convention is to place the odds in the control or unexposed group in the denominator. The parameter of interest is the odds ratio in the population, OR, and the point estimate is the odds ratio in the sample,

$$\hat{OR} = \frac{\hat{p}_1/(1-\hat{p}_1)}{\hat{p}_2/(1-\hat{p}_2)}.$$

The odds ratio can also be expressed as

$$\hat{OR} = \frac{x_1/(n_1-x_1)}{x_2/(n_2-x_2)}.$$

Similar to the RR, the OR does not follow a normal distribution and so we use the log transformation to promote normality. We first generate a CI for ln(OR) and then take the antilog of the upper and lower limits of the CI for ln(OR) to determine the upper and lower limits of the CI for the OR. The two steps are detailed in **Table 6–21**.

The null, or no difference, value for the OR is 1. If a 95% CI for the OR does not include 1, then the odds are said to be statistically significantly different. We again reconsider Example 6.9 and Example 6.10 and produce estimates of ORs and compare these to our estimates of RDs and RRs.

Example 6.13. Consider again the data shown in Table 6–18 reflecting the prevalence of CVD among participants who were and were not currently smoking cigarettes at the time of the fifth examination in the Framingham Offspring Study. The outcome is prevalent CVD and the two comparison groups are defined by current smoking status. In Example 6.9, we generated a point estimate for the difference in prevalent CVD between smokers and nonsmokers of 0.0114 and a 95% CI of (−0.0133, 0.0361). Because the 95% CI included 0, we could not conclude that there was a statistically significant difference in prevalent

---

TABLE 6–21 Confidence Interval for OR

CI for ln(OR)	$\ln(\hat{OR}) \pm z\sqrt{\frac{1}{x_1} + \frac{1}{(n_1-x_1)} + \frac{1}{x_2} + \frac{1}{(n_2-x_2)}}$	(Find z in Table 1B)
CI for OR	exp(Lower limit), exp(Upper limit)	

CVD between smokers and nonsmokers. In Example 6.11, we generated a point estimate for the relative risk,

$$\hat{RR} = \frac{\hat{p}_1}{\hat{p}_2} = \frac{0.1089}{0.0975} = 1.12,$$

and a 95% CI of (0.89, 1.41), and because the CI for the RR included 1, we concluded that there was no statistically significant difference in prevalent CVD between smokers and nonsmokers.

We now use these data to estimate an OR. A point estimate for the odds ratio is

$$\hat{OR} = \frac{x_1/(n_1 - x_1)}{x_2/(n_2 - x_2)} = \frac{81/663}{298/2757} = 1.13,$$

and the interpretation is that smokers have 1.13 times the odds of prevalent CVD as compared to nonsmokers. The 95% CI estimate for the OR is computed using the two-step procedure outlined in Table 6–21,

$$\ln(\hat{OR}) \pm z\sqrt{\frac{1}{x_1} + \frac{1}{n_1 - x_1} + \frac{1}{x_2} + \frac{1}{n_2 - x_2}},$$

$$\ln(1.13) \pm 1.96\sqrt{\frac{1}{81} + \frac{1}{663} + \frac{1}{298} + \frac{1}{2757}},$$

$$0.122 \pm 0.260,$$

$$(-0.138, 0.382).$$

A 95% CI for ln(OR) is (−0.138, 0.382). To generate the CI for the OR, we take the antilog (exp) of the lower and upper limits,

$$(\exp(-0.138), \exp(0.382)),$$

$$(0.87, 1.47).$$

Thus, we are 95% confident that the odds of prevalent CVD in smokers is between 0.87 and 1.47 times the odds of prevalent CVD in nonsmokers. The null value is 1 and because this CI includes 1, we cannot conclude that there is a statistically significant difference in the odds of prevalent CVD between smokers and nonsmokers.

Example 6.14. Consider again the data shown in Table 6–19 from the randomized trial assessing the effectiveness of a newly developed pain reliever as compared to the standard of care. The primary outcome is reduction in pain of 3 or more scale points (defined by clinicians as a clinically meaningful reduction).

We generated a point estimate for the difference in proportions of patients reporting a clinically meaningful reduction in pain between pain relievers as 0.46 − 0.22 = 0.24, and the 95% CI was (0.06, 0.42). Because the 95% CI for the RD did not contain 0 (the null value), we concluded that there was a statistically significant difference between pain relievers. We then generated a point estimate for the relative risk,

$$\hat{RR} = \frac{\hat{p}_1}{\hat{p}_2} = \frac{0.46}{0.22} = 2.09.$$

and a 95% CI of (1.14, 3.82). Because this CI for the RR did not include 1, we concluded that this difference was statistically significant.

We now use these data to generate a point estimate and 95% CI estimate for the OR. The point estimate for the odds ratio is

$$\hat{OR} = \frac{x_1/(n_1 - x_1)}{x_2/(n_2 - x_2)} = \frac{23/27}{11/39} = 3.02.$$

This suggests that the odds that a patient on the new pain reliever reports a meaningful reduction in pain is 3.02 times the odds that a patient on the standard pain reliever reports a meaningful reduction. The 95% CI estimate for the OR is computed using the two-step procedure outlined in Table 6–21,

$$\ln(3.02) \pm 1.96\sqrt{\frac{1}{23} + \frac{1}{27} + \frac{1}{11} + \frac{1}{39}},$$

$$1.105 \pm 0.870,$$

$$(0.235, 1.975).$$

A 95% CI for ln(OR) is (0.235, 1.975). To generate the CI for the OR, we take the antilog (exp) of the lower and upper limits,

$$(\exp(0.235), \exp(1.975)),$$

$$(1.26, 7.21).$$

Thus, we are 95% confident that the odds that a patient on the new pain reliever reports a meaningful reduction in pain is between 1.26 and 7.21 times the odds that a patient on the standard pain reliever reports a meaningful reduction. Because this CI does not include 1, we conclude that this difference is statistically significant.

When the study design allows for the calculation of an RR, it is the preferred measure as it is far more interpretable than an OR. However, the OR is extremely important as it is the only measure of effect that can be computed in a case-control study design. When the outcome of interest is relatively rare (e.g., fewer than 10% successes), then the OR and RR will be very close in magnitude. In such a case, investigators often interpret the OR as if it were a RR—i.e., as a comparison of risks rather than a comparison of odds, which is less intuitive.

6.7 SUMMARY

In this chapter, we presented various formulas for estimating different unknown population parameters. In each application, a random sample or two independent random samples were selected from the population and sample statistics (e.g., sample sizes, means, and standard deviations, or sample sizes and proportions) were generated. Point estimates are the best single-valued estimates of unknown population parameters. Because these can vary from sample to sample, most investigations start with a point estimate and build in a margin of error. The margin of error quantifies sampling variability and includes a value from the z or t distribution reflecting the selected confidence level as well as the standard error of the point estimate.

It is important to remember that the confidence interval contains a range of likely values for the unknown population parameter—a range of values for the population parameter consistent with the data. It is also possible that the confidence interval does not contain the true population parameter. This is important to remember in interpreting intervals. Confidence intervals are also very useful for comparing means or proportions and can be used to assess whether there is a statistically meaningful difference in parameters. This is based on whether the confidence interval includes the null value (e.g., 0 for the difference in means, mean difference, and risk difference, or 1 for the relative risk and odds ratio). The precision of a confidence interval is defined by the margin of error (or the width of the interval). A larger margin of error (wider interval) is indicative of a less precise estimate. **Table 6–22** summarizes key formulas for confidence interval estimates.

TABLE 6–22 Summary of Key Formulas

Number of Groups, Outcome: Parameter	Confidence Interval*
One sample, continuous: CI for μ	$\overline{X} \pm z \dfrac{s}{\sqrt{n}}$
Two independent samples, continuous: CI for $(\mu_1 - \mu_2)$	$(\overline{X}_1 - \overline{X}_2) \pm zS_p \sqrt{\dfrac{1}{n_1} + \dfrac{1}{n_2}}$
Two matched samples, continuous: CI for μ_d	$\overline{X}_d \pm z \dfrac{s_d}{\sqrt{n}}$
One sample, dichotomous: CI for p	$\hat{p} \pm z \sqrt{\dfrac{\hat{p}(1-\hat{p})}{n}}$
Two independent samples, dichotomous: CI for RD = $(p_1 - p_2)$	$\hat{p}_1 - \hat{p}_2 \pm z \sqrt{\dfrac{\hat{p}_1(1-\hat{p}_1)}{n_1} + \dfrac{\hat{p}_2(1-\hat{p}_2)}{n_2}}$
Two independent samples, dichotomous: CI for RR = $\dfrac{p_1}{p_2}$	$\ln(\hat{RR}) \pm z \sqrt{\dfrac{(n_1 - x_1)/x_1}{n_1} + \dfrac{(n_2 - x_2)/x_2}{n_2}}$ exp(Lower limit), exp(Upper limit)
Two independent samples, dichotomous: CI for OR = $\dfrac{x_1/(n_1 - x_1)}{x_2/(n_2 - x_2)}$	$\ln(\hat{OR}) \pm z \sqrt{\dfrac{1}{x_1} + \dfrac{1}{(n_1 - x_1)} + \dfrac{1}{x_2} + \dfrac{1}{(n_2 - x_2)}}$ exp(Lower limit), exp(Upper limit)

*See Tables 6–3, 6–8, and 6–12 for alternative formulas that are appropriate for small samples.

6.8 PRACTICE PROBLEMS

1. A study is run to estimate the mean total cholesterol level in children 2 to 6 years of age. A sample of nine participants is selected and their total cholesterol levels are measured as follows:

   185 225 240 196 175 180 194 147 223

   Generate a 95% CI for the true mean total cholesterol levels in children.

2. A clinical trial is planned to compare an experimental medication designed to lower blood pressure to a placebo. Before starting the trial, a pilot study is conducted involving ten participants. The objective of the study is to assess how systolic blood pressure changes over time untreated. Systolic blood pressures are measured at baseline and again 4 weeks later. Compute a 95% CI for the difference in blood pressures over 4 weeks.

   Baseline: 120 145 130 160 152 143 126 121 115 135

   4 Weeks: 122 142 135 158 155 140 130 120 124 130

3. The main trial for the medication in Problem 2 is conducted and involves a total of 200 patients. Patients are enrolled and randomized to receive either the experimental medication or the placebo. The data shown in **Table 6–23** are data collected at the end of the study after 6 weeks on the assigned treatment.
   a. Generate a 95% CI for the difference in mean systolic blood pressures between groups.
   b. Generate a 95% CI for the difference in proportions of patients with hypertension between groups.
   c. Generate a point estimate for the RR of side effects in patients assigned to the experimental group as compared to placebo. Generate a 95% CI for the RR.

4. The following data are collected as part of a study of coffee consumption among undergraduate students. The following values reflect cups per day consumed:

   3 4 6 8 2 1 0 2

   a. Compute the sample mean.
   b. Compute the sample standard deviation.
   c. Construct a 95% CI for the mean number of cups of coffee consumed among all undergraduates.

5. A clinical trial is conducted to evaluate a new pain medication for arthritis. Participants are randomly assigned to receive the new medication or a placebo. The outcome is pain relief within 30 minutes, and the data are shown in **Table 6–24**.
   a. Estimate the RD in pain relief between treatments.
   b. Estimate the RR in pain relief between treatments.
   c. Estimate the OR in pain relief between treatments.
   d. Construct a 95% CI for the OR.

6. Data are collected in a clinical trial evaluating a new compound designed to improve wound healing in trauma patients. The new compound is compared against a placebo. After treatment for 5 days with the new compound or placebo, the extent of wound healing is measured and the data are shown in **Table 6–25**. Suppose that clinicians feel that if the percent reduction in the size of the wound is greater than 50%, then the treatment is a success.
   a. Generate a 95% CI for the percent success in patients receiving the new compound.

TABLE 6–24 Pain Relief by Treatment

	Pain Relief	No Pain Relief
New medication	44	76
Placebo	21	99

TABLE 6–23 Outcome Data by Treatment

	Experimental ($n=100$)	Placebo ($n=100$)
Mean (SD) systolic blood pressure	120.2 (15.4)	131.4 (18.9)
Hypertensive (%)	14	22
Side effects (%)	6	8

TABLE 6–25 Wound Healing by Treatment

Treatment	Number of Patients with Percent Reduction in Size of Wound				
	None	1–25	26–50	51–75	76–100
New compound ($n=125$)	4	11	37	32	41
Placebo ($n=125$)	12	24	45	34	10

b. Generate a 95% CI for the difference in the percent success between the new compound and placebo.
c. Generate a 95% CI for the RR of treatment success between treatments.
d. Generate a 95% CI for the OR of treatment success between treatments.

7. A clinical trial is conducted to compare an experimental medication to placebo to reduce the symptoms of asthma. Two hundred participants are enrolled in the study and randomized to receive either the experimental medication or placebo. The primary outcome is a self-reported reduction of symptoms. Among 100 participants who receive the experimental medication, 38 report a reduction of symptoms as compared to 21 participants of 100 assigned to the placebo.
   a. Generate a 95% CI for the difference in proportions of participants reporting a reduction of symptoms between the experimental and placebo groups.
   b. Estimate the RR for reduction in symptoms between groups.
   c. Estimate the OR for reduction in symptoms between groups.
   d. Generate a 95% CI for the RR.

8. **Table 6–26** displays descriptive statistics on the participants involved in the study described in Problem 7.
   a. Generate a 95% CI for the mean age among participants assigned to the placebo.
   b. Generate a 95% CI for the difference in mean ages in participants assigned to the experimental versus the placebo groups.
   c. Generate a 95% CI for the difference in mean BMI in participants assigned to the experimental versus the placebo groups.

TABLE 6–26 Descriptive Statistics by Treatment

	Experimental Medication (n=100)	Placebo (n=100)
Mean (SD) age, (years)	47.2 (4.3)	46.1 (5.1)
Men (%)	46%	58%
Mean (SD) educational level (years)	13.1 (2.9)	14.2 (3.1)
Mean (SD) annual income	$36,560 ($1054)	$37,470 ($998)
Mean (SD) body mass index (BMI)	24.7 (2.7)	25.1 (2.4)

9. A crossover trial is conducted to compare an experimental medication for migraine headaches to a currently available medication. A total of 50 patients are enrolled in the study and each patient receives both treatments. The outcome is the time, in minutes, until the headache pain resolves. Following each treatment, patients record the time it takes until pain is resolved. Treatments are assigned in random order (i.e., some patients receive the currently available medication first and then the experimental medication, and others receive the experimental medication first and then the currently available medication). The mean difference in times between the experimental and currently available medication is −9.4 minutes with a standard deviation of 2.8 minutes. Construct a 95% CI for the mean difference in times between the experimental and currently available medication.

10. Suppose in the study described in Problem 9 each participant is also asked if the assigned medication causes any stomach upset. Among the 50 participants, 12 reported stomach upset with the experimental medication. Construct a 90% CI for the proportion of participants who experience stomach upset with the experimental medication.

11. A 95% confidence interval for the mean diastolic blood pressure in women is 62 to 110. What is the margin of error? What is the standard error?

12. The following data are collected from 10 randomly selected patients undergoing physical therapy following knee surgery. The data represent the percentage gain in range of motion after 3 weeks of therapy. Generate a 95% confidence interval for the mean percentage gain in range of motion.

| 24% | 32% | 50% | 62% | 21% |
| 45% | 80% | 24% | 30% | 10% |

13. The data in **Table 6–27** are collected in a randomized trial to test the efficacy of a new drug for migraine headaches. The following are characteristics of study participants overall and then organized by the treatment to which they are assigned.
   a. Generate a 95% confidence interval for the difference in mean ages between groups.
   b. Generate a 95% confidence interval for the difference in proportions of males between groups.
   c. Generate a 95% confidence interval for the difference in proportions of patients with severe migraine headaches between groups.

TABLE 6–27 Characteristics in Participants in Study of Treatment for Migraine Headaches

	New Drug (*n* = 100)	Placebo (*n* = 100)	All (*n* = 200)
Mean (SD) age, years	32.8 (4.7)	31.9 (5.1)	32.0 (4.9)
% Male	54%	48%	51%
% High school graduate	76%	80%	78%
Severity of migraine headaches			
% Mild	22%	20%	21%
% Moderate	38%	42%	39%
% Severe	40%	38%	39%
Median (Q1–Q3) number of days missed work in past year due to migrane	5 (3–12)	6 (2–18)	6 (3–17)
Min–Max number of days missed work in past year due to migraine	0–35	0–48	0–48

TABLE 6–28 Baseline Characteristics of Participants in Pacemaker Study

Baseline Characteristics	New Pacemaker (*n* = 100)	Available Pacemaker (*n* = 100)
Mean (SD) age, years	67.3 (5.9)	66.9 (5.6)
% Male	48%	52%
Outcomes		
Mean (SD) number of days with AF event	8.4 (3.2)	14.9 (3.9)
% hospitalized for AF	4%	9%

14. A clinical trial is run to assess the efficacy of a new pacemaker device in patients with atrial fibrillation (AF). Two hundred participants are randomized to receive the new pacemaker or a currently available pacemaker. There are two primary outcomes of interest: the number of days in a 3-month period with an atrial fibrillation event, and hospitalization for atrial fibrillation over the 3-month follow-up period. Data on baseline characteristics and the outcomes are shown in **Table 6–28**.
    a. Compute a 95% confidence interval for the difference in mean number of days with an AF event between participants receiving the new pacemaker as compared to the available pacemaker.
    b. Compute a 95% confidence interval for the mean number of days with an AF event among participants receiving the new pacemaker.
15. A pilot study is run to investigate the feasibility of recruiting pregnant women into a study of risk factors for preterm delivery. Women are invited to participate at their first clinical visit for prenatal care. The following represent the gestational ages in weeks of women who consented to participate in the study. Compute a 95% confidence interval for the mean gestational age of women enrolling in the study.

    11 14 21 22 9 10 13 18

16. A clinical trial is run comparing a new drug for high cholesterol to a placebo. A total of 40 participants are randomized (with equal assignment to treatments) to receive either the new drug or placebo. Their total serum cholesterol levels are measured after eight weeks on the assigned treatment. Participants receiving the new drug reported a mean total serum cholesterol level of 209.5 (SD = 21.6) and participants receiving the placebo reported a mean total serum cholesterol level of 228.1 (SD = 19.7).
    a. Construct a 95% confidence interval for the difference in mean total serum cholesterol levels between participants receiving the new drug versus placebo.
    b. Is the new drug effective? Justify briefly.

REFERENCES

1. Newcomb, R.G. "Two-sided confidence intervals for the single proportion: Comparison of seven methods." *Statistics in Medicine* 1998; 17(8): 857–872.
2. StatXact Version 7. © 2006 by Cytel, Inc., Cambridge, MA.
3. D'Agostino, R.B., Sullivan, L.M., and Beiser, A. *Introductory Applied Biostatistics*. Belmont, CA: Duxbury-Brooks/Cole, 2004.
4. Rosner, B. *Fundamentals of Biostatistics*. Belmont, CA: Duxbury-Brooks/ Cole, 2006.
5. Agresti, A. *Categorical Data Analysis* (2nd ed.). New York: John Wiley & Sons, 2002.
6. Rothman, K.J. and Greenland, S. *Modern Epidemiology* (2nd ed.). Philadelphia: Lippincott-Raven, 1998.

CHAPTER 7

Hypothesis Testing Procedures

When and Why

Key Questions

- What does it mean when a study is statistically significant but not clinically meaningful?
- What does a test of hypothesis tell you that a confidence interval does not?
- When testing a hypothesis, is it better to run a matched pairs test or an independent samples test?

In the News

The Physicians' Health Study, which was sponsored by the National Cancer Institute and the National, Heart, Lung, and Blood Institute, was a randomized, controlled trial involving more than 20,000 male physicians that investigated the effects of low-dose aspirin and beta-carotene on cardiovascular mortality and cancer incidence, respectively.[1] The aspirin component of the trial was stopped early because there was statistically significant evidence of a reduction in myocardial infarction (heart attack) risk in participants who took aspirin. Participants who took aspirin had a 44% reduction in risk of heart attack as compared to their counterparts who received placebo (relative risk = 0.56; 95% confidence interval, 0.45 to 0.70; $p < 0.0001$). More than 11,000 participants were assigned to each arm of the trial (aspirin or placebo), and 139 participants taking aspirin experienced heart attack as compared to 239 who took placebo.[2]

For some time, low-dose aspirin has been considered a safe and effective preventive measure.[3] However, there may be risks for some groups of patients.[4]

---

[1] Physicians' Health Study. Available at http://phs.bwh.harvard.edu/index.html.
[2] Steering Committee of the Physicians' Health Study research group. "Final report on the aspirin component of the ongoing Physicians' Health Study." *New England Journal of Medicine* 1989; 321: 129–135.
[3] CardioSmart. American College of Cardiology. "Heart disease: aspirin." Available at https://www.cardiosmart.org/~/media/Documents/Fact%20Sheets/en/tb1597.ashx. Accessed July 21, 2016.
[4] *Everyday Health* guest columnist. "6 essential facts about aspirin therapy for your heart." *Everyday Health*. Available at http://www.everydayhealth.com/columns/health-answers/essential-facts-about-aspirin-therapy-your-heart/. Accessed July 21, 2016.

Dig In

- What percentage of patients suffered heart attacks in the Physicians' Health Study in each arm of the trial?
- How substantial is the evidence supporting the use of low-dose aspirin from the Physicians' Health Study?
- Are the recommendations for use of low-dose aspirin clear to consumers (all of whom may not understand statistics)?

Learning Objectives

By the end of this chapter, the reader will be able to

- Define null and research hypothesis, test statistic, level of significance, and decision rule
- Distinguish between Type I and Type II errors and discuss the implications of each
- Explain the difference between one- and two-sided tests of hypothesis
- Estimate and interpret *p*-values
- Explain the relationship between confidence interval estimates and *p*-values in drawing inferences
- Perform analysis of variance by hand
- Appropriately interpret results of analysis of variance tests
- Distinguish between one- and two-factor analysis of variance tests
- Perform chi-square tests by hand
- Appropriately interpret results of chi-square tests
- Identify the appropriate hypothesis testing procedure based on type of outcome variable and number of samples

The second area of statistical inference is hypothesis testing. In **hypothesis testing**, a specific statement or hypothesis is generated about a population parameter, and sample statistics are used to assess the likelihood that the hypothesis is true.

This statement or hypothesis is based on available information and the investigator's belief about the parameter. The process of hypothesis testing involves setting up two competing hypotheses: one reflects no difference, no association, or no effect (called the **null hypothesis**) and the other reflects the investigator's belief (called the research or **alternative hypothesis**). We select a random sample (or multiple samples when there are more comparison groups) and generate summary statistics. We then assess the likelihood that the sample data support the research or alternative hypothesis. Similar to estimation, the process of hypothesis testing is based on probability theory and the Central Limit Theorem.

The techniques for hypothesis testing again depend on the appropriate classification of the key study outcome variable or endpoint. The number of comparison groups in the investigation must also be specified. It is again important to determine whether the comparison groups are independent (i.e., physically separate, such as men versus women or participants assigned to receive a new drug or placebo in a clinical trial) or dependent (i.e., matched or paired, such as pre- and post-assessments on the same participants). These issues dictate the appropriate hypothesis testing technique. In estimation, we focused explicitly on techniques for one and two samples. We discussed estimation for a specific parameter (e.g., the mean μ or proportion p of a population), for differences (e.g., difference in means $\mu_1 - \mu_2$, the risk difference $p_1 - p_2$) and for ratios [(e.g., the relative risk $RR = p_1/p_2$ and odds ratio $OR = (p_1/(1-p_1))/(p_2/(1-p_2))$)]. Here we focus on procedures for one, two, and more than two samples. Hypothesis testing can generalize to the situation of more than two groups, whereas estimation is not intuitive when there are more than two groups. **Table 7–1** outlines hypothesis testing procedures that we consider here.

7.1 INTRODUCTION TO HYPOTHESIS TESTING

Before discussing the different procedures that are appropriate for each scenario outlined in Table 7–1, we first present the general approach using a simple example. The Centers for Disease Control and Prevention (CDC) reported on trends in weight, height, and body mass index (BMI) from the 1960s through 2002.[1] The data for the report were collected in the National Health Examination and the National Health and Nutrition Examination Surveys, which were conducted between 1960 and 2002. The general trend was that Americans were much heavier and slightly taller in 2002 as compared to 1960. The report indicated that both men and women gained approximately 24 pounds, on average, between 1960 and 2002. In 2002, the mean weight for American men was reported at 191 pounds and the mean weight for American women was reported at 163 pounds.

TABLE 7–1 Hypothesis Testing Techniques

Number of Samples	Outcome Variable
One sample	Continuous
Two independent samples	Continuous
Two dependent, matched samples	Continuous
More than two independent samples	Continuous
One sample	Dichotomous
Two independent samples	Dichotomous
More than two independent samples	Dichotomous
One sample	Categorical or ordinal (more than 2 response options)
Two or more independent samples	Categorical or ordinal

Suppose we focus on the mean weight in men. In 2002, the mean weight for men was reported at 191 pounds. Suppose that an investigator hypothesizes that weights are even higher in 2006 (i.e., that the trend continued over the next 4 years). In hypothesis testing, we set up competing hypotheses about the unknown parameter. One hypothesis is called the null hypothesis and the other is called the alternative or research hypothesis. The research hypothesis is that the mean weight in men in 2006 is more than 191 pounds. The null hypothesis is that there is no change in weight and therefore the mean weight is still 191 pounds in 2006. The null and research hypotheses are denoted as

Null hypothesis, $H_0: \mu = 191$ (no change),

Research hypothesis, $H_1: \mu > 191$ (investigator's belief).

To test the hypotheses, we select a random sample of American males in 2006 and measure their weights. Suppose we have resources available to recruit $n = 100$ men into our sample. We weigh each participant and compute summary statistics on the sample data. Suppose in the sample we determine the following: $n = 100$, $\overline{X} = 197.1$, $s = 25.6$.

Do the sample data support the null or research hypothesis? The sample mean of 197.1 pounds is numerically higher than 191 pounds. However, is this difference more than would be expected by chance? In hypothesis testing, we assume that the null hypothesis holds until proven otherwise.

We therefore need to determine the likelihood of observing a sample mean of 197.1 or higher when the true population mean is 191 (i.e., if the null hypothesis is true or under the null hypothesis). We compute this probability using the Central Limit Theorem. Specifically,

$$P(\overline{X} > 197.1) = P\left(z > \frac{197.1 - 191}{25.6/\sqrt{100}}\right) =$$
$$P(z > 2.38) = 1 - 0.9913 = 0.0087.$$

(Notice that we use the sample standard deviation in computing the z score. This is generally an appropriate substitution as long as the sample size is large, $n > 30$.) Thus, there is less than a 1% chance of observing a sample mean as large as 197.1 when the true population mean is 191. Do you think that the null hypothesis is likely true? Based on how unlikely it is to observe a sample mean of 197.1 under the null hypothesis (i.e., less than 1% chance), we might infer from our data that the null hypothesis is probably not true.

Suppose that the sample data had turned out differently. Suppose that we instead observed the following in 2006: $n = 100$, $\overline{X} = 192.1$, $s = 25.6$. How likely is it to observe a sample mean of 192.1 or higher when the true population mean is 191 (i.e., if the null hypothesis is true)? We again compute this probability using the Central Limit Theorem. Specifically,

$$P(\overline{X} > 192.1) = P\left(z > \frac{192.1 - 191}{25.6/\sqrt{100}}\right) =$$
$$P(z > 0.43) = 1 - 0.6664 = 0.3336.$$

There is a 33.4% chance of observing a sample mean as large as 192.1 when the true population mean is 191. Do you think that the null hypothesis is likely true?

We need to determine a threshold or cutoff point (called the *critical value*) to decide when to believe the null hypothesis and when to believe the research hypothesis. It is important to note that although it is possible to observe any sample mean when the true population mean is 191, some values are very unlikely. Based on the previous two samples, it would seem reasonable to believe the research hypothesis when $\overline{X} = 197.1$, but to believe the null hypothesis when $\overline{X} = 192.1$. What we need is a threshold value such that if $\overline{X}$ is above that threshold, we believe that H_1 is true and if $\overline{X}$ is below that threshold, we believe that H_0 is true. The difficulty in determining a threshold for $\overline{X}$ is that it depends on the scale of measurement. In this example, the critical value might be 195 pounds (i.e., if the sample mean is 195 pounds or more, we believe that H_1 is true and if the sample mean is less than 195 pounds, we believe that H_0 is true). Suppose we are interested in assessing an increase in blood pressure over time. The critical value would be different because blood pressures are measured in millimeters of mercury (mmHg) as opposed to in pounds. In the following, we explain how the critical value is determined and how we handle the issue of scale.

First, to address the issue of scale in determining the critical value we convert our sample data (in particular, the sample mean) into a z score. We know from Chapter 5 that the center of the z distribution is 0 and extreme values are those that exceed 2 or fall below -2 (values above 2 and below -2 represent approximately 5% of all z values). If the observed sample mean is close to the mean specified in H_0 ($\mu = 191$), then z is close to 0. If the observed sample mean is much larger than the mean specified in H_0, then z is large.

In hypothesis testing, we select a critical value from the z distribution. This is done by first determining what is called the level of significance, denoted α. Remember that if the null hypothesis is true, it is possible to observe any sample mean. What we are doing here is drawing a line at extreme values. The **level of significance** is the probability that we reject the null hypothesis (in favor of the alternative) when it is actually true:

$$\alpha = \text{Level of significance} = P(\text{Reject } H_0 \mid H_0 \text{ is true}).$$

Because α is a probability, it ranges between 0 and 1. The usual value for α is 0.05, or 5%. If an investigator selects $\alpha = 0.05$, he or she is allowing a 5% probability of incorrectly rejecting the null hypothesis in favor of the alternative when the null is true. The typical values for α are 0.01, 0.05, and 0.10, with $\alpha = 0.05$ the most commonly used value.

Suppose in our weight study we select $\alpha = 0.05$. We need to determine the value of z that holds 5% of the values above it (**Figure 7-1**). The critical value of z for $\alpha = 0.05$ is $z = 1.645$ (i.e., 5% of the distribution is above 1.645). With this value, we can set up what is called our *decision rule* for the test. The rule is to reject H_0 if the z score is 1.645 or higher.

With the first sample, we have $\overline{X} = 197.1$ and $z = 2.38$. Because $2.38 > 1.645$, we reject the null hypothesis. (The same conclusion can be drawn by comparing the 0.0087 probability of observing a sample mean as extreme as 197.1 to the level of significance of 0.05. If the observed probability is smaller than the level of significance, we reject H_0.) Because the z score exceeds the critical value ($2.38 > 1.645$), we conclude that the mean weight for men in 2006 is more than 191 pounds, the value reported in 2002. If we observed the second sample ($\overline{X} = 192.1$), we would not reject the null hypothesis because the z score is 0.43, which is not in the rejection region (i.e., the region in the tail end of the curve at or above 1.645). With the second sample we do not have evidence to conclude that weights have increased. (Again, the same conclusion can be reached by comparing probabilities.

FIGURE 7–1 Critical Value of Z for $\alpha = 0.05$

$\alpha = 0.05$

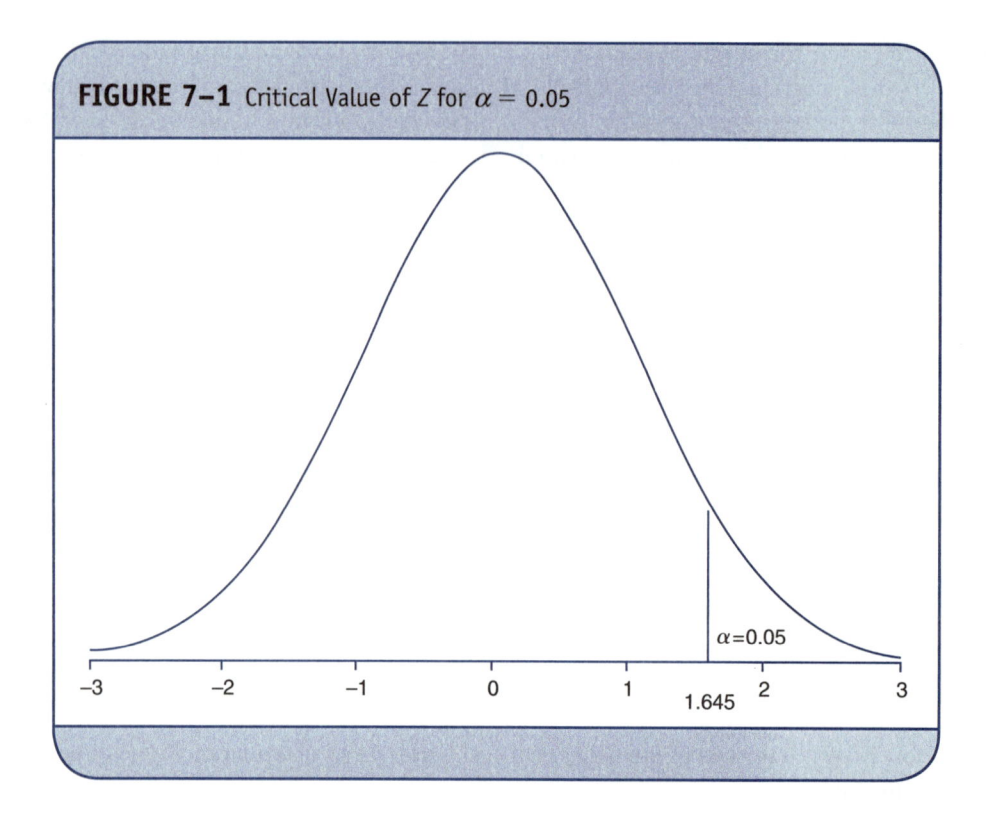

−3 −2 −1 0 1 1.645 2 3

The probability of observing a sample mean as extreme as 192.1 is 33.4%, which is not below our 5% level of significance.)

The procedure for hypothesis testing is based on the ideas described previously. Specifically, we set up competing hypotheses, select a random sample from the population of interest, and compute summary statistics. We then determine whether the sample data supports the null or alternative hypotheses. The procedure can be broken down into the following five steps. We use this five-step approach in performing tests of hypotheses for all of the scenarios in Table 7–1.

Step 1: Set up hypotheses and determine the level of significance.

H_0: Null hypothesis (no change, no difference),

H_1: Research hypothesis (investigator's belief),

$\alpha = 0.05$.

The research or alternative hypothesis can take one of three forms. An investigator might believe that the parameter has increased, decreased, or changed. For example, an investigator might hypothesize:

1. $H_1: \mu > \mu_0$, where μ_0 is the comparator or null value (e.g., $\mu_0 = 191$ in our example about weight in men in 2006) and an increase is hypothesized—this type of test is called an **upper-tailed test**.

2. $H_1: \mu < \mu_0$, where a decrease is hypothesized—this is called a **lower-tailed test**.

3. $H_1: \mu \neq \mu_0$, where a difference is hypothesized—this is called a **two-tailed test**.

The exact form of the research hypothesis depends on the investigator's belief about the parameter of interest and whether it has possibly increased, decreased, or is different from the null value. The research hypothesis is set up by the investigator before any data are collected.

Step 2: Select the appropriate test statistic.

The test statistic is a single number that summarizes the sample information. An example of a test statistic is the z statistic computed as

$$z = \frac{\overline{X} - \mu_0}{s/\sqrt{n}}.$$

When the sample size is small, we use t statistics (just as we did in estimation in Chapter 6). As we present each scenario, alternative test statistics are provided along with conditions for their appropriate use.

Step 3: Set up the decision rule.

The *decision rule* is a statement that tells under what circumstances to reject the null hypothesis. The decision rule is based on specific values of the test statistic (e.g., reject

H₀ if $z \geq 1.645$). The decision rule for a specific test depends on three factors: the research hypothesis, the test statistic, and the level of significance.

The decision rule depends on whether an upper-tailed, lower-tailed, or two-tailed test is proposed. In an upper-tailed test, the decision rule has investigators reject H₀ if the test statistic is greater than or equal to the critical value. In a lower-tailed test, the decision rule has investigators reject H₀ if the test statistic is less than or equal to the critical value. In a two-tailed test, the decision rule has investigators reject H₀ if the test statistic is extreme—either greater than or equal to an upper critical value or less than or equal to a lower critical value.

The exact form of the test statistic is also important in determining the decision rule. If the test statistic follows the standard normal distribution (z), then the decision rule is based on the standard normal distribution. If the test statistic follows the t distribution, then the decision rule is based on the t distribution. The appropriate critical value is selected from the t distribution again depending on the specific alternative hypothesis and the level of significance.

The third factor is the level of significance, which is selected in Step 1 (e.g., $\alpha = 0.05$). For example, in an upper-tailed z test, if $\alpha = 0.05$, then the critical value is $z = 1.645$.

Figures 7–2 through **Figure 7–4** illustrate the rejection regions defined by the decision rule for upper-, lower-, and two-tailed z tests with $\alpha = 0.05$. Notice that the rejection regions are in the upper, lower, and both tails of the curves, respectively. Notice in the two-tailed test that the rejection region is split into two equal parts. The total area in the rejection region is still equal to α. In Figure 7–4, $\alpha = 0.05$ and the area in each tail is 0.025.

Critical values of z for upper-, lower-, and two-tailed tests can be found in Table 1C in the Appendix. Critical values of t for upper, lower, and two-tailed tests can be found in Table 2 in the Appendix.

Step 4: Compute the test statistic.
Here we compute the test statistic by substituting the observed sample data into the test statistic identified in Step 2.

Step 5: Conclusion.
The final conclusion is made by comparing the test statistic (which is a summary of the information observed in the sample) to the decision rule. The final conclusion is either to reject the null hypothesis (because the sample data are

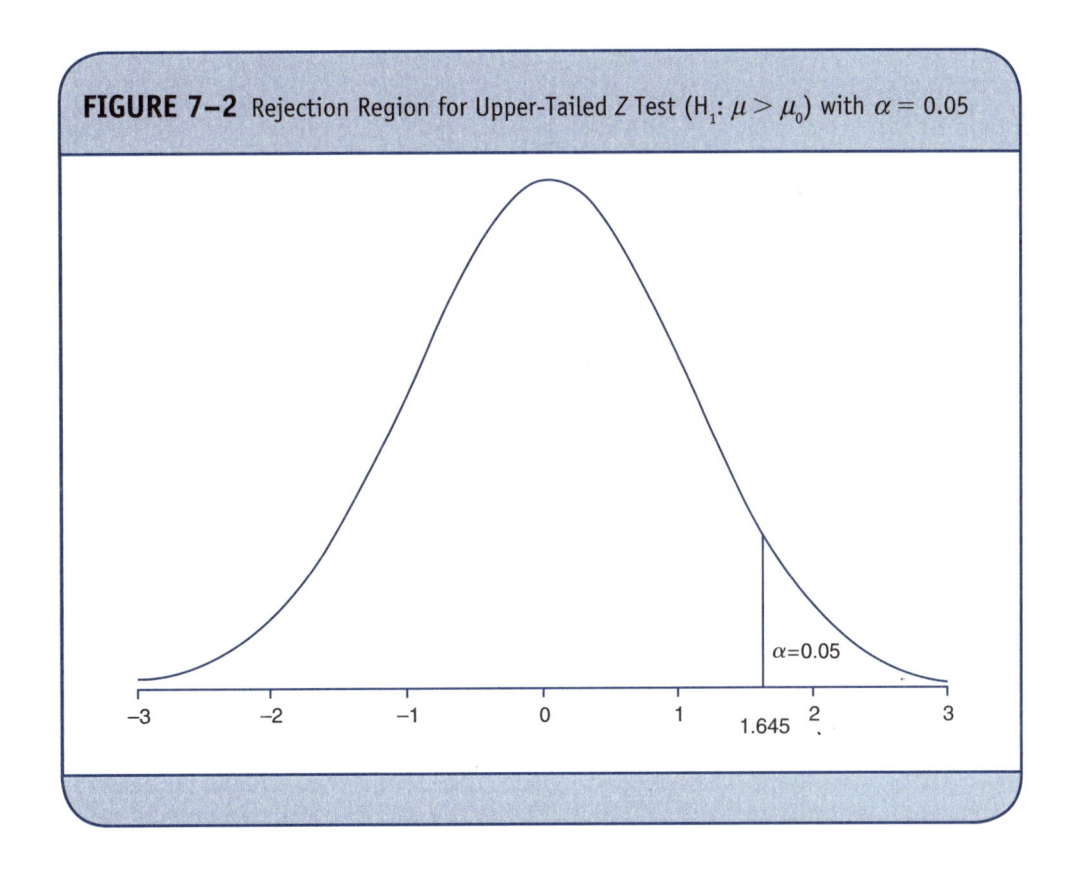

FIGURE 7–2 Rejection Region for Upper-Tailed Z Test (H₁: $\mu > \mu_0$) with $\alpha = 0.05$

FIGURE 7–3 Rejection Region for Lower-Tailed Z Test (H$_1$: $\mu < \mu_0$) with α = 0.05

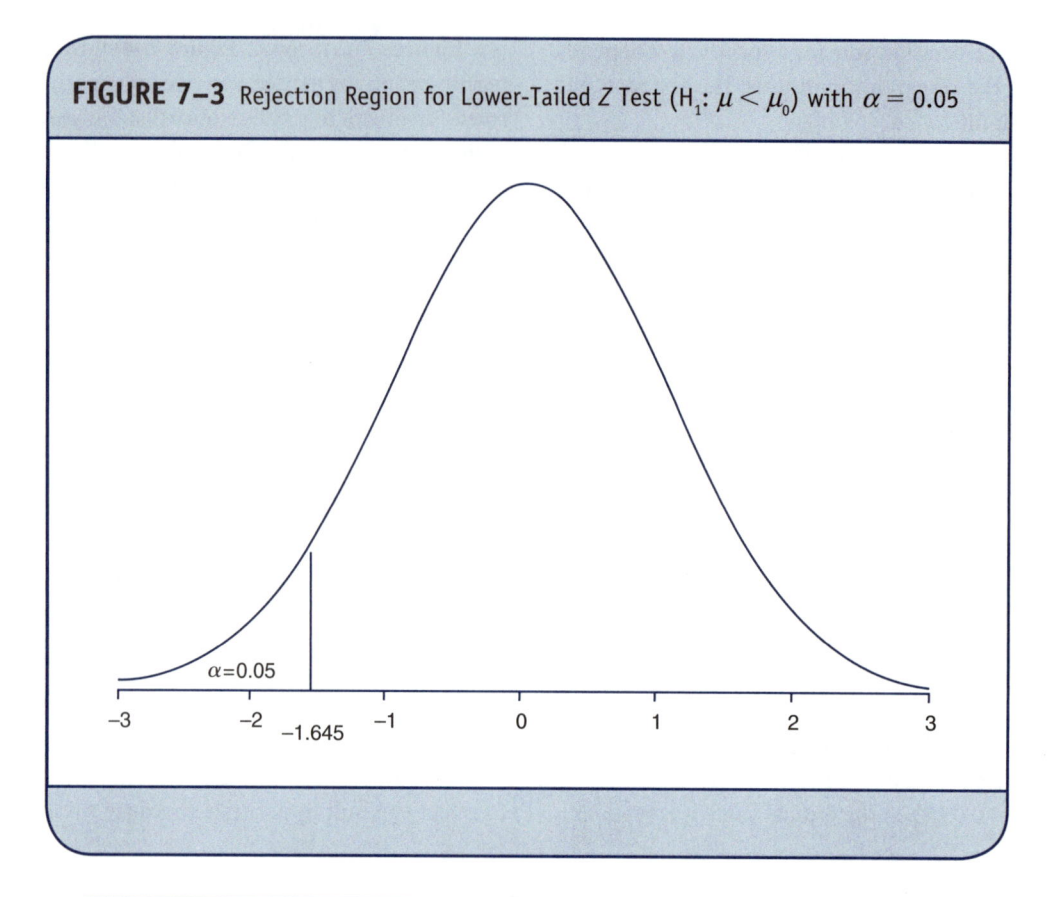

FIGURE 7–4 Rejection Region for Two-Tailed Z Test (H$_1$: $\mu \neq \mu_0$) with α = 0.05

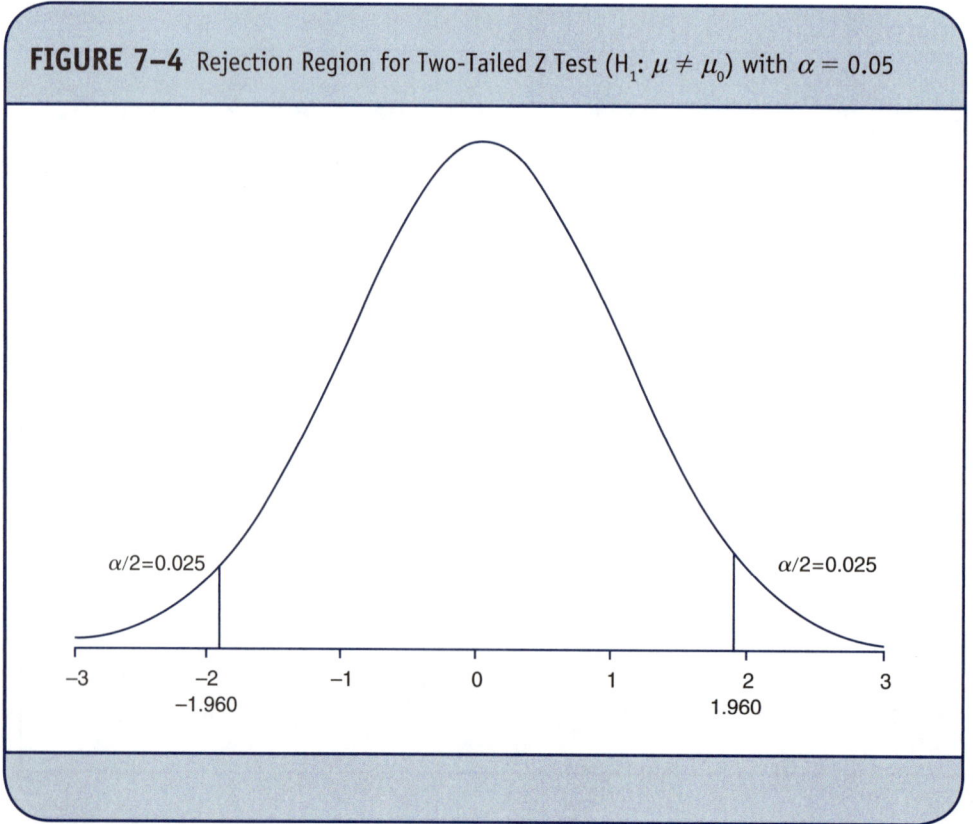

TABLE 7–2 Critical Values for Upper-Tailed Z Tests

Upper-Tailed Test α	Critical Value of z
0.10	1.282
0.05	1.645
0.025	1.960
0.010	2.326
0.005	2.576
0.001	3.090
0.0001	3.719

very unlikely if the null hypothesis is true) or not to reject the null hypothesis (because the sample data are not very unlikely).

If the null hypothesis is rejected, an exact significance level is computed to describe the likelihood of observing the sample data assuming that the null hypothesis is true. The exact level of significance is called the **p-value** and it will be less than the chosen level of significance. Statistical computing packages provide exact *p*-values as part of their standard output for hypothesis tests. We approximate *p*-values using Table 1C for tests involving *z* statistics and Table 2 for tests involving *t* statistics.

We now use the five-step procedure to test the research hypothesis that the mean weight in men in 2006 is more than 191 pounds. We assume the observed sample data are as follows: $n = 100$, $\overline{X} = 197.1$, and $s = 25.6$.

Step 1: Set up hypotheses and determine the level of significance.

$$H_0: \mu = 191,$$
$$H_1: \mu > 191,$$
$$\alpha = 0.05.$$

The research hypothesis is that weights have increased, and therefore an upper-tailed test is used.

Step 2: Select the appropriate test statistic.
Because the sample size is large ($n > 30$), the appropriate test statistic is

$$z = \frac{\overline{X} - \mu_0}{s/\sqrt{n}}.$$

(In Section 7.2, we present alternative test statistics appropriate for small samples.)

Step 3: Set up the decision rule.
In this example, we are performing an upper-tailed test ($H_1: \mu > 191$), with a *z* test statistic and selected $\alpha = 0.05$. The decision rule is shown in Figure 7–2,

$$\text{Reject } H_0 \text{ if } z \geq 1.645.$$

Step 4: Compute the test statistic.
We now substitute the sample data into the formula for the test statistic identified in Step 2,

$$z = \frac{\overline{X} - \mu_0}{s/\sqrt{n}} = \frac{197.1 - 191}{25.6/\sqrt{100}} = 2.38.$$

Step 5: Conclusion.
We reject H_0 because $2.38 > 1.645$. We have statistically significant evidence at $\alpha = 0.05$ to show that the mean weight in men in 2006 is more than 191 pounds. Because we reject the null hypothesis, we now approximate the *p*-value, which is the likelihood of observing data as more extreme under the assumed statistical model. An alternative definition of the *p*-value is the smallest level of significance where we still reject H_0.

In this example, we observe $z = 2.38$, and for $\alpha = 0.05$ the critical value is 1.645. Because 2.38 exceeds 1.645, we reject H_0. In our conclusion, we report a statistically significant increase in mean weight at a 5% level of significance. The data actually provide stronger evidence. **Table 7–2** is a copy of Table 1C in the Appendix that contains critical values for upper-tailed tests.

What is the smallest level of significance we could choose and still reject H_0? If we select $\alpha = 0.025$, the critical value is 1.96, and we still reject H_0 because $2.38 > 1.960$. If we select $\alpha = 0.010$, the critical value is 2.326, and we still reject H_0 because $2.38 > 2.326$. However, if we select $\alpha = 0.005$, the critical value is 2.576 and we cannot reject H_0 because $2.38 < 2.576$. Therefore, the smallest α where we still reject H_0 is 0.010. This is the *p*-value. A statistical computing package produces a more precise *p*-value, which would be between 0.005 and 0.010. Here we are approximating the *p*-value using Table 1C in the Appendix, and we report $p < 0.010$.

p-values reflect the exact significance of tests of hypothesis. In this example, we find $p < 0.010$, indicating that there is less than a 1% chance that we are incorrectly rejecting the null hypothesis if the null hypothesis is true. Suppose in this example that the test statistic is $z = 1.70$. We still reject H_0 at $\alpha = 0.05$ because $1.70 > 1.645$. However, with $z = 1.70$, the *p*-value would be reported as $p < 0.05$. The *p*-value is shown graphically in **Figure 7–5**.

FIGURE 7–5 *p*-Value, Exact Significance Level

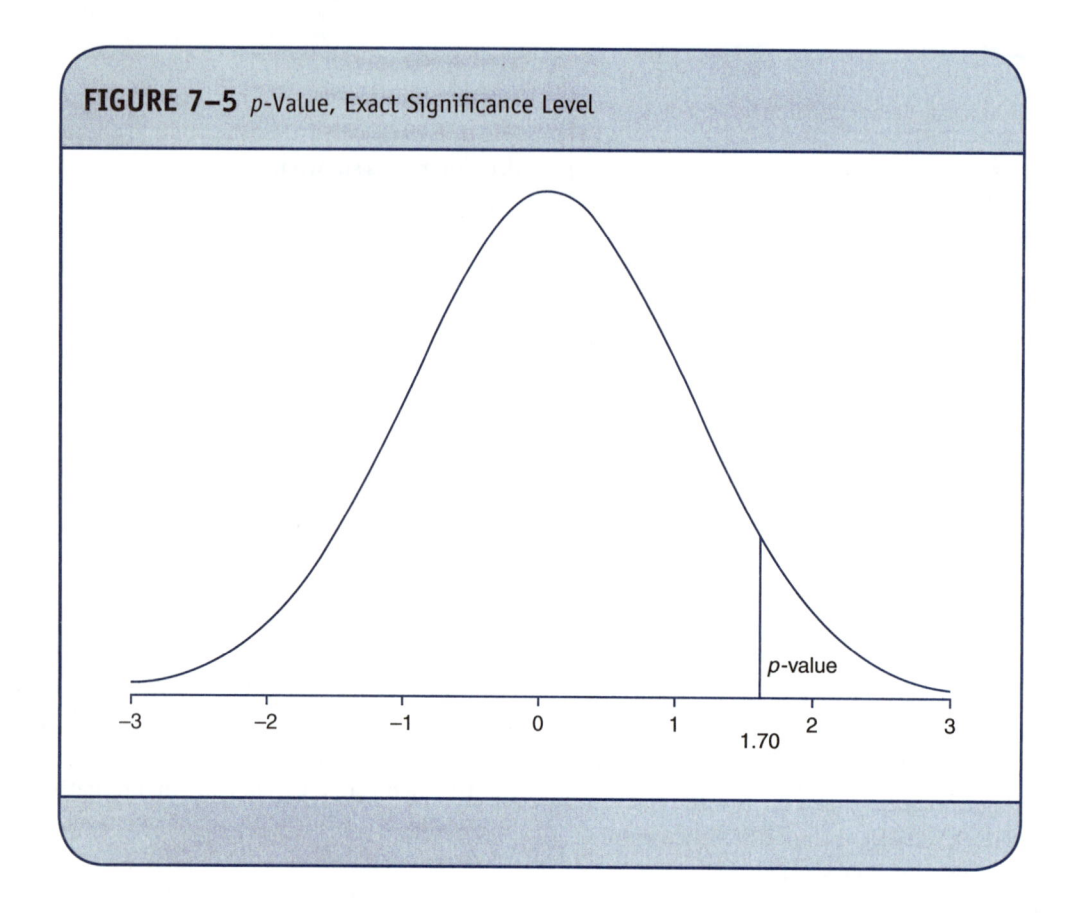

Smaller *p*-values are indicative of more incompatibility of the data with the assumed statistical model. In the literature, investigators often report *p*-values to summarize the significance of tests of hypothesis. The following rule can be used to interpret *p*-values:

$$\text{Reject } H_0 \text{ if } p \leq \alpha,$$

where *p* is the *p*-value and α is the level of significance selected by the investigator.

For example, suppose we wish to test $H_0: \mu = 100$ versus $H_1: \mu > 100$ at $\alpha = 0.05$. Data are collected and analyzed with a statistical computing package that reports $p = 0.0176$. Because $p = 0.0176 < \alpha = 0.05$, we reject H_0. However, if we had selected $\alpha = 0.01$, we would not reject H_0. In the examples that follow, we approximate *p*-values using Table 1C in the Appendix for tests involving *z* statistics and Table 2 in the Appendix for tests involving *t* statistics. We also discuss appropriate interpretation.

In all tests of hypothesis (for continuous, dichotomous, categorical and ordinal outcomes, with one, two, or more than two samples), there are two errors that can be committed. The first is called a **Type I error** and refers to the situation where we incorrectly reject H_0 when, in fact, it is true. This is also called a false positive result (as we incorrectly conclude that the research hypothesis is true when it is not). When we run a test of hypothesis and decide to reject H_0 (e.g., because the test statistic is greater than or equal to the critical value in an upper-tailed test), either we make a correct decision because the research hypothesis is true or we commit a Type I error. The different conclusions are summarized in **Table 7–3**. Note that we never know whether the null hypothesis is really true or false (i.e., we never know which row of Table 7–3 reflects reality).

TABLE 7–3 Errors in Tests of Hypothesis

	Conclusion in Test of Hypothesis	
	Do Not Reject H_0	**Reject H_0**
H_0 is true	Correct decision	Type I error
H_0 is false	Type II error	Correct decision

In the first step of a hypothesis test, we select a level of significance, α, and $\alpha =$ P(Type I error). Because we purposely select a small value for α, we control the probability of committing a Type I error. For example, if we select $\alpha = 0.05$ and our test tells us to reject H_0, then there is a 5% probability that we commit a Type I error. Most investigators are very comfortable with this and are confident when rejecting H_0 that the research hypothesis is true, as it is the more likely scenario when we reject H_0.

When we run a test of hypothesis and decide not to reject H_0 (e.g., because the test statistic is below the critical value in an upper-tailed test), then either we make a correct decision because the null hypothesis is true or we commit a **Type II error**. β represents the probability of a Type II error and is defined as

$$\beta = \text{P(Type II error)} = \text{P(Do not reject } H_0 \mid H_0 \text{ is false)}.$$

Unfortunately, we cannot choose β to be small (e.g., 0.05) to control the probability of committing a Type II error because β depends on several factors, including the sample size, the level of significance (α), and the research hypothesis. These issues will be discussed in much further detail in Chapter 8. For now, we must recognize that when we do not reject H_0, it may be very likely that we are committing a Type II error (i.e., failing to reject H_0 when it is false). Therefore, when tests are run and the null hypothesis is not rejected, we often make a weak concluding statement allowing for the possibility that we might be committing a Type II error. If we do not reject H_0, we conclude that we do not have significant evidence to show that H_1 is true. We do not conclude that H_0 is true. The most common reason for a Type II error is a small sample size.

7.2 TESTS WITH ONE SAMPLE, CONTINUOUS OUTCOME

Hypothesis testing applications with a continuous outcome variable in a single population are performed according to the five-step procedure outlined earlier. A key component is setting up the null and research hypotheses. The objective is to compare the mean in a single population (μ) to a known mean (μ_0). The known value is generally derived from another study or report—for example, a study in a similar but not identical population or a study performed some years ago. The latter is called a historical control ("control" here referring to the fact that the historical study is the comparator). It is important in setting up the hypotheses in a one-sample test that the mean specified in the null hypothesis is a fair and reasonable comparator. This is discussed in the examples that follow.

In one-sample tests for a continuous outcome, we set up our hypotheses against an appropriate comparator. We select a sample and compute descriptive statistics on the sample data using the techniques described in Chapter 4. Specifically, we compute the sample size (n), the sample mean ($\overline{X}$), and the sample standard deviation (s). We then determine the appropriate test statistic (Step 2) for the hypothesis test. The formulas for test statistics depend on the sample size and are given in **Table 7–4**. Appropriate use of the t distribution assumes that the outcome of interest is approximately normally distributed.

The National Center for Health Statistics (NCHS) published a report in 2005 entitled "Health, United States," and it contains extensive information on major trends in the health of Americans.[2] Data are provided for the U.S. population as a whole and for specific ages, by sex and race. We use some of the national statistics here as comparative values in one-sample tests in this section for means of continuous outcomes and in Section 7.3 and Section 7.4 for proportions of dichotomous, categorical, and ordinal variables respectively.

Example 7.1. The NCHS report indicated that in 2002, Americans paid an average of $3302 per year on health care and prescription drugs. An investigator hypothesizes that in 2005, expenditures are lower primarily due to the availability of generic drugs. To test the hypothesis, a sample of 100 Americans is selected and their expenditures on health care and prescription drugs in 2005 are measured. The sample data are summarized as follows: $n = 100$, $\overline{X} = \$3190$, and $s = \$890$. Is there statistical evidence of a reduction in expenditures on health care and prescription drugs in 2005? Is the sample mean of $3190 evidence of a true reduction in the mean or is it within chance fluctuation? We run the test using the five-step approach.

TABLE 7–4 Test Statistics for Testing $H_0: \mu = \mu_0$

$n \geq 30$	$z = \dfrac{\overline{X} - \mu_0}{s/\sqrt{n}}$	(Find critical value in Table 1C)
$n < 30$	$t = \dfrac{\overline{X} - \mu_0}{s/\sqrt{n}}$	(Find critical value in Table 2, $df = n - 1$)

Step 1: Set up hypotheses and determine the level of significance.

$$H_0: \mu = 3302,$$
$$H_1: \mu < 3302,$$
$$\alpha = 0.05.$$

The research hypothesis is that expenditures have decreased, and therefore a lower-tailed test is used.

Step 2: Select the appropriate test statistic.
Because the sample size is large ($n > 30$), the appropriate test statistic is

$$z = \frac{\overline{X} - \mu_0}{s/\sqrt{n}}.$$

Step 3: Set up the decision rule.
This is a lower-tailed test, using a z statistic and a 5% level of significance. The appropriate critical value can be found in Table 1C in the Appendix and the decision rule is

$$\text{Reject } H_0 \text{ if } z \leq -1.645.$$

Step 4: Compute the test statistic.
We now substitute the sample data into the formula for the test statistic identified in Step 2:

$$z = \frac{\overline{X} - \mu_0}{s/\sqrt{n}} = \frac{3190 - 3302}{890/\sqrt{100}} = -1.26$$

Step 5: Conclusion.
We do not reject H_0 because $-1.26 > -1.645$. We do not have statistically significant evidence at $\alpha = 0.05$ to show that the mean expenditures on health care and prescription drugs are lower in 2005 than the mean of $3302 reported in 2002.

Recall that when we fail to reject H_0 in a test of hypothesis that either the null hypothesis is true (the mean expenditures in 2005 are the same as those in 2002 and equal to $3302) or we are committing a Type II error (we fail to reject H_0 when in fact it is false). In summarizing this test, we conclude that we do not have sufficient evidence to reject H_0. We do not conclude that H_0 is true because there may be a moderate to high probability that we are committing a Type II error. It is possible that the sample size is not large enough to detect a difference in mean expenditures (see Chapter 8 for more details).

Example 7.2. The NCHS reported that the mean total cholesterol level in 2002 for all adults was 203. In Chapter 4, we presented data on $n = 3539$ participants who attended the seventh examination of the Offspring in the Framingham Heart Study. Descriptive statistics on variables measured in the sample were presented in Table 4.20 and included the following statistics on total cholesterol levels of participants: $n = 3310$, $\overline{X} = 200.3$, and $s = 36.8$. Is there statistical evidence of a difference in mean cholesterol levels in the Framingham Offspring as compared to the national mean? Here we want to assess whether the sample mean of 200.3 in the Framingham sample is statistically significantly different from 203 (i.e., beyond what we would expect by chance). We run the test using the five-step approach.

Step 1: Set up hypotheses and determine the level of significance.

$$H_0: \mu = 203,$$
$$H_1: \mu \neq 203,$$
$$\alpha = 0.05.$$

The research hypothesis is that cholesterol levels are different in the Framingham Offspring, and therefore a two-tailed test is used.

Step 2: Select the appropriate test statistic.
Because the sample size is large ($n > 30$), the appropriate test statistic is

$$z = \frac{\overline{X} - \mu_0}{s/\sqrt{n}}.$$

Step 3: Set up the decision rule.
This is a two-tailed test, using a z statistic and a 5% level of significance. The appropriate critical values can be found in Table 1C in the Appendix and the decision rule is

$$\text{Reject } H_0 \text{ if } z \leq -1.960 \text{ or if } z \geq 1.960.$$

Step 4: Compute the test statistic.
We now substitute the sample data into the formula for the test statistic identified in Step 2:

$$z = \frac{\overline{X} - \mu_0}{s/\sqrt{n}} = \frac{200.3 - 203}{36.8/\sqrt{3310}} = -4.22.$$

Step 5: Conclusion.
We reject H_0 because $-4.22 < -1.960$. We have statistically significant evidence at $\alpha = 0.05$ to show that the mean total cholesterol level in the Framingham Offspring is different from the national mean of 203 reported in 2002. Because we reject

H_0, we also approximate a *p*-value. Using the two-sided significance levels in Table 1C in the Appendix, $p < 0.0001$.

Example 7.2 raises an important concept of statistical versus clinical or practical significance. From a statistical standpoint, the mean total cholesterol level in the Framingham sample is highly statistically significantly different from the national mean with $p < 0.0001$ (i.e., there is less than a 0.01% chance that we are incorrectly rejecting the null hypothesis given the observed data). However, the sample mean in the Framingham Offspring study is 200.3, less than 3 units different from the national mean of 203. The reason the data are so highly statistically significant is the very large sample size. It is always important to assess both statistical and practical significance of data. This is particularly relevant when the sample size is large. The five-step procedure allows for an assessment of statistical significance. Investigators must also assess practical or clinical significance. Is a 3-unit difference in total cholesterol a meaningful difference?

Example 7.3. Consider again the NCHS-reported mean total cholesterol level of 203 in 2002 for all adults. Suppose a new drug is proposed to lower total cholesterol, and a study is designed to evaluate the efficacy of the drug in lowering cholesterol. Fifteen patients are enrolled in the study and asked to take the new drug for 6 weeks. At the end of 6 weeks, each patient's total cholesterol level is measured and the sample statistics are as follows: $n = 15$, $\overline{X} = 195.9$, and $s = 28.7$. Is there statistical evidence of a reduction in mean total cholesterol in patients after using the new drug for 6 weeks? We run the test using the five-step approach.

Step 1: Set up hypotheses and determine the level of significance.

$$H_0: \mu = 203,$$
$$H_1: \mu < 203,$$
$$\alpha = 0.05.$$

Step 2: Select the appropriate test statistic.

Because the sample size is small ($n < 30$), the appropriate test statistic is

$$t = \frac{\overline{X} - \mu_0}{s/\sqrt{n}}.$$

Step 3: Set up the decision rule.

This is a lower-tailed test, using a *t* statistic and a 5% level of significance. The appropriate critical value can be found in Table 2 in the Appendix. To determine the critical value of *t*, we need degrees of freedom, *df*, defined as $df = n - 1$. In this example, $df = 15 - 1 = 14$. The critical value for a lower-tailed test with $df = 14$ and $\alpha = 0.05$ is -1.761 and the decision rule is

$$\text{Reject } H_0 \text{ if } t \leq -1.761.$$

Step 4: Compute the test statistic.

We now substitute the sample data into the formula for the test statistic identified in Step 2.

$$t = \frac{\overline{X} - \mu_0}{s/\sqrt{n}} = \frac{195.9 - 203}{28.7/\sqrt{15}} = -0.96.$$

Step 5: Conclusion.

We do not reject H_0 because $-0.96 > -1.761$. We do not have statistically significant evidence at $\alpha = 0.05$ to show that the mean total cholesterol level in patients taking the new drug for 6 weeks is lower than the national mean. Again, because we fail to reject the null hypothesis, we make a weaker concluding statement, allowing for the possibility that we may be committing a Type II error (i.e., fail to reject H_0 when the drug is efficacious).

Example 7.3 raises an important issue in terms of study design. In this example, we assume in the null hypothesis that the mean cholesterol level is 203. This is taken to be the mean cholesterol level in patients without treatment. Is this an appropriate comparator? Alternative and potentially more efficient study designs to evaluate the effect of the new drug could involve two treatment groups, where one group receives the new drug and the other does not, or we could measure each patient's baseline or pre-treatment cholesterol level and then assess changes from baseline to 6 weeks post-treatment. These designs are discussed in Section 7.5 and Section 7.6, respectively.

7.3 TESTS WITH ONE SAMPLE, DICHOTOMOUS OUTCOME

Hypothesis testing applications with a dichotomous outcome variable in a single population are also performed according to the five-step procedure. Similar to tests for means described in Section 7.2, a key component is setting up the null and research hypotheses. The objective is to compare the proportion of successes in a single population to a known proportion (p_0). That known proportion is generally derived from another study or report and is sometimes called a historical control. It is important in setting up the hypotheses in a one-sample test that the proportion specified in the null hypothesis is a fair and reasonable comparator.

In one-sample tests for a dichotomous outcome, we set up our hypotheses against an appropriate comparator. We select a sample and compute descriptive statistics on the sample data using the techniques described in Chapter 4. Specifically, we compute the sample size (n) and the sample proportion ($\hat{p}$), which is computed by taking the ratio of the number of successes to the sample size, $\hat{p} = x/n$. We then determine the appropriate test statistic (Step 2) for the hypothesis test. The formula for the test statistic is given in **Table 7–5**.

The preceding formula is appropriate for large samples, defined when the smaller of np_0 and $n(1 - p_0)$ is at least 5. This is similar, but not identical, to the condition required for appropriate use of the confidence interval formula for a population proportion specified in Table 6–6—i.e., $\min[n\hat{p}, n(1 - \hat{p})] \geq 5$. Here we use the proportion specified in the null hypothesis (p_0) as the true proportion of successes rather than the sample proportion ($\hat{p}$). If we fail to satisfy the condition, then alternative procedures called exact methods must be used to test the hypothesis about the population proportion.[3,4]

Example 7.4. The NCHS report indicated that in 2002 the prevalence of cigarette smoking among American adults was 21.1%. In Chapter 4, we presented data on prevalent smoking in $n = 3536$ participants who attended the seventh examination of the offspring in the Framingham Heart Study. Suppose we want to assess whether the prevalence of smoking is lower in the Framingham offspring sample due to the focus on cardiovascular health in that community. Data from the Framingham Offspring Study were summarized in Table 4–11 and indicated that 482 / 3536 = 13.6% of the respondents were currently smoking at the time of the exam. Is there evidence of a statistically lower prevalence of smoking in the Framingham Offspring Study as compared to the prevalence among all Americans?

Step 1: Set up hypotheses and determine the level of significance.

$$H_0: p = 0.211,$$
$$H_1: p < 0.211,$$
$$\alpha = 0.05.$$

TABLE 7–5 Test Statistic for Testing $H_0: p = p_0$

$\min[np_0, n(1-p_0)] \geq 5$ $\quad z = \dfrac{\hat{p} - p_0}{\sqrt{p_0(1-p_0)/n}}$ $\quad$ (Find critical value in Table 1C)

Step 2: Select the appropriate test statistic.
The formula for the test statistic is in Table 7–5. We must first check that the sample size is adequate. Specifically, we need to check

$$\min[np_0, n(1-p_0)] = \min[3536(0.211), 3536(1 - 0.211)] = \min(746, 2790) = 746$$

The sample size is more than adequate so the following formula can be used:

$$z = \frac{\hat{p} - p_0}{\sqrt{p_0(1 - p_0)/n}}.$$

Step 3: Set up the decision rule.
This is a lower-tailed test, using a z statistic and a 5% level of significance. The appropriate critical value can be found in Table 1C in the Appendix, and the decision rule is

$$\text{Reject } H_0 \text{ if } z \leq -1.645.$$

Step 4: Compute the test statistic.
We now substitute the sample data into the formula for the test statistic identified in Step 2:

$$z = \frac{\hat{p} - p_0}{\sqrt{p_0(1-p_0)/n}} = \frac{0.136 - 0.211}{\sqrt{0.211(1 - 0.211)/3536}} = -10.93.$$

Step 5: Conclusion.
We reject H_0 because $-10.93 < -1.645$. We have statistically significant evidence at $\alpha = 0.05$ to show that the prevalence of smoking in the Framingham Offspring is lower than the national prevalence (21.1%). Using Table 1C in the Appendix, the p-value is $p < 0.0001$.

Example 7.5. The NCHS report indicated that in 2002, 75% of children aged 2 to 17 saw a dentist in the past year. An investigator wants to assess whether use of dental services is different in children living in the city of Boston. A sample of 125 children aged 2 to 17 living in Boston is surveyed and 64 reported seeing a dentist over the past 12 months. Is there a significant difference in the proportion of children living in Boston who use dental services as compared to the national proportion?

Step 1: Set up hypotheses and determine the level of significance.

$$H_0: p = 0.75,$$
$$H_1: p \neq 0.75,$$
$$\alpha = 0.05.$$

Step 2: Select the appropriate test statistic.
The formula for the test statistic is in Table 7–5. We must first check that the sample size is adequate. Specifically, we need to check

$$\min[np_0, n(1-p_0)] = \min[125(0.75), 125(1-0.75)] = \min(94, 31) = 31.$$

The sample size is more than adequate, so the following formula can be used:

$$z = \frac{\hat{p} - p_0}{\sqrt{p_0(1-p_0)/n}}.$$

Step 3: Set up the decision rule.
This is a two-tailed test, using a z statistic and a 5% level of significance. The appropriate critical value can be found in Table 1C in the Appendix and the decision rule is

Reject H_0 if $z \leq -1.960$ or if $z \geq 1.960$.

Step 4: Compute the test statistic.
We now substitute the sample data into the formula for the test statistic identified in Step 2. The sample proportion is $\hat{p} = 64/125 = 0.512$, and

$$z = \frac{\hat{p} - p_0}{\sqrt{p_0(1-p_0)/n}} = \frac{0.512 - 0.75}{\sqrt{0.75(1-0.75)/125}} = -6.15.$$

Step 5: Conclusion.
We reject H_0 because $-6.15 < -1.960$. We have evidence at $\alpha = 0.05$ to show that there is a statistically significant difference in the proportion of children living in Boston who use dental services as compared to the national proportion. Using Table 1C in the Appendix, the p-value is $p < 0.0001$.

7.4 TESTS WITH ONE SAMPLE, CATEGORICAL AND ORDINAL OUTCOMES

Hypothesis testing with a categorical or ordinal outcome variable in a single population is again performed according to the five-step procedure. Similar to tests for means and proportions described in Section 7.2 and Section 7.3, a key component is setting up the null and research hypotheses. Categorical and ordinal variables are variables that take on more than two distinct responses or categories, and responses can be ordered or unordered (i.e., ordinal or categorical). The procedure we describe here can be used for ordinal or categorical outcomes, and the objective is to compare the distribution of responses—or the proportions of participants in each response category—to a known distribution. The known distribution is derived from another study or report and it is again important in setting up the hypotheses that the comparator distribution specified in the null hypothesis is a fair comparison.

In one-sample tests for a categorical or ordinal outcome, we set up our hypotheses against an appropriate comparator. We select a sample and compute descriptive statistics on the sample data using the techniques described in Chapter 4. Specifically, we compute the sample size (n) and the proportions of participants in each response category ($\hat{p}_1, \hat{p}_2, ..., \hat{p}_k$) where k represents the number of response categories. We then determine the appropriate test statistic (Step 2) for the hypothesis test. The formula for the test statistic is in **Table 7–6**.

With the χ^2 statistic, we compare the observed frequencies in each response category (O) to the frequencies we would expect (E) if the null hypothesis were true. These expected frequencies are determined by allocating the sample to the response categories according to the distribution specified in H_0. This is done by multiplying the observed sample size (n) by the proportions specified in the null hypothesis ($p_{10}, p_{20}, ..., p_{k0}$). To ensure that the sample size is appropriate for the use of the test statistic in Table 7–6, we need to ensure that the expected frequency in each response category is at least 5, or

$$\min(np_{10}, np_{20}, ..., np_{k0}) \geq 5.$$

The test of hypothesis with a categorical or ordinal outcome measured in a single sample, where the goal is to assess whether the distribution of responses follows a known distribution, is called the χ^2 goodness-of-fit test. As the name indicates, the idea is to assess whether the distribution of responses in the sample "fits" a specified population distribution. In the next example, we illustrate the test using the five-step approach. As we work through the example, we provide additional details related to the use of this new test statistic.

TABLE 7–6 Test Statistic for Testing
$H_0: p_1 = p_{10}, p_2 = p_{20}, ..., p_k = p_{k0}$

$$\chi^2 = \sum \frac{(O-E)^2}{E}$$

(Find critical value in Table 3, $df = k - 1$)

Example 7.6. A university conducted a survey of its recent graduates to collect demographic and health information for future planning purposes, as well as to assess students' satisfaction with their undergraduate experiences. The survey revealed that a substantial proportion of students were not engaging in regular exercise, many felt their nutrition was poor, and a substantial number were smoking. In response to a question on regular exercise, 60% of all graduates reported getting no regular exercise, 25% reported exercising sporadically, and 15% reported exercising regularly as undergraduates. The next year, the university launched a health promotion campaign on campus in an attempt to increase health behaviors among undergraduates. The program included modules on exercise, nutrition, and smoking cessation. To evaluate the impact of the program, the university again surveyed graduates and asked the same questions. The survey is completed by 470 graduates and the data shown in **Table 7–7** are collected on the exercise question. Based on the data, is there evidence of a shift in the distribution of responses to the exercise question following the implementation of the health promotion campaign on campus? Run the test at a 5% level of significance.

In this example, we have one sample and an ordinal outcome variable (with three response options). We specifically want to compare the distribution of responses in the sample to the distribution reported the previous year (i.e., 60%, 25%, and 15% reporting no, sporadic, and regular exercise, respectively). We now run the test using the five-step approach.

Step 1: Set up hypotheses and determine the level of significance.

The null hypothesis again represents the "no change" or "no difference" situation. If the health promotion campaign has no impact, then we expect the distribution of responses to the exercise question to be the same as that measured prior to the implementation of the program.

H_0: $p_1 = 0.60, p_2 = 0.25, p_3 = 0.15$ or, equivalently,

H_0: Distribution of responses is 0.60, 0.25, 0.15,

H_1: H_0 is false,

$\alpha = 0.05$.

Notice that the research hypothesis is written in words rather than in symbols. The research hypothesis as stated captures any difference in the distribution of responses from that specified in the null hypothesis. We do not specify a specific alternative distribution; instead, we are testing whether the sample data "fit" the distribution in H_0 or not. With the χ^2 goodness-of-fit test, there is no upper- or lower-tailed version of the test.

Step 2: Select the appropriate test statistic.

The formula for the test statistic is in Table 7–6 and is

$$\chi^2 = \Sigma \frac{(O-E)^2}{E}.$$

We must first assess whether the sample size is adequate. Specifically, we need to check $\min(np_{10},...,np_{k0}) \geq 5$. The sample size here is $n = 470$ and the proportions specified in the null hypothesis are 0.60, 0.25, and 0.15. Thus,

$$\min[470(0.60), 470(0.25), 470(0.15)] = \min(282, 117.5, 70.5) = 70.5.$$

The sample size is more than adequate, so the test statistic can be used.

Step 3: Set up the decision rule.

The decision rule for the χ^2 test is set up in a similar way to decision rules we established for z and t tests. The decision rule depends on the level of significance and the degrees of freedom, defined as $df = k - 1$, where k is the number of response categories. Again, with χ^2 tests there are no upper- or lower-tailed versions of the test. If the null hypothesis is true, the observed and expected frequencies are close in value and the χ^2 statistic is close to 0. If the null hypothesis is false, then the χ^2 statistic is large. The rejection region for the χ^2 test is always in the upper-tail (right), as shown in **Figure 7–6**.

Table 3 in the Appendix contains critical values for the χ^2 test, indexed by degrees of freedom and the desired level of significance. Here we have $df = k-1 = 3-1 = 2$ and a 5% level of significance. The appropriate critical value from Table 3 is 5.99, and the decision rule is

Reject H_0 if $\chi^2 \geq 5.99$.

TABLE 7–7 Survey Results on Regular Exercise

	No Regular Exercise	Sporadic Exercise	Regular Exercise	Total
Number of students	255	125	90	470

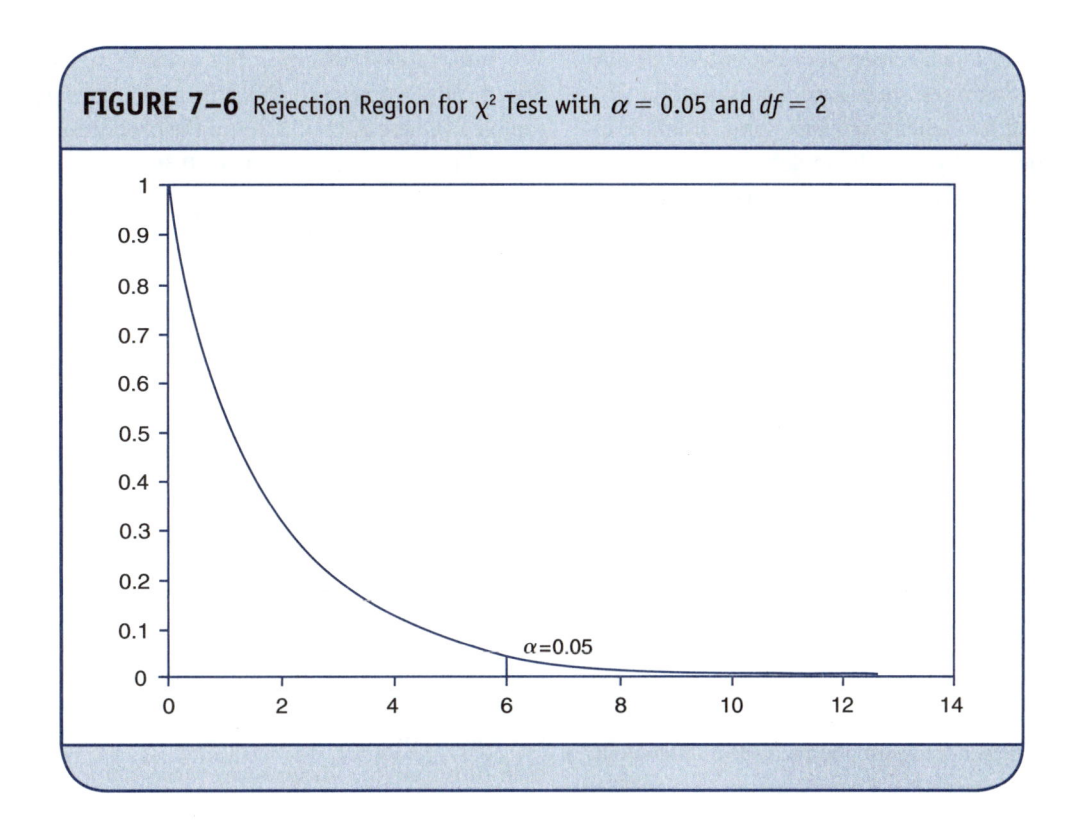

FIGURE 7–6 Rejection Region for χ^2 Test with $\alpha = 0.05$ and $df = 2$

α=0.05

Step 4: Compute the test statistic.

We now compute the expected frequencies using the sample size and the proportions specified in the null hypothesis. We then substitute the sample data (observed frequencies) and the expected frequencies into the formula for the test statistic identified in Step 2. The computations can be organized as shown in **Table 7–8**. Notice that the expected frequencies are taken to one decimal place and that the sum of the observed frequencies is equal to the sum of the expected frequencies. The test statistic is computed as

$$\chi^2 = \frac{(255-282)^2}{282} + \frac{(125-177.5)^2}{117.5} + \frac{(90-70.5)^2}{70.5},$$

$$\chi^2 = 2.59 + 0.48 + 5.39 = 8.46.$$

Step 5: Conclusion.

We reject H_0 because $8.46 > 5.99$. We have statistically significant evidence at $\alpha = 0.05$ to show that H_0 is false, or that the distribution of responses is not 0.60, 0.25, and 0.15. Using Table 3 in the Appendix, we can approximate the *p*-value. We need to look at critical values for smaller levels of significance with $df = 2$. Using Table 3 in the Appendix, the *p*-value is $p < 0.025$.

In the χ^2 goodness-of-fit test, we conclude that either the distribution specified in H_0 is false (when we reject H_0) or that we do not have sufficient evidence to show that the distribution specified in H_0 is false (when we fail to reject H_0). In Example 7.6, we reject H_0 and conclude that the distribution of responses to the exercise question following the implementation of the health promotion campaign is not the same as

TABLE 7–8 Computing Expected Frequencies

	No Regular Exercise	Sporadic Exercise	Regular Exercise	Total
Observed frequencies (O)	255	125	90	470
Expected frequencies (E)	470(0.60) = 282	470(0.25) = 117.5	470(0.15) = 70.5	470

the distribution prior. The test itself does not provide details of how the distribution has shifted. A comparison of the observed and expected frequencies provides some insight into the shift (when the null hypothesis is rejected). In Example 7.6, we observe the data shown in the first row of Table 7–8. If the null hypothesis is true, we would have expected more students to fall in the *No Regular Exercise* category and fewer in the *Regular Exercise* category. In the sample, 255 / 470 = 54% report no regular exercise and 90 / 470 = 19% report regular exercise. Thus, there is a shift toward more regular exercise following the implementation of the health promotion campaign.

Example 7.7. The NCHS provided data on the distribution of weight (in categories) among Americans in 2002. The distribution was based on specific values of body mass index (BMI) computed as weight in kilograms over height in meters squared. Underweight was defined as a BMI less than 18.5, normal weight as a BMI between 18.5 and 24.9, overweight as a BMI between 25 and 29.9, and obese as a BMI of 30 or greater. Americans in 2002 were distributed as follows: 2% underweight, 39% normal weight, 36% overweight, and 23% obese. Suppose we want to assess whether the distribution of BMI is different in the Framingham offspring sample. Using data from the $n = 3536$ participants who attended the seventh examination of the offspring in the Framingham Heart Study, we create the BMI categories as defined and are the data shown in **Table 7–9**.

Step 1: Set up hypotheses and determine the level of significance.

H_0: $p_1 = 0.02$, $p_2 = 0.39$, $p_3 = 0.36$, $p_4 = 0.23$ or, equivalently,

H_0: Distribution of responses is 0.02, 0.39, 0.36, 0.23,

H_1: H_0 is false,

$\alpha = 0.05$.

Step 2: Select the appropriate test statistic.

The formula for the test statistic is in Table 7–6 and is

$$\chi^2 = \sum \frac{(O - E)^2}{E}.$$

We must assess whether the sample size is adequate. Specifically, we need to check $\min(np_{10}, ..., np_{k0}) \geq 5$. The sample size here is $n = 3326$ and the proportions specified in the null hypothesis are 0.02, 0.39, 0.36, and 0.23. Thus,

$$\min[3326(0.02), 3326(0.39), 3326(0.36), 3326(0.23)] = \min(66.5, 1297.1, 1197.4, 765.0) = 66.5.$$

The sample size is more than adequate, so the test statistic can be used.

Step 3: Set up the decision rule.

Here we have $df = k - 1 = 4 - 1 = 3$ and a 5% level of significance. The appropriate critical value from Table 3 in the Appendix is 7.81, and the decision rule is

Reject H_0 if $\chi^2 \geq 7.81$.

Step 4: Compute the test statistic.

We now compute the expected frequencies using the sample size and the proportions specified in the null hypothesis. We then substitute the sample data (observed frequencies) into the formula for the test statistic identified in Step 2. We organize the computations in **Table 7–10**.

The test statistic is computed as

$$\chi^2 = \frac{(20 - 66.5)^2}{66.5} + \frac{(932 - 1297.1)^2}{1297.1} + \frac{(1374 - 1197.4)^2}{1197.4} + \frac{(1000 - 765.0)^2}{765.0},$$

$\chi^2 = 32.52 + 102.77 + 26.05 + 72.19 = 233.53$.

Step 5: Conclusion.

We reject H_0 because 233.53 > 7.81. We have statistically significant evidence at $\alpha = 0.05$ to show that H_0 is false or that the distribution of BMI in the Framingham offspring is different from the national data reported in 2002. Using Table 3 in the Appendix, we can approximate the *p*-value. We need

TABLE 7–9 Distribution of BMI in Framingham Offspring

	Underweight BMI < 18.5	Normal Weight BMI 18.5–24.9	Overweight BMI 25.0–29.9	Obese BMI 30+	Total
Number of participants	20	932	1374	1000	3326

TABLE 7–10 Observed and Expected Frequencies

	Underweight BMI < 18.5	Normal Weight BMI 18.5–24.9	Overweight BMI 25.0–29.9	Obese BMI 30+	Total
Observed frequencies (O)	20	932	1374	1000	3326
Expected frequencies (E)	66.5	1297.1	1197.4	765.0	3326

to look at critical values for smaller levels of significance for $df = 3$. Using Table 3, the p-value is $p < 0.005$.

Again, the χ^2 goodness-of-fit test allows us to assess whether the distribution of responses "fits" a specified distribution. In Example 7.7, we show that the distribution of BMI in the Framingham Offspring Study is different from the national distribution. To understand the nature of the difference, we compare observed and expected frequencies or observed and expected proportions (or percentages). In Example 7.7, the frequencies are large because of the large sample size—the observed percentages of patients in the Framingham sample are 0.6% underweight, 28% normal weight, 41% overweight, and 30% obese. In the Framingham Offspring sample, there are higher percentages of overweight and obese persons (41% and 30% in Framingham as compared to 36% and 23% nationally) and lower percentages of underweight and normal weight persons (0.6% and 28% in Framingham as compared to 2% and 39% nationally).

7.5 TESTS WITH TWO INDEPENDENT SAMPLES, CONTINUOUS OUTCOME

There are many applications where it is of interest to compare two independent groups with respect to their mean scores on a continuous outcome. In Chapter 6, we presented techniques to estimate the difference in means. Here we again compare means between groups, but rather than generating an estimate of the difference, we test whether the observed difference (increase, decrease, or difference) is statistically significant or not.

In this section, we discuss the comparison of means when the two comparison groups are independent or physically separate. The two groups might be determined by a particular attribute (e.g., sex, history of cardiovascular disease) or might be set up by the investigator (e.g., participants assigned to receive an experimental drug or placebo). The first step in the analysis involves computing descriptive statistics on each of the two samples using the techniques described in Chapter 4. Specifically, we compute the sample size, mean, and standard deviation in each sample and we denote these summary statistics as follows: n_1, $\bar{X}_1$, and s_1 for Sample 1 and n_2, $\bar{X}_2$, and s_2 for Sample 2. The designation of Sample 1 and Sample 2 is essentially arbitrary. In a clinical trial setting, the convention is to call the treatment Group 1 and the control Group 2. However, when comparing men and women, either group can be 1 or 2.

In the two independent samples application with a continuous outcome, the parameter of interest in the test of hypothesis is the difference in population means, $\mu_1 - \mu_2$. The null hypothesis is always that there is no difference between groups with respect to means, i.e., H_0: $\mu_1 - \mu_2 = 0$. The null hypothesis can also be written as H_0: $\mu_1 = \mu_2$. In the research hypothesis, an investigator can hypothesize that the first mean is larger than the second (H_1: $\mu_1 > \mu_2$), that the first mean is smaller than the second (H_1: $\mu_1 < \mu_2$), or that the means are different (H_1: $\mu_1 \neq \mu_2$). The three different alternatives represent upper-, lower-, and two-tailed tests, respectively. **Table 7–11** contains the formulas for test statistics for the difference in population means.

In the formulas in Table 7–11, $\bar{X}_1$ and $\bar{X}_2$ are the means of the outcome in the independent samples, and S_p is the pooled estimate of the common standard deviation (again assuming that the variances in the populations are similar) computed as the weighted average of the standard deviations in the samples,

$$S_p = \sqrt{\frac{(n_1-1)s_1^2 + (n_2-1)s_2^2}{n_1 + n_2 - 2}}.$$

TABLE 7–11 Test Statistics for Testing $H_0: \mu_1 = \mu_2$

$n_1 \geq 30$ and $n_2 \geq 30$	$z = \dfrac{\overline{X}_1 - \overline{X}_2}{S_p\sqrt{1/n_1 + 1/n_2}}$	(Find critical value of z in Table 1C)
$n_1 < 30$ or $n_2 < 30$	$t = \dfrac{\overline{X}_1 - \overline{X}_2}{S_p\sqrt{1/n_1 + 1/n_2}}$	(Find critical value of t in Table 2, $df = n_1 + n_2 - 2$)

Because we are assuming equal variances between groups, we pool the information on variability (sample variances) to generate an estimate of the variability in the population. As a guideline, if the ratio of the sample variances is between 0.5 and 2, the assumption of equality of population variances is taken to be appropriate. (Note that because S_p is a weighted average of the standard deviations in the sample, S_p is always between s_1 and s_2.)

Example 7.8. In Example 6.5, we used data presented in Chapter 4 on $n = 3539$ participants who attended the seventh examination of the offspring in the Framingham Heart Study and constructed a 95% confidence interval for the difference in mean systolic blood pressures between men and women. Table 7–12 contains summary statistics on the characteristics measured in men and women. Suppose we now wish to assess whether there is a statistically significant difference in mean systolic blood pressures between men and women using a 5% level of significance.

Step 1: Set up hypotheses and determine the level of significance.

$$H_0: \mu_1 = \mu_2$$
$$H_1: \mu_1 \neq \mu_2$$
$$\alpha = 0.05.$$

Step 2: Select the appropriate test statistic.
Because both samples are large ($n_1 \geq 30$ and $n_2 \geq 30$), we use the z test statistic as opposed to t. Before implementing the

TABLE 7–12 Summary Statistics in Men and Women

	Men			Women		
	n	$\overline{X}$	s	n	$\overline{X}$	s
Systolic blood pressure	1623	128.2	17.5	1911	126.5	20.1
Diastolic blood pressure	1622	75.6	9.8	1910	72.6	9.7
Total serum cholesterol	1544	192.4	35.2	1766	207.1	36.7
Weight (lbs)	1612	194.0	33.8	1894	157.7	34.6
Height (in.)	1545	68.9	2.7	1781	63.4	2.5
Body mass index (BMI)	1545	28.8	4.6	1781	27.6	5.9

formula, we first check whether the assumption of equality of population variances is reasonable. The guideline suggests investigating the ratio of the sample variances, s_1^2/s_2^2. Suppose we call the men Group 1 and the women Group 2. Again, this is arbitrary; it needs to be noted only when interpreting the results. The ratio of the sample variances is $17.5^2/20.1^2 = 0.76$, which falls in between 0.5 and 2, suggesting that the assumption of equality of population variances is reasonable. The appropriate test statistic is

$$z = \frac{\overline{X}_1 - \overline{X}_2}{S_p\sqrt{1/n_1 + 1/n_2}}.$$

Step 3: Set up the decision rule.
This is a two-tailed test, using a z statistic and a 5% level of significance. The appropriate critical values can be found in Table 1C in the Appendix and the decision rule is

Reject H_0 if $z \leq -1.960$ or if $z \geq 1.960$.

Step 4: Compute the test statistic.
We now substitute the sample data into the formula for the test statistic identified in Step 2. Before substituting, we first compute S_p, the pooled estimate of the common standard deviation.

$$S_p = \sqrt{\frac{(n_1-1)s_1^2 + (n_2-1)s_2^2}{n_1 + n_2 - 2}},$$

$$S_p = \sqrt{\frac{(1623-1)(17.5)^2 + (1911-1)(20.1)^2}{1623 + 1911 - 2}} = \sqrt{359.12} = 19.0.$$

Notice that the pooled estimate of the common standard deviation, S_p, falls between the standard deviations in the comparison groups (i.e., 17.5 and 20.1). S_p is slightly closer in value to the standard deviation in women (20.1) as there are slightly more women in the sample. Recall that S_p is a weighted average of the standard deviations in the comparison groups, weighted by the respective sample sizes.

We now calculate the test statistic,

$$z = \frac{128.2 - 126.5}{19.0\sqrt{1/1623 + 1/1911}} = \frac{1.7}{0.64} = 2.66.$$

Step 5: Conclusion.
We reject H_0 because $2.66 > 1.960$. We have statistically significant evidence at $\alpha = 0.05$ to show that there is a difference in mean systolic blood pressures between men and women. The p-value can be found in Table 1C in the Appendix and is equal to $p < 0.010$.

In Example 7.8, we find that there is a statistically significant difference in mean systolic blood pressures between men and women at $p < 0.010$. Notice that there is a very small difference in the sample means $(128.2 - 126.5 = 1.7$ units) but this difference is beyond what would be expected by chance. The large sample sizes in this example are driving the statistical significance. In Example 6.5, we computed a 95% confidence interval for the difference in mean systolic blood pressures as 1.7 ± 1.26, or (0.44, 2.96). The confidence interval provides an assessment of the magnitude of the difference between means, whereas the test of hypothesis and p-value provides an assessment of the statistical significance of the difference. From the confidence interval in Example 6.5, we see that the difference in means is significant at the 5% level of significance because the 95% confidence interval does not include the null value of 0. The formal test is needed to compute the exact statistical significance of the difference or the p-value.

In Example 7.3, we analyzed data from a study to evaluate a new drug designed to lower total cholesterol. The study involved one sample of patients, each patient took the new drug for 6 weeks and had his or her cholesterol measured. As a means of evaluating the efficacy of the new drug, the mean total cholesterol following 6 weeks of treatment was compared to a national mean of 203 based on the NCHS. At the end of the example, we discussed the appropriateness of the historical comparator as well as an alternative study design to evaluate the effect of the new drug involving two treatment groups, where one group receives the new drug and the other does not. In Example 7.9, we revisit this example with a concurrent or parallel control group, which is very typical in randomized controlled trials or clinical trials (for more details, see Section 2.3.1).

Example 7.9. A new drug is proposed to lower total cholesterol. A randomized controlled trial is designed to evaluate the efficacy of the drug in lowering cholesterol. Thirty participants are enrolled in the trial and are randomly assigned to receive either the new drug or a placebo. The participants do not know which treatment they are assigned. Each participant is asked to take the assigned treatment for 6 weeks. At the end of 6 weeks, each patient's total cholesterol level is measured and the sample statistics are shown in **Table 7-13**. Is there statistical evidence of a reduction in mean total cholesterol in patients taking the new drug for 6 weeks as compared to participants taking a placebo? We run the test using the five-step approach and call the new drug Group 1 and the placebo Group 2.

TABLE 7–13 Cholesterol Levels by Treatment

	Sample Size	Mean	Standard Deviation
New drug	15	195.9	28.7
Placebo	15	227.4	30.3

Step 1: Set up hypotheses and determine the level of significance.

$$H_0: \mu_1 = \mu_2,$$
$$H_1: \mu_1 < \mu_2,$$
$$\alpha = 0.05.$$

Step 2: Select the appropriate test statistic.

Because both samples are small ($n_1 < 30$ and $n_2 < 30$), we use the t test statistic. Before implementing the formula, we first check whether the assumption of equality of population variances is reasonable. The ratio of the sample variances, $s_1^2 / s_2^2 = 28.7^2 / 30.3^2 = 0.90$, which falls between 0.5 and 2, suggesting that the assumption of equality of population variances is reasonable. The appropriate test statistic is

$$t = \frac{\overline{X}_1 - \overline{X}_2}{S_p \sqrt{1/n_1 + 1/n_2}}.$$

Step 3: Set up the decision rule.

This is a lower-tailed test, using a t statistic and a 5% level of significance. The appropriate critical value can be found in Table 2 in the Appendix. To determine the critical value of t, we need degrees of freedom, df, defined as $df = n_1 + n_2 - 2$. In this example, $df = 15 + 15 - 2 = 28$. The critical value for a lower-tailed test with $df = 28$ and $\alpha = 0.05$ is -1.701, and the decision rule is

$$\text{Reject } H_0 \text{ if } t \leq -1.701.$$

Step 4: Compute the test statistic.

We now substitute the sample data into the formula for the test statistic identified in Step 2. Before substituting, we first compute S_p, the pooled estimate of the common standard deviation:

$$S_p = \sqrt{\frac{(15-1)(28.7)^2 + (15-1)(30.3)^2}{15+15-2}} = \sqrt{870.89} = 29.5.$$

Now calculate the test statistic,

$$t = \frac{195.9 - 227.4}{29.5\sqrt{1/15 + 1/15}} = \frac{-31.5}{10.77} = -2.92.$$

Step 5: Conclusion.

We reject H_0 because $-2.92 < -1.701$. We have statistically significant evidence at $\alpha = 0.05$ to show that the mean total cholesterol level is lower in patients taking the new drug for 6 weeks as compared to patients taking a placebo. Using Table 2 in the Appendix, the p-value is $p < 0.005$.

The clinical trial in Example 7.9 finds a statistically significant reduction in total cholesterol, whereas in Example 7.3 we did not demonstrate the efficacy of the new drug. Notice that the mean total cholesterol level in patients taking a placebo is 227.4, which is different from the mean cholesterol of 203 reported among all Americans in 2002 and used as the comparator. The historical control value may not have been the most appropriate comparator as cholesterol levels have been increasing over time. In Section 7.6, we present another design that can be used to assess the efficacy of the new drug.

7.6 TESTS WITH MATCHED SAMPLES, CONTINUOUS OUTCOME

An alternative study design to that described in the previous section (to compare two groups with respect to their mean scores on a continuous outcome) is one based on *matched* or *paired* samples. The two comparison groups are said to be *dependent* (matched or paired) and the data can arise from a single sample of participants where each participant is measured twice, possibly before and after an intervention, or from two samples that are matched or paired on one or more specific characteristics (e.g., siblings). When the samples are dependent, we focus on difference scores in each participant or between members of a pair, and the test of hypothesis is based on the mean difference, μ_d. The null hypothesis again reflects "no difference" and is stated as $\mu_d = 0$. Note that there are some instances where it is of interest to test whether there is a difference of a particular magnitude (e.g., $\mu_d = 5$) but in most instances, the null hypothesis reflects no difference (i.e., $\mu_d = 0$). The appropriate formula for the test of hypothesis depends on the sample size. The formulas are shown in **Table 7–14** and are identical to those we presented for hypothesis testing with one sample and a continuous outcome presented in Section 7.2, except here we focus on difference scores.

TABLE 7–14 Test Statistics for Testing $H_0: \mu_d = 0$

$n \geq 30$	$z = \dfrac{\overline{X}_d - \mu_d}{s_d/\sqrt{n}}$	(Find critical value of z in Table 1C)
$n < 30$	$t = \dfrac{\overline{X}_d - \mu_d}{s_d/\sqrt{n}}$	(Find critical value of t in Table 2, $df = n - 1$)

Example 7.10. In Example 6.7, we compared systolic blood pressures measured at the sixth and seventh examinations (approximately 4 years apart) of the Framingham Offspring Study in a subsample of $n = 15$ randomly selected participants. The data are shown in Example 6.7, where we generated a 95% confidence interval for the mean difference in systolic blood pressures over a 4-year period. Using the same data, we now test whether there is a statistically significant difference in systolic blood pressures over 4 years using the five-step approach.

Step 1: Set up hypotheses and determine the level of significance.

$$H_0: \mu_d = 0,$$

$$H_1: \mu_d \neq 0,$$

$$\alpha = 0.05.$$

Step 2: Select the appropriate test statistic.
Because the sample is small ($n < 30$), we use the t test statistic,

$$t = \dfrac{\overline{X}_d - \mu_d}{s_d/\sqrt{n}}$$

Step 3: Set up the decision rule.
This is a two-tailed test, using a t statistic and a 5% level of significance. The appropriate critical value can be found in Table 2 in the Appendix with degrees of freedom, df, defined as $df = n - 1 = 15 - 1 = 14$. The critical value is 2.145 and the decision rule is

Reject H_0 if $t \leq -2.145$ or if $t \geq 2.145$.

Step 4: Compute the test statistic.
We now substitute the sample data into the formula for the test statistic identified in Step 2. In Example 6.7, we had: $n = 15$, $\overline{X}_d = -5.3$, and $s_d = 12.8$. The test statistic is

$$t = \dfrac{\overline{X}_d - \mu_d}{s_d/\sqrt{n}} = \dfrac{-5.3 - 0}{12.8/\sqrt{15}} = -1.60.$$

Step 5: Conclusion.
We do not reject H_0 because $-2.145 < -1.60 < 2.145$. We do not have statistically significant evidence at $\alpha = 0.05$ to show that there is a difference in systolic blood pressures over time.

In Example 6.7, we estimated a confidence interval and were 95% confident that the mean difference in systolic blood pressures between Examination 6 and Examination 7 (approximately 4 years apart) was between -12.4 and 1.8. Because the null value of the confidence interval for the mean difference is 0, we concluded that there was no statistically significant difference in blood pressures over time because the confidence interval for the mean difference included 0. The test of hypothesis gives the same result.

In Example 7.11, we revisit Example 7.3 and Example 7.9 where we evaluated a new drug designed to lower total cholesterol. In Example 7.3, we collected data on a sample of patients who took the new drug and we compared their mean total cholesterol level to a historical control. In Example 7.9, we evaluated the efficacy of the new drug using a clinical trial with a concurrent or parallel placebo control group. In Example 7.11, we again evaluate the efficacy of the new drug using a matched design.

Example 7.11. A new drug is proposed to lower total cholesterol and a study is designed to evaluate the efficacy of the drug in lowering cholesterol. Fifteen patients agree to participate in the study and each is asked to take the new drug for 6 weeks. However, before starting the treatment, each patient's total cholesterol level is measured. The initial measurement is a pre-treatment or baseline value. After taking the drug for 6 weeks, each patient's total cholesterol level is measured again and the data are shown in **Table 7–15**. The right column contains difference scores for each patient, computed by subtracting the 6-week cholesterol level from the baseline level. The differences represent the reduction in total cholesterol over 6 weeks. (The differences could have been computed by subtracting the baseline total cholesterol level from the level measured at 6 weeks. The way in which the differences are computed does not affect the outcome of the analysis, only the interpretation.)

TABLE 7–15 Differences in Cholesterol over 6 Weeks

Subject Identification Number	Baseline	6 Weeks	Difference
1	215	205	10
2	190	156	34
3	230	190	40
4	220	180	40
5	214	201	13
6	240	227	13
7	210	197	13
8	193	173	20
9	210	204	6
10	230	217	13
11	180	142	38
12	260	262	−2
13	210	207	3
14	190	184	6
15	200	193	7

TABLE 7–16 Summary Statistics on Difference Scores

Subject Identification Number	Difference	Difference^2
1	10	100
2	34	1156
3	40	1600
4	40	1600
5	13	169
6	13	169
7	13	169
8	20	400
9	6	36
10	13	169
11	38	1444
12	−2	4
13	3	9
14	6	36
15	7	49
Total	254	7110

Because the differences are computed by subtracting the cholesterol levels measured at 6 weeks from the baseline values, positive differences indicate reductions and negative differences indicate increases (e.g., Participant 12 increases by 2 units over 6 weeks). The goal here is to test whether there is a statistically significant reduction in cholesterol. Because of the way in which we compute the differences, we want to look for an increase in the mean difference (i.e., a positive reduction). To conduct the test, we need to summarize the differences. In this sample, we have $n = 15$, $\overline{X}_d = 16.9$, and $s_d = 14.2$, respectively. The data necessary to compute the sample statistics is shown in **Table 7–16**.

$$\overline{X}_d = \frac{\sum \text{Differences}}{n} = \frac{254}{15} = 16.9$$

and

$$s_d = \sqrt{\frac{\sum \text{Differences}^2 - (\sum \text{Differences})^2 / n}{n-1}},$$

$$s_d = \sqrt{\frac{7110 - (254)^2 / 15}{15 - 1}} = \sqrt{200.64} = 14.2$$

Is there statistical evidence of a reduction in mean total cholesterol in patients after using the new medication for 6 weeks? We run the test using the five-step approach.

Step 1: Set up hypotheses and determine the level of significance.

$$H_0: \mu_d = 0,$$
$$H_1: \mu_d > 0,$$
$$\alpha = 0.05.$$

Note that if we had computed differences by subtracting the baseline level from the level measured at 6 weeks, then negative differences would have reflected reductions and the research hypothesis would have been $H_1: \mu_d < 0$.

Step 2: Select the appropriate test statistic.

Because the sample size is small ($n < 30$), the appropriate test statistic is

$$t = \frac{\overline{X}_d - \mu_d}{s_d / \sqrt{n}}.$$

Step 3: Set up the decision rule.
This is an upper-tailed test, using a t statistic and a 5% level of significance. The appropriate critical value can be found in Table 2 in the Appendix. To determine the critical value of t we need degrees of freedom, df, defined as $df = n - 1$. In this example, $df = 15 - 1 = 14$. The critical value for an upper-tailed test with $df = 14$ and $\alpha = 0.05$ is 1.761, and the decision rule is

$$\text{Reject } H_0 \text{ if } t \geq 1.761.$$

Step 4: Compute the test statistic.
We now substitute the sample data into the formula for the test statistic identified in Step 2:

$$t = \frac{\overline{X}_d - \mu_d}{s_d/\sqrt{n}} = \frac{16.9 - 0}{14.2/\sqrt{15}} = 4.61.$$

Step 5: Conclusion.
We reject H_0 because $4.61 > 1.761$. We have statistically significant evidence at $\alpha = 0.05$ to show that there is a reduction in cholesterol levels over 6 weeks. Using Table 2 in the Appendix, the p-value is $p < 0.005$.

In Example 7.9 and Example 7.11, using parallel samples and matched designs, respectively, we find statistically significant reductions in total cholesterol. In Example 7.3, using a historical comparator, we do not. It is extremely important to design studies that are best suited to detect a meaningful difference when one exists. There are often several alternatives, and investigators work with biostatisticians to determine the best design for each application. It is worth noting that the matched design used in Example 7.11 can be problematic in that observed differences may only reflect a "placebo" effect. All participants took the assigned medication but is the observed reduction attributable to the medication or a result of participation in a study?

7.7 TESTS WITH TWO INDEPENDENT SAMPLES, DICHOTOMOUS OUTCOME

Techniques presented in Section 7.5 are appropriate when there are two independent comparison groups and the outcome of interest is continuous (e.g., blood pressure, total cholesterol, weight loss). Here we consider the situation where there are two independent comparison groups and the outcome of interest is dichotomous (e.g., success/failure). The goal of the analysis is to compare proportions of successes between the two groups. The relevant sample data are the sample sizes in each comparison group (n_1 and n_2) and the sample proportions ($\hat{p}_1$ and $\hat{p}_2$), which are computed by taking the ratios of the numbers of successes to the sample sizes in each group, e.g., $\hat{p}_1 = x_1/n_1$ and $\hat{p}_2 = x_2/n_2$.

In Chapter 3, we introduced the risk difference, relative risk, and odds ratio, which are different measures to compare proportions in two independent groups. We developed confidence intervals for each of these measures in Chapter 6 in Section 6.6.1, Section 6.6.2, and Section 6.6.3, respectively. The confidence interval formulas were different and depended on the specific measure.

In tests of hypothesis comparing proportions between two independent groups, one test is performed and it can be interpreted relative to a risk difference, relative risk, or odds ratio. As a reminder, the risk difference is computed by taking the difference in proportions between comparison groups, the relative risk is computed by taking the ratio of proportions, and the odds ratio is computed by taking the ratio of the odds of success in the comparison groups (see Section 3.4.2 for more details). Because the null values for the risk difference, the relative risk, and the odds ratio are different, the hypotheses in tests of hypothesis look slightly different depending on which measure is used. When performing tests of hypothesis for the risk difference, relative risk, or odds ratio, the convention is to label the exposed or treated Group 1 and the unexposed or control Group 2.

For example, suppose a study is designed to assess whether there is a significant difference in proportions in two independent comparison groups. The test of interest is

$$H_0: p_1 = p_2 \text{ versus } H_1: p_1 \neq p_2.$$

The following are the hypotheses for testing for a difference in proportions using the risk difference (RD), the relative risk (RR), and the odds ratio (OR) as measures of effect. First, the previous hypotheses are equivalent to

$$H_0: p_1 - p_2 = 0 \text{ versus } H_1: p_1 - p_2 \neq 0,$$

which are, by definition, equal to

$$H_0: RD = 0 \text{ versus } H_1: RD \neq 0.$$

If an investigator wants to focus on the relative risk, the equivalent hypotheses are

$$H_0: RR = 1 \text{ versus } H_1: RR \neq 1,$$

and if the investigator wants to focus on the odds ratio, the equivalent hypotheses are

$$H_0: \text{OR} = 1 \text{ versus } H_1: \text{OR} \neq 1.$$

Suppose a test is performed to test

$$H_0: \text{RD} = 0 \text{ versus } H_1: \text{RD} \neq 0,$$

and the test rejects H_0 at $\alpha = 0.05$. Based on this test, we can conclude that there is significant evidence ($\alpha = 0.05$) of a difference in proportions, significant evidence that the risk difference is not 0, and significant evidence that the relative risk and odds ratio are not 1.

The risk difference is analogous to the difference in means when the outcome is continuous. The test for the difference in means was described in Section 7.5. Here the parameter of interest is the difference in proportions in the population, $\text{RD} = p_1 - p_2$, and the null value for the risk difference is 0. In a test of hypothesis for the risk difference, the null hypothesis is always

$$H_0: \text{RD} = 0.$$

This is equivalent to

$$H_0: \text{RR} = 1 \text{ and } H_0: \text{OR} = 1.$$

In the research hypothesis, the investigator can hypothesize that the first proportion is larger than the second,

$$H_1: p_1 > p_2, \text{ which is equivalent to } H_1: \text{RD} > 0,$$
$$H_1: \text{RR} > 1, \text{ and } H_1: \text{OR} > 1,$$

that the first proportion is smaller than the second,

$$H_1: p_1 < p_2, \text{ which is equivalent to } H_1: \text{RD} < 0,$$
$$H_1: \text{RR} < 1, \text{ and } H_1: \text{OR} < 1,$$

or that the proportions are different,

$$H_1: p_1 \neq p_2, \text{ which is equivalent to } H_1: \text{RD} \neq 0, H_1: \text{RR} \neq 1,$$
$$\text{and } H_1: \text{OR} \neq 1.$$

The three different alternatives represent upper-, lower-, and two-tailed tests, respectively.

The formula for the test statistic for the difference in proportions is given in **Table 7–17**. Note that $\hat{p}_1$ and $\hat{p}_2$ are the proportions of successes in Groups 1 and 2, respectively. $\hat{p}$ is the overall proportion of successes, which is computed as $\hat{p} = \dfrac{x_1 + x_2}{n_1 + n_2}$. The preceding formula is appropriate for large samples, defined as at least five successes ($n\hat{p}$) and at least five failures [$n(1 - \hat{p})$] in each of the two samples. If there are fewer than five successes or failures in either comparison group, then alternative procedures called exact methods must be used to test whether there is a difference in population proportions.[3,4]

Example 7.12. In Example 6.9, we analyzed data from $n = 3799$ participants who attended the fifth examination of the offspring in the Framingham Heart Study. The outcome of interest was prevalent CVD and we compared prevalent CVD between participants who were and were not currently smoking cigarettes at the time of the fifth examination of the Framingham Offspring Study. The data are shown in **Table 7–18**. We now use the data to test if the prevalence of CVD is significantly different in smokers as compared to nonsmokers.

The prevalence of CVD (or proportion of participants with prevalent CVD) among nonsmokers is $298 / 3055 = 0.0975$, and the prevalence of CVD among current smokers is $81 / 744 = 0.1089$. Here smoking status defines the comparison groups, and we call the current smokers Group 1 (exposed) and the nonsmokers (unexposed) Group 2. The test of hypothesis is conducted on the following page using the five-step approach.

TABLE 7–17 Test Statistic for Testing $H_0: p_1 = p_2$

$\min[n_1 \hat{p}_1, n_1(1 - \hat{p}_1), n_2 \hat{p}_2, n_2(1 - \hat{p}_2)] \geq 5$

$$z = \dfrac{\hat{p}_1 - \hat{p}_2}{\sqrt{\hat{p}(1 - \hat{p})\left(\dfrac{1}{n_1} + \dfrac{1}{n_2}\right)}}$$

(Find critical value of z in Table 1C)

TABLE 7-18 Prevalent CVD in Smokers and Nonsmokers

	Free of CVD	History of CVD	Total
Nonsmoker	2757	298	3055
Current smoker	663	81	744
Total	3420	379	3799

Step 1: Set up hypotheses and determine the level of significance.

$$H_0: p_1 = p_2,$$
$$H_1: p_1 \neq p_2,$$
$$\alpha = 0.05.$$

Step 2: Select the appropriate test statistic.

The formula for the test statistic is in Table 7–17. We must first check that the sample size is adequate. Specifically, we need to ensure that we have at least five successes and five failures in each comparison group. In this example, we have more than enough successes (cases of prevalent CVD) and failures (persons free of CVD) in each comparison group. The sample size is more than adequate, so the following formula can be used:

$$z = \frac{\hat{p}_1 + \hat{p}_2}{\sqrt{\hat{p}(1-\hat{p})\left(\frac{1}{n_1} + \frac{1}{n_2}\right)}}.$$

Step 3: Set up the decision rule.

This is a two-tailed test, using a z statistic and a 5% level of significance. The appropriate critical value can be found in Table 1C in the Appendix and the decision rule is

Reject H_0 if $z \leq -1.960$ or if $z \geq 1.960$.

Step 4: Compute the test statistic.

We now substitute the sample data into the formula for the test statistic identified in Step 2. We first compute the overall proportion of successes:

$$\hat{p} = \frac{x_1 + x_2}{n_1 + n_2} = \frac{81 + 298}{744 + 3055} = \frac{379}{3799} = 0.0998.$$

We now substitute to compute the test statistic,

$$z = \frac{0.1089 - 0.0975}{\sqrt{0.0988(1 - 0.0988)(1/744 + 1/3055)}} =$$

$$\frac{0.0114}{0.0123} = 0.927.$$

Step 5: Conclusion.

We do not reject H_0 because $-1.960 < 0.927 < 1.960$. We do not have statistically significant evidence at $\alpha = 0.05$ to show that there is a difference in prevalent CVD between smokers and nonsmokers.

In Example 6.9, we estimated the 95% confidence interval for the difference in prevalent CVD (or risk difference) between smokers and nonsmokers as 0.0114 ± 0.0247, or between -0.0133 and 0.0361. Because the 95% confidence interval for the risk difference included 0, we could not conclude that there was a statistically significant difference in prevalent CVD between smokers and nonsmokers. This is consistent with the test of hypothesis result.

In Example 6.11, we used the same data and estimated the relative risk along with a 95% confidence interval for the relative risk. We estimated that among smokers the risk of CVD was 1.12 times that among nonsmokers with a 95% confidence interval of 0.89 to 1.41. Because this confidence interval included the null value of 1, we could not conclude that there was a difference in prevalent CVD between smokers and nonsmokers.

In Example 6.13, we again used the same data and estimated the odds ratio along with a 95% confidence interval for the odds ratio. We estimated that among smokers the odds of CVD was 1.13 times that among nonsmokers with a 95% confidence interval estimate of 0.87 to 1.47. Because this confidence interval included the null value of 1, we could not conclude that there was a difference in prevalent CVD between smokers and nonsmokers.

Smoking has been shown to be a risk factor for cardiovascular disease. What might explain the fact that we did not observe a statistically significant difference using data from the Framingham Heart Study? (Hint: Here we consider prevalent CVD. Would the results have been different if we considered incident CVD?)

Example 7.13. In Example 6.10, we analyzed data from a randomized trial designed to evaluate the effectiveness of a newly developed pain reliever designed to reduce pain in patients following joint replacement surgery. The trial compared the new pain reliever to the pain reliever currently in use (called the standard of care). A total of 100 patients undergoing joint replacement surgery agreed to participate in the trial. Patients were randomly assigned to receive either the new pain reliever or the standard pain reliever following surgery and were blind to the treatment assignment. Before receiving the assigned treatment, patients were asked to rate their pain on a scale of 0 to 10, with higher scores indicative of more pain. Each patient was then given the assigned treatment and after 30 minutes was again asked to rate his or her pain on the same scale. The primary outcome was a reduction in pain of 3 or more scale points (defined by clinicians as a clinically meaningful reduction). The data shown in **Table 7–19** were observed in the trial. We now test whether there is a statistically significant difference in the proportions of patients reporting a meaningful reduction (i.e., a reduction of 3 or more scale points) using the five-step approach.

Step 1: Set up hypotheses and determine the level of significance.

$$H_0: p_1 = p_2,$$
$$H_1: p_1 \neq p_2,$$
$$\alpha = 0.05.$$

Here the new pain reliever is represented as Group 1 and the standard pain reliever as Group 2.

Step 2: Select the appropriate test statistic.

The formula for the test statistic is in Table 7–17. We must first check that the sample size is adequate. Specifically, we need to ensure that we have at least five successes and five failures in each comparison group or that

$$\min[n_1\hat{p}_1, n_1(1-\hat{p}_1), n_2\hat{p}_2, n_2(1-\hat{p}_2)] \geq 5.$$

In this example, we have

$$\min[50(0.46), 50(1-0.46), 50(0.22), 50(1-0.22)] =$$
$$\min(23, 27, 11, 39) = 11.$$

TABLE 7–19 Pain Reduction by Treatment

Treatment Group	n	Reduction of 3+ Points Number	Proportion
New pain reliever	50	23	0.46
Standard pain reliever	50	11	0.22

The sample size is adequate, so the following formula can be used:

$$z = \frac{\hat{p}_1 + \hat{p}_2}{\sqrt{\hat{p}(1-\hat{p})\left(\frac{1}{n_1} + \frac{1}{n_2}\right)}}.$$

Step 3: Set up the decision rule.

This is a two-tailed test, using a z statistic and a 5% level of significance. The appropriate critical value can be found in Table 1C in the Appendix and the decision rule is

Reject H_0 if $z \leq -1.960$ or if $z \geq 1.960$.

Step 4: Compute the test statistic.

We now substitute the sample data into the formula for the test statistic identified in Step 2. We first compute the overall proportion of successes:

$$\hat{p} = \frac{x_1 + x_2}{n_1 + n_2} = \frac{23 + 11}{50 + 50} = \frac{34}{100} = 0.34.$$

We now substitute to compute the test statistic,

$$z = \frac{0.46 - 0.22}{\sqrt{0.34(1-0.34)\left(\frac{1}{50} + \frac{1}{50}\right)}} = \frac{0.24}{0.095} = 2.526.$$

Step 5: Conclusion.

We reject H_0 because 2.526 > 1960. We have statistically significant evidence at $\alpha = 0.05$ to show that there is a difference in the proportions of patients on the new pain reliever reporting a meaningful reduction (i.e., a reduction of 3 or more scale points) as compared to patients on the standard pain reliever.

In Example 6.10, we estimated a 95% confidence interval for the difference in proportions of patients on the new pain reliever reporting a meaningful reduction (i.e., a reduction of 3 or more scale points) as compared to patients on the standard pain reliever, and it was 0.24 ± 0.18, or between 0.06 and 0.42. Because the 95% confidence interval did not include 0, we concluded that there was a statistically significant difference in proportions, which is consistent with the test of hypothesis result.

In Example 6.12, we used the same data and estimated the relative risk along with a 95% confidence interval for the relative risk. We estimated that patients receiving the new pain reliever had 2.09 times the likelihood of reporting a meaningful reduction in pain as compared to patients receiving the standard pain reliever, with a 95% confidence interval of 1.14 to 3.82. Because this confidence interval did not include the null value of 1, we concluded that there was a significant difference in the relative risk of a meaningful reduction in pain in patients on the new and standard pain relievers.

In Example 6.14, we used the same data and estimated the odds ratio along with a 95% confidence interval for the odds ratio. We estimated that the odds that patients receiving the new pain reliever reported a meaningful reduction in pain was 3.02 times the odds that patients on standard pain reliever reported a meaningful reduction in pain, with a 95% confidence interval of 1.26 to 7.21. Because this confidence interval did not include the null value of 1, we concluded that there was a significant difference in the odds of a meaningful reduction in pain in patients on the new and standard pain relievers.

The procedures discussed here apply to applications where there are two independent comparison groups and a dichotomous outcome. There are other applications in which it is of interest to compare a dichotomous outcome in matched or paired samples. For example, in a clinical trial we might wish to test the effectiveness of a new antibiotic eyedrop for the treatment of bacterial conjunctivitis. Participants use the new antibiotic eyedrop in one eye and a comparator (placebo or active control treatment) in the other. The success of the treatment (yes/no) is recorded for each participant for each eye. Because the two assessments (success or failure) are paired, we cannot use the procedures discussed here. The appropriate test is called *McNemar's test* (sometimes called McNemar's test for dependent proportions). For more details, see Agresti[3] or Rosner.[6]

7.8 TESTS WITH MORE THAN TWO INDEPENDENT SAMPLES, CONTINUOUS OUTCOME

There are many applications where it is of interest to compare more than two independent groups with respect to their mean scores on a continuous outcome. For example, in some clinical trials there are more than two comparison groups. Suppose a clinical trial is designed to evaluate a new medication for asthma and investigators compare an experimental medication to a placebo and to a standard treatment (i.e., a medication currently being used). In an observational study such as the Framingham Heart Study, it might be of interest to compare mean blood pressures or mean cholesterol levels in persons who are underweight, normal weight, overweight, and obese.

The technique to test for a difference in more than two independent means is an extension of the two independent samples procedure discussed in Section 7.5 and is called **analysis of variance (ANOVA)**. The ANOVA technique applies when there are more than two independent comparison groups. The ANOVA procedure is used to compare the means of the comparison groups and is conducted using the same five-step approach used in the scenarios discussed in previous sections. However, because there are more than two groups, the computation of the test statistic is more involved. The test statistic must take into account the sample sizes, sample means, and sample standard deviations in each of the comparison groups. Before illustrating the computation of the test statistic, we first present the logic of the procedure.

Consider an example with four independent groups and a continuous outcome measure. The independent groups might be defined by a particular characteristic of the participants, such as BMI (e.g., underweight, normal weight, overweight, or obese), or by the investigator (e.g., by randomizing participants to one of four competing treatments, call them A, B, C, and D). Suppose that the outcome is systolic blood pressure and we wish to test whether there is a statistically significant difference in mean systolic blood pressures among the four groups. The sample data are summarized as shown in **Table 7–20**.

The hypotheses of interest in an ANOVA are

$$H_0: \mu_1 = \mu_2 = \cdots = \mu_k,$$

H_1: Means are not all equal,

where k is the number of independent comparison groups. For the example just described, the hypotheses are

$$H_0: \mu_1 = \mu_2 = \mu_3 = \mu_4,$$

H_1: Means are not all equal.

The null hypothesis in ANOVA is always that there is no difference in the means. The research or alternative hypothesis is always that the means are not all equal and is usually written in words rather than in mathematical symbols. The research hypothesis captures any difference in means and includes the situation where all four means are unequal, where one is different from the other three, where two are different, and so

TABLE 7–20 Summary Statistics for ANOVA

	Group 1	Group 2	Group 3	Group 4
Sample size	n_1	n_2	n_3	n_4
Sample mean	$\overline{X}_1$	$\overline{X}_2$	$\overline{X}_3$	$\overline{X}_4$
Sample standard deviation	s_1	s_2	s_3	s_4

on. The alternative hypothesis (as shown earlier) captures all possible situations other than equality of all means specified in the null hypothesis.

The test statistic for ANOVA is given in **Table 7–21**. The F statistic is computed by taking the ratio of what is called the "between-treatment" variability to the "residual or error" variability. This is where the name of the procedure originates. In analysis of variance, we are testing for a difference in means (H_0: means are all equal versus H_1: means are not all equal) by evaluating variability in the data. The numerator captures between-treatment variability (i.e., differences among the sample means), and the denominator contains an estimate of the variability in the outcome. The test statistic is a measure that allows us to assess whether the differences among the sample means (numerator) are larger than would be expected by chance if the null hypothesis is true. Recall in the two-independent samples test (see test statistics in Table 7–11), the test statistics (z or t) were computed by taking the ratio of the difference in sample means (numerator) to the variability in the outcome (estimated by $S_p\sqrt{\frac{1}{n_1} + \frac{1}{n_2}}$).

The decision rule for the F test in ANOVA is set up in a similar way to decision rules we established for z and t tests. The decision rule again depends on the level of significance and the degrees of freedom. The F statistic has two degrees of freedom. These are denoted df_1 and df_2, and called the numerator and denominator degrees of freedom, respectively. The degrees of freedom are defined as $df_1 = k - 1$ and $df_2 = N - k$, where k is the number of comparison groups and N is the total number of observations in the analysis. Similar to χ^2 tests, there is no upper- or lower-tailed version of the test. If the null hypothesis is true, the between-treatment variation (numerator) is close in value to the residual or error variation (denominator) and the F statistic is small. If the null hypothesis is false, then the F statistic is large. The rejection region for the F test is always in the upper tail (right) of the distribution, as shown in **Figure 7–7**.

Table 4 in the Appendix contains critical values for the F distribution for tests when $\alpha = 0.05$, indexed by df_1 and df_2. Figure 7–7 is an example for the situation with $\alpha = 0.05$, $df_1 = 3$, and $df_2 = 36$. The degrees of freedom are based on an application with four comparison groups (k = 4) and a sample size of 40 (N = 40). The appropriate critical value from Table 4 in the Appendix is 2.87, and the decision rule is

$$\text{Reject } H_0 \text{ if } F \geq 2.87.$$

We next illustrate the ANOVA procedure using the five-step approach. Because the computation of the test statistic is involved, the computations are often organized in an ANOVA table. The ANOVA table breaks down the components of variation in the data into variation between treatments and error or residual variation. The ANOVA table is set up as shown in **Table 7–22**.

The ANOVA table is organized as follows. The first column is entitled *Source of Variation* and delineates the between-treatment and error or residual variation. The total variation is the sum of the between-treatment and error variation. The second column is entitled *Sums of Squares* (SS). The between-treatment sums of squares is

$$SSB = \sum n_j(\overline{X}_j - \overline{X})^2,$$

TABLE 7–21 Test Statistic for Testing $H_0: \mu_1 = \mu_2 = \ldots = \mu_k$

$$F = \frac{\sum n_j(\overline{X}_j - \overline{X})^2 / (k-1)}{\sum\sum(X - \overline{X}_j)^2 / (N-k)}$$

(Find critical value in Table 4, $df_1 = k - 1$, $df_2 = N - k$)

where n_j = the sample size in the jth group (e.g., j = 1, 2, 3, and 4 when there are 4 comparison groups), $\overline{X}_j$ is the sample mean in the jth group, and $\overline{X}$ is the overall mean. k represents the number of independent groups (k > 2), and N represents the total number of observations in the analysis.

Note that N does not refer to a population size, but instead to the total sample size in the analysis (the sum of the sample sizes in the comparison groups, e.g., $N = n_1 + n_2 + n_3 + n_4$). The test statistic is complicated because it incorporates all of the sample data. While it is not easy to see the extension, the F statistic is a generalization of the test statistic shown in Table 7–11, which is appropriate for exactly two groups.

NOTE: Appropriate use of the F statistic is based on several assumptions. First, the outcome is assumed to follow a normal distribution in each of the comparison groups. Second, the data are assumed to be equally variable in each of the k populations (i.e., the population variances are equal, or $\sigma_1^2 = \sigma_2^2 = \ldots = \sigma_k^2$). This means that the outcome is equally variable in each of the comparison populations. This assumption is the same as that assumed for appropriate use of the test statistics in Table 7–11 when there were two comparison groups. It is possible to assess the likelihood that the assumption of equal variances is true and the test can be conducted in most statistical computing packages.[7] If the variability in the k comparison groups is not similar, then alternative techniques must be used.[8]

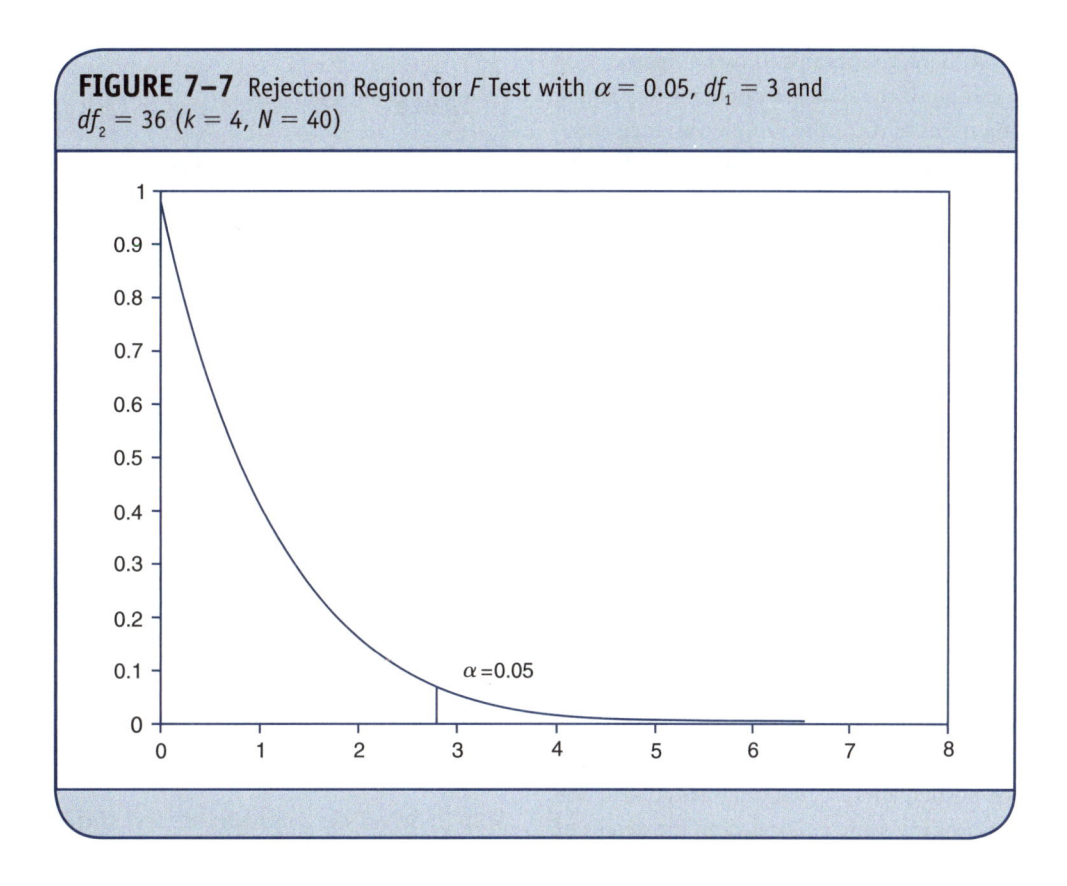

FIGURE 7–7 Rejection Region for F Test with $\alpha = 0.05$, $df_1 = 3$ and $df_2 = 36$ ($k = 4$, $N = 40$)

(Figure: F-distribution curve with $\alpha = 0.05$ rejection region marked near 2.8)

and is computed by summing the squared differences between each treatment (or group) mean ($\bar{X}_j$) and the overall mean ($\bar{X}$). The squared differences are weighted by the sample sizes per group (n_j). The error sums of squares is

$$SSE = \sum\sum(X - \bar{X}_j)^2,$$

and is computed by summing the squared differences between each observation (X) and its group mean ($\bar{X}_j$) (i.e., the squared differences between each observation in Group 1 and the Group 1 mean, the squared differences between each observation in Group 2 and the Group 2 mean, and so on). The double summation ($\sum\sum$) indicates summation of the squared differences within each treatment and then summation of these totals across treatments to produce a single value. (This is illustrated in the following examples.) The total sums of squares is

$$SST = \sum\sum(X - \bar{X})^2,$$

and is computed by summing the squared differences between each observation (X) and the overall sample mean

TABLE 7–22 ANOVA Table

Source of Variation	Sums of Squares (SS)	Degrees of Freedom (df)	Mean Squares (MS)	F
Between treatments	$SSB = \sum n_j(\bar{X}_j - \bar{X})^2$	$k - 1$	$MSB = \dfrac{SSB}{k-1}$	$F = \dfrac{MSB}{MSE}$
Error or residual	$SSE = \sum\sum(X - \bar{X}_j)^2$	$N - k$	$MSE = \dfrac{SSB}{N-k}$	
Total	$SST = \sum\sum(X - \bar{X})^2$	$N - 1$		

($\overline{X}$). In an ANOVA, data are organized by comparison or treatment groups. If all of the data are pooled into a single sample, SST is the numerator of the sample variance computed on the pooled or total sample. SST does not figure into the F statistic directly. However, SST = SSB + SSE; thus if two sums of squares are known, the third can be computed from the other two.

The third column contains degrees of freedom. The between-treatment degrees of freedom is $df_1 = k - 1$. The error degrees of freedom is $df_2 = N - k$. The total degrees of freedom is $N - 1$, and it is also true that $(k - 1) + (N - k) = N - 1$. The fourth column contains *Mean Squares* (MS), which are computed by dividing sums of squares (SS) by degrees of freedom (df), row by row. Specifically,

$$\text{MSB} = \frac{\text{SSB}}{k-1} \text{ and } \text{MSE} = \frac{\text{SSE}}{N-k}.$$

Dividing SST / ($N - 1$) produces the variance of the total sample. The F statistic is in the right column of the ANOVA table and is computed by taking the ratio of MSB / MSE.

Example 7.14. A clinical trial is run to compare weight-loss programs, and participants are randomly assigned to one of the comparison programs and counseled on the details of the assigned program. Participants follow the assigned program for 8 weeks. The outcome of interest is weight loss, defined as the difference in weight measured at the start of the study (baseline) and weight measured at the end of the study (8 weeks), in pounds.

Three popular weight-loss programs are considered. The first is a low-calorie diet. The second is a low-fat diet and the third is a low-carbohydrate diet. For comparison purposes, a fourth group is considered as a control group. Participants in the fourth group are told that they are participating in a study of healthy behaviors with weight loss being only one component of interest. The control group is included here to assess the placebo effect (i.e., weight loss due to simply participating in the study). A total of 20 patients agree to participate in the study and are randomly assigned to one of the four diet groups. Weights are measured at baseline and patients are counseled on the proper implementation of the assigned diet (with the exception of the control group). After 8 weeks, each participant's weight is again measured and the difference in weights is computed by subtracting the 8-week weight from the baseline weight. Positive differences indicate weight loss and negative differences indicate weight gain. For interpretation purposes, we refer to the difference in weights as weight loss, and the observed weight losses are shown in **Table 7–23**. Is there a statistically

TABLE 7–23 Weight Loss in Each Treatment

Low-Calorie	Low-Fat	Low-Carbohydrate	Control
8	2	3	2
9	4	5	2
6	3	4	−1
7	5	2	0
3	1	3	3

significant difference in the mean weight loss among the four diets? We run the ANOVA using the five-step approach.

Step 1: Set up hypotheses and determine the level of significance.

$$H_0: \mu_1 = \mu_2 = \mu_3 = \mu_4,$$

$$H_1: \text{Means are not all equal},$$

$$\alpha = 0.05.$$

Step 2: Select the appropriate test statistic.
The test statistic is the F statistic for ANOVA,

$$F = \frac{\text{MSB}}{\text{MSE}}.$$

Step 3: Set up the decision rule.
The appropriate critical value can be found in Table 4 in the Appendix. To determine the critical value of F, we need **degrees of freedom**, $df_1 = k - 1$ and $df_2 = N - k$. In this example, $df_1 = k - 1 = 4 - 1 = 3$ and $df_2 = N - k = 20 - 4 = 16$. The critical value is 3.24, and the decision rule is

$$\text{Reject } H_0 \text{ if } F \geq 3.24.$$

Step 4: Compute the test statistic.
To organize our computations, we complete the ANOVA table. To compute the sums of squares, we first compute the sample means for each group (see **Table 7–24**). The overall mean based on the total sample ($n = 20$) is $\overline{X} = 3.6$. We can now compute SSB:

$$\text{SSB} = \sum n_j(\overline{X}_j - \overline{X})^2,$$

$$\text{SSB} = 5(6.6 - 3.6)^2 + 5(3.0 - 3.6)^2 +$$

$$5(3.4 - 3.6)^2 + 5(1.2 - 3.6)^2,$$

$$\text{SSB} = 45.0 + 1.8 + 0.2 + 28.8 = 75.8.$$

Next, we compute SSE. SSE requires computing the squared differences between each observation and its group mean. We compute SSE in parts (see **Tables 7-25, 7-26, 7-27, 7-28**), and then sum.

$$SSE = \sum\sum(X - \overline{X}_j)^2 = 21.4 + 10.0 + 5.4 + 10.6 = 47.4.$$

TABLE 7-24 Summary Statistics on Weight Loss by Treatment

Low-Calorie	Low-Fat	Low-Carbohydrate	Control
$n_1 = 5$	$n_2 = 5$	$n_3 = 5$	$n_4 = 5$
$\overline{X}_1 = 6.6$	$\overline{X}_2 = 3.0$	$\overline{X}_3 = 3.4$	$\overline{X}_4 = 1.2$

TABLE 7-25 Deviations from Mean Weight Loss on Low-Calorie Diet

Low-Calorie	$(X - 6.6)$	$(X - 6.6)^2$
8	1.4	2.0
9	2.4	5.8
6	−0.6	0.4
7	0.4	0.2
3	−3.6	13.0
Total	0	21.4

TABLE 7-26 Deviations from Mean Weight Loss on Low-Fat Diet

Low-Fat	$(X - 3.0)$	$(X - 3.0)^2$
2	−1.0	1.0
4	1.0	1.0
3	0.0	0.0
5	2.0	4.0
1	−2.0	4.0
Total	0	10.0

TABLE 7-27 Deviations from Mean Weight Loss on Low-Carbohydrate Diet

Low-Carbohydrate	$(X - 3.4)$	$(X - 3.4)^2$
3	−0.4	0.2
5	1.6	2.6
4	0.6	0.4
2	−1.4	2.0
3	−0.4	0.2
Total	0	5.4

TABLE 7-28 Deviations from Mean Weight Loss in Control Group

Control	$(X - 1.2)$	$(X - 1.2)^2$
2	0.8	0.6
2	0.8	0.6
−1	−2.2	4.8
0	−1.2	1.4
3	1.8	3.2
Total	0	10.6

We can now construct the ANOVA table (see **Table 7-29**).

Step 5: Conclusion.

We reject H_0 because 8.43 > 3.24. We have statistically significant evidence at $\alpha = 0.05$ to show that there is a difference in mean weight loss among the four diets.

Note that Table 4 in the Appendix provides critical values for the F test only when $\alpha = 0.05$. Without tables for other levels of significance, we cannot approximate p-values for ANOVA by hand. Because ANOVA is a tedious technique to implement by hand, it is usually performed with the use of a statistical computing package that produces, along with the ANOVA table, an exact p-value.

ANOVA is a test that provides a global assessment of the statistical significance in more than two independent means. In this example, we find that there is a statistically significant difference in mean weight loss among the four diets considered. In addition to reporting the results of the statistical test of hypothesis (i.e., that there is a statistically

TABLE 7-29 ANOVA Table

Source of Variation	Sums of Squares (SS)	Degrees of Freedom (df)	Mean Squares (MS)	F
Between treatments	75.8	4 − 1 = 3	75.8 / 3 = 25.3	25.3 / 3.0 = 8.43
Error or residual	47.4	20 − 4 = 16	47.4 / 16 = 3.0	
Total	123.2	20 − 1 = 19		

significant difference in mean weight loss at $\alpha = 0.05$), investigators should also report the observed sample means to facilitate interpretation of the results. In this example, participants in the low-calorie diet lost a mean of 6.6 pounds over 8 weeks, as compared to 3 and 3.4 pounds in the low-fat and low-carbohydrate groups, respectively. Participants in the control group lost a mean of 1.2 pounds, which could be called a placebo effect because these participants were not participating in an active arm of the trial specifically targeted for weight loss.

Example 7.15. Calcium is an essential mineral that regulates the heart and is important for blood clotting and for building healthy bones. The National Osteoporosis Foundation recommends a daily calcium intake of 1000 to 1200 mg/day for adult men and women.[9] Whereas calcium is contained in some foods, most adults do not get enough calcium in their diets and take supplements. Unfortunately, some of the supplements have side effects such as gastric distress, making them difficult for some patients to take on a regular basis.

A study is designed to test whether there is a difference in mean daily calcium intake in adults with normal bone density, adults with osteopenia (a low bone density, which may lead to osteoporosis), and adults with osteoporosis. Adults 60 years of age with normal bone density, osteopenia, and osteoporosis are selected at random from hospital records and invited to participate in the study. Each participant's daily calcium intake is measured based on reported food intake and supplements. The data are shown in **Table 7–30**.

Is there a statistically significant difference in mean calcium intake in patients with normal bone density as compared to patients with osteopenia and osteoporosis? We run the ANOVA using the five-step approach.

Step 1: Set up hypotheses and determine the level of significance.

$$H_0: \mu_1 = \mu_2 = \mu_3,$$

TABLE 7-30 Calcium Intake in Each Treatment

Normal Bone Density	Osteopenia	Osteoporosis
1200	1000	890
1000	1100	650
980	700	1100
900	800	900
750	500	400
800	700	350

H_1: Means are not all equal,

$\alpha = 0.05$.

Step 2: Select the appropriate test statistic.
The test statistic is the F statistic for ANOVA,

$$F = \frac{\text{MSB}}{\text{MSE}}.$$

Step 3: Set up the decision rule.
The appropriate critical value can be found in Table 4 in the Appendix. To determine the critical value of F, we need degrees of freedom, $df_1 = k - 1$ and $df_2 = N - k$. In this example, $df_1 = k - 1 = 3 - 1 = 2$ and $df_2 = N - k = 18 - 3 = 15$. The critical value is 3.68, and the decision rule is

Reject H_0 if $F \geq 3.68$.

Step 4: Compute the test statistic.
To organize our computations, we complete the ANOVA table. To compute the sums of squares, we must first compute the sample means for each group (see **Table 7–31**). The overall mean is $\overline{X} = 817.8$. We can now compute SSB:

TABLE 7–31 Summary Statistics on Calcium Intake by Treatment

Normal Bone Density	Osteopenia	Osteoporosis
$n_1 = 6$	$n_2 = 6$	$n_3 = 6$
$\bar{X}_1 = 938.3$	$\bar{X}_2 = 800.0$	$\bar{X}_3 = 715.0$

$$SSB = \sum n_j(\bar{X}_j - \bar{X})^2,$$

$$SSB = 6(938.3 - 817.8)^2 + 6(715.0 - 817.8)^2 + 6(800.0 - 817.8)^2,$$

$$SSB = 87{,}121.5 + 63{,}407.0 + 1{,}901.0 = 152{,}429.5.$$

Next, we calculate SSE. SSE requires computing the squared differences between each observation and its group mean. We compute SSE in parts (see **Tables 7–32, 7–33,** and **7–34**), and then sum.

$$SSE = \sum\sum(X - \bar{X}_j)^2 = 130{,}083.4 + 240{,}000.0 + 449{,}750.0 = 819{,}833.4.$$

TABLE 7–32 Deviations from Mean Calcium Intake in Patients with Normal Bone Density

Normal Bone Density	$(X - 938.3)$	$(X - 938.3)^2$
1200	261.7	68,486.9
1000	61.7	3806.9
980	41.7	1738.9
900	−38.3	1466.9
750	−188.3	35,456.9
800	−138.3	19,126.9
Total	0	130,083.4

TABLE 7–33 Deviations from Mean Calcium Intake in Patients with Osteopenia

Osteopenia	$(X - 800.0)$	$(X - 800.0)^2$
1000	200.0	40,000.0
1100	300.0	90,000.0
700	−100.0	10,000.0
800	0.0	0.0
500	−300.0	90,000.0
700	200.0	10,000.0
Total	0	240,000.0

TABLE 7–34 Deviations from Mean Calcium Intake in Patients with Osteoporosis

Osteoporosis	$(X - 715.0)$	$(X - 715.0)^2$
890	175.0	30,625.0
650	−65.0	4,225.0
1100	385.0	148,225.0
900	185.0	34,225.0
400	−315.0	99,225.0
350	−365.0	133,225.0
Total	0	449,750.0

We can now construct the ANOVA table (see **Table 7–35**).

Step 5: Conclusion.
We do not reject H_0 because 1.39 < 3.68. We do not have statistically significant evidence at $\alpha = 0.05$ to show that there is a difference in mean calcium intake in patients with normal bone density as compared to osteopenia and osteoporosis.

TABLE 7–35 ANOVA Table

Source of Variation	Sums of Squares (SS)	Degrees of Freedom (df)	Mean Squares (MS)	F
Between treatments	152,429.5	2	76,214.8	1.39
Error or residual	819,833.4	15	54,655.6	
Total	972,262.9	17		

When sample sizes in each comparison group are equal, the design is called a **balanced design**. Balanced designs are preferred over unbalanced designs (unequal numbers of participants in the comparison groups) because they are more robust (e.g., they ensure a 5% Type I error rate when $\alpha = 0.05$) when the assumptions, such as normality of the outcome, are violated.

The ANOVA tests described here are called one-way or one-factor ANOVAs. There is one treatment or grouping factor with $k > 2$ levels, and we wish to compare the means of a continuous outcome across the different categories of this factor (also called different treatments). The factor might represent different diets, different classifications of risk for disease (e.g., osteoporosis), different medical treatments, different age groups, or different racial or ethnic groups. There are situations where it may be of interest to compare means of a continuous outcome across two or more factors. For example, suppose a clinical trial is designed to compare three different treatments for joint pain in patients with osteoarthritis. Investigators might also hypothesize that there are differences in the outcome by sex. This is an example of a **two-factor ANOVA** where the factors are treatment (with three levels) and sex (with two levels). In the two-factor ANOVA, investigators can assess whether there are differences in means due to the treatments, due to the sex of the participant, or due to the combination or interaction of treatment and sex. If there are differences in the means of the outcome by treatment, we say there is a main effect of treatment. If there are differences in the means of the outcome by sex, we say there is a main effect of sex. If there are differences in the means of the outcome among treatments but these vary by sex, we say there is an interaction effect. Higher-order ANOVAs are conducted in the same way as one-factor ANOVAs presented here, and the computations are again organized in ANOVA tables, with more rows to distinguish the different sources of variation (e.g., between treatments, between men and women). An example of a two-factor ANOVA is given in Example 7.16. More details on higher-order ANOVA can be found in Snedecor and Cochran.[10]

Example 7.16. Consider the clinical trial outlined above, in which three competing treatments for joint pain are compared in terms of their mean time to pain relief in patients with osteoarthritis. Because investigators hypothesize that there may be a difference in time to pain relief in men versus women, they randomly assign 15 participating men to one of the three competing treatments and randomly assign 15 participating women to one of the three competing treatments (i.e., stratified randomization). Participating men and women do not know to which treatment they are assigned. They are instructed to take the assigned medication when they experience joint pain and to record the time, in minutes, until the pain subsides. The data (times to pain relief) are shown in **Table 7–36** and are organized by the assigned treatment and gender of the participant.

The analysis in two-factor ANOVA is similar to that illustrated in Section 7.8 for one-factor ANOVA. The computations are again organized in an ANOVA table, but the total variation is partitioned into that due to the main effect of treatment, the main effect of gender, and the interaction effect. The results of the analysis are shown in **Table 7–37** (see Snedecor and Cochran for technical details.[10])

There are four statistical tests in the ANOVA table. The first test is an overall test to assess whether there is a difference among the six cell means (cells are defined by treatment and sex). The *F*-statistic is 20.7 and is highly statistically significant with $p = 0.0001$. When the overall test is significant, focus then turns to the factors that may be driving the significance (in this example, treatment, sex, or the interaction between the two). The next three statistical tests assess the significance of the main effect of treatment, the main effect of sex, and the interaction effect. In this example, there is a highly significant main effect of treat-

TABLE 7–36 Time to Pain Relief by Treatment and Sex

Treatment	Male	Female
A	12	21
	15	19
	16	18
	17	24
	14	25
B	14	21
	17	20
	19	23
	20	27
	17	25
C	25	37
	27	34
	29	36
	24	26
	22	29

TABLE 7-37 ANOVA Table for Two-Factor ANOVA

Source of Variation	Sums of Squares SS	Degrees of Freedom df	Mean Squares MS	F	p-value
Model	967.0	5	193.4	20.7	0.0001
Treatment	651.5	2	325.7	34.8	0.0001
Sex	313.6	1	313.6	33.5	0.0001
Treatment * Sex	1.9	2	0.9	0.1	0.9054
Error or Residual	224.4	24	9.4		
Total	1191.4	29			

ment ($p = 0.0001$) and a highly significant main effect of sex ($p = 0.0001$). The interaction between the two does not reach statistical significance. **Table 7-38** contains the mean times to pain relief in each of the treatments for men and women (note that each sample mean is computed on the 5 observations measured under that experimental condition).

Treatment A appears to be the most efficacious treatment for both men and women. The mean times to relief are lower in Treatment A for both men and women and highest in Treatment C for both men and women. In each treatment, women report longer times to pain relief.

Suppose that the same clinical trial is replicated in a second clinical site and the data in **Table 7-39** are observed. The ANOVA table for the data measured in clinical site 2 are summarized in **Table 7-40**.

Notice that the overall test is significant ($F = 19.4$, $p = 0.0001$) and there is a significant treatment effect and sex effect and a highly significant interaction effect. **Table 7-41** contains the mean times to relief in each of the treatments for men and women.

Notice that now the differences in mean time to pain relief among the treatments depend on sex. Among men, the mean time to pain relief is highest in Treatment A and lowest in Treatment C. Among women, the reverse is true. This is an interaction effect. When interaction effects are present, some investigators do not examine main effects. We discuss interaction effects in more detail in Chapter 9.

7.9 TESTS FOR TWO OR MORE INDEPENDENT SAMPLES, CATEGORICAL AND ORDINAL OUTCOMES

In Section 7.4, we presented the χ^2 goodness-of-fit test, which was used to test whether the distribution of responses to a categorical or ordinal variable measured in a single sample followed

TABLE 7-38 Mean Time to Pain Relief by Treatment and Sex

Treatment	Male	Female
A	14.8	21.4
B	17.4	23.2
C	25.4	32.4

TABLE 7-39 Time to Pain Relief by Treatment and Sex: Clinical Site 2

Treatment	Male	Female
A	22	21
	25	19
	26	18
	27	24
	24	25
B	14	21
	17	20
	19	23
	20	27
	17	25
C	15	37
	17	34
	19	36
	14	26
	12	29

TABLE 7-40 ANOVA Table for Two-Factor ANOVA: Clinical Site 2

Source of Variation	Sums of Squares SS	Degrees of Freedom df	Mean Squares MS	F	p-value
Model	907.0	5	181.4	19.4	0.0001
Treatment	71.5	2	35.7	3.8	0.0362
Sex	313.6	1	313.6	33.5	0.0001
Treatment * Sex	521.9	2	260.9	27.9	0.0001
Error or Residual	224.4	24	9.4		
Total	1131.4	29			

TABLE 7-41 Mean Time to Pain Relief by Treatment and Sex: Clinical Site 2

Treatment	Male	Female
A	24.8	21.4
B	17.4	23.2
C	15.4	32.4

a known distribution. Here we extend that application to the two or more independent samples case. Specifically, the outcome of interest has two or more responses, and the responses are ordered or unordered (i.e., ordinal or categorical). We now consider the situation where there are two or more independent comparison groups, and the goal of the analysis is to compare the distribution of responses to the categorical or ordinal outcome variable among several independent comparison groups.

The test is called the χ^2 test of independence and the null hypothesis is that there is no difference in the distribution of responses to the outcome across comparison groups. This is often stated as: The outcome variable and the grouping variable (e.g., the comparison treatments or comparison groups) are independent (hence the name of the test). Independence here implies homogeneity in the distribution of the outcome among comparison groups. The null hypothesis in the χ^2 test of independence is often stated in words as

H_0: The distribution of the outcome is independent of the groups.

The research hypothesis is that there is a difference in the distribution of the outcome variable among the comparison groups (i.e., that the distribution of responses "depends" on the group). To test the hypothesis, we measure the categorical or ordinal outcome variable in each participant in each comparison group. The data of interest are the observed frequencies (or number of participants in each response category in each group). The formula for the test statistic for the χ^2 test of independence is given in **Table 7-42**.

The data for the χ^2 test of independence are organized in a two-way table. The outcome and grouping variable are shown in the rows and columns of the table. **Table 7-43** illustrates the sample data layout. The table entries (blank) are the numbers of participants in each group responding to each response category of the outcome variable.

In Table 7-43, the grouping variable is shown in the rows of the table; r denotes the number of independent groups. The outcome variable is shown in the columns of the table; c denotes the number of response options in the outcome variable. Each combination of a row (group) and column (response) is called a cell of the table. The table has $r \times c$ cells and is sometimes called an $r \times c$ ("r by c") table. For example, if there are four groups and five categories in the outcome variable, the data are organized in a 4 × 5 table. The row and column totals are shown along the right margin and the bottom of the table, respectively. The total sample size, N, is computed by summing the row totals or the column totals. Similar to ANOVA, N here does not refer to a population size but rather to the total sample size in the analysis. The sample data are organized into a table like that shown in Table 7-43.

TABLE 7-42 Test Statistic for Testing H_0: Distribution of Outcome Is Independent of Groups

$$\chi^2 = \sum \frac{(O-E)^2}{E}$$

(Find critical value in Table 3, $df = (r-1)(c-1)$)

TABLE 7–43 Data Layout for Chi-Square Test of Independence

Grouping Variable	Response Option 1	Response Option 2	...	Response Option c	Row Totals
Group 1					
Group 2					
...					
Group r					
Column totals					N

The numbers of participants within each group who select each response option are shown in the cells of the table, and these are the observed frequencies used in the test statistic.

The test statistic for the χ^2 test of independence involves comparing observed (sample data) and expected frequencies in each cell of the table. The expected frequencies are computed assuming that the null hypothesis is true. The null hypothesis states that the two variables (the grouping variable and the outcome) are independent. In Chapter 5, we introduced the concept of independence. The definition of independence is

Two events, A and B, are independent if $P(A|B) = P(A)$ or if $P(B|A) = P(B)$, or, equivalently,

Two events, A and B, are independent if $P(A \text{ and } B) = P(A) P(B)$.

The last statement indicates that if two events, A and B, are independent, then the probability of their intersection can be computed by multiplying the probability of each individual event. To conduct the χ^2 test of independence, we must compute expected frequencies in each cell of the table. Expected frequencies are computed by assuming that the grouping variable and outcome are independent (i.e., under the null hypothesis). Thus, if the null hypothesis is true, using the definition of independence:

P(Group 1 and Response Option 1) = P(Group 1) P(Response Option 1).

The preceding states that the probability that an individual is in Group 1 and his or her outcome is Response Option 1 is computed by multiplying the probability that person is in Group 1 by the probability that a person gives Response Option 1. This is true if Group and Response are independent. To conduct the χ^2 test of independence, we need expected frequencies and not expected probabilities.

To convert the probability to a frequency, we multiply by N (the total sample size). Consider the following example.

The data shown in **Table 7–44** are measured in a sample of size $N = 150$. The frequencies in the cells of the table are the observed frequencies. If Group and Response are independent, then we can compute the probability that a person in the sample is in Group 1 and Response Option 1 using

P(Group 1 and Response 1) = P(Group 1) P(Response 1),

P(Group 1 and Response 1) = (25 / 150)(62 / 150) = 0.069.

Thus, if Group and Response are independent, we would expect 6.9% of the sample to be in the top-left cell of the table (Group 1 and Response 1). The expected frequency is 150(0.069) = 10.4. We could do the same for Group 2 and Response 1:

P(Group 2 and Response 1) = P(Group 2) P(Response 1),

P(Group 2 and Response 1) = (50 / 150) (62 / 150) = 0.138.

The expected frequency in Group 2 and Response 1 is 150(0.138) = 20.7.

TABLE 7–44 Observed Responses by Group

	Response 1	Response 2	Response 3	Total
Group 1	10	8	7	25
Group 2	22	15	13	50
Group 3	30	28	17	75
Total	62	51	37	150

The formula for determining the expected cell frequencies in the χ^2 test of independence is

$$\text{Expected cell frequency} = \frac{\text{Row total} \times \text{column total}}{N}.$$

The preceding equation produces the expected frequency in one step rather than computing the expected probability first and then converting to a frequency.

Example 7.17. In Example 7.6, we examined data from a survey of university graduates that assessed (among other things) how frequently they exercised. The survey was completed by 470 graduates. We used the χ^2 goodness-of-fit test to assess whether there was a shift in the distribution of responses to the exercise question following the implementation of a health promotion campaign on campus. We specifically considered one sample (all students) and compared the observed distribution of responses to the exercise question to the distribution of responses the prior year (a historical control). Suppose we wish to assess whether there is a relationship between exercise on campus and students' living arrangement. As part of the same survey, graduates were asked where they lived their senior year. The response options were dormitory, on-campus apartment, off-campus apartment, and at home (i.e., commuted to and from the university). The data are shown in **Table 7–45**.

Based on the data, is there a relationship between exercise and a student's living arrangement? Here we have four independent comparison groups (living arrangements) and an ordinal outcome variable with three response options. We specifically want to test whether living arrangement and exercise are independent. We run the test using the five-step approach.

TABLE 7–45 Exercise by Living Arrangement

	No Regular Exercise	Sporadic Exercise	Regular Exercise	Total
Dormitory	32	30	28	90
On-campus apartment	74	64	42	180
Off-campus apartment	110	25	15	150
At home	39	6	5	50
Total	255	125	90	470

Step 1: Set up hypotheses and determine the level of significance.

H_0: Living arrangement and exercise are independent,

H_1: H_0 is false,

$\alpha = 0.05$.

The null and research hypotheses are written in words rather than in symbols. The research hypothesis is that the grouping variable (living arrangement) and the outcome variable (exercise) are dependent or related.

Step 2: Select the appropriate test statistic.

The formula for the test statistic is in Table 7–42 and is given as

$$\chi^2 = \sum \frac{(O-E)^2}{E}.$$

The condition for appropriate use of the preceding test statistic is that each expected frequency is at least five. In Step 4, we compute the expected frequencies and ensure that the condition is met.

Step 3: Set up the decision rule.

The decision rule for the χ^2 test of independence is set up in a similar way to decision rules we established for z and t tests. The decision rule depends on the level of significance and the degrees of freedom, defined as $df = (r-1)(c-1)$, where r and c are the numbers of rows and columns in the two-way data table. The row variable is the living arrangement, and there are four arrangements considered; thus $r = 4$. The column variable is exercise, and three responses are considered; thus $c = 3$. For this test, $df = (4-1)(3-1) = 3(2) = 6$. Again, with χ^2 tests there are no upper- or lower-tailed versions of the test. If the null hypothesis is true, the observed and expected frequencies are close in value and the χ^2 statistic is close to 0. If the null hypothesis is false, then the χ^2 statistic is large. The rejection region for the χ^2 test of independence is always in the upper tail (right) of the distribution, as shown in Figure 7–6.

Table 3 in the Appendix contains critical values for the χ^2 test indexed by degrees of freedom and the desired level of significance. For $df = 6$ and a 5% level of significance, the appropriate critical value from Table 3 is 12.59, and the decision rule is

Reject H_0 if $\chi^2 \geq 12.59$.

Step 4: Compute the test statistic.
We now compute the expected frequencies using the formula,

$$\text{Expected cell frequency} = \frac{\text{Row total} \times \text{column total}}{N}.$$

The computations are organized in a two-way table. The expected frequencies are taken to one decimal place and the sums of the observed frequencies are equal to the sums of the expected frequencies in each row and column of the table (see **Table 7–46**).

Recall in Step 2 that a condition for the appropriate use of the test statistic was that each expected frequency is at least five. This is true for this sample (the smallest expected frequency is 9.6), and therefore it is appropriate to use the test statistic. The test statistic is computed as follows:

$$\chi^2 = \frac{(32-48.8)^2}{48.8} + \frac{(30-23.9)^2}{23.9} + \frac{(28-17.2)^2}{17.2} +$$

$$\frac{(74-97.7)^2}{97.7} + \frac{(64-47.9)^2}{47.9} + \frac{(42-34.5)^2}{34.5} +$$

$$\frac{(110-81.4)^2}{81.4} + \frac{(25-39.9)^2}{39.9} + \frac{(15-28.7)^2}{28.7} +$$

$$\frac{(39-27.1)^2}{27.1} + \frac{(6-13.3)^2}{13.3} + \frac{(5-9.6)^2}{9.6},$$

$$\chi^2 = 5.78 + 1.56 + 6.78 + 5.75 + 5.41 + 1.63 + 10.05 + 5.56 + 6.54 + 5.23 + 4.01 + 2.20 = 60.5.$$

Step 5: Conclusion.
We reject H_0 because $60.5 > 12.59$. We have statistically significant evidence at $\alpha = 0.05$ to show that H_0 is false or that living arrangement and exercise are not independent (i.e., they are dependent or related). Using Table 3 in the Appendix, we can approximate the p-value. We need to look at critical values for smaller levels of significance with $df = 6$. Using Table 3, the p-value is $p < 0.005$.

The χ^2 test of independence is used to test whether the distribution of the outcome variable is different across the comparison groups. In Example 7.17, we reject H_0 and conclude that the distribution of exercise is not independent of living arrangement, or that there is a relationship between living arrangement and exercise. The test provides an overall assessment of statistical significance. When the null hypothesis is rejected, it is important to review the sample data to understand the nature of the relationship. Consider again the data in Example 7.17.

Because there are different numbers of students in each living situation, it makes the comparisons of exercise patterns difficult on the basis of the frequencies alone. **Table 7–47** displays the percentages of students in each exercise category by living arrangement. The percentages sum to 100% in each row of the table. For comparison purposes, percentages are also shown for the total sample along the bottom row of the table. From Table 7–47, it is clear that higher percentages of students living in dormitories and in on-campus apartments report regular exercise (31% and 23%) as compared to students living in off-campus apartments and at home (10% each).

Example 7.18. In Example 7.13, we analyzed data from a randomized trial designed to evaluate the effectiveness of a newly developed pain reliever to reduce pain in patients following joint replacement surgery. The trial compared a new pain reliever to the pain reliever currently in use (called the standard of care). Suppose there was a third arm of the trial and

TABLE 7–46 Expected Frequencies

	No Regular Exercise	Sporadic Exercise	Regular Exercise	Total
Dormitory	48.8	23.9	17.2	90
On-campus apartment	97.7	47.9	34.5	180
Off-campus apartment	81.4	39.9	28.7	150
At home	27.1	13.3	9.6	50
Total	255	125	90	470

TABLE 7–47 Percentages of Students Exercising by Living Arrangement

	No Regular Exercise	Sporadic Exercise	Regular Exercise
Dormitory	36	33	31
On-campus apartment	41	36	23
Off-campus apartment	73	17	10
At home	78	12	10
Total	54	27	19

patients assigned to the third arm received a higher dose of the newly developed pain reliever. Suppose that $N = 150$ patients agreed to participate in the trial and were randomly assigned to one of the three treatments. Before receiving the assigned treatment, patients were asked to rate their pain on a scale of 0 to 10, with higher scores indicative of more pain. Each patient was then given the assigned treatment and after 30 minutes was again asked to rate his or her pain on the same scale. The primary outcome was a reduction in pain of 3 or more scale points (defined by clinicians as a clinically meaningful reduction). The data shown in **Table 7–48** are observed in the trial.

Here we have three independent comparison groups and a categorical (dichotomous) outcome variable. We want to test whether there is a difference in the proportions of patients reporting a meaningful reduction in pain among the three treatments, and we run the test using the five-step approach.

Step 1: Set up hypotheses and determine the level of significance.

H_0: Treatment and reduction in pain are independent,

H_1: H_0 is false,

$\alpha = 0.05$.

Step 2: Select the appropriate test statistic.

The formula for the test statistic is in Table 7–42 and is given as

$$\chi^2 = \sum \frac{(O-E)^2}{E}.$$

The condition for appropriate use of the preceding test statistic is that each expected frequency is at least five. In Step 4, we compute the expected frequencies and ensure that the condition is met.

Step 3: Set up the decision rule.

The row variable is the treatment and there are three considered here ($r = 3$). The column variable is the outcome and two responses are considered ($c = 2$). For this test, $df = (3 - 1)(2 - 1) = 2(1) = 2$. For $df = 2$ and a 5% level of significance, the appropriate critical value from Table 3 in the Appendix is 5.99, and the decision rule is

Reject H_0 if $\chi^2 \geq 5.99$.

Step 4: Compute the test statistic.

We now compute the expected frequencies using the formula,

$$\text{Expected cell frequency} = \frac{\text{Row total} \times \text{column total}}{N}.$$

The expected frequencies are shown in **Table 7–49**.

The test statistic is computed as

$$X^2 = \frac{(23-21.3)^2}{21.3} + \frac{(27-28.7)^2}{28.7} + \frac{(30-21.3)^2}{21.3} +$$

$$\frac{(20-28.7)^2}{28.7} + \frac{(11-21.3)^2}{21.3} + \frac{(39-28.7)^2}{28.7},$$

$$\chi^2 = 0.14 + 0.10 + 3.56 + 2.64 + 4.98 + 3.70 = 15.12.$$

Step 5: Conclusion.

We reject H_0 because $15.12 > 5.99$. We have statistically significant evidence at $\alpha = 0.05$ to show that H_0 is false, or that there is a difference in the proportions of patients reporting a meaningful reduction in pain among the three treatments. Using Table 3 in the Appendix, we can approximate the p-value. We need to

TABLE 7–48 Pain Reduction by Treatment

	Meaningful Reduction (3 + Points)	No Meaningful Reduction (< 3 Points)	Total
New pain reliever	23	27	50
Higher dose of new pain reliever	30	20	50
Standard pain reliever	11	39	50
Total	64	86	150

TABLE 7–49 Expected Frequencies

	Meaningful Reduction (3 + Points)	No Meaningful Reduction (< 3 Points)	Total
New pain reliever	21.3	28.7	50
Higher dose of new pain reliever	21.3	28.7	50
Standard pain reliever	21.3	28.7	50
Total	64	86	150

look at critical values for smaller levels of significance for $df = 2$. Using Table 3 in the Appendix, the p-value is $p < 0.005$.

What is the nature of the relationship between meaningful reduction in pain and the three treatments?

7.10 SUMMARY

In this chapter, we presented hypothesis testing techniques. Tests of hypothesis involve several steps, including specifying the null hypothesis and the alternative or research hypothesis, selecting and computing an appropriate test statistic, setting up a decision rule, and drawing a conclusion. There are many details to consider in hypothesis testing. The first is to determine the appropriate test. We discussed z, t, χ^2, and F tests here for different applications. The appropriate test depends on the distribution of the outcome variable (continuous, dichotomous, categorical, or ordinal), the number of comparison groups (one, two, or more than two), and whether the comparison groups are independent or dependent. **Table 7–50** summarizes the different tests of hypothesis discussed here.

TABLE 7–50 Summary of Key Formulas for Tests of Hypothesis

Outcome Variable, Number of Groups: Null Hypothesis	Test Statistic*
Continuous outcome, one sample: $H_0: \mu = \mu_0$	$z = \dfrac{\overline{X} - \mu_0}{s/\sqrt{n}}$
Continuous outcome, two independent samples: $H_0: \mu_1 = \mu_2$	$z = \dfrac{\overline{X}_1 - \overline{X}_2}{S_p\sqrt{1/n_1 + 1/n_2}}$
Continuous outcome, two matched samples: $H_0: \mu_d = 0$	$z = \dfrac{\overline{X}_d - \mu_d}{s_d/\sqrt{n}}$
Continuous outcome, more than two independent samples: $H_0: \mu_1 = \mu_2 = \ldots = \mu_k$	$F = \dfrac{\sum n_j (\overline{X}_j - \overline{X})^2 / (k-1)}{\sum\sum (X - \overline{X}_j)^2 / (N-k)}$
Dichotomous outcome, one sample: $H_0: p = p_0$	$z = \dfrac{\hat{p} - p_0}{\sqrt{\dfrac{p_0(1-p_0)}{n}}}$
Dichotomous outcome, two independent samples: $H_0: p_1 = p_2$, RD = 0, RR = 1, OR = 1	$z = \dfrac{\hat{p}_1 - \hat{p}_2}{\sqrt{\hat{p}(1-\hat{p})(1/n_1 + 1/n_2)}}$
Categorical or ordinal outcome, one sample: $H_0: p_1 = p_{10}, p_2 = p_{20}, \ldots, p_k = p_{k0}$	$\chi^2 = \sum \dfrac{(O-E)^2}{E}$, $df = k-1$
Categorical or ordinal outcome, two or more independent samples: H_0: Outcome and groups are independent	$\chi^2 = \sum \dfrac{(O-E)^2}{E}$, $df = (r-1)(c-1)$

*See Table 7–4, 7–11, and 7–14 for alternative formulas that are appropriate for small samples.

Once the type of test is determined, the details of the test must be specified. Specifically, the null and research hypotheses must be clearly stated. The null hypothesis always reflects the "no change" or "no difference" situation. The alternative or research hypothesis reflects the investigator's belief. The investigator might hypothesize that a parameter (e.g., a mean, proportion, difference in means, or difference in proportions) will increase, will decrease, or will be different under specific conditions (sometimes the conditions are different experimental conditions and at other times the conditions are defined by participant attributes). Once the hypotheses are specified, data are collected and summarized. The appropriate test is then conducted according to the five-step approach. If the test leads to rejection of the null hypothesis, an approximate p-value is computed to summarize the statistical significance of the findings. When tests of hypothesis are conducted using statistical computing packages, exact p-values are computed. Because the statistical tables in this textbook are limited, we only approximate p-values. If the test fails to reject the null hypothesis, then a weaker concluding statement is made.

In hypothesis testing, there are two types of errors that can be committed. A Type I error occurs when a test incorrectly rejects the null hypothesis. This is referred to as a false positive result, and the probability that this occurs is equal to the level of significance, α. The investigator chooses the level of significance and purposely chooses a small value, such as $\alpha = 0.05$, to control the probability of committing a Type I error. A Type II error occurs when a test fails to reject the null hypothesis when, in fact, it is false. The probability that this occurs is equal to β. Unfortunately, the investigator cannot specify β because it depends on several factors, including the sample size (smaller samples have higher β), the level of significance and the difference in the parameter under the null and alternative hypothesis. (For more details, see D'Agostino, Sullivan, and Beiser.[5])

We noted in several examples the relationship between confidence intervals and tests of hypothesis. While the approaches are somewhat different, they are clearly related. It is possible to draw a conclusion about statistical significance by examining a confidence interval. For example, if a 95% confidence interval does not contain the null value of the parameter of interest (e.g., 0 when analyzing a difference in means or risk difference, 1 when analyzing relative risks or odds ratios), then we conclude that a two-sided test of hypothesis is significant at $\alpha = 0.05$. It is important to note that the correspondence between a confidence interval and a test of hypothesis relates to a two-sided test, and that the confidence level corresponds to a specific two-sided level of significance (e.g., 95% to $\alpha = 0.05$, 90% to $\alpha = 0.10$, and so on). The exact significance of the test, the p-value, can be determined only by using the hypothesis testing approach.

7.11 PRACTICE PROBLEMS

1. A clinical trial evaluates a new compound designed to improve wound healing in trauma patients. The new compound is compared against a placebo. After treatment for 5 days with the new compound or placebo, the extent of wound healing is measured and the data are shown in **Table 7–51**. Is there a difference in the extent of wound healing by treatment? Run the appropriate test at a 5% level of significance. (Hint: Are treatment and the percent of wound healing independent?)

2. Use the data in Problem 1 and pool the data across the treatments into one sample of size $n = 250$. Use the pooled data to test whether the distribution of the percent of wound healing is approximately normal. Specifically, use the following distribution: 30%, 40%, 20%, and 10%, and $\alpha = 0.05$ to run the appropriate test.

3. **Table 7–52** displays summary statistics on the participants involved in the study described in Problem 1. Are any of the characteristics significantly different between groups? Justify briefly. (Hint: No calculations, just an interpretation.)

4. An investigator hypothesizes that cholesterol levels in children might be affected by educating their parents on proper nutrition and exercise. A sample of 40 families with a child between the ages of 10 to 15 who has been diagnosed with high cholesterol agrees to participate in the study. All parents are provided educational information on nutrition and exercise. After following the prescribed program, their child's total cholesterol level is measured. The childrens' mean cholesterol level is 175 with a standard deviation of 19.5. Is there significant evidence of a reduction in total cholesterol in the

TABLE 7–51 Wound Healing by Treatment

	Percent Wound Healing			
Treatment	0–25	26–50	51–75	76–100
New compound ($n = 125$)	15	37	32	41
Placebo ($n = 125$)	36	45	34	10

Practice Problems

TABLE 7–52 Summary Statistics by Treatment

	New Compound	Placebo	*p*-value
Mean age (years)	47.2	46.1	0.7564
Men (%)	44	59	0.0215
Mean educational level (years)	13.1	14.2	0.6898
Mean annual income	$36,560	$37,470	0.3546
Mean body mass index (BMI)	24.7	25.1	0.0851

children? Run the appropriate test at the 5% level of significance and assume that the null value for total cholesterol is 191.

5. An experiment is designed to investigate the impact of different positions of the mother during ultrasound on fetal heart rate. Fetal heart rate is measured by ultrasound in beats per minute. The study includes 20 women who are assigned to one position and have the fetal heart rate measured in that position. Each woman is between 28 and 32 weeks gestation. The data are shown in **Table 7–53**. Is there a significant difference in mean fetal heart rates by position? Run the test at a 5% level of significance.

6. A clinical trial is conducted comparing a new pain reliever for arthritis to a placebo. Participants are randomly assigned to receive the new treatment or a placebo and the outcome is pain relief within 30 minutes. The data are shown in **Table 7–54**. Is there a significant difference in the proportions of patients reporting pain relief? Run the test at a 5% level of significance.

7. A clinical trial is planned to compare an experimental medication designed to lower blood pressure to a placebo. Before starting the trial, a pilot study is conducted involving seven participants. The objective of the study is to assess how systolic blood pressures change over time if left untreated. Systolic blood pressures are measured at baseline and again 4 weeks later. Is there a statistically significant difference in blood pressures over time? Run the test at a 5% level of significance.

Baseline: 120 145 130 160 152 143 126

4 Weeks: 122 142 135 158 155 140 130

8. The main trial in Problem 7 is conducted and involves a total of 200 patients. Patients are enrolled

TABLE 7–54 Pain Relief by Treatment

	Pain Relief	No Pain Relief
New medication	44	76
Placebo	21	99

TABLE 7–53 Fetal Heart Rate by Position

Back	Side	Sitting	Standing
140	141	144	147
144	143	145	145
146	145	147	148
141	144	148	149
139	136	144	145
Mean = 142.0	Mean = 141.8	Mean = 145.6	Mean = 146.8

CHAPTER 7 Hypothesis Testing Procedures

TABLE 7-55 Outcomes by Treatment

	Experimental (n = 100)	Placebo (n = 100)
Mean (SD) systolic blood pressure	120.2 (15.4)	131.4 (18.9)
Hypertensive (%)	14	22
Side effects (%)	6	8

TABLE 7-56 Hypertensive Status by Treatment

	Site 1	Site 2	Site 3
Hypertensive	10	14	12
Not hypertensive	68	56	40

and randomized to receive either the experimental medication or the placebo. The data shown in **Table 7-55** are collected at the end of the study after 6 weeks on the assigned treatment.
   a. Test if there is a significant difference in mean systolic blood pressures between groups using $\alpha = 0.05$.
   b. Test if there is a significant difference in the proportions of hypertensive patients between groups using $\alpha = 0.05$.

9. Suppose in the trial described in Problem 8 that patients were recruited from three different clinical sites. Use the data in **Table 7-56** to test if there is a difference in the proportions of hypertensive patients across clinical sites. Run the test at a 5% level of significance.

10. A clinical trial is conducted to compare an experimental medication to a placebo to reduce the symptoms of asthma. Two hundred participants are enrolled in the study and randomized to receive either the experimental medication or a placebo. The primary outcome is self-reported reduction of symptoms. Among 100 participants who receive the experimental medication, 38 report a reduction of symptoms as compared to 21 participants of 100 assigned to placebo. Test if there is a significant difference in the proportions of participants reporting a reduction of symptoms between the experimental and placebo groups. Use $\alpha = 0.05$.

11. Suppose more detail is recorded in the primary outcome in the clinical trial described in Problem 10. The data are shown in **Table 7-57**. Is there a difference in the change in symptoms by treatment group? Run the appropriate test at a 5% level of significance.

12. Suppose a secondary outcome is recorded in the trial described in Problem 10 reflecting asthma symptom severity measured on a scale of 0 to 100, with higher scores indicating more severe symptoms. In the participants who receive the experimental medication, the mean symptom score is 74 with a standard deviation of 5.6, and in the placebo group, the mean symptom score is 85 with a standard deviation of 6. Is there a significant difference in mean symptom scores between groups? Run the appropriate test at a 5% level of significance.

13. Recent recommendations suggest 60 minutes of physical activity per day. A sample of 50 adults in a study of cardiovascular risk factors report exercising a mean of 38 minutes per day with a standard deviation of 19 minutes. Based on the sample data, is the physical activity significantly less than recommended? Run the appropriate test at a 5% level of significance.

14. Suppose a hypertension trial is mounted and 18 participants are randomly assigned to one of three comparison treatments. Each participant takes the assigned medication and his or her systolic blood

TABLE 7-57 Change in Symptoms by Treatment

	Change in Symptoms				
Treatment	Much Worse	Worse	No Change	Better	Much Better
Experimental	10	17	35	28	10
Placebo	12	25	42	12	9

TABLE 7–58 Systolic Blood Pressure by Treatment

Standard Treatment	Placebo	New Treatment
124	134	114
111	143	117
133	148	121
125	142	124
128	150	122
115	160	128

TABLE 7–59 Cholesterol Levels by Treatment

Diet Program	Sample Size	Mean Cholesterol	Std Dev Cholesterol
Low carbohydrate	50	225.4	24.5
Conventional	75	203.8	21.6

pressure is recorded after 6 months on the assigned treatment. The data are shown in **Table 7–58**. Is there a difference in mean systolic blood pressure among treatments? Run the appropriate test at $\alpha = 0.05$.

15. A study is conducted to compare mean cholesterol levels for individuals following a low-carbohydrate diet for at least 6 months to individuals following a conventional (low-fat, low-calorie) diet for at least 6 months. The data are summarized in **Table 7–59**. Test if there is a significant difference in mean cholesterol levels between the diet programs using a 5% level of significance.

16. Another outcome variable in the study described in Problem 15 is hypercholesterolemia, defined as total cholesterol over 220. Among the individuals who follow the low-carbohydrate diet, 56% are hypercholesterolemic, and among the individuals who follow the conventional diet, 40% are hypercholesterolemic. Test if there is a significant difference in the proportions using a 5% level of significance.

17. **Table 7–60** compares background characteristics of the participants involved in the study described in Problem 15. Are there any statistically significant differences in patient characteristics between the different diet programs? Justify briefly.

18. Suppose the results of the analyses in Problem 15 through Problem 17 are reported and criticized because the participants were not randomized to different diets, and that there may be other factors associated with changes in cholesterol. A third study is run to estimate the effect of the low-carbohydrate diet on cholesterol levels. In the third study, participants' cholesterol levels are measured before starting the program and then again after 6 months on the program. The data are shown below:

Before Program   210   230   190   215   260   200

After 6 Months   215   240   190   200   280   210

Is there a significant increase in cholesterol after 6 months on the low-carbohydrate diet? Run the appropriate test at a 5% level of significance.

19. A study is conducted to compare three new appetite suppressants (A, B, and C) to a placebo in terms of their effects on weight reduction. A total of 80 participants are involved and are randomly assigned to the comparison groups in equal numbers. The outcome of interest is weight reduction, measured in pounds. The data shown in **Table 7–61** are observed after 3 months on treatment. Is there a significant difference

TABLE 7–60 Background Characteristics by Treatment

	Low Carbohydrate ($n = 50$)	Conventional ($n = 75$)	p-value
Mean age (years)	52.1	53.4	0.7564
Men (%)	42	34	0.0145
Mean educational level (years)	15.3	12.9	0.0237
Mean family income	$39,540	$47,980	0.0576

TABLE 7-61 Weight Reduction by Treatment

	A	B	C	Placebo
Mean (SD) weight reduction	6.4 (4.1)	8.9 (4.0)	2.2 (3.9)	2.5 (4.3)

in mean weight reduction among the four treatments? Use a 5% level of significance. (Hint: SST = 1889.)

20. The mean lifetime for cardiac stents is 8.9 years. A medical device company has implemented some improvements in the manufacturing process and hypothesizes that the lifetime is now longer. A study of 40 new devices reveals a mean lifetime of 9.7 years with a standard deviation of 3.4 years. Is there statistical evidence of a prolonged lifetime of the stents? Run the test at a 5% level of significance.

21. A study is conducted in 100 children to assess risk factors for obesity. Children are enrolled and undergo a complete physical examination. At the examination, the height and weight of the child, their mother, and their father are measured and are converted to body mass index scores (weight(kg)/height(m)^2). Data on self-reported health behaviors are captured by interview and merged with the physical examination data. The primary outcome variable is child's obesity: for analysis, children are classified as normal weight (BMI < 25) or overweight/obese (BMI ≥ 25). Data on key study variables are summarized in **Table 7-62**. Is there a statistically significant difference in mean age between normal and overweight/obese children? Run the test at a 5% level of significance.

22. Use the data shown in Problem 21 and test if there is an association between mother's BMI and the child's obesity status (i.e., normal versus overweight/obese)? Run the test at a 5% level of significance.

23. Use the data shown in Problem 21 and test if there is a significant difference in the proportions of normal versus overweight/obese children who are male. Run the test at a 5% level of significance.

24. A study is run to compare body mass index (BMI) in participants assigned to different diet programs and the data are analyzed in Excel®. Use the Excel results to answer the questions below.
    a. Complete the ANOVA table (**Table 7-63**).
    b. Write the hypotheses to be tested.
    c. Write the conclusion of the test.

25. A clinical trial is conducted to test the efficacy of a new drug for hypertension. The new drug is compared to a standard drug and to a placebo in a study involving $n = 120$ participants. The primary outcome is systolic blood pressure measured after 4 weeks on

TABLE 7-62 Key Study Variables in Normal and Overweight/Obese Children

Characteristics	Normal Weight ($n = 62$)	Overweight/Obese ($n = 38$)	Total ($n = 100$)
Mean (SD) age, years	13.4 (2.6)	11.1 (2.9)	12.5 (2.7)
% Male	45%	51%	47%
Mother's BMI			
Normal (BMI < 25)	40 (65%)	16 (41%)	56 (56%)
Overweight (BMI 25–29.9)	15 (24%)	14 (38%)	29 (29%)
Obese (BMI ≥ 30)	7 (11%)	8 (21%)	15 (15%)
Father's BMI			
Normal (BMI < 25)	34 (55%)	16 (41%)	50 (50%)
Overweight (BMI 25–29.9)	20 (32%)	14 (38%)	34 (34%)
Obese (BMI ≥ 30)	8 (13%)	8 (21%)	16 (16%)
Mean (SD) systolic blood pressure	123 (15)	139 (12)	129 (14)
Mean (SD) total cholesterol	186 (25)	211 (28)	196 (26)

TABLE 7-63 ANOVA Table

ANOVA

Source of Variation	SS	df	MS	F	p-value	F crit
Between Groups	40.791667	3			0.0012733	3.10
Within Groups	35.166666					
Total	75.958327	23				

the assigned drug. **Table 7–64** shows characteristics of study participants measured at baseline (prior to randomization). Test if the mean SBP at baseline is significantly higher in participants assigned to the standard drug as compared to those assigned to the new drug. Use a 5% level of significance.

26. Use the data shown in Problem 25 and test if there is a significant difference in the proportions of men assigned to each of the three treatments. (HINT: Are gender and treatment independent?) Use a 5% level of significance.

27. Use the data shown in Problem 25 and test whether there is a significant difference in the proportion of diabetic participants in the placebo group as compared to the standard drug group. Use a 5% level of significance.

28. Use the data shown in Problem 25 and test whether there is a significant difference in mean age among the three groups. (HINT: SStotal = 2893.) Use a 5% level of significance.

29. Some scientists believe that alcoholism is linked to social isolation. One measure of social isolation is marital status. A study of 280 adults is conducted, and each participant is classified as not alcoholic, diagnosed alcoholic, or undiagnosed alcoholic, and also by marital status (see **Table 7–65**). Is there significant evidence of an association? Run the appropriate test at a 5% level of significance.

30. A study is performed to examine the relationship between the concentration of plasma antioxidant vitamins and cancer risk. **Table 7–66** shows data for

TABLE 7-65 Social Isolation and Alcohol

	Diagnosed Alcoholic	Undiagnosed Alcoholic	Not Alcoholic
Married	21	37	58
Not Married	59	63	42

TABLE 7-66 Plasma Antioxidant Vitamins in Cancer Patients and in Controls

	N	Mean	SD
Stomach cancer patients	20	2.41	0.15
Controls	50	2.78	0.19

TABLE 7-64 Baseline Characteristics of Study Participants

Baseline Characteristic	Placebo (n = 40)	Standard Drug (n = 40)	New Drug (n = 40)
Mean (SD) age	75.2 (4.4)	75.6 (4.8)	74.7 (5.6)
n (%) Male	12 (30.0%)	13 (32.5%)	9 (22.5%)
Mean (SD) BMI	26.1 (4.9)	25.1 (3.0)	26.9 (4.1)
Mean (SD) SBP	142.1 (19.2)	150.4 (19.8)	144.5 (19.7)
n (%) Diabetic	13 (32.5%)	11 (27.5%)	8 (20.0%)
n (%) Current smokers	4 (10.0%)	2 (5.0%)	1 (2.5%)

plasma vitamin-A concentration in stomach cancer patients and in controls (participants similar to the cancer patients but free of disease). Is there a significant difference in the mean concentration of plasma antioxidant vitamins between patients with stomach cancer and controls? Run the appropriate test at a 5% level of significance.

REFERENCES

1. Ogden, C.L., Fryar, C.D., Carroll, M.D., and Flegal, K.M. "Mean body weight, height, and body mass index, United States 1960–2002." *Advance Data from Vital and Health Statistics* 2004; 347.
2. National Center for Health Statistics. *Health, United States, 2005 with Chartbook on Trends in the Health of Americans*. Hyattsville, MD: U.S. Government Printing Office, 2005.
3. Agresti, A. *Categorical Data Analysis* (2nd ed.). New York: John Wiley & Sons, 2002.
4. StatXact Version 7. © 2006 by Cytel, Inc., Cambridge, MA.
5. D'Agostino, R.B., Sullivan, L.M., and Beiser, A. *Introductory Applied Biostatistics*. Belmont, CA: Duxbury-Brooks/Cole, 2004.
6. Rosner, B. *Fundamentals of Biostatistics*. Belmont, CA: Duxbury-Brooks/ Cole, 2006.
7. SAS version 9.1. © 2002–2003 by SAS Insitute, Inc., Cary, NC.
8. Glantz, S.A. and Slinker, B.K. *Primer of Applied Regression and Analysis of Variance*. New York: McGraw-Hill, 2001.
9. National Osteoporosis Foundation. Available at *http://www.nof.org*.
10. Snedecor, G.W. and Cochran, W.G. *Statistical Methods*. Ames, IA: Iowa State University Press, 1980.

CHAPTER 8

Power and Sample Size Determination

WHEN AND WHY

Key Questions

- Why do some studies have 100 participants and others 10,000? What is the right size for samples?
- How can some studies with very small samples produce statistically significant results?
- What are the ethical issues when a study is under-powered?

In the News

Clinical trials are very carefully designed to ensure that the sample size is sufficiently large to ensure a high probability of observing a clinically meaningful difference in the outcome between treatments, if one exists. Such trials usually include planned interim analyses, to be conducted at various points before enrollment is complete, so as to ensure participant welfare. In some instances, clinical trials are stopped early, before the target enrollment numbers are reached. Among the various reasons that a trial might be stopped are that the treatments are found to be hugely different or not different at all, the participants suffer extensive side effects, the participants have poor compliance, or study misconduct occurs. Stopping a study early is a complex decision, and one that is never taken lightly. Oftentimes, studies that are stopped early remain controversial, such as the trial that investigated the efficacy of hormone replacement therapy in healthy postmenopausal women.[1,2] The women in the trial who were taking a combination of estrogen and progestin did experience relief from symptoms of menopause and had fewer hip fractures and a lower risk of developing colon cancer. However, compared to women who did not take hormones, they had higher risks of developing breast cancer, heart disease, blood clots, and stroke.

---

[1]Writing Group for the Women's Health Initiative Investigators. "Risks and benefits of estrogen plus progestin in health postmenopausal women." *Journal of the American Medical Association* 2002; 288: 321–333.
[2]National Institutes of Health. Available at *http://www.nhlbi.nih.gov/whi/pr_02-7-9.pdf*.

Dig In

Imagine that a clinical trial is designed to compare a new treatment to a current available treatment (standard of care). Suppose that the new treatment costs less and has fewer adverse events than the standard of care. The investigators run the test of hypothesis comparing the primary outcomes of the new treatment and the standard of care and fail to reject the null hypothesis at a 5% level of significance.

- Which treatment would you recommend for patients and why?
- What kinds of errors are possible with hypothesis testing, and what are the implications for patients when we commit errors?
- How can we best guard against making such errors in tests of hypothesis?

LEARNING OBJECTIVES

By the end of this chapter, the reader will be able to

- Provide examples demonstrating how the margin of error, effect size, and variability of the outcome affect sample size computations
- Compute the sample size required to estimate population parameters with precision
- Interpret statistical power in tests of hypothesis
- Compute the sample size required to ensure high power in tests of hypothesis

In Chapter 6 and Chapter 7, we presented techniques for estimation and hypothesis testing, respectively. In Chapter 6, we saw that confidence interval estimates based on larger samples had smaller margins of error. In Chapter 7, we saw that tests of hypothesis based on larger samples were more likely to detect small increases, decreases, or differences in the parameter of interest. Larger samples produce more precise analyses. However, there is a point at which a larger sample size does not substantially improve the precision in the analysis.

Studies should be designed to include a sufficient number of participants to adequately address the research question. Realizing at the end of a study that the sample was simply too small to answer the research question is wasteful in terms of participant and investigator time, resources to conduct the assessments, analytic efforts, and so on. It can also be viewed as unethical as participants may have been put at risk as part of a study that was unable to answer an important question. Alternatively, studies should not be too large because again resources can be wasted and some participants may be unnecessarily placed at risk.

A critically important aspect of study design is determining the appropriate sample size to answer the research question. There are formulas that are used to estimate the sample size needed to produce a confidence interval estimate with a specified margin of error, or to ensure that a test of hypothesis has a high probability of detecting a meaningful difference in the parameter if one exists. Ideally, these formulas are used to generate estimates of the sample size needed to answer the study question before any data are collected.

The formulas we present here generate the sample sizes required to satisfy statistical criteria. In many studies, the sample size is determined by financial or logistical constraints. For example, suppose a study is proposed to evaluate a new screening test for Down Syndrome. Suppose that the screening test is based on the analysis of a blood sample taken from women early in pregnancy. To evaluate the properties of the screening test (e.g., the sensitivity and specificity), each pregnant woman will be asked to provide a blood sample and in addition to undergo an amniocentesis. The amniocentesis is included as the gold standard, and the plan is to compare the results of the screening test to the results of the amniocentesis. Suppose that the collection and processing of the blood sample costs $250 per participant and that the amniocentesis costs $900 per participant. These financial constraints alone might substantially limit the number of women who can be enrolled. Just as it is important to consider both statistical and practical issues in interpreting the results of a statistical analysis, it is also important to weigh both statistical and logistical issues in determining the sample size for an analysis. Our focus here is on statistical considerations. Investigators must evaluate whether the sample size determined to be sufficient from a statistical standpoint is realistic and feasible.

8.1 ISSUES IN ESTIMATING SAMPLE SIZE FOR CONFIDENCE INTERVALS ESTIMATES

In Chapter 6, we presented confidence intervals for various parameters—e.g., μ, p, $(\mu_1 - \mu_2)$, μ_d, $(p_1 - p_2)$. Confidence intervals for every parameter take the following general form:

$$\text{Point estimate} \pm \text{Margin of error}.$$

In Chapter 6, when we introduced the concept of confidence intervals, we derived the expression for the confidence interval for μ as

$$\overline{X} \pm z \frac{\sigma}{\sqrt{n}}.$$

In practice we use the sample standard deviation, s, to estimate the population standard deviation, σ. Note also that there is an alternative formula for estimating the mean of a continuous outcome in a single population, and it is used when the sample size is small ($n < 30$). It involves a value from the t distribution, as opposed to one from the standard normal distribution, to reflect the desired level of confidence. When performing sample-size computations, we use the large sample formula shown here. The resultant sample size might be small and in the analysis stage; the appropriate confidence interval formula must be used.

The point estimate for the population mean is the sample mean and the margin of error is $z \frac{\sigma}{\sqrt{n}}$. In planning studies, we want to determine the sample size needed to ensure that the margin of error is sufficiently small to be informative. For example, suppose we want to estimate the mean weight of female college students. We conduct a study and generate a 95% confidence interval as 125 ± 40 lbs, or 85 to 165 lbs. The margin of error is so wide that the confidence interval is uninformative. To be informative, an investigator might want the margin of error to be no more than 5 or 10 lbs (meaning that the 95% confidence interval would have a width—lower limit to upper limit—of 10 or 20 lbs). To determine the sample size needed, the investigator must specify the desired margin of error. It is important to note that this is not a statistical issue but a clinical or a practical one. The sample size needed also depends on the outcome variable under consideration. For example, suppose we want to estimate the mean birth weight of infants born to mothers who smoke cigarettes during pregnancy. Birth weights in infants clearly have a much more restricted range than weights of female college students. Therefore, we would probably want to generate a confidence interval for the mean birth weight that has a margin of error not exceeding 1 or 2 lbs.

The margin of error in the one sample confidence interval for μ can be written as

$$E = z\frac{\sigma}{\sqrt{n}}.$$

We want to determine the sample size, n, that ensures that the margin of error, E, does not exceed a specified value. We can take the preceding formula and, with some algebra, solve for n:

$$n = \left(\frac{z\sigma}{E}\right)^2.$$

This formula generates the minimum sample size, n, required to ensure that the margin of error, E, does not exceed a specified value. To solve for n, we must input z, σ, and E. z is the value from Table 1B in the Appendix for the desired confidence level (e.g., $z = 1.96$ for 95% confidence) and E is the margin of error that the investigator specifies as important from a practical standpoint. The last input is σ, the standard deviation of the outcome of interest. Sometimes it is difficult to estimate σ. When we use the prior sample size formula (or one of the other formulas that we present in the sections that follow), we are planning a study to estimate the unknown mean of a particular outcome variable in a population. It is unlikely that we know the standard deviation of that variable. In sample size computations, investigators often use a value for the standard deviation from a previous study or a study done in a different but comparable population.

The sample size computation is not an application of statistical inference, and therefore it is reasonable to use an appropriate estimate for the standard deviation. The estimate can be derived from a different study that was reported in the literature; some investigators perform a small pilot study to estimate the standard deviation. A pilot study usually involves a small number of participants (e.g., $n = 10$) who are selected by convenience as opposed to random sampling. Data from the participants in the pilot study are used to compute the sample standard deviation, s, which serves as a good estimate for σ in the sample size formula. Regardless of how the estimate of the variability of the outcome is derived, it should always be conservative (i.e., as large as is reasonable), so that the resultant sample size is not too small.

The formula for n produces the minimum sample size to ensure that the margin of error in a confidence interval will not exceed E. In planning studies, investigators should also consider attrition or loss to follow-up. Sample size formulas for confidence intervals give the number of participants needed with complete data to ensure that the margin of error in the confidence interval does not exceed E. We illustrate how attrition is addressed in planning studies through examples in the following sections.

8.1.1 Sample Size for One Sample, Continuous Outcome

In studies where the plan is to estimate the mean of a continuous outcome variable in a single population (μ), the formula for determining sample size is

$$n = \left(\frac{z\sigma}{E}\right)^2$$

where z is the value from the standard normal distribution reflecting the confidence level that will be used (e.g., $z = 1.96$ for 95%), σ is the standard deviation of the outcome variable, and E is the desired margin of error. The preceding formula generates the minimum number of subjects required to ensure that the margin of error in the confidence interval for μ does not exceed E.

Example 8.1. An investigator wants to estimate the mean systolic blood pressure in children with congenital heart disease who are between the ages of 3 and 5 years of age. How many children should be enrolled in the study?

To determine the requisite sample size, the investigator must provide the necessary inputs for the sample size formula,

$$n = \left(\frac{z\sigma}{E}\right)^2.$$

Suppose that a 95% confidence interval is planned; thus, $z = 1.96$. The investigator must then specify the margin of error. Suppose the investigator decides that a margin of error of 5 units will be sufficiently precise. The final input is the standard deviation of systolic blood pressure in children with congenital heart disease. A suitable estimate might be found in the literature. There are certainly data on systolic blood pressure in the Framingham Heart Study, but those data might not be sufficiently comparable as they are based on adults, most of whom are probably free of congenital heart disease.

Suppose the investigators conduct a literature search and find that the standard deviation of systolic blood pressure in children with other cardiac defects is between 15 and 20. To estimate the sample size, we consider the larger standard deviation. This will produce the most conservative (largest) sample size,

$$n = \left(\frac{z\sigma}{E}\right)^2 = \left(\frac{1.96 \times 20}{5}\right)^2 = 61.5.$$

To ensure that the 95% confidence interval estimate of the mean systolic blood pressure in children between the ages of 3 and 5 with congenital heart disease is within 5 units of the true mean, a sample of size 62 is needed. Note that we always round up, as the sample size formulas always generates the minimum number of subjects needed to ensure the specified precision.

Had we assumed a standard deviation of 15, the sample size would have been $n = 35$. Because the estimates of the standard deviation are derived from studies of children with other cardiac defects, it is advisable to use the larger standard deviation and plan for a study with 62 children. Selecting the smaller sample size could potentially produce a confidence interval estimate with a much larger margin of error.

Example 8.2. An investigator wants to estimate the mean birth weight of infants born full term (approximately 40 weeks gestation) to mothers who are 19 years of age and under. The mean birth weight of infants born full term to mothers 20 years of age and older is 3510 g with a standard deviation of 385 g.[1] How many women 19 years of age or younger must be enrolled in the study to ensure that a 95% confidence interval estimate of the mean birth weight of their infants has a margin of error not exceeding 100 g?

$$n = \left(\frac{z\sigma}{E}\right)^2 = \left(\frac{1.96 \times 385}{100}\right)^2 = 56.9.$$

To ensure that the 95% confidence interval estimate of the mean birth weight is within 100 g of the true mean, a sample of size 57 is needed. In planning the study, the investigator must consider the fact that some women may deliver prematurely. If women are enrolled into the study during pregnancy, then more than 57 women will need to be enrolled so that after excluding those who deliver prematurely, 57 who deliver full term will be available for analysis. For example, if 5% of the women are expected to deliver prematurely (i.e., 95% will deliver full term), then 60 women must be enrolled to ensure that 57 deliver full term. The number of women who must be enrolled, N, is computed as

N (number to enroll) × (% retained) = Desired sample size,

$$N \times 0.95 = 57,$$

$$N = \frac{57}{0.95} = 60.$$

8.1.2 Sample Size for One Sample, Dichotomous Outcome

In studies where the plan is to estimate the proportion of successes in a dichotomous outcome variable in a single population, the formula for determining sample size is

$$n = p(1-p)\left(\frac{z}{E}\right)^2.$$

where z is the value from the standard normal distribution reflecting the confidence level that will be used (e.g., $z = 1.96$ for 95%), E is the desired margin of error, and p is the proportion of successes in the population. Here we are planning a study to generate a 95% confidence interval for the unknown population proportion, p, and the formula to estimate the sample size needed requires p! Obviously, this is a circular problem—if we knew the proportion of successes in the population, then a study would not be necessary. To estimate the sample size, we need an approximate value of p. The range of p is 0 to 1 and the range of $p(1 - p)$ is 0 to 0.25. The value of p that maximizes $p(1 - p)$ is $p = 0.5$. Thus, if there is no information available to approximate p, then $p = 0.5$ can be used to generate the most conservative, or largest, sample size.

Example 8.3. An investigator wants to estimate the proportion of freshmen at his university who currently smoke cigarettes (i.e., the prevalence of smoking). How many freshmen should be involved in the study to ensure that a 95% confidence interval estimate of the proportion of freshmen who smoke is within 0.05 or 5% of the true proportion?

If we have no information on the proportion of freshmen who smoke, we use the following to estimate the sample size:

$$n = p(1-p)\left(\frac{z}{E}\right)^2 = 0.5(1-0.5)\left(\frac{1.96}{0.05}\right)^2 = 384.2.$$

To ensure that the 95% confidence interval estimate of the proportion of freshmen who smoke is within 5% of the true proportion, a sample of size 385 is needed.

Suppose that a similar study was conducted two years ago and found that the prevalence of smoking was 27% among freshmen. If the investigator believes that this is a reasonable estimate of prevalence today, it can be used to plan the study:

$$n = p(1-p)\left(\frac{z}{E}\right)^2 = 0.27(1-0.27)\left(\frac{1.96}{0.05}\right)^2 = 302.9.$$

To ensure that the 95% confidence interval estimate of the proportion of freshmen who smoke is within 5% of the true proportion, a sample of size 303 is needed. Notice that this sample size is substantially smaller than the one estimated previously. Having some information on the magnitude of the proportion in the population always produces a sample size that is less than or equal to the one based on a population proportion of 0.5. However, the estimate must be realistic.

Example 8.4. An investigator wants to estimate the prevalence of breast cancer among women who are between 40 and 45 years of age living in Boston. How many women must be involved in the study to ensure that the estimate is precise? National data suggest[2] that 1 in 235 women are diagnosed with breast cancer by age 40. This translates to a proportion of 0.0043 (0.43%), or a prevalence of 43 per 10,000 women. Suppose the investigator wants the estimate to be within 10 per 10,000 women with 95% confidence. The sample size is computed as

$$n = p(1-p)\left(\frac{z}{E}\right)^2 =$$

$$0.0043(1-0.0043)\left(\frac{1.96}{0.0010}\right)^2 = 16,447.8.$$

A sample of size $n = 16,448$ ensures that a 95% confidence interval estimate of the prevalence of breast cancer is within 0.0010 (or within 10 women per 10,000) of its true value.

This is a situation where investigators might decide that a sample of this size is not feasible. Suppose that the investigators believe that a sample of size 5000 is reasonable from a practical point of view. How precisely can they estimate the prevalence with a sample of size $n = 5000$? Recall from Chapter 6 that the confidence interval formula to estimate prevalence is

$$\hat{p} \pm z\sqrt{\frac{\hat{p}(1-\hat{p})}{n}}.$$

Assuming that the prevalence of breast cancer in the sample is close to that based on national data, we would expect the margin of error to be approximately equal to

$$z\sqrt{\frac{\hat{p}(1-\hat{p})}{n}} = 1.96\sqrt{\frac{0.0043(1-0.0043)}{5000}} = 0.0018.$$

With $n = 5000$ women, a 95% confidence interval would be expected to have a margin of error of 0.0018 (or 18 per 10,000). The investigators must decide if this is sufficiently precise to answer the research question. Note that this calculation is based on the assumption that the prevalence of breast cancer in Boston is similar to that reported nationally. This may or may not be a reasonable assumption. In fact, it is the objective of the current study to estimate the prevalence in Boston. With input from clinical investigators and biostatisticians, the research team must carefully evaluate the implications of selecting a sample of size $n = 5000$, $n = 16,448$, or any size in between.

8.1.3 Sample Sizes for Two Independent Samples, Continuous Outcome

In studies where the plan is to estimate the difference in means between two independent populations ($\mu_1 - \mu_2$), the formula for determining the sample sizes required in each comparison group is

$$n_i = 2\left(\frac{z\sigma}{E}\right)^2$$

where n_i is the sample size required in each group ($i = 1, 2$), z is the value from the standard normal distribution reflecting the confidence level that will be used (e.g., $z = 1.96$ for 95%), and E is the desired margin of error. σ again reflects the standard deviation of the outcome variable. Recall from Chapter 6 that when we generated a confidence interval estimate for the difference in means, we used S_p (the pooled estimate of the common standard deviation) as a measure of variability in the outcome, where S_p is computed as

$$S_p = \sqrt{\frac{(n_1-1)s_1^2 + (n_2-1)s_2^2}{n_1+n_2-2}}$$

If data are available on the variability of the outcome in each comparison group, then S_p can be computed and used in the sample size formula. However, it is more often the case that data on the variability of the outcome are available from only one group, often the untreated (e.g., placebo control) or unexposed group. When planning a clinical trial to investigate a new drug or procedure, data are often available from other trials that involved a placebo or an active control group (i.e., a standard medication or treatment given for the condition under study). The standard deviation of the outcome variable measured in patients assigned to the placebo, control, or unexposed group can be used to plan a future trial. This situation is illustrated below.

Note that the formula shown here generates sample size estimates for samples of equal size. If a study is planned where different numbers of patients will be assigned or different numbers of patients will comprise the comparison groups, then alternative formulas can be used. Interested readers can see Howell for more details.[3]

Example 8.5. An investigator wants to plan a clinical trial to evaluate the efficacy of a new drug designed to increase HDL cholesterol (the "good" cholesterol). The plan is to enroll participants and to randomly assign them to receive either the new drug or a placebo. HDL cholesterol will be measured in each participant after 12 weeks on the assigned treatment. Based on prior experience with similar trials, the investigator expects that 10% of all participants will be lost to follow-up or will drop out of the study. A 95% confidence interval will be estimated to quantify the difference in mean HDL levels between patients taking the new drug as compared to the placebo. The investigator would like the margin of error to be no more than 3 units. How many patients should be recruited into the study?

The sample sizes (i.e., the number of participants who must receive the new drug and the number who must receive a placebo) are computed as

$$n_i = 2\left(\frac{z\sigma}{E}\right)^2.$$

A major issue is determining the variability in the outcome of interest, here the standard deviation (σ) of HDL cholesterol. To plan this study, we can use data from the Framingham Heart Study. In participants who attended the seventh examination of the Framingham Offspring Study and were not on treatment for high cholesterol, the standard deviation of HDL cholesterol is 17.1. We use this value and the other inputs to compute the sample sizes:

$$n_i = 2\left(\frac{z\sigma}{E}\right)^2 = 2\left(\frac{1.96 \times 17.1}{3}\right)^2 = 249.6.$$

Samples of size $n_1 = 250$ and $n_2 = 250$ will ensure that the 95% confidence interval for the difference in mean HDL levels will have a margin of error of no more than 3 units. Again, these sample sizes refer to the numbers of participants with complete data. The investigators hypothesized a 10% attrition (or drop-out) rate in both groups. To ensure that a total sample size of 500 is available at 12 weeks, the investigator needs to recruit more participants to allow for attrition.

N (number to enroll) $\times$ (% retained) = Desired sample size.

$$N \times 0.90 = 500.$$

$$N = \frac{500}{0.90} = 556.$$

The investigator must enroll 556 participants. Each patient will be randomly assigned to receive either the new drug or a placebo; assuming that 10% are lost to follow-up, 500 will be available for analysis.

Example 8.6. An investigator wants to compare two diet programs in children who are obese. One diet is a low-fat diet and the other is a low-carbohydrate diet. The plan is to enroll children and weigh them at the start of the study. Each child will then be randomly assigned to either the low-fat or the low-carbohydrate diet. Each child will follow the assigned diet for 8 weeks, at which time he or she will again be weighed. The number of pounds lost will be computed for each child. Based on data reported from diet trials in adults, the investigator expects that 20% of all children will not complete the study. A 95% confidence interval will be estimated to quantify the difference in weight lost between the two diets, and the investigator would like the margin of error to be no more than 3 lbs. How many children should be recruited into the study?

The sample sizes (i.e., the number of children who must follow the low-fat diet and the number of children who must follow the low-carbohydrate diet) are computed as

$$n_i = 2\left(\frac{z\sigma}{E}\right)^2.$$

Again, the issue is determining the variability in the outcome of interest, here the standard deviation (σ) in pounds lost over 8 weeks. To plan this study, investigators use data from adult studies. Suppose one such study compared the same diets in adults and involved 100 participants in each diet group. The study reported a standard deviation in weight lost over 8 weeks on a low-fat diet of 8.4 lbs and a standard deviation in weight lost over 8 weeks on a low-carbohydrate diet of 7.7 lbs. These data can be used to estimate the common standard deviation in weight lost as

$$S_p = \sqrt{\frac{(n_1-1)s_1^2 + (n_2-1)s_2^2}{n_1+n_2-2}} =$$

$$\sqrt{\frac{(100-1)(8.4)^2 + (100-1)(7.7)^2}{100+100-2}} = 8.1.$$

We now use this value and the other inputs to compute the sample sizes:

$$n_i = 2\left(\frac{z\sigma}{E}\right)^2 = 2\left(\frac{1.96 \times 8.1}{3}\right)^2 = 56.0.$$

Samples of size $n_1 = 56$ and $n_2 = 56$ will ensure that the 95% confidence interval for the difference in weight lost between diets will have a margin of error of no more than 3 lbs. Again, these sample sizes refer to the numbers of children with complete data. The investigators hypothesized a 20% attrition rate. To ensure that the total sample size of 112 is available at 8 weeks, the investigator needs to recruit more participants to allow for attrition.

N (number to enroll) $\times$ (% retained) = Desired sample size.

$$N \times 0.80 = 112.$$

$$N = \frac{112}{0.80} = 140.$$

The investigator must enroll 140 children. Each child will be randomly assigned to either the low-fat or low-carbohydrate diet; assuming that 20% are lost to follow-up, 112 will be available for analysis.

8.1.4 Sample Size for Matched Samples, Continuous Outcome

In studies where the plan is to estimate the mean difference of a continuous outcome (μ_d) based on matched data, the formula for determining sample size is

$$n = \left(\frac{z\sigma_d}{E}\right)^2$$

where z is the value from the standard normal distribution reflecting the confidence level that will be used (e.g., $z = 1.96$ for 95%), E is the desired margin of error, and σ_d is the standard deviation of the difference scores. It is extremely important that the standard deviation of the difference scores (e.g., the differences based on measurements over time, or the differences between matched pairs) is used here to appropriately estimate the sample size.

Example 8.7. Consider again the diet study proposed in Example 8.6. Suppose the investigator is considering an alternative design—a crossover trial. In the crossover trial, each child follows each diet for 8 weeks. At the end of each 8-week period, the weight lost during that period is measured. Children will be randomly assigned to the first diet (i.e., some will follow the low-fat diet first while others will follow the low-carbohydrate diet first). The investigators expect to lose 30% of the participants over the course of the 16-week study. The difference in weight lost on the low-fat diet and the low-carbohydrate diet will be computed for each child, and a confidence interval for the mean difference in weight lost will be computed. How many children are required to ensure that a 95% confidence interval estimate for the mean difference in weight lost has a margin of error of no more than 3 lbs?

To compute the sample size, we need an estimate of the variability in the difference in weight lost between diets as opposed to the variability in weight loss in either diet program, as was required in Example 8.6. Suppose that the standard deviation of the difference in weight loss between a low-fat diet and a low-carbohydrate diet is approximately 9.1 lbs based on a crossover trial conducted in adults. The sample size is computed as

$$n = \left(\frac{z\sigma_d}{E}\right)^2 = \left(\frac{1.96 \times 9.1}{3}\right)^2 = 35.3.$$

To ensure that the 95% confidence interval estimate of the mean difference in weight lost between diets is within 3 lbs of the true mean, a sample size of 36 children with complete data is needed. The investigators hypothesized a 30% attrition rate. To ensure that the total sample size of 36 is available at 16 weeks, the investigator needs to recruit 52 participants to allow for attrition:

N (number to enroll) $\times$ (% retained) = Desired sample size.

$$N \times 0.70 = 36.$$

$$N = \frac{36}{0.70} = 52.$$

Notice that fewer patients are needed in the crossover trial as compared to the trial with two concurrent or parallel comparison groups considered in Example 8.6. In the crossover trial, each child is given each treatment (diet). Crossover trials can be efficient in terms of sample size. However, the crossover can be challenging in other respects. The crossover trial is problematic when there is loss to follow-up. Because each participant contributes data on each treatment, when a participant is lost after completing the first treatment, his or her data are unusable without information on the second treatment. Diet studies are particularly problematic in terms of participant loss to follow-up. Sometimes participants start to lose weight and feel they no longer need the study, and so they drop out. Other participants do not lose weight and become frustrated and drop out. These and other issues must be considered carefully in determining the most appropriate design for any study.

8.1.5 Sample Sizes for Two Independent Samples, Dichotomous Outcome

In studies where the plan is to estimate the difference in proportions between two independent populations (i.e., to estimate the risk difference $p_1 - p_2$), the formula for determining the sample sizes required in each comparison group is given as

$$n_i = \left[p_1(1-p_1) + p_2(1-p_2)\right]\left(\frac{z}{E}\right)^2$$

where n_i is the sample size required in each group ($i = 1, 2$), z is the value from the standard normal distribution reflecting the confidence level that will be used (e.g., $z = 1.96$ for 95%), E is the desired margin of error, and p_1 and p_2 are the proportions of successes in each comparison group. Again, here we are planning a study to generate a 95% confidence interval for the difference in unknown proportions, and the formula to estimate the sample sizes needed requires p_1 and p_2. To estimate the sample size, we need approximate values of p_1 and p_2. The values of p_1 and p_2 that maximize the sample sizes are $p_1 = p_2 = 0.5$. Thus, if there is no information available to approximate p_1 and p_2, 0.5 can be used to generate the most conservative, or largest, sample sizes.

Similar to the situation we presented in Section 8.1.3 for two independent samples and a continuous outcome, it may be the case that data are available on the proportion of successes in one group, usually the untreated (e.g., placebo control) or unexposed group. If so, that known proportion can be used for both p_1 and p_2 in the formula shown here.

This formula generates sample size estimates for samples of equal size. If a study is planned where different numbers of patients will be assigned or different numbers of patients will comprise the comparison groups, then alternative formulas can be used. Interested readers can see Fleiss for more details.[4]

Example 8.8. An investigator wants to estimate the impact of smoking during pregnancy on premature delivery. Normal pregnancies last approximately 40 weeks and premature deliveries are those that occur before 37 weeks. The 2005 National Vital Statistics report indicates that approximately 12% of infants are born prematurely in the United States.[5] The investigator plans to collect data through medical record review and generate a 95% confidence interval for the difference in proportions of infants born prematurely to women who smoked during pregnancy as compared to those who did not. How many women should be enrolled in the study to ensure that the 95% confidence interval for the difference in proportions has a margin of error of no more than 0.04, or 4%?

The sample sizes (i.e., numbers of women who smoked and did not smoke during pregnancy) are computed using the previous formula. National data suggest that 12% of infants are born prematurely. We use this estimate for both groups in the sample size computation:

$$n_i = \left[p_1(1-p_1) + p_2(1-p_2)\right]\left(\frac{z}{E}\right)^2 =$$

$$\left[0.12(1-0.12) + 0.12(1-0.12)\right]\left(\frac{1.96}{0.04}\right)^2 = 507.1.$$

Samples of size $n_1 = 508$ women who smoked during pregnancy and $n_2 = 508$ women who did not smoke during pregnancy will ensure that the 95% confidence interval for the difference in proportions who deliver prematurely will have a margin of error of no more than 4%. Is attrition an issue here?

Example 8.9. An investigator wants to estimate the impact of smoking on the incidence of prostate cancer. The incidence of prostate cancer by age 70 is approximately 1 in 6 men (17%). A United States study reported that heavy smokers (defined as men smoking more than a pack a day for 40 years) had twice the risk of developing prostate cancer as compared to nonsmokers.[6] An investigator plans to replicate the study in England using a cohort study design. Men who are free of prostate cancer will be enrolled at age 50 and followed for 30 years. The plan is to enroll approximately equal numbers of smokers and nonsmokers in the study and follow them prospectively for the outcome of interest—a diagnosis of prostate cancer. The plan is to generate a 95% confidence interval for the difference in proportions of smoking and nonsmoking men who develop prostate cancer. The study will follow men for 30 years, and investigators expect to lose approximately 20% of the participants over the course of the follow-up period. How many men should be enrolled in the study to ensure that the 95% confidence interval for the difference in proportions has a margin of error of no more than 0.05, or 5%?

The sample sizes (i.e., numbers of men who smoke and do not smoke) are computed using the previous formula. The estimates of the incidence of prostate cancer from the United States study are used to design the study planned for England (i.e., $p_1 = 0.34$ and $p_2 = 0.17$ are the estimates of incidence of prostate cancer in smokers versus nonsmokers, respectively):

$$n_i = \left[p_1(1-p_1) + p_2(1-p_2)\right]\left(\frac{z}{E}\right)^2 =$$

$$\left[0.34(1-0.34) + 0.17(1-0.17)\right]\left(\frac{1.96}{0.05}\right)^2 = 561.6.$$

Samples of size $n_1 = 562$ men who smoke and $n_2 = 562$ men who do not smoke will ensure that the 95% confidence interval for the difference in the incidence of prostate cancer will have a margin of error of no more than 5%.

The sample sizes here refer to the numbers of men with complete data. The investigators hypothesized 20% attrition (in each group). To ensure that the total sample size of 1124 is available for analysis, the investigator needs to recruit more participants to allow for attrition:

N (number to enroll) $\times$ (% retained) = Desired sample size.

$$N \times 0.80 = 1124.$$

$$N = \frac{1124}{0.80} = 1405.$$

The investigator must enroll 1405 men. Because enrollment will be on the basis of smoking status at baseline, the investigator will enroll approximately 703 men who smoke and 703 men who do not.

8.2 ISSUES IN ESTIMATING SAMPLE SIZE FOR HYPOTHESIS TESTING

In Chapter 7, we introduced hypothesis testing techniques for means, proportions, differences in means, and differences in proportions. Whereas each test involves details that are specific to the outcome of interest (continuous, dichotomous, categorical, or ordinal) and to the number of comparison groups (one, two, or more than two), there are elements common to each test.

For example, in each test of hypothesis there are two errors that can be committed. The first is called a **Type I error** and refers to the situation where we incorrectly reject H_0 when, in fact, it is true. In the first step of any test of hypothesis, we select a level of significance, α, and

$$\alpha = P(\text{Type I error}) = P(\text{Reject } H_0 \mid H_0 \text{ is true}).$$

Because we purposely select a small value for α, we control the probability of committing a Type I error. The second type of error is called a **Type II error** and it is defined as the probability we do not reject H_0 when it is false. The probability of a Type II error is denoted β, and

$$\beta = P(\text{Type II error}) = P(\text{Do not reject } H_0 \mid H_0 \text{ is false}).$$

In hypothesis testing, we usually focus on **power**, which is defined as the probability that we reject H_0 when it is false,

$$\text{Power} = 1 - \beta = P(\text{Reject } H_0 \mid H_0 \text{ is false}).$$

Power is the probability that a test correctly rejects a false null hypothesis. A good test is one with low probability of committing a Type I error (i.e., small α) and high power (i.e., small β, high power).

In the following sections, we present formulas to determine the sample size required to ensure that a test has high power. The sample size computations depend on the level of significance, α, the desired power of the test (equivalent to $1 - \beta$), the variability of the outcome, and the effect size. The effect size is the difference in the parameter of interest that represents a clinically meaningful difference. Similar to the margin of error in confidence interval applications, the effect size is determined based on clinical or practical criteria and not statistical criteria. The concept of statistical power can be difficult to grasp. Before presenting the formulas to determine the sample size required to ensure high power in a test, we first discuss power from a conceptual point of view.

Suppose we want to test the following hypotheses at $\alpha = 0.05$,

$$H_0: \mu = 90,$$
$$H_1: \mu \neq 90.$$

To test the hypotheses, suppose we select a sample of size $n = 100$. For this example, assume that the standard deviation of the outcome is $\sigma = 20$. We compute the sample mean and then must decide whether the sample mean provides evidence to support the research hypothesis or not. This is done by computing a test statistic and comparing the test statistic to an appropriate critical value.

If the null hypothesis is really true ($\mu = 90$), then we are likely to select a sample whose mean is close in value to 90. However, it is also possible to select a sample whose mean is much larger or much smaller than 90. Recall from the Central Limit Theorem (see Chapter 5), that for large n (here $n = 100$ is sufficiently large) the distribution of the sample means is approximately normal with a mean of $\mu_{\bar{X}} = \mu = 90$ and a standard deviation of $\sigma_{\bar{X}} = \sigma/\sqrt{n} = 20/\sqrt{100} = 2.0$. If the null hypothesis is true, it is possible to observe any sample mean shown in **Figure 8–1**, as all are possible under $H_0: \mu = 90$.

When we set up the decision rule for our test of hypothesis, we determine critical values based on $\alpha = 0.05$ and a two-sided test. When we run tests of hypothesis, we usually standardize the data (e.g., convert to z or t) and the critical values are appropriate values from the probability distribution used in the test. To facilitate interpretation, we continue this discussion with $\bar{X}$ as opposed to z. The critical values of $\bar{X}$ for a two-sided test with $\alpha = 0.05$ are 86.08 and 93.92 (these values correspond to -1.96 and 1.96, respectively, on the z scale), so the decision rule is

$$\text{Reject } H_0 \text{ if } \bar{X} \leq 86.08 \text{ or if } \bar{X} \geq 93.92.$$

The rejection region is shown in the tails of **Figure 8–2**. The areas in the two tails of the curve represent the probability of a Type I error, $\alpha = 0.05$. This concept was discussed in some detail in Chapter 7.

Now suppose that the alternative hypothesis, H_1, is true ($\mu \neq 90$) and that the true mean is actually 94. **Figure 8–3** shows the distributions of the sample mean under the null and alternative hypotheses. The values of the sample mean are shown along the horizontal axis.

If the true mean is 94, then the alternative hypothesis is true. In our test, we selected $\alpha = 0.05$ and reject H_0 if the observed sample mean exceeds 93.92 (focusing on the upper tail of the rejection region for now). The critical value (93.92) is indicated by the vertical line. The probability of a Type II error is denoted β, and

$$\beta = P(\text{Do not reject } H_0 \mid H_0 \text{ is false}).$$

FIGURE 8–1 Distribution of $\bar{X}$ Under $H_0: \mu = 90$

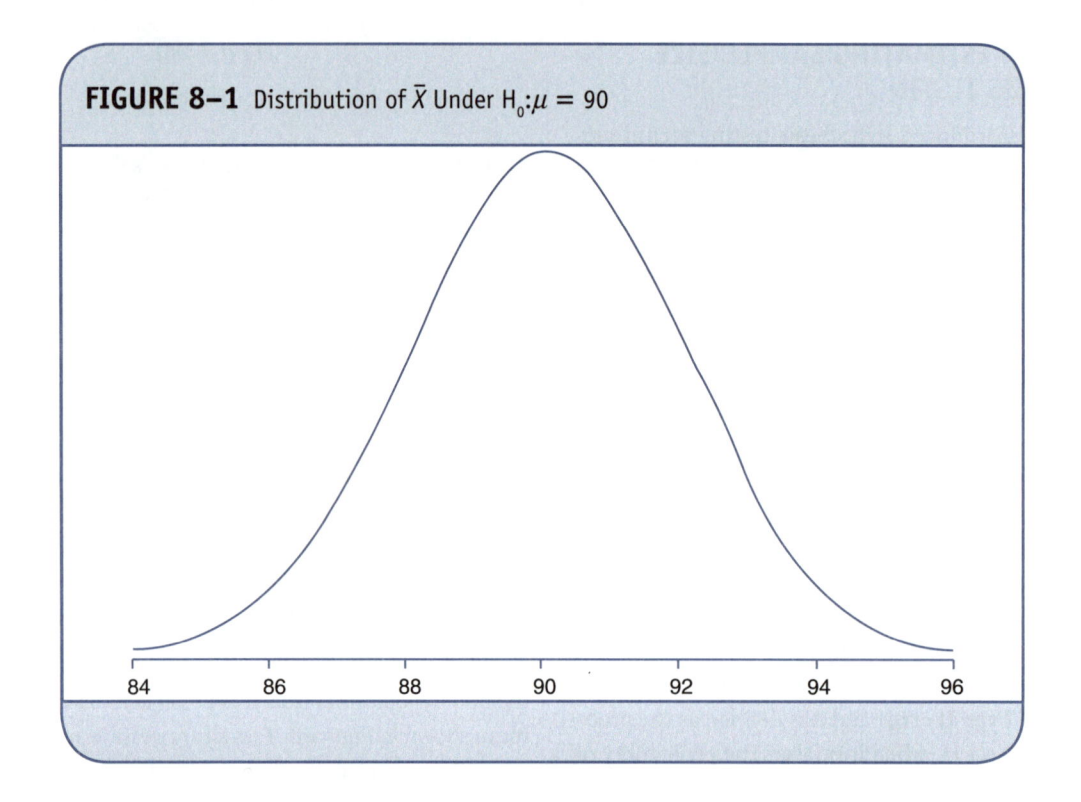

FIGURE 8–2 Rejection Region for Test $H_0: \mu = 90$ Versus $H_1: \mu \neq 90$ at $\alpha = 0.05$

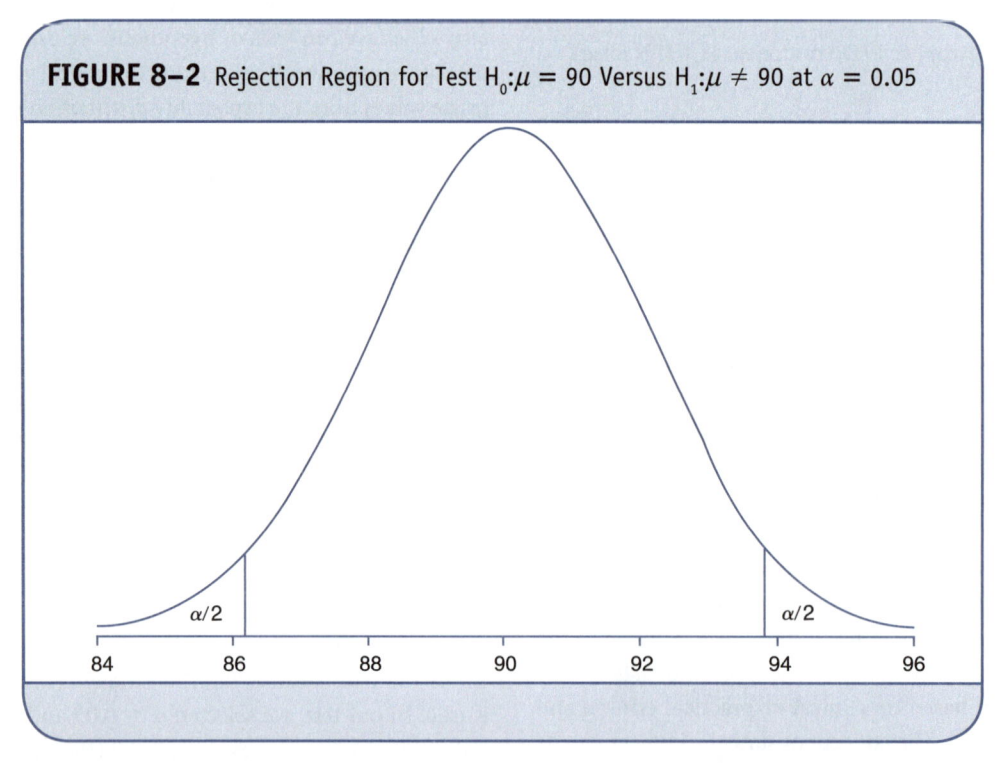

β is shown in Figure 8–3 as the area under the rightmost curve (H_1) to the left of the vertical line (where we do not reject H_0). Power is defined as

$$1 - \beta = P(\text{Reject } H_0 \mid H_0 \text{ is false})$$

and is shown in Figure 8–3 as the area under the rightmost curve (H_1) to the right of the vertical line (where we reject H_0).

We previously noted that β and power were related to α, the variability of the outcome and the effect size. From Figure 8–3 we can see what happens to β and power if we

FIGURE 8–3 Distribution of $\bar{X}$ Under $H_0: \mu = 90$ and Under $H_1: \mu = 94$

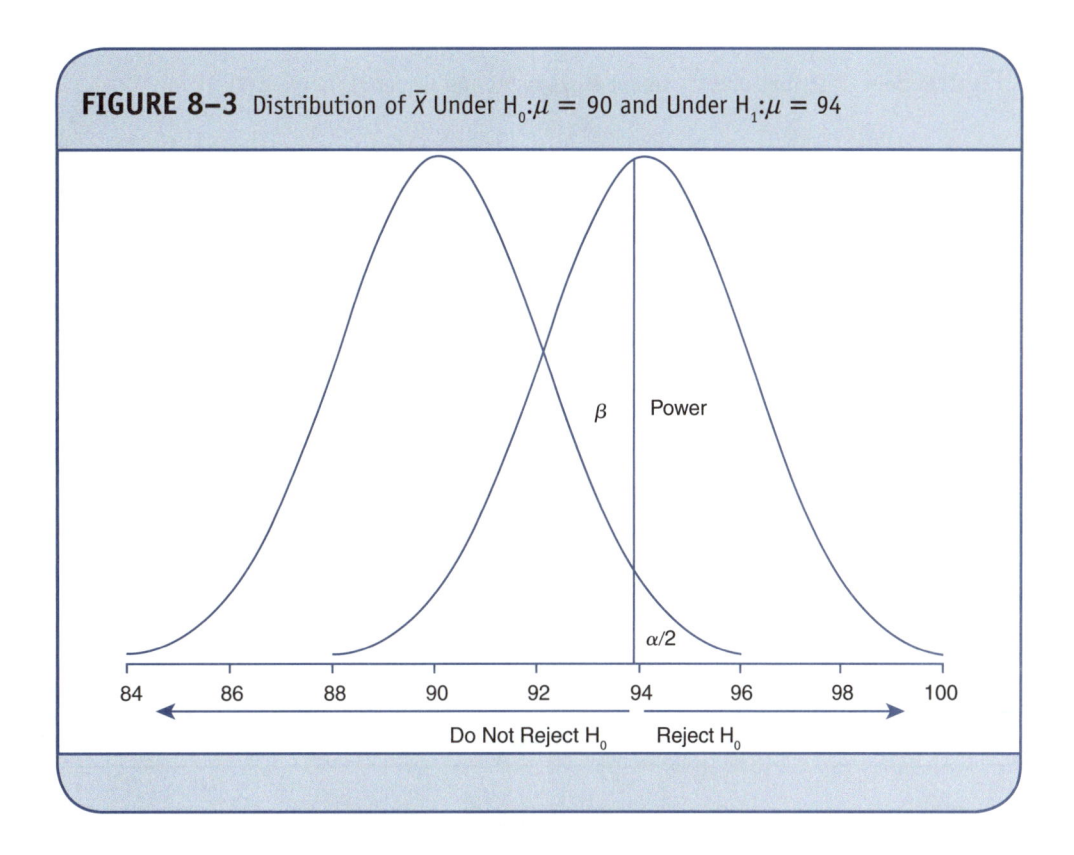

increase α. For example, suppose we increase α to $\alpha = 0.10$. The upper critical value would be 92.56 instead of 93.92. The vertical line would shift to the left, increasing α, decreasing β, and increasing power. Whereas a better test is one with higher power, it is not advisable to increase α as a means to increase power. Nonetheless, there is a direct relationship between α and power (as α increases, so does power).

β and power are also related to the variability of the outcome and to the effect size. The effect size is the difference in the parameter of interest (e.g., μ) that represents a clinically meaningful difference. Figure 8–3 graphically displays α, β, and power when the difference in the mean under the null as compared to the alternative hypothesis is 4 units (i.e., 90 versus 94). **Figure 8–4** shows the same components for the situation where the mean under the alternative hypothesis is 98.

Notice that there is much higher power when there is a larger difference between the mean under H_0 as compared to H_1 (i.e., 90 versus 98). A statistical test is much more likely to reject the null hypothesis in favor of the alternative if the true mean is 98 than if the true mean is 94. Notice in Figure 8–4 that there is little overlap in the distributions under the null and alternative hypotheses. If a sample mean of 97 or higher is observed, it is very unlikely that it came from a distribution whose mean is 90. In Figure 8–3, if we observed a sample mean of 93, it would not be as clear as to whether it came from a distribution whose mean is 90 or one whose mean is 94.

In the following sections, we provide formulas to determine the sample size needed to ensure that a test has high power. In designing studies, most people consider power of 80% or 90% (just as we generally use 95% as the confidence level for confidence interval estimates). The inputs for the sample size formulas include the desired power, the level of significance, and the effect size. The effect size is selected to represent a clinically meaningful or practically important difference in the parameter of interest. This will be illustrated in examples that follow.

The formulas we present produce the minimum sample size to ensure that the test of hypothesis will have a specified probability of rejecting the null hypothesis when it is false (i.e., a specified power). In planning studies, investigators again must account for attrition or loss to follow-up. The formulas shown produce the minimum number of participants needed with complete data. In the examples, we also illustrate how attrition is addressed in planning studies.

8.2.1 Sample Size for One Sample, Continuous Outcome

In studies where the plan is to perform a test of hypothesis comparing the mean of a continuous outcome variable in

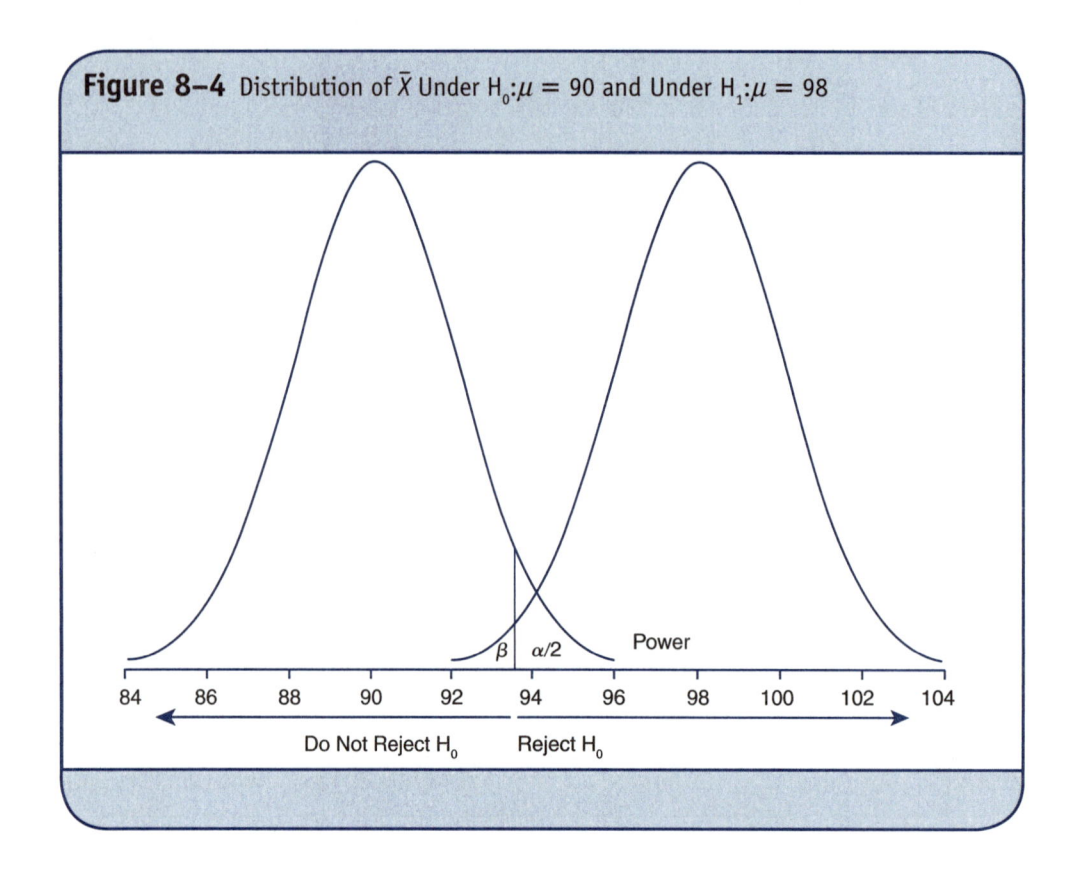

Figure 8-4 Distribution of $\bar{X}$ Under $H_0: \mu = 90$ and Under $H_1: \mu = 98$

a single population to a known mean, the hypotheses of interest are

$$H_0: \mu = \mu_0,$$
$$H_1: \mu \neq \mu_0$$

where μ_0 is the known mean (e.g., a historical control). Formulas to determine the sample size required to ensure high power in tests of hypothesis are often based on two-sided tests. Here we consider two-sided tests. If a one-sided test is planned, the sample size forumulas must be modified. The formula for determining sample size to ensure that the test has a specified power is

$$n = \left(\frac{z_{1-\alpha/2} + z_{1-\beta}}{ES} \right)^2$$

where α is the selected level of significance and $z_{1-\alpha/2}$ is the value from the standard normal distribution holding $1-\alpha/2$ below it. For example, if $\alpha = 0.05$, then $1-\alpha/2 = 0.975$ and $z = 1.960$. $1-\beta$ is the selected power and $z_{1-\beta}$ is the value from the standard normal distribution holding $1-\beta$ below it. Sample size estimates for hypothesis testing are often based on achieving 80% or 90% power. The $z_{1-\beta}$ values for these popular scenarios are

80% power $\quad z_{0.8} = 0.84$

90% power $\quad z_{0.9} = 1.282$

ES is the *effect size*, defined as

$$ES = \frac{|\mu_1 - \mu_0|}{\sigma},$$

where μ_0 is the mean under H_0, μ_1 is the mean under H_1, and σ is the standard deviation of the outcome of interest. The numerator of the effect size, the absolute value of the difference in means $|\mu_1 - \mu_0|$, represents what is considered a clinically meaningful or practically important difference in means. Similar to the issue we faced when determining sample size to estimate confidence intervals, it can sometimes be difficult to estimate the standard deviation. In sample size computations, investigators often use a value for the standard deviation from a previous study or a study performed in a different population. Regardless of how the estimate of the variability of the outcome is derived, it should always be conservative (i.e., as large as is reasonable) so that the resultant sample size will not be too small.

Example 8.10. An investigator hypothesizes that in people free of diabetes, fasting blood glucose—a risk factor for coronary heart disease—is higher in those who drink at least two cups of coffee per day. A cross-sectional study is planned to assess the mean fasting blood glucose levels in people who drink at least two cups of coffee per day. The mean fasting blood glucose level in people free of diabetes is reported as 95 mg/dl with a standard deviation of 9.8 mg/dl.[7] If the mean blood glucose level in people who drink at least two cups of coffee per day is 100 mg/dl, this would be important clinically. How many patients should be enrolled in the study to ensure that the power of the test is 80% to detect this difference? A two-sided test will be used with a 5% level of significance.

We first compute the effect size,

$$ES = \frac{|\mu_1 - \mu_0|}{\sigma} = \frac{|100 - 95|}{9.8} = 0.51.$$

We now substitute the effect size and the appropriate z values for the selected α and power to compute the sample size.

$$n = \left(\frac{z_{1-\alpha/2} + z_{1-\beta}}{ES}\right)^2 = \left(\frac{1.96 + 0.84}{0.51}\right)^2 = 30.1.$$

A sample of size $n = 31$ will ensure that a two-sided test with $\alpha = 0.05$ has 80% power to detect a 5-mg/dl difference in mean fasting blood glucose levels.

In the planned study, participants will be asked to fast overnight and to provide a blood sample for the analysis of glucose levels. Based on prior experience, the investigators hypothesize that 10% of the participants will fail to fast or will refuse to submit the blood sample (i.e., they will fail to follow the study protocol). Therefore, a total of 35 participants will be enrolled in the study to ensure that 31 are available for analysis:

$$N \text{ (number to enroll)} \times (\% \text{ following protocol}) = \text{Desired sample size}$$

$$N \times 0.90 = 31$$

$$N = \frac{31}{0.90} = 35.$$

Example 8.11. In Example 7.1, we conducted a test of hypothesis to assess whether there was a reduction in expenditures on health care and prescription drugs in 2005 as compared to 2002. In 2002, Americans paid an average of $3302 per year on health care and prescription drugs.[8] An investigator sampled 100 Americans and found that they spent a mean of $3190 on health care and prescription drugs with a standard deviation of $890. The test failed to reject the null hypothesis, and the investigator is concerned that the sample size was too small. How many participants are needed to ensure that the test has 80% power to detect a difference of $150 in expenditures (a difference the investigator feels is meaningful)? A two-sided test will be used with a 5% level of significance.

We first compute the effect size,

$$ES = \frac{|\mu_1 - \mu_0|}{\sigma} = \frac{150}{890} = 0.17.$$

Notice that the numerator is 150, the difference specified by the investigator. We also use the standard deviation measured in the previous study ($s = 890$) to inform the design of the next study.

We now substitute the effect size and the appropriate z values for the selected α and power to compute the sample size,

$$n = \left(\frac{z_{1-\alpha/2} + z_{1-\beta}}{ES}\right)^2 = \left(\frac{1.96 + 0.84}{0.17}\right)^2 = 271.3.$$

A sample of size $n = 272$ will ensure that a two-sided test with $\alpha = 0.05$ has 80% power to detect a $150 difference in expenditures. Thus, the study with $n = 100$ participants was underpowered to detect a meaningful difference ($150) in expenditures.

8.2.2 Sample Size for One Sample, Dichotomous Outcome

In studies where the plan is to perform a test of hypothesis comparing the proportion of successes in a dichotomous outcome variable in a single population to a known proportion, the hypotheses of interest are

$$H_0: p = p_0,$$
$$H_1: p \neq p_0$$

where p_0 is the known proportion (e.g., a historical control). The formula for determining the sample size to ensure that the test has a specified power is

$$n = \left(\frac{z_{1-\alpha/2} + z_{1-\beta}}{ES}\right)^2$$

where α is the selected level of significance and $z_{1-\alpha/2}$ is the value from the standard normal distribution holding $1 - \alpha/2$ below it. $1 - \beta$ is the selected power, $z_{1-\beta}$ is the value from the standard normal distribution holding $1 - \beta$ below it, and ES is the effect size, defined as

$$ES = \frac{|p_1 - p_0|}{\sqrt{p_0(1 - p_0)}},$$

where p_0 is the proportion under H_0 and p_1 is the proportion under H_1. The numerator of the effect size, the absolute value of the difference in proportions $|p_1 - p_0|$, again represents what is considered a clinically meaningful or practically important difference in proportions.

Example 8.12. A recent report from the Framingham Heart Study indicated that 26% of people free of cardiovascular disease had elevated LDL cholesterol levels, defined as LDL of more than 159 mg/dl.[9] An investigator hypothesizes that a higher proportion of patients with a history of cardiovascular disease will have elevated LDL cholesterol. How many patients should be studied to ensure that the power of the test is 90% to detect an absolute difference of 5% in the proportion with elevated LDL cholesterol? A two-sided test will be used with a 5% level of significance.

We first compute the effect size,

$$ES = \frac{|p_1 - p_0|}{\sqrt{p_0(1-p_0)}} = \frac{0.05}{\sqrt{0.26(1-0.26)}} = 0.11.$$

We now substitute the effect size and the appropriate z values for the selected α and power to compute the sample size,

$$n = \left(\frac{z_{1-\alpha/2} + z_{1-\beta}}{ES}\right)^2 = \left(\frac{1.96 + 1.282}{0.11}\right)^2 = 868.6.$$

A sample of size $n = 869$ will ensure that a two-sided test with $\alpha = 0.05$ has 90% power to detect a difference of 5% in the proportion of patients with a history of cardiovascular disease who have an elevated LDL cholesterol level.

Example 8.13. A medical device manufacturer produces implantable stents. During the manufacturing process, approximately 10% of the stents are deemed to be defective. The manufacturer wants to test whether the percentage of defective stents is more than 10% in one of his plants. If the process produces more than 15% defective stents, then corrective action must be taken. Therefore, the manufacturer wants the test to have 90% power to detect a difference in proportions of this magnitude. How many stents must be evaluated? A two-sided test will be used with a 5% level of significance.

We first compute the effect size,

$$ES = \frac{|p_1 - p_0|}{\sqrt{p_0(1-p_0)}} = \frac{|0.15 - 0.10|}{\sqrt{0.10(1-0.10)}} = 0.17.$$

We now substitute the effect size and the appropriate z values for the selected α and power to compute the sample size,

$$n = \left(\frac{z_{1-\alpha/2} + z_{1-\beta}}{ES}\right)^2 = \left(\frac{1.96 + 1.282}{0.17}\right)^2 = 363.7.$$

A sample of size $n = 364$ stents will ensure that a two-sided test with $\alpha = 0.05$ has 90% power to detect a difference of 5% in the proportion of defective stents.

8.2.3 Sample Sizes for Two Independent Samples, Continuous Outcome

In studies where the plan is to perform a test of hypothesis comparing the means of a continuous outcome variable in two independent populations, the hypotheses of interest are

$$H_0: \mu_1 = \mu_2$$
$$H_1: \mu_1 \neq \mu_2,$$

where μ_1 and μ_2 are the means in the two comparison populations. The formula for determining the sample sizes to ensure that the test has a specified power is

$$n_i = 2\left(\frac{z_{1-\alpha/2} + z_{1-\beta}}{ES}\right)^2$$

where n_i is the sample size required in each group ($i = 1, 2$), α is the selected level of significance, $z_{1-\alpha/2}$ is the value from the standard normal distribution holding $1 - \alpha/2$ below it, $1 - \beta$ is the selected power, and $z_{1-\beta}$ is the value from the standard normal distribution holding $1 - \beta$ below it. ES is the effect size, defined as

$$ES = \frac{|\mu_1 - \mu_2|}{\sigma},$$

where $|\mu_1 - \mu_2|$ is the absolute value of the difference in means between the two groups expected under the alternative hypothesis, H_1. σ is the standard deviation of the outcome of interest.

Recall from Chapter 6 when we performed tests of hypothesis comparing the means of two independent groups, we used S_p (the pooled estimate of the common standard deviation) as a measure of variability in the outcome. If data are available on variability of the outcome in each comparison group, then S_p can be computed and used to generate the sample sizes. However, it is more often the case that data on the variability of the outcome are available from only one group, usually the untreated (e.g., placebo control) or unexposed group. When planning a clinical trial to investigate a new drug or procedure, data are often available from other trials that may have involved a placebo or an active control group (i.e., a standard medication or treatment given for the condition under study). The standard deviation of the outcome variable measured in patients assigned to the placebo, control, or unexposed group can be used to plan a future trial. This situation is illustrated in the following examples.

Note that the formula shown previously generates sample size estimates for samples of equal size. If a study is planned where different numbers of patients will be assigned or the comparison groups will comprise different numbers of patients, then alternative formulas can be used. Interested readers can see Howell for more details.[3]

Example 8.14. An investigator is planning a clinical trial to evaluate the efficacy of a new drug designed to reduce systolic blood pressure. The plan is to enroll participants and to randomly assign them to receive either the new drug or a placebo. Systolic blood pressures will be measured in each participant after 12 weeks on the assigned treatment. Based on prior experience with similar trials, the investigator expects that 10% of all participants will be lost to follow-up or will drop from the study. If the new drug shows a 5-unit reduction in mean systolic blood pressure, this would represent a clinically meaningful reduction. How many patients should be enrolled in the trial to ensure that the power of the test is 80% to detect this difference? A two-sided test will be used with a 5% level of significance.

To compute the effect size, an estimate of the variability in systolic blood pressures is needed. Analysis of data from the Framingham Heart Study (see Table 4.20), showed that the standard deviation of systolic blood pressure is $s = 19.0$. This value is used to plan the trial.

The effect size is

$$ES = \frac{|\mu_1 - \mu_2|}{\sigma} = \frac{5}{19.0} = 0.26.$$

We now substitute the effect size and the appropriate z values for the selected α and power to compute the sample size,

$$n_i = 2\left(\frac{z_{1-\alpha/2} + z_{1-\beta}}{ES}\right)^2 = 2\left(\frac{1.96 + 0.84}{0.26}\right)^2 = 232.0.$$

Samples of size $n_1 = 232$ and $n_2 = 232$ will ensure that the test of hypotheses will have 80% power to detect a 5-unit difference in mean systolic blood pressures in patients receiving the new drug as compared to patients receiving the placebo. These sample sizes refer to the numbers of participants required with complete data. The investigators hypothesized a 10% attrition rate in both groups. To ensure that the total sample size of 464 is available at 12 weeks, the investigator needs to recruit more participants to allow for attrition:

N (number to enroll) $\times$ (% retained) = Desired sample size

$$N \times 0.90 = 464$$

$$N = \frac{464}{0.90} = 516$$

The investigator must enroll 516 participants. Each patient will be randomly assigned to receive either the new drug or placebo; assuming that 10% are lost to follow-up, 464 will be available for analysis.

Example 8.15. An investigator is planning a study to assess the association between alcohol consumption and grade point average among college seniors. The plan is to categorize students as heavy drinkers or not using five or more drinks on a typical drinking day as the criterion for heavy drinking. Mean grade point averages will be compared between students classified as heavy drinkers versus not using a two independent samples test of means. The standard deviation in grade point averages is assumed to be 0.42 and a meaningful difference in grade point averages (relative to drinking status) is 0.25 unit.

How many college seniors should be enrolled in the study to ensure that the power of the test is 80% to detect a 0.25 unit difference in mean grade point averages? A two-sided test will be used with a 5% level of significance.

We first compute the effect size,

$$ES = \frac{|\mu_1 - \mu_2|}{\sigma} = \frac{0.25}{0.42} = 0.60.$$

We now substitute the effect size and the appropriate z values for the selected α and power to compute the sample size,

$$n_i = 2\left(\frac{z_{1-\alpha/2} + z_{1-\beta}}{ES}\right)^2 = 2\left(\frac{1.96 + 0.84}{0.60}\right)^2 = 43.6.$$

Samples of size $n_1 = 44$ heavy drinkers and $n_2 = 44$ who drink fewer than five drinks per typical drinking day will ensure that the test of hypothesis has 80% power to detect a 0.25 unit difference in mean grade point averages.

This computation assumes that approximately equal numbers of students will be classified as heavy drinkers versus not. A study by Wechsler et al. showed that approximately 44% of college students were binge drinkers (defined as drinking five or more drinks per occasion for men and four or more per occasion for women).[9] To ensure that this study has at least 44 heavy drinkers, a total of $N = 100$ college students should be selected. This will ensure that approximately 44 (44% of 100) are heavy drinkers and 56 (56% of 100) are not.

8.2.4 Sample Size for Matched Samples, Continuous Outcome

In studies where the plan is to perform a test of hypothesis on the mean difference in a continuous outcome variable based on matched data, the hypotheses of interest are

$$H_0: \mu_d = 0,$$
$$H_1: \mu_d \neq 0,$$

where μ_d is the mean difference in the population. The formula for determining the sample size to ensure that the test has a specified power is

$$n = \left(\frac{z_{1-\alpha/2} + z_{1-\beta}}{ES} \right)^2,$$

where α is the selected level of significance and $z_{1-\alpha/2}$ is the value from the standard normal distribution holding $1-\alpha/2$ below it. $1-\beta$ is the selected power, $z_{1-\beta}$ is the value from the standard normal distribution holding $1-\beta$ below it and ES is the effect size, defined as

$$ES = \frac{\mu_d}{\sigma_d},$$

where μ_d is the mean difference expected under the alternative hypothesis, H_1, and σ_d is the standard deviation of the difference in the outcome (e.g., the difference based on measurements over time or the difference between matched pairs).

Example 8.16. In Example 8.7, we generated sample size requirements for a crossover trial to compare two diet programs for their effectiveness in promoting weight loss. The planned analysis was a confidence interval estimate for the mean difference in weight loss on the low-fat as compared to the low-carbohydrate diet. Suppose we design the same study with a test of hypothesis as the primary analysis. The proposed study will have each child follow each diet for 8 weeks. At the end of each 8-week period, the weight lost during that period will be measured. Children will be randomly assigned to the first diet (i.e., some will follow the low-fat diet first while others will follow the low-carbohydrate diet first). The investigators expect to lose 30% of the participants over the course of the 16-week study. The difference in weight lost on the low-fat diet and the low-carbohydrate diet will be computed for each child; here we will test if there is a statistically significant difference in weight loss between the diets. How many children are required to ensure that a two-sided test with a 5% level of significance has 80% power to detect a mean difference in weight loss between the two diets of 3 lbs?

To compute the sample size, we again need an estimate of the variability of the difference in weight loss between diets. In Example 8.7, we used a value from a previous study of 9.1 lbs (this was based on a similar trial conducted in adults).

We first compute the effect size,

$$ES = \frac{\mu_d}{\sigma_d} = \frac{3}{9.1} = 0.33.$$

We now substitute the effect size and the appropriate z values for the selected α and power to compute the sample size,

$$n = \left(\frac{z_{1-\alpha/2} + z_{1-\beta}}{ES} \right)^2 = \left(\frac{1.96 + 0.84}{0.33} \right)^2 = 72.0.$$

A sample of size $n = 72$ children will ensure that a two-sided test with $\alpha = 0.05$ has 80% power to detect a mean difference of 3 lbs between diets using a crossover trial. The investigators hypothesized a 30% attrition rate. To ensure that the total sample size of 72 is available at 16 weeks, the investigator needs to recruit 103 participants:

$$N \text{ (number to enroll)} \times (\% \text{ retained}) = \text{Desired sample size}$$

$$N \times 0.70 = 72$$

$$N = \frac{72}{0.70} = 103$$

Notice that different numbers of children are needed depending on the analysis plan. In Example 8.7, we determined that 36 children would be needed to ensure that a 95% confidence interval for the mean difference has a margin of error of no more than 3 lbs. With a test of hypothesis, to ensure 80% power in detecting a mean difference of 3 lbs between diets we need 72 children. The analysis plan is key in the design of the study as it clearly affects the sample size.

Example 8.17. An investigator wants to evaluate the efficacy of an acupuncture treatment for reducing pain in patients with chronic migraine headaches. The plan is to enroll patients who suffer from migraine headaches. Each will be asked to rate the severity of the pain they experience with their next migraine before any treatment is administered. Pain will be recorded on a scale of 1 to 100 with higher scores indicative of more severe pain. Each patient will then undergo the acupuncture treatment. On their next migraine (post-treatment), each patient will again be asked to rate the severity of the pain. The difference in pain will be computed for each patient. A two-sided test of hypothesis will be conducted at $\alpha = 0.05$ to assess whether there is a statistically significant difference in pain scores before and after treatment. How many patients should be involved in the study to ensure that the test has 80% power to detect a difference of 10 units on the pain scale? Assume that the standard deviation in the difference scores is approximately 20 units.

We first compute the effect size,

$$ES = \frac{\mu_d}{\sigma_d} = \frac{10}{20} = 0.50.$$

We now substitute the effect size and the appropriate z values for the selected α and power to compute the sample size,

$$n = \left(\frac{z_{1-\alpha/2} + z_{1-\beta}}{ES}\right)^2 = \left(\frac{1.96 + 0.84}{0.50}\right)^2 = 31.4.$$

A sample of size $n = 32$ patients with migraines will ensure that a two-sided test with $\alpha = 0.05$ has 80% power to detect a mean difference of 10 points in pain before and after treatment. (Note that this assumes that all 32 patients complete the treatment and have available pain measurements before and after treatment.)

8.2.5 Sample Sizes for Two Independent Samples, Dichotomous Outcome

In studies where the plan is to perform a test of hypothesis comparing the proportions of successes in two independent populations, the hypotheses of interest are

$$H_0: p_1 = p_2,$$

$$H_1: p_1 \neq p_2,$$

where p_1 and p_2 are the proportions in the two comparison populations. The formula for determining the sample sizes to ensure that the test has a specified power is

$$n_i = 2\left(\frac{z_{1-\alpha/2} + z_{1-\beta}}{ES}\right)^2$$

where n_i is the sample size required in each group ($i = 1, 2$), α is the selected level of significance, and $z_{1-\alpha/2}$ is the value from the standard normal distribution holding $1-\alpha/2$ below it, $1-\beta$ is the selected power, and $z_{1-\beta}$ is the value from the standard normal distribution holding $1-\beta$ below it. ES is the effect size, defined as

$$ES = \frac{|p_1 - p_2|}{\sqrt{p(1-p)}},$$

where $|p_1 - p_2|$ is the absolute value of the difference in proportions between the two groups expected under the alternative hypothesis, H_1, and p is the overall proportion, based on pooling the data from the two comparison groups.

Example 8.18. In Example 8.14, we determined the sample size needed for a clinical trial proposed to evaluate the efficacy of a new drug designed to reduce systolic blood pressure. The primary outcome was systolic blood pressure measured after 12 weeks on the assigned treatment, and the proposed analysis was a comparison of mean systolic blood pressures between treatments. We now consider a similar trial with a different primary outcome. Suppose we consider a diagnosis of hypertension (yes/no) as the primary outcome. Hypertension will be defined as a systolic blood pressure above 140 or a diastolic blood pressure above 90. Suppose we expect that 30% of the participants will meet the criteria for hypertension in the placebo group. The new drug will be considered efficacious if there is a 20% reduction in the proportion of patients receiving the new drug who meet the criteria for hypertension (i.e., if 24% of the patients receiving the new drug meet the criteria for hypertension). How many patients should be enrolled in the trial to ensure that the power of the test is 80% to detect this difference in the proportions of patients with hypertension? A two-sided test will be used with a 5% level of significance.

We first compute the effect size by substituting the proportions of hypertensive patients expected in each group, $p_1 = 0.24$ and $p_2 = 0.30$, and the overall proportion, $p = 0.27$,

$$ES = \frac{|p_1 - p_2|}{\sqrt{p(1-p)}} = \frac{|0.24 - 0.30|}{\sqrt{0.27(1-0.27)}} = 0.135.$$

We now substitute the effect size and the appropriate z values for the selected α and power to compute the sample size,

$$n_i = 2\left(\frac{z_{1-\alpha/2} + z_{1-\beta}}{ES}\right)^2 = 2\left(\frac{1.96 + 0.84}{0.135}\right)^2 = 860.4.$$

Samples of size $n_1 = 861$ patients on the new drug and $n_2 = 861$ patients on placebo will ensure that the test of hypothesis will have 80% power to detect a 20% reduction in the proportion of patients on the new drug who meet the criteria for hypertension (assuming that 30% of the participants on placebo meet the criteria for hypertension). In the trial proposed in Example 8.14, the investigators hypothesized a 10% attrition rate in each group. To ensure that the total sample size of 1722 is available at 12 weeks, the investigator needs to recruit more participants to allow for attrition:

N (number to enroll) $\times$ (% retained) = Desired sample size

$$N \times 0.90 = 1722$$

$$N = \frac{1722}{0.90} = 1914$$

The investigator must enroll 1914 participants. Each participant will be randomly assigned to receive either the new medication or a placebo. Assuming that 10% are lost to follow-up, 1722 will be available for analysis.

Example 8.19. An investigator hypothesizes that a higher proportion of students who use their athletic facility regularly develop flu as compared to their counterparts who do not. The study will be conducted in the spring semester, and each student will be asked if they used the athletic facility regularly over the past 6 months and they will be followed for 3 months to assess whether they develop the flu. A test of hypothesis will be conducted to compare the students who used the athletic facility regularly and those who did not in terms of developing the flu. During a typical year, approximately 35% of the students experience flu. The investigators feel that a 30% increase in flu among those who used the athletic facility regularly would be clinically meaningful. How many students should be enrolled in the study to ensure that the power of the test is 80% to detect this difference in the proportions? A two-sided test will be used with a 5% level of significance.

We first compute the effect size by substituting the proportions of students in each group who are expected to develop flu, $p_1 = 0.46$ and $p_2 = 0.35$, and the overall proportion, $p = 0.41$,

$$ES = \frac{|p_1 - p_2|}{\sqrt{p(1-p)}} = \frac{|0.46 - 0.35|}{\sqrt{0.41(1 - 0.41)}} = 0.22.$$

We now substitute the effect size and the appropriate z values for the selected α and power to compute the sample size,

$$n_i = 2\left(\frac{z_{1-\alpha/2} + z_{1-\beta}}{ES}\right)^2 = 2\left(\frac{1.96 + 0.84}{0.22}\right)^2 = 324.0.$$

Samples of sizes $n_1 = 324$ and $n_2 = 324$ ensure that the test of hypothesis will have 80% power to detect a 30% increase in the proportion of students who use the athletic facility regularly and develop flu, assuming that 35% of the students who do not will develop flu.

8.3 SUMMARY

Determining the appropriate design of a study is more important than the statistical analysis. A poorly designed study can never be salvaged, whereas a poorly analyzed study can always be re-analyzed. A critical component in study design is the determination of the appropriate sample size. The sample size must be large enough to adequately answer the research question, yet not be so large as to involve too many patients when fewer would have sufficed. The determination of the appropriate sample size involves statistical criteria as well as clinical or practical considerations. Sample size determination involves teamwork; biostatisticians must work closely with clinical investigators to determine the sample size to address the research question of interest with adequate precision or power to produce results that are clinically meaningful.

Table 8–1 summarizes the sample size formulas for each scenario described here. The formulas are organized by the proposed analysis, a confidence interval estimate, or a test of hypothesis.

8.4 PRACTICE PROBLEMS

1. Suppose we want to design a new placebo-controlled trial to evaluate an experimental medication to increase lung capacity. The primary outcome is peak expiratory flow rate, a continuous variable measured in liters per minute. The primary outcome will be measured after 6 months of treatment. The mean peak expiratory flow rate in adults is 300 l/min with a standard deviation of 50 l/min. How many subjects should be enrolled to ensure 80% power to detect a difference of 15 l/min with a two-sided test and $\alpha = 0.05$? The investigators expect to lose 10% of the participants over the course of follow-up.

2. An investigator wants to estimate caffeine consumption in high school students. How many students would be required to ensure that a 95% confidence interval estimate for the mean caffeine intake (measured in mg) is within 15 mg of the true mean? Assume that the standard deviation in caffeine intake is 68 mg.

3. Consider the study proposed in Problem 2. How many students would be required to estimate the proportion of students who consume coffee? Suppose we want the estimate to be within 5 percentage points of the true proportion with 95% confidence.

4. A clinical trial was conducted comparing a new compound designed to improve wound healing in trauma patients to a placebo. After treatment for five days, 58% of the patients taking the new compound had a substantial reduction in the size of their wound as compared to 44% in the placebo group. The trial failed to show significance. How many subjects would be required to detect the difference in proportions observed in the trial with 80% power? A two-sided test is planned at $\alpha = 0.05$.

5. A crossover trial is planned to evaluate the impact of an educational intervention program to reduce alcohol consumption in patients determined to be at risk for alcohol problems. The plan is to measure alcohol consumption (the number of drinks on a typical drinking day) before the intervention

TABLE 8–1 Sample Size Formulas

Outcome Variable, Number of Groups: Planned Analysis	Sample Size to Estimate Confidence Interval (CI)	Sample Size for Test of Hypothesis
Continuous outcome, one sample: CI for μ, H_0: $\mu = \mu_0$	$n = \left(\dfrac{Z\sigma}{E}\right)^2$	$n = \left(\dfrac{Z_{1-\alpha/2} + Z_{1-\beta}}{ES}\right)^2$ $ES = \dfrac{|\mu_1 - \mu_0|}{\sigma}$
Continuous outcome, two independent samples: CI for $(\mu_1 - \mu_2)$, H_0: $\mu_1 = \mu_2$	$n_i = 2\left(\dfrac{Z\sigma}{E}\right)^2$	$n_i = 2\left(\dfrac{Z_{1-\alpha/2} + Z_{1-\beta}}{ES}\right)^2$ $ES = \dfrac{|\mu_1 - \mu_2|}{\sigma}$
Continuous outcome, two matched samples: CI for μ_d, H_0: $\mu_d = 0$	$n = \left(\dfrac{Z\sigma_d}{E}\right)^2$	$n = \left(\dfrac{Z_{1-\alpha/2} + Z_{1-\beta}}{ES}\right)^2$ $ES = \dfrac{\mu_d}{\sigma_d}$
Dichotomous outcome, one sample: CI for p, H_0: $p = p_0$	$n = p(1-p)\left(\dfrac{Z}{E}\right)^2$	$n = \left(\dfrac{Z_{1-\alpha/2} + Z_{1-\beta}}{ES}\right)^2$ $ES = \dfrac{|p_1 - p_0|}{\sqrt{p_0(1-p_0)}}$
Dichotomous outcome, two independent samples: CI for $(p_1 - p_2)$, H_0: $p_1 = p_2$	$n_i = [p_1(1-p_1) + p_2(1-p_2)]\left(\dfrac{Z}{E}\right)^2$	$n_i = 2\left(\dfrac{Z_{1-\alpha/2} + Z_{1-\beta}}{ES}\right)^2$ $ES = \dfrac{|p_1 - p_2|}{\sqrt{p(1-p)}}$

and then again after participants complete the educational intervention program. How many participants would be required to ensure that a 95% confidence interval for the mean difference in the number of drinks is within 2 drinks of the true mean difference? Assume that the standard deviation of the difference in the mean number of drinks is 6.7 drinks and that 20% of the participants will drop out over the course of follow-up.

6. An investigator wants to design a study to estimate the difference in the proportions of men and women who develop early onset cardiovascular disease (defined as cardiovascular disease before age 50). A study conducted 10 years ago found that 15% and 8% of men and women, respectively, developed early onset cardiovascular disease. How many men and women are needed to generate a 95% confidence interval estimate for the difference in proportions with a margin of error not exceeding 4%?

7. The mean body mass index (BMI) for boys of age 12 years is 23.6 kg/m². An investigator wants to test if the BMI is higher in 12-year-old boys living in New York City. How many boys are needed to ensure that a two-sided test of hypothesis has 80% power to detect a difference in BMI of 2 kg/m²? Assume that the standard deviation in BMI is 5.7 kg/m².

8. An investigator wants to design a study to estimate the difference in the mean BMI between 12-year-old boys and girls living in New York City. How many boys and girls are needed to ensure that a 95% confidence interval estimate for the difference in mean BMI between boys and girls has a margin of error not exceeding 2 kg/m^2? Use the estimate of the variability in BMI from Problem 7.

9. How many stomach cancer patients and controls would be required to estimate the difference in proportions of patients with sufficient concentrations of plasma antioxidant vitamins with a margin of error not exceeding 5% using a 95% CI? The data in **Table 8–2** were reported recently from a similar study investigating the sufficiency of plasma antioxidant vitamins in throat cancer patients.

10. Recently it has been observed that HIV-infected patients develop peripheral lypoatrophy while on potent antiretroviral therapy. A clinical trial is planned to determine whether a new chemical will improve this condition. In the trial, participants will be randomized to receive the new chemical or a placebo, and changes from baseline in subcutaneous adipose cross-sectional area as measured by CT scan will be calculated after 24 weeks of treatment. Investigators hope to show that the increases in patients receiving the chemical will be greater than the increases in patients receiving the placebo. They hypothesize that the participants assigned to the chemical arm will exhibit a mean change of 30% and that the participants assigned to the placebo arm will exhibit a mean change of 0%. Prior literature suggests that the standard deviation of the changes will be 57% in both arms. How many participants are needed to ensure 80% power? Assume that $\alpha = 0.05$, that there are equal numbers in each group, and that 20% of the participants will drop out over the course of follow-up.

11. We wish to design a study to assess risk factors for stroke and dementia. There are two primary outcome measures. The first is a measure of neurologic function. Patients with "normal" function typically score 70 with an SD of 15. The second outcome is incident stroke. In patients over the age of 65, approximately 14% develop stroke over 25 years. How many participants over the age of 65 must be enrolled to ensure the following?
    a. That the margin of error in a 95% CI for the difference in mean neurologic function scores between men and women does not exceed 2 mg/dL.
    b. That we have 80% power to detect a 20% reduction in incidence of stroke between participants who participate in a healthy-aging program compared to those who do not. Assume that the incidence of stroke is 14% in those who do not participate.
    c. How many men and women over the age of 65 should be enrolled to satisfy parts (a) and (b) assuming that 15% will be lost to follow-up over the course of the 25-year follow-up?

12. Suppose we want to design a new study to assess the implications of childhood obesity on poor health outcomes in adulthood. In the study we will consider total cholesterol at age 50 and incident cardiovascular disease by age 50 as the key outcomes. We will compare normal versus overweight/obese children. Assume that the standard deviation in total cholesterol is 35.
    a. How many normal and overweight/obese children are required to ensure that a 95% CI estimate for the difference in total cholesterol has a margin of error not exceeding 5 mg/dL?
    b. How many normal and overweight/obese children are required to ensure 80% power to detect a 4% difference in incident cardiovascular disease between groups? Assume that approximately 3% of normal-weight children will develop cardiovascular disease by age 50 and that 7% of overweight/obese children will develop cardiovascular disease by age 50 and that a two-sided test with $\alpha = 0.05$ will be used.
    c. How many normal and overweight/obese children should be enrolled to satisfy parts (a) and (b) assuming that 20% will be lost to follow-up over the course of the study?

TABLE 8–2 Plasma Antioxidant Vitamins in Throat Cancer Patients

	N	% with Sufficient Concentrations of Plasma Antioxidant Vitamins
Throat cancer patients	25	47%
Controls	25	56%

REFERENCES

1. Buschman, N.A., Foster, G., and Vickers, P. "Adolescent girls and their babies: Achieving optimal birth weight. Gestational weight gain and pregnancy outcome in terms of gestation at delivery and infant birth weight: A comparison between adolescents under 16 and adult women." *Child: Care, Health and Development* 2001; 27(2): 163–171.
2. Feuer, E.J. and Wun, L.M. DEVCAN: Probability of Developing or Dying of Cancer, Version 4.0. Bethesda, MD: National Cancer Institute, 1999.
3. Howell, D.C. *Statistical Methods for Psychology.* Boston: Duxbury Press, 1982.
4. Fleiss, J.L. *Statistical Methods for Rates and Proportions.* New York: John Wiley & Sons, 1981.
5. National Center for Health Statistics. *Health, United States, 2005 with Chartbook on Trends in the Health of Americans.* Hyattsville, MD: US Government Printing Office, 2005.
6. Plaskon, L.A., Penson, D.F., Vaughan, T.L., and Stanford, J.L. "Cigarette smoking and risk of prostate cancer in middle-aged men." *Cancer Epidemiology Biomarkers & Prevention* 2003; 12: 604–609.
7. Rutter, M.K., Meigs, J.B., Sullivan, L.M., D'Agostino, R.B., and Wilson, P.W. "C-reactive protein, the metabolic syndrome, and prediction of cardiovascular events in the Framingham Offspring Study." *Circulation* 2004; 110: 380–385.
8. Vasan, R.S., Sullivan, L.M., Wilson, P.W., Sempos, C.T., Sundstrom, J., Kannel, W.B., Levy, D., and D'Agostino, R.B. "Relative importance of borderline and elevated levels of coronary heart disease risk factors." *Annals of Internal Medicine* 2005; 142: 393–402.
9. Wechsler, H., Lee, J.E., Kuo, M., and Lee, H. "College binge drinking in the 1990s: A continuing problem. Results of the Harvard School of Public Health, 1999 College Alcohol Study." *Journal of American College Health* 2000; 48: 199–210.

CHAPTER 9

Multivariable Methods

When and Why

Key Questions
- What does it mean when investigators adjust for other variables?
- How can you be sure that an association is real?
- How can you be sure that you accounted for all other possible explanations of associations?

In the News
There is much discussion about childhood and adult obesity in the United States, and increasingly around the globe. In the United States, approximately 35% of adults and 17% of children are obese, with these rates being higher among blacks (48% of adults, 20% of children) and Latinos (43% of adults, 22% of children) as compared to whites (33% of adults, 14% of children).[1] The prevalence of overweight and obesity in the United States varies widely by state, from a low of 22.1% in Utah to a high of 39.8% in Louisiana.[2]

The World Health Organization (WHO) reports an increase globally in the number of overweight and obese young children (ages 0 to 5 years) from 32 million in 1990 to 42 million in 2013.[3] WHO has suggested a number of preventive measures for these conditions, including increased rates of breastfeeding, consumption of healthier foods, and regular physical activity.

There are a number of health risks associated with obesity, such as hypertension, diabetes, and mortality.[4]

Mortality associated with overweight and obesity is reportedly three times higher among white men compared to black men, and approximately 20% higher among black women compared to white women.[5]

---

[1] Available at http://stateofobesity.org/disparities/.
[2] Available at https://www.healthiergeneration.org/about_childhood_obesity/in_your_state/.
[3] Available at http://www.who.int/end-childhood-obesity/facts/en/.
[4] Available at https://www.cdc.gov/obesity/adult/causes.html
[5] Masters, R.K., Reither, E.N., Powers, D.A., Yang, Y.C., Burger, A.E., and Link, B.G. "The impact of obesity on US mortality levels: the importance of age and cohort factors in population estimates." *American Journal of Public Health* 2013; 103: 1895–1901.

Dig In
- Are some groups more likely to be obese? Why?
- What might explain some of the differences in health risks attributable to obesity among racial/ethnic groups?
- What factors might mask or enhance the associations between obesity and hypertension, diabetes, or mortality?

Learning Objectives
By the end of this chapter, the reader will be able to
- Define and provide examples of dependent and independent variables in a study of a public health problem
- Explain the principle of statistical adjustment to a lay audience
- Organize data for regression analysis
- Define and provide an example of confounding
- Define and provide an example of effect modification
- Interpret coefficients in multiple linear and multiple logistic regression analysis

In Chapter 6 and Chapter 7, we presented statistical inference procedures for estimation and hypothesis testing. We discussed many scenarios that differed in terms of the nature of the outcome variable (continuous, dichotomous, categorical, or ordinal), the number of comparison groups (one, two, or more than two), and whether the groups were independent (physically separate) or dependent (matched). In each scenario, we considered one outcome variable, and we investigated whether there were differences in that outcome variable among the comparison groups or whether there was an association between an exposure or risk factor and an outcome. What we have not considered up to this point is the possibility that observed differences or associations might be due to other characteristics or variables. Multivariable statistical methods are used to assess the interrelationships among several risk factors or exposure variables and a single outcome. The topic of multivariable analysis is extensive. We describe only general principles and concepts here. Interested readers should see Kleinbaum, Kupper, and Muller; Jewell; and Hosmer and Lemeshow for more in-depth discussion.[1-3]

We specifically discuss the use of multivariable modeling techniques to address **confounding** and effect modification. In Chapter 2, we defined confounding as a distortion of the effect of an exposure or risk factor on an outcome by another characteristic or variable. **Effect modification** occurs when there is a different relationship between the exposure or risk factor and the outcome depending on the level of another characteristic or variable. In both situations, the third variable can exaggerate or mask the association between the risk factor and the outcome. Effect modification is also called statistical interaction.

Analytically, confounding and effect modification are handled differently. When there is confounding, multivariable methods can be used to generate an estimate of the association between an exposure or risk factor and an outcome adjusting for or taking into account the impact of the confounder. In contrast, with effect modification there is a different relationship between the risk factor and outcome depending on the level of a third variable, and therefore an overall estimate of association does not accurately convey the information in the data. In the presence of effect modification, results are generally presented separately for each level of the third variable (using the techniques presented in Chapter 6 and Chapter 7). To formalize these concepts, we first present some examples and then move into the methods.

Suppose we want to assess the association between smoking and cardiovascular disease in a cohort study. We may find that smokers in the cohort are much more likely to develop cardiovascular disease. Suppose that we estimate a relative risk of $\hat{RR} = 2.6$ with a 95% confidence interval of (1.5, 4.1). In our study, smokers have 2.6 times the risk of developing cardiovascular disease as compared to nonsmokers. However, it may also be the case that the smokers are less likely to exercise and have higher cholesterol levels. Without examining these other characteristics, we could incorrectly infer that there is a strong association between smoking and cardiovascular disease when the relationship may be due, in part, to lack of exercise or to high cholesterol. Multivariable methods can be used to address these complex relationships and to tease out what proportion of the association might be due to smoking, as opposed to other risk factors. In other words, we use these methods to adjust the magnitude of association for the impact of other variables (e.g., the association between smoking and cardiovascular disease adjusted for exercise and cholesterol). Methods are also discussed that allow us to assess whether an association remains statistically significant after adjusting for the impact of other variables.

Multivariable methods can also be used to assess effect modification—specifically, the situation in which the relationship between a risk factor and the outcome of interest varies by a third variable. For example, suppose we are interested in the efficacy of a new drug designed to lower total cholesterol. A clinical trial is conducted and the drug is shown to be effective, with a statistically significant reduction in total cholesterol in patients who receive the new drug as compared to patients who receive a placebo. Suppose that the investigators look more closely at the data and find that the reduction (or effect of the drug) is only present in participants with a specific genetic marker and that there is no reduction in persons who do not possess the marker. This is an example of effect modification, or statistical interaction. The effect of the treatment is different depending on the presence or absence of the genetic marker. Multivariable methods can be used to identify effect modification. When effect modification is present, the strategy is to present separate results according to the third variable (i.e., to report the effect of the treatment separately in persons with and without the marker).

Other uses of multivariable methods include the consideration of several risk factors simultaneously, where the goal is to assess the relative importance of each with regard to a single outcome variable. For example, the Framingham Heart Study has a long history of developing multivariable risk functions.[4,5] These risk functions are used to predict the likelihood that a person will develop cardiovascular disease over a fixed period (e.g., the next 10 years) as a function of his or her risk factors. The risk factors include age, sex, systolic blood pressure, total and HDL cholesterol levels, current smoking, and diabetes status. Multivariable techniques are used in this setting to account for all of the risk factors simultaneously. This application is similar to the first, except that rather than focusing on the association between one risk factor and the outcome (adjusting for others), we are interested in the relationships between each of the risk factors and the outcome.

There are many other uses of multivariable methods. We focus here on only a few specific applications. Because the computations are intensive, our presentation is more general with particular emphasis on interpretation rather than computation. In practice, the computations are performed using a statistical computing package.[6]

It is important to note that the statistical analysis of any study should begin with a complete description of the study data using the methods described in Chapter 4. The primary analyses should proceed as planned to either generate estimates of unknown parameters or to perform tests of hypothesis using the techniques described in Chapter 6 and Chapter 7, respectively. The analyses we presented in Chapter 6 and Chapter 7 are called unadjusted or **crude analyses** as they focused exclusively on the associations between one risk factor or exposure and the outcome. Multivariable methods are used after the study data are described and after unadjusted analyses are performed. In the clinical trials setting, the unadjusted analyses are generally the final analyses due primarily to the randomization component that (in theory) eliminates the possibility of confounding. While the multivariable methods we describe are a means to account for confounding, they should not be relied on to "correct" problems in a study.

Multivariable models are used for the statistical adjustment of confounding, whereas careful study design can provide much more in the way of minimizing confounding. Randomization is a very effective means of minimizing confounding. Unfortunately, not all studies are suitable for randomization. There are other options to minimize confounding (e.g., matching), but in all cases it is important to recognize that multivariable models can only adjust or account for differences in confounding variables that are measured in the study. Therefore, investigators must carefully plan studies not only with an eye toward minimizing confounding whenever possible, but also with forethought to measure variables that might be potential confounders.

9.1 CONFOUNDING AND EFFECT MODIFICATION

Confounding is present when the relationship between a risk factor and an outcome is modified by a third variable, called the confounding variable. A **confounding variable** or confounder is one that is related to the exposure or risk factor of interest and also to the outcome. Identification of confounding variables requires a strong working knowledge of the substantive area under investigation. (An additional consideration in determining whether the third variable is a confounder is assessing whether or not the third variable is part of a causal chain. A confounder is not in the causal chain from exposure to outcome. This is a complex issue; for more details, see Aschengrau and Seage.[7]) Confounding can be assessed in a number of different ways. Some investigators use formal statistical tests and others use clinical or practical judgment. We discuss both approaches.

Effect modification, also called statistical interaction, occurs when there is a different association between the risk factor and outcome depending on a third variable. Effect modification is generally assessed by statistical significance using regression techniques (discussed in Section 9.4 and Section 9.5). We illustrate confounding and effect modification in Example 9.1 and Example 9.2, respectively.

Example 9.1. Consider a study designed to assess the association between obesity—defined as a body mass index (BMI) of 30 or more—and the incidence of cardiovascular disease. Data are collected in a cohort study in which participants between the ages of 35 and 65 who are free of cardiovascular disease (CVD) are enrolled and followed over 10 years. Each participant's BMI is assessed at baseline and participants are followed for 10 years for incident cardiovascular disease. A summary of the data is contained in **Table 9-1**.

The proportion of obese persons at baseline who develop CVD (or the incidence of CVD among obese persons) is 46 / 300 = 0.153, and the proportion of non-obese persons who develop CVD (or the incidence of CVD among non-obese persons) is 60 / 700 = 0.086. The 10-year incidence of CVD is shown graphically in **Figure 9-1**. The estimate of the relative risk of CVD in obese as compared to non-obese persons is

$$\hat{RR} = \frac{\hat{p}_1}{\hat{p}_2} = \frac{0.153}{0.086} = 1.78.$$

Obese persons have 1.78 times the risk of developing CVD as compared to non-obese persons in this study.

Many studies have shown an increase in the risk of CVD with advancing age. Could age be a confounder in this analysis?

TABLE 9-1 Baseline Obesity and Incident CVD Over 10-Year Follow-Up

	Incident CVD	No CVD	Total
Obese	46	254	300
Not obese	60	640	700
Total	106	894	1000

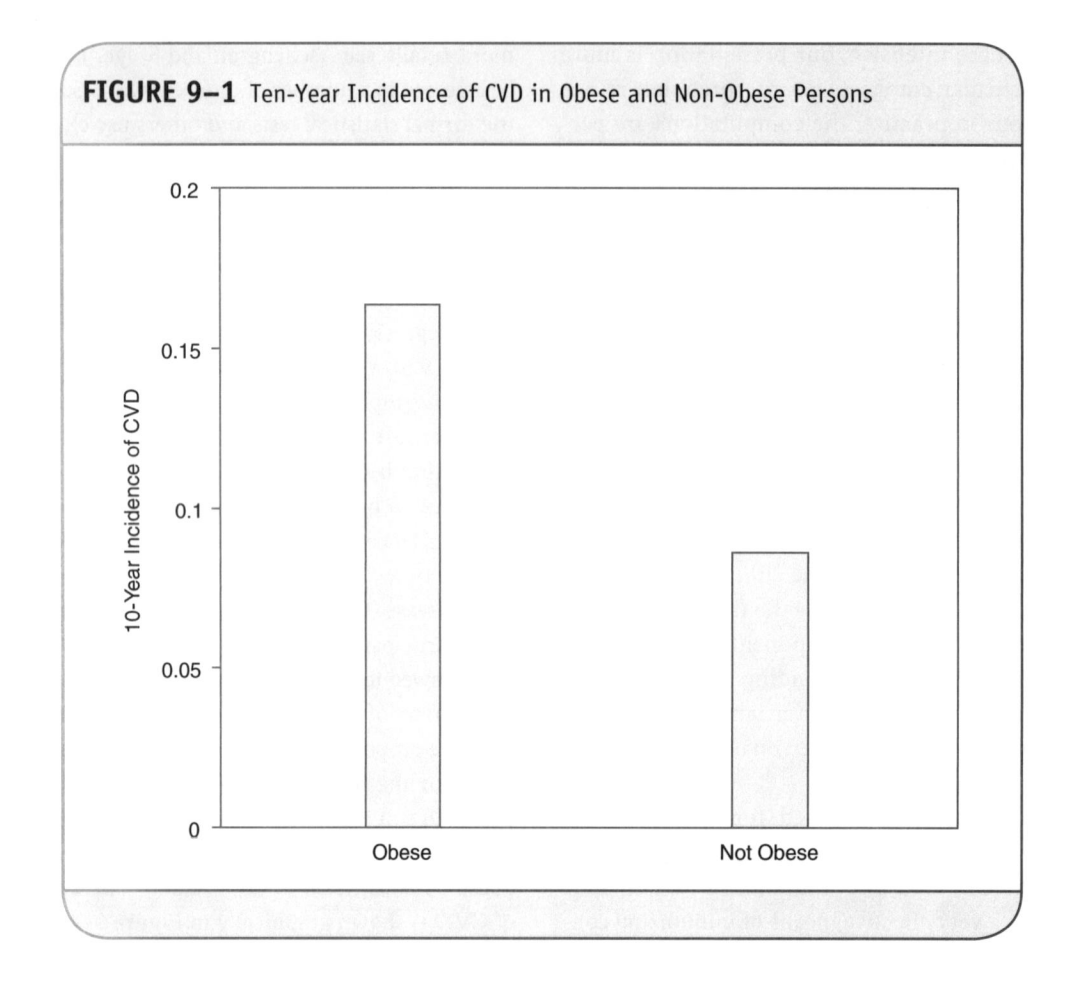

FIGURE 9–1 Ten-Year Incidence of CVD in Obese and Non-Obese Persons

Could any of the observed association between obesity and incident CVD be attributable to age? A confounder is a variable that is associated with the risk factor and with the outcome. Is age associated with obesity? Is age associated with incident CVD?

In this study, age is measured as a continuous variable (in years). For interpretation purposes, suppose we dichotomize age at 50 years (less than 50 years of age versus 50 years and older). In this study, $n = 600$ (60%) participants are less than 50 years of age and $n = 400$ (40%) are 50 years of age and older. **Table 9–2** shows the relationship between age and obesity.

Age is related to obesity. A higher proportion of persons 50 years of age and older are obese (200 / 400 = 0.500) as compared to persons less than 50 years of age (100 / 600 = 0.167). Looking at the data another way, a much higher proportion of obese persons are 50 years of age and older (200 / 300 = 0.667) as compared to non-obese persons (200 / 700 = 0.286). Could the fact that the majority of the obese persons are older explain the association between obesity and incident CVD? Is age driving the association?

TABLE 9–2 Baseline Age and Baseline Obesity

Age	Obese	Not Obese	Total
< 50	100	500	600
50+	200	200	400
Total	300	700	1000

Table 9–3 shows the relationship between age and incident CVD. Age is related to incident CVD. A much higher proportion of persons 50 years of age and older develop CVD (61 / 400 = 0.153) as compared to persons less than 50 years of age (45 / 600 = 0.075).

A higher proportion of the obese persons are older, and age is related to incident CVD. Thus, age meets the definition of a confounder (i.e., it is associated with the risk factor, obesity, and the outcome, incident CVD). We now examine the relationship between obesity and incident CVD in persons less

TABLE 9–3 Baseline Age and Incident CVD Over 10-Year Follow-Up

Age	Incident CVD	No CVD	Total
< 50	45	555	600
50+	61	339	400
Total	106	894	1000

TABLE 9–4 Baseline Obesity and Incident CVD by Age Group

Age < 50	Incident CVD	No CVD	Total
Obese	10	90	100
Not obese	35	465	500
Total	45	555	600

Age 50+	Incident CVD	No CVD	Total
Obese	36	164	200
Not obese	25	175	200
Total	61	339	400

than 50 years of age and in persons 50 years of age and older separately (see **Table 9–4**).

Among people less than 50 years of age, the proportion of obese persons who develop CVD is 10 / 100 = 0.100, and the proportion of non-obese persons who develop CVD is 35 / 500 = 0.070. The relative risk of CVD in persons less than 50 years of age who are obese as compared to non-obese is

$$\hat{RR} = \frac{\hat{p}_1}{\hat{p}_2} = \frac{0.100}{0.070} = 1.43.$$

Among people 50 years of age and older, the proportion of obese persons who develop CVD is 36 / 200 = 0.180, and the proportion of non-obese persons who develop CVD is 25 / 200 = 0.125. The relative risk of CVD in persons 50 years of age and older who are obese as compared to non-obese is

$$\hat{RR} = \frac{\hat{p}_1}{\hat{p}_2} = \frac{0.180}{0.125} = 1.44.$$

The relative risk based on the combined sample suggests a much stronger association between obesity and incident CVD ($\hat{RR}$ = 1.78). **Figure 9–2** shows the 10-year incidence of CVD for obese and non-obese persons in each age group.

There are different methods to determine whether a variable is a confounder or not. Some investigators perform formal tests of hypothesis to assess whether the variable is associated with the risk factor and with the outcome; other investigators do not conduct statistical tests but instead inspect the data, and if there is a practically important or clinically meaningful relationship between the variable and the risk factor and between the variable and the outcome (regardless of whether that relationship reaches statistical significance), then the variable is said to be a confounder. The data in Table 9–2 through **Table 9–5** would suggest that age is a confounder by the latter method. (In Section 9.3, we provide yet another method to assess confounding that again is based on evaluation of associations from a practical or clinical point of view.) We now illustrate the approach to assessing confounding using tests of hypothesis. We specifically test whether there is a statistically significant association between age and obesity and between age and incident CVD.

The tests of hypothesis can be performed using the χ^2 test of independence. In the first test, we determine whether age and obesity are related. The data for the test are shown in Table 9–2.

Step 1. Set up hypotheses and determine the level of significance.

H_0: Age and obesity are independent

H_1: H_0 is false

$\alpha = 0.05$

Step 2. Select the appropriate test statistic.
The formula for the test statistic is in Table 7–42 and is given as

$$\chi^2 = \sum \frac{(O-E)^2}{E}$$

The condition for appropriate use of the preceding test statistic is that each expected frequency in each cell of the table is at least 5. In Step 4, we compute the expected frequencies and ensure that the condition is met.

Step 3. Set up the decision rule.
The row variable is age ($r = 2$) and the column variable is obesity ($c = 2$), so for this test, $df = (2 - 1)(2 - 1) = 1(1) = 1$. For $df = 1$ and a 5% level of significance, the appropriate

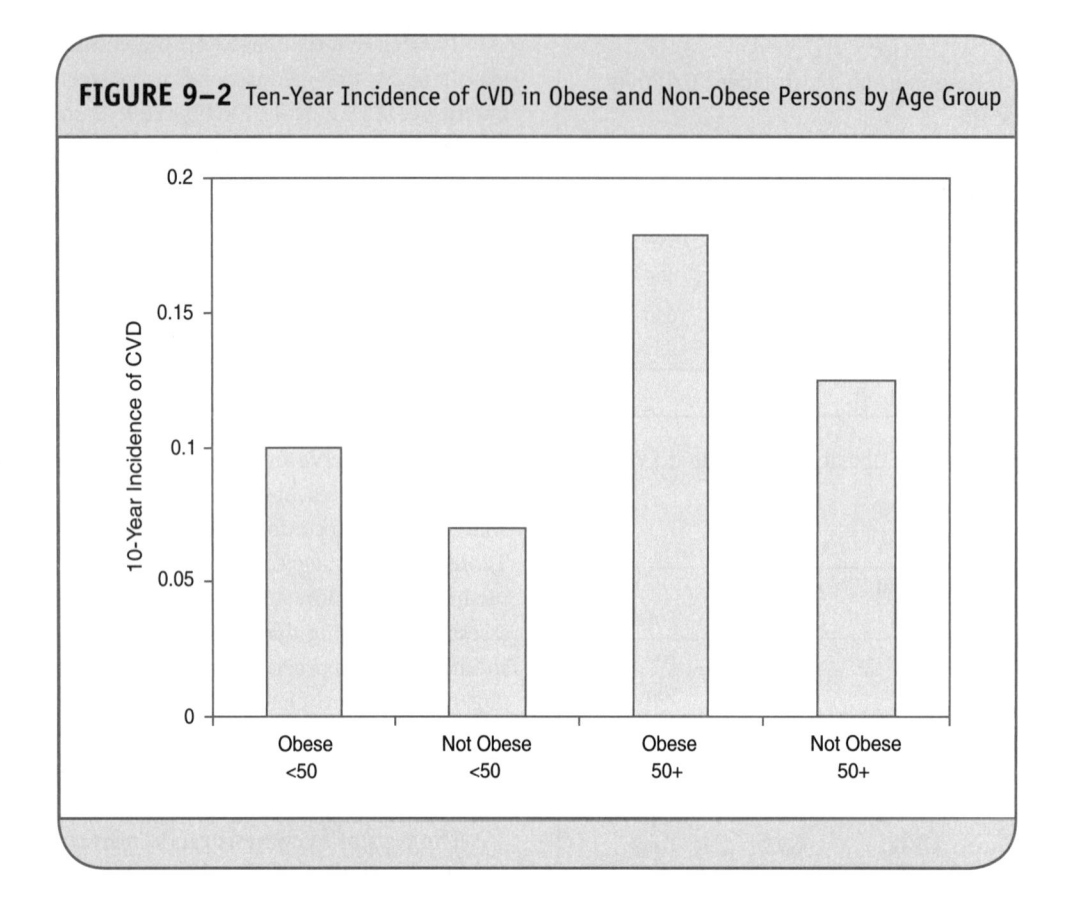

FIGURE 9–2 Ten-Year Incidence of CVD in Obese and Non-Obese Persons by Age Group

TABLE 9–5 Age and Obesity Frequencies

Age	Obese	Not Obese	Total
< 50	100 (180)	500 (420)	600
50+	200 (120)	200 (280)	400
Total	300	700	1000

critical value from Table 3 in the Appendix is 3.84, and the decision rule is

$$\text{Reject } H_0 \text{ if } \chi^2 \geq 3.84.$$

Step 4. Compute the test statistic.
We now compute the expected frequencies using the formula,

$$\text{Expected frequency} = \frac{\text{Row total} \times \text{column total}}{N}.$$

The numbers in each cell of the table are the observed frequencies and the expected frequencies, shown in parentheses (see Table 9–5). The test statistic is computed as

$$\chi^2 = \frac{(100-180)^2}{180} + \frac{(500-420)^2}{420} + \frac{(200-120)^2}{120} + \frac{(200-280)^2}{280},$$

$$\chi^2 = 35.56 + 15.24 + 53.33 + 22.86 = 126.99.$$

Step 5. Conclusion.
We reject H_0 because $126.99 > 3.84$. We have statistically significant evidence at $\alpha = 0.05$ to show that H_0 is false; age and obesity are not independent (i.e., they are related). Using Table 3 in the Appendix, the p-value is $p < 0.005$.

The second test is to assess whether age is associated with incident CVD. We again use the χ^2 test of independence, and the data for the test are shown in Table 9–3.

Step 1. Set up hypotheses and determine the level of significance.

H_0: Age and incident CVD are independent

H_1: H_0 is false

$\alpha = 0.05$

Step 2. Select the appropriate test statistic.
The formula for the test statistic is in Table 7–42 and is given below:

$$\chi^2 = \Sigma \frac{(O-E)^2}{E}$$

The condition for appropriate use of the preceding test statistic is that each expected frequency is at least 5. In Step 4, we compute the expected frequencies and ensure that the condition is met.

Step 3. Set up the decision rule.
The row variable is age ($r = 2$) and the column variable is incident CVD ($c = 2$), so for this test, $df = (2 - 1)(2 - 1) = 1(1) = 1$. For $df = 1$ and a 5% level of significance, the decision rule is

Reject H_0 if $\chi^2 \geq 3.84$.

Step 4. Compute the test statistic.
We now compute the expected frequencies using the formula,

$$\text{Expected frequency} = \frac{\text{Row total} \times \text{column total}}{N}.$$

The numbers in each cell of the table are the observed frequencies and the expected frequencies, shown in parentheses (see **Table 9–6**). The test statistic is computed as

$$\chi^2 = \frac{(45-63.6)^2}{63.6} + \frac{(555-536.4)^2}{536.4} + \frac{(61-42.4)^2}{42.4} + \frac{(339-357.6)^2}{357.6},$$

$$\chi^2 = 5.44 + 0.64 + 8.16 + 0.97 = 15.21.$$

Step 5. Conclusion.
We reject H_0 because $15.21 > 3.84$. We have statistically significant evidence at $\alpha = 0.05$ to show that H_0 is false; age and incident CVD are not independent. Using Table 3 in the Appendix, the p-value is $p < 0.005$.

Because age is related to obesity (17% of younger persons are obese as compared to 50% of older persons, and 67% of obese persons are 50 years of age and older as compared to 29% of non-obese persons) and age is related to incident CVD (7.5% of younger persons develop CVD as compared to 15.3% of older persons), age is a confounder. Therefore, some of the association between obesity and incident CVD can be attributed to age. In Section 9.2 to Section 9.5, we discuss multivariable methods that can be used to estimate the magnitude of the association between obesity and incident CVD adjusting for age.

Example 9.2. A clinical trial is conducted to evaluate the efficacy of a new drug to increase HDL cholesterol (the "good" cholesterol). One hundred patients are enrolled in the trial and randomized to receive either the new drug or a placebo. Background characteristics (e.g., age, sex, educational level, income) and clinical characteristics (e.g., height, weight, blood pressure, total and HDL cholesterol levels) are measured at baseline. Suppose that the background and clinical characteristics (including baseline HDL levels) are comparable between groups (data not shown). Each patient is instructed to take the assigned medication for 8 weeks, at which time his or her HDL cholesterol is measured. Summary statistics on the primary outcome, HDL cholesterol, are shown in **Table 9–7**.

On average, the mean HDL levels are 0.95 mg/dL higher in patients treated with the new drug. A two-sample test to compare mean HDL levels between treatments (i.e., $H_0: \mu_1 = \mu_2$ versus $H_1: \mu_1 \neq \mu_2$) has a test statistic of $z = 1.13$, which is not statistically significant at $\alpha = 0.05$. **Figure 9–3** shows the mean HDL levels in each group. The error bars represent the standard errors (SE) in each group.

Suppose that the lack of effect concerns investigators as they had expected a significant result on the basis of preliminary studies. They wonder whether other variables might be masking the effect of the treatment. Suppose other studies had shown differential effectiveness of a similar drug in men and women. In this study, there are 19 men and 81 women. **Table 9–8** shows the number and percentage of men assigned to each treatment.

TABLE 9–6 Age and CVD Frequencies

Age	Incident CVD	No CVD	Total
< 50	45 (63.6)	555 (536.4)	600
50+	61 (42.4)	339 (357.6)	400
Total	106	894	1000

TABLE 9–7 HDL by Treatment

	Sample Size	Mean HDL	Standard Deviation of HDL
New drug	50	40.16	4.46
Placebo	50	39.21	3.91

FIGURE 9–3 Mean (SE) HDL Levels by Treatment

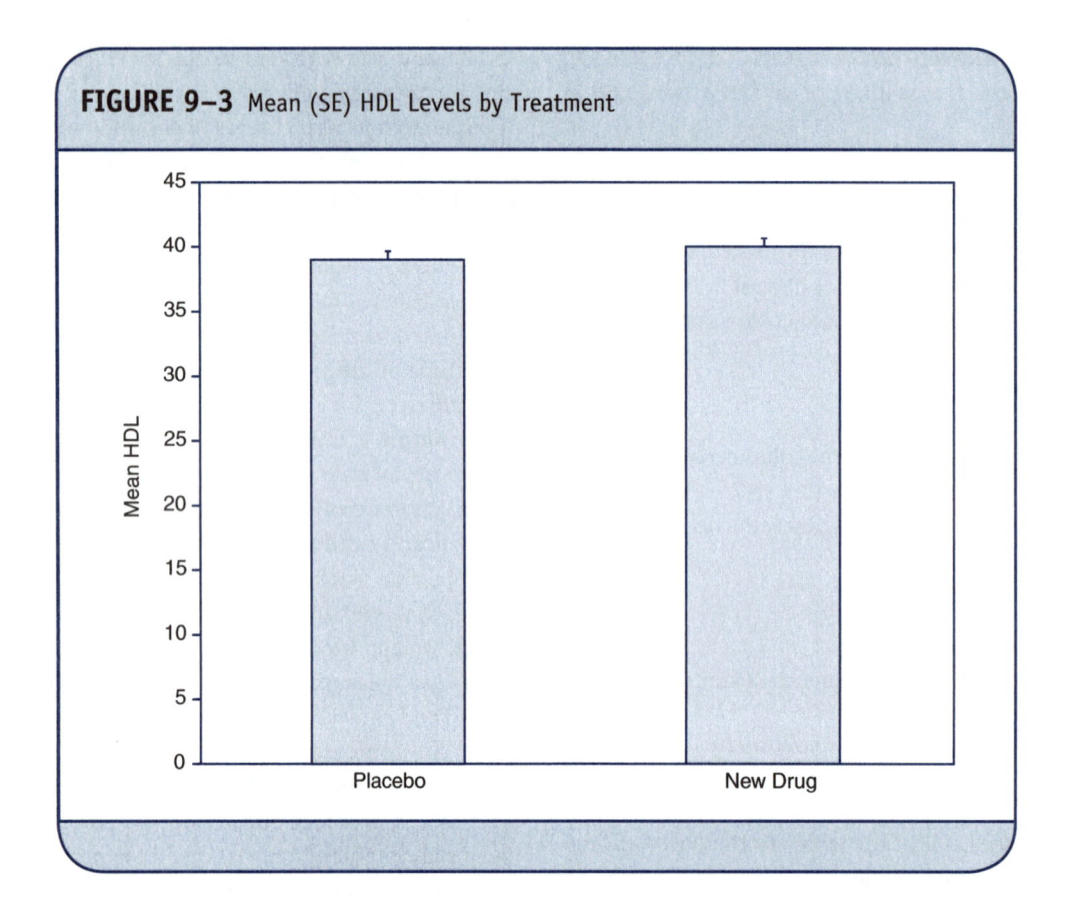

TABLE 9–8 Number (%) of Men in Each Treatment Group

	Sample Size	Number	Percentage
New drug	50	10	20
Placebo	50	9	18

TABLE 9–9 HDL by Treatment and Sex

	Women		
	Sample Size	Mean HDL	Std Dev of HDL
New drug	40	38.88	3.97
Placebo	41	39.24	4.21
	Men		
	Sample Size	Mean HDL	Std Dev of HDL
New drug	10	45.25	1.89
Placebo	9	39.06	2.22

There does not appear to be a difference in the proportions of men assigned to the treatments (20% vs 18%). A two-sample test of independent proportions or an χ^2 test of independence can be used to compare the proportions of men between treatments (i.e., $H_0: p_1 = p_2$ versus $H_1: p_1 \neq p_2$, or H_0: Sex and treatment are independent versus H_1: H_0 is false). The test of proportions has a test statistic of $z = 0.25$ (and the χ^2 test has $\chi^2 = 0.06$), which is not statistically significant at $\alpha = 0.05$.

Thus, sex is not a confounder as it is not related to the treatment. There is no meaningful difference in the proportions of men assigned to receive the new drug or the placebo; the difference in proportions is also not statistically significant. (Recall that a confounder is related to both the risk factor—the treatment variable in this case—and the outcome.) The investigators then wonder whether there might be effect modification by sex. Effect modification is a difference in the effect of treatment depending on the participant's sex. The mean HDL levels for men and women assigned to the new drug and placebo are shown in **Table 9-9**.

On average, the mean HDL levels are 0.36 mg/dL lower in women treated with the new drug as compared to a placebo. On average, the mean HDL levels are 6.19 mg/dL higher in men treated with the new drug. This is an example of effect modification by sex. There is a different relationship between treatment and outcome depending on the sex of the participant. There is very little, if any, effect of treatment among women and a large effect among men.

Figure 9–4 shows the mean HDL levels in each treatment group, stratified (or presented separately) by sex. Among women, there is almost no difference in mean HDL levels by treatment. However, among men there is a substantial difference in mean HDL, with men assigned to the new drug having a much higher mean HDL than men assigned to the placebo. It is important to note that there are very few men in the analysis. Thus, the comparison of the mean HDL levels in men assigned to the new drug versus the placebo should be interpreted carefully.

It is also important to note that in the clinical trials setting, the only analyses that should be conducted are those that are planned *a priori* and specified in the study protocol. In contrast, in epidemiologic studies there are often exploratory analyses that are conducted to fully understand associations (or lack thereof). However, these analyses should also be restricted to those that are biologically sensible.

When there is effect modification, analysis of the pooled data can be misleading. In this example, the pooled data (men and women combined) show no effect of the treatment. Because there is effect modification by sex, it is important to look at the differences in HDL levels among men and women, considered separately (or stratified by sex). However, in stratified analyses investigators must be careful to ensure that the sample size is adequate to provide a meaningful analysis. In Section 9.3 to Section 9.5, we discuss methods to formally assess effect modification.

9.2 THE COCHRAN–MANTEL–HAENSZEL METHOD

In Example 9.1 and Example 9.2, we illustrated the concepts of confounding and effect modification, respectively. Confounding occurs when a third variable, the confounder, affects the relationship between the risk factor or exposure and the outcome. In Example 9.1, age was the confounder. Age was related to obesity and to incident CVD. Stated differently, there was an imbalance in the age distribution among obese and non-obese persons. The association between obesity and incident CVD was similar among younger and older persons (in stratified analysis). When the data were pooled (all ages combined), the effect of obesity was magnified. In Example 9.2, we found that sex was an effect modifier. Stated differently, there was a statistical interaction between sex and treatment on mean HDL levels

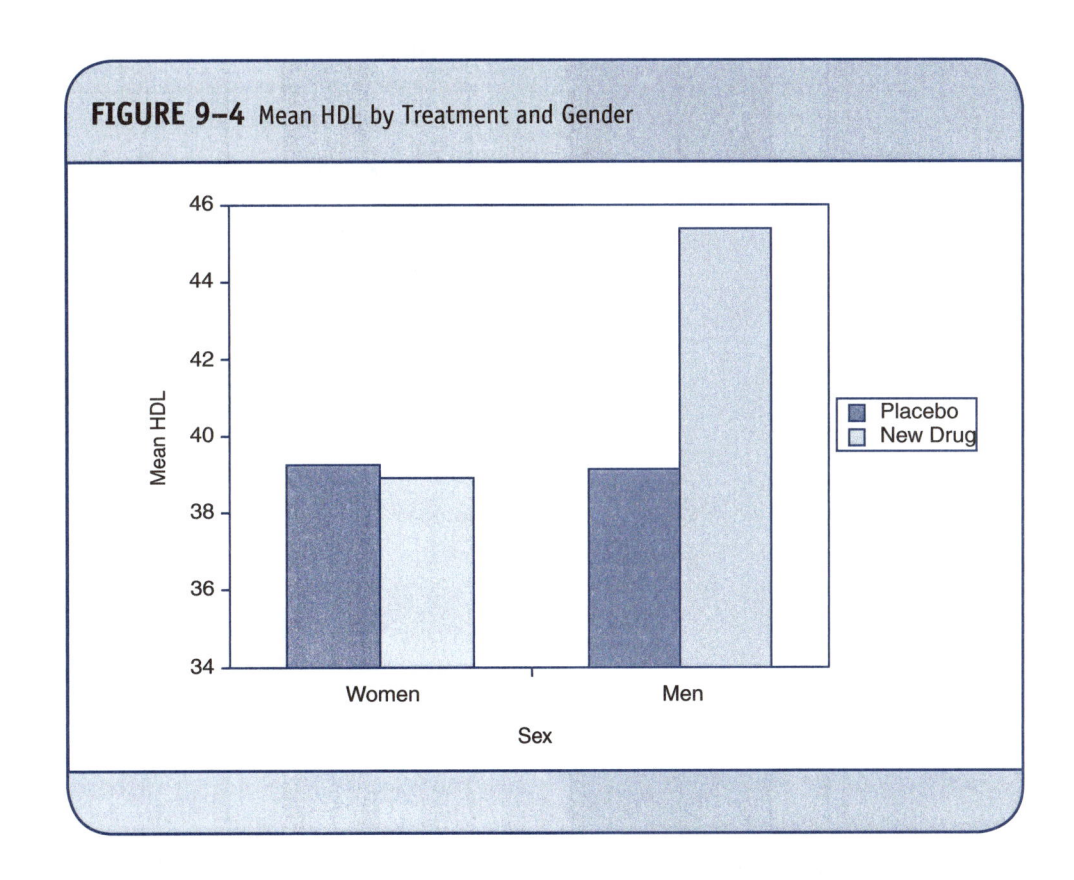

FIGURE 9–4 Mean HDL by Treatment and Gender

(the primary outcome). The effect of treatment was different in men and women.

The **Cochran–Mantel–Haenszel method** is a technique that generates an estimate of an association between a risk factor or exposure and an outcome accounting for confounding. The method is used with a dichotomous outcome variable and a dichotomous risk factor and essentially computes a weighted average of the relative risks (or odds ratios, whichever measure is used to quantify association) across the stratum (or groups) defined by the confounding variable. To implement the Cochran–Mantel–Haenszel method, the confounder must be categorized so that a series of two-by-two tables can be generated showing the association between the risk factor and outcome in each stratum.

In Example 9.1, we found that the association between obesity and incident CVD was much stronger in the pooled sample (all ages combined) than in either age group. The estimate based on the pooled sample is sometimes called an unadjusted or crude estimate. We generated a crude estimate of the relative risk of 1.78. The crude or unadjusted odds ratio is 1.93. We now want to estimate the association between obesity and incident CVD that accounts for confounding by age.

Table 9–10 is a two-by-two table summarizing the association between a dichotomous risk factor and a dichotomous outcome. In previous chapters, we estimated the relative risk and the odds ratios as

$$\hat{RR} = \frac{\hat{p}_1}{\hat{p}_2} \text{ and } \hat{OR} = \frac{\hat{p}_1/(1-\hat{p}_1)}{\hat{p}_2/(1-\hat{p}_2)}.$$

Using the notation in Table 9–10, the relative risk and odds ratio are equivalent to

$$\hat{RR} = \frac{a/(a+b)}{c/(c+d)} \text{ and } \hat{OR} = \frac{a/b}{c/d} = \frac{ad}{bc}.$$

When there is a confounding variable, we set up a series of two-by-two tables, one for each stratum (category) of the confounding variable. For example, in Example 9.1 we considered two age groups and thus we set up two tables showing the association between obesity and incident CVD in each of the two age strata.

The Cochran–Mantel–Haenszel estimates of the relative risk and odds ratio adjusted for confounding are

$$\hat{RR}_{CMH} = \frac{\sum a_i(c_i+d_i)/n_i}{\sum c_i(a_i+b_i)/n_i} \text{ and } \hat{OR}_{CMH} = \frac{\sum a_i d_i/n_i}{\sum b_i c_i/n_i},$$

where a_i, b_i, c_i, and d_i are the numbers of participants in the cells of the two-by-two table in the ith stratum of the confounding variable (see Table 9–10). n_i represents the number of participants in the ith stratum. We illustrate the use of the Cochran–Mantel–Haenszel method in Example 9.3.

Example 9.3. Consider again the data in Example 9.1. We use those data to estimate the relative risk and odds ratio describing the association between obesity and incident CVD adjusting for age. We again consider two age groups or stratum, less than 50 years of age and 50 years of age and older. The data are shown in **Table 9–11** for each age stratum.

The Cochran–Mantel–Haenszel method produces a single, summary measure of association that accounts for the fact that there is a different association in each age stratum. The adjusted relative risk and adjusted odds ratio are

TABLE 9–10 Data Layout for Cochran–Mantel–Haenszel Estimates

	Outcome Present	Outcome Absent	Total
Risk factor present (exposed)	a	b	a + b
Risk factor absent (unexposed)	c	d	c + d
Total	a + c	b + d	n

TABLE 9–11 Obesity and Incident CVD by Age Group

Age < 50	Incident CVD	No CVD	Total
Obese	10	90	100
Not obese	35	465	500
Total	45	555	600

Age 50+	Incident CVD	No CVD	Total
Obese	36	164	200
Not obese	25	175	200
Total	61	339	400

$$\hat{RR}_{CMH} = \frac{\sum a_i(c_i+d_i)/n_i}{\sum c_i(a_i+b_i)/n_i} =$$

$$\frac{10(35+465)/600 + 36(25+175)/400}{35(10+90)/600 + 25(36+164)/400} = 1.44$$

and

$$\hat{OR}_{CMH} = \frac{\sum a_i d_i/n_i}{\sum b_i c_i/n_i} =$$

$$\frac{10(465)/600 + 36(175)/400}{90(35)/600 + 164(25)/400} = 1.52$$

Table 9–12 summarizes the relative risks and odds ratios we have computed thus far to summarize the association between obesity and incident CVD.

Notice that the adjusted relative risk and adjusted odds ratio, 1.44 and 1.52, are not equal to the unadjusted or crude relative risk and odds ratio, 1.78 and 1.93. The adjustment for age produces estimates of the relative risk and odds ratio that are much closer to the stratum-specific estimates (the adjusted estimates are weighted averages of the stratum-specific estimates).

9.3 INTRODUCTION TO CORRELATION AND REGRESSION ANALYSIS

Regression analysis is a technique to assess the relationship between an outcome variable and one or more risk factors or confounding variables. The outcome variable is also called the response or **dependent variable** and the risk factors and confounders are called the predictors, explanatory, or **independent variables**. There is one potentially misleading aspect to this nomenclature. The term predictor can be interpreted as the ability to predict even beyond the limits of the data. The term explanatory might give an impression of an effect (when inferences should be limited to associations). The terms independent and dependent variable are less subject to these interpretations as they do not strongly imply cause and effect. In regression analysis, the dependent variable is denoted y and the independent variable is denoted x.

Before we discuss regression analysis, we first describe a related technique, called **correlation analysis**. Correlation analysis is used to quantify the association between two continuous variables (e.g., between an independent and a dependent variable or between two independent variables). In correlation analysis, we estimate a sample correlation coefficient, more specifically the Pearson Product Moment correlation coefficient. The sample correlation coefficient, denoted r, ranges between –1 and +1 and quantifies the direction and strength of the linear association between the two variables. The correlation between two variables can be positive (i.e., higher levels of one variable are associated with higher levels of the other) or negative (i.e., higher levels of one variable are associated with lower levels of the other). The sign of the correlation coefficient indicates the direction of the association. The magnitude of the correlation coefficient indicates the strength of the association. For example, a correlation of $r = 0.9$ suggests a strong, positive association between two variables, whereas a correlation of $r = -0.2$ suggests a weak, negative association. A correlation close to zero suggests no linear association between two continuous variables. It is important to note that there may be a nonlinear association between two continuous variables. The correlation coefficient does not detect this. Thus, it is always important to evaluate the data carefully in addition to computing summary statistics such as the correlation coefficient. Graphical displays are particularly useful to explore associations between variables. For example, **Figure 9–5** shows four different scenarios, and in each scenario one continuous variable is plotted along the x-axis and the other along the y-axis. Scenario 1 suggests a strong positive association between the two variables (e.g., $r = 0.9$); scenario 2 suggests a weak positive association between the two variables (e.g., $r = 0.2$); scenario 3 suggests no association (or no correlation) between the two variables (e.g., $r = 0$); and scenario 4 suggests a strong negative association between the two variables (e.g., $r = -0.9$).

Scenario 1 in Figure 9–5 might depict the strong positive association generally observed between infant birth weight and birth length, or between systolic and diastolic blood pressures. Scenario 2 might depict the weaker association seen between age and body mass index (which tends to increase with age). Scenario 3 might depict the lack of association between the extent of media exposure in adolescence and age at which adolescents initiate sexual activity. Scenario 4 might depict the strong negative association generally observed

TABLE 9–12 Estimates of Association Between Obesity and CVD

	Relative Risk	Odds Ratio
Crude, unadjusted	1.78	1.93
Age < 50	1.43	1.48
Age 50+	1.44	1.52
Adjusted for age	1.44	1.52

FIGURE 9-5 Varying Correlations Between Variables

1. Strong, Positive
2. Weak, Positive
3. No correlation
4. Strong, Negative

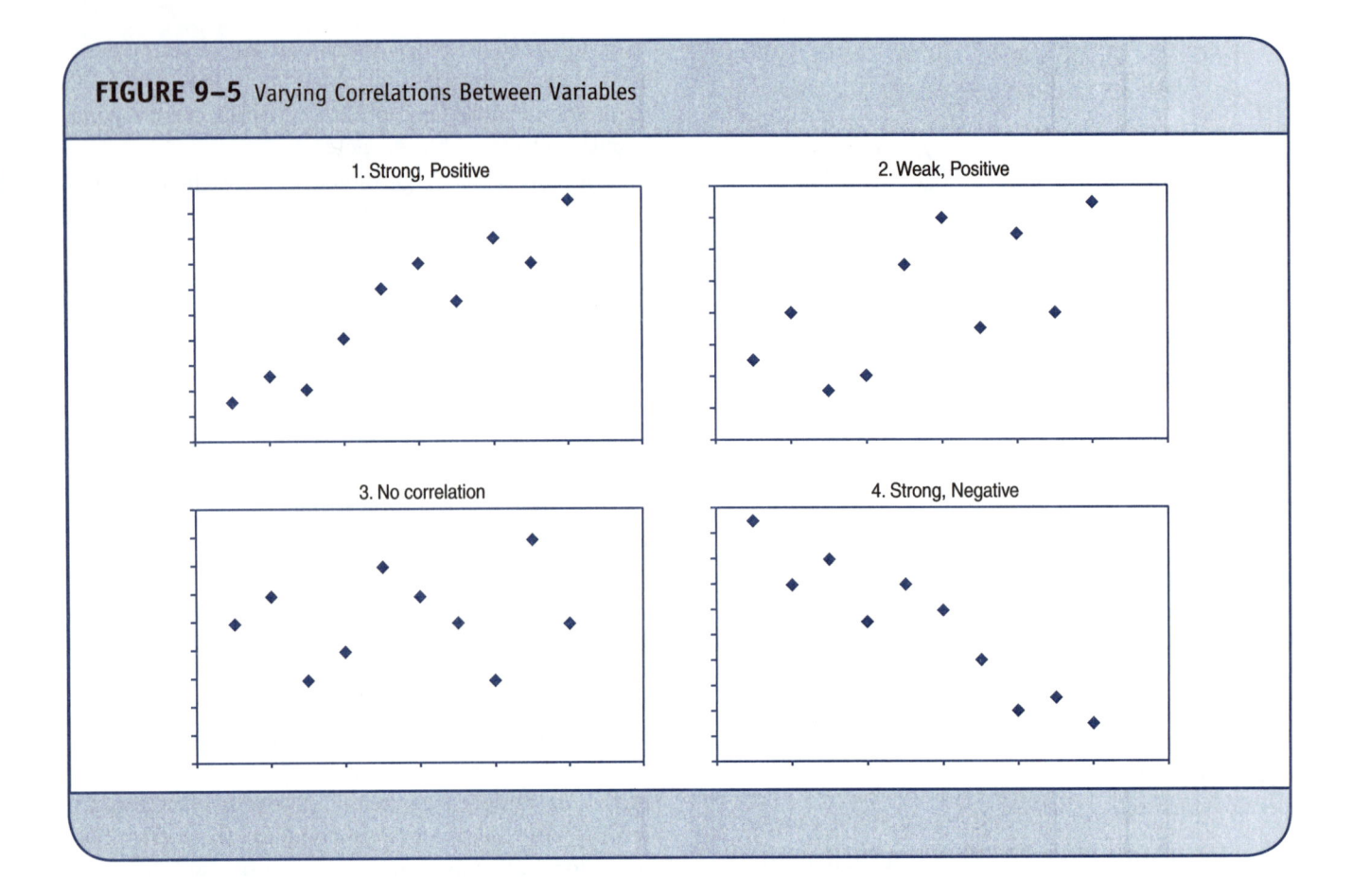

between the number of hours of aerobic exercise per week and percent body fat. In Example 9.4 we illustrate the computation of the sample correlation coefficient.

Example 9.4. A small study is conducted involving 17 infants to investigate the association between gestational age at birth, measured in weeks, and birth weight, measured in grams. The data are shown in **Table 9–13**.

We wish to estimate the association between gestational age and birth weight. In this example, birth weight is the dependent variable, and gestational age is the independent variable. Thus y = birth weight and x = gestational age. The data are displayed in a scatter diagram in **Figure 9–6**. Each point represents an (x, y) pair (in this case the gestational age, measured in weeks, and the birth weight, measured in grams). Scatter diagrams display the independent variable on the horizontal axis (or x-axis) and the dependent variable on the vertical axis (or y-axis).

Figure 9–6 shows a positive or direct association between gestational age and birth weight. Infants with shorter gestational ages are more likely to be born with lower weights, and infants with longer gestational ages are more likely to be born with higher weights. We now estimate the correlation between gestational age and birth weight using the sample data.

TABLE 9–13 Gestational Age and Birth Weight

Infant Identification Number	Gestational Age (weeks)	Birth Weight (grams)
1	34.7	1895
2	36.0	2030
3	29.3	1440
4	40.1	2835
5	35.7	3090
6	42.4	3827
7	40.3	3260
8	37.3	2690
9	40.9	3285
10	38.3	2920
11	38.5	3430
12	41.4	3657
13	39.7	3685
14	39.7	3345
15	41.1	3260
16	38.0	2680
17	38.7	2005

FIGURE 9–6 Scatter Diagram of Gestational Age and Birth Weight

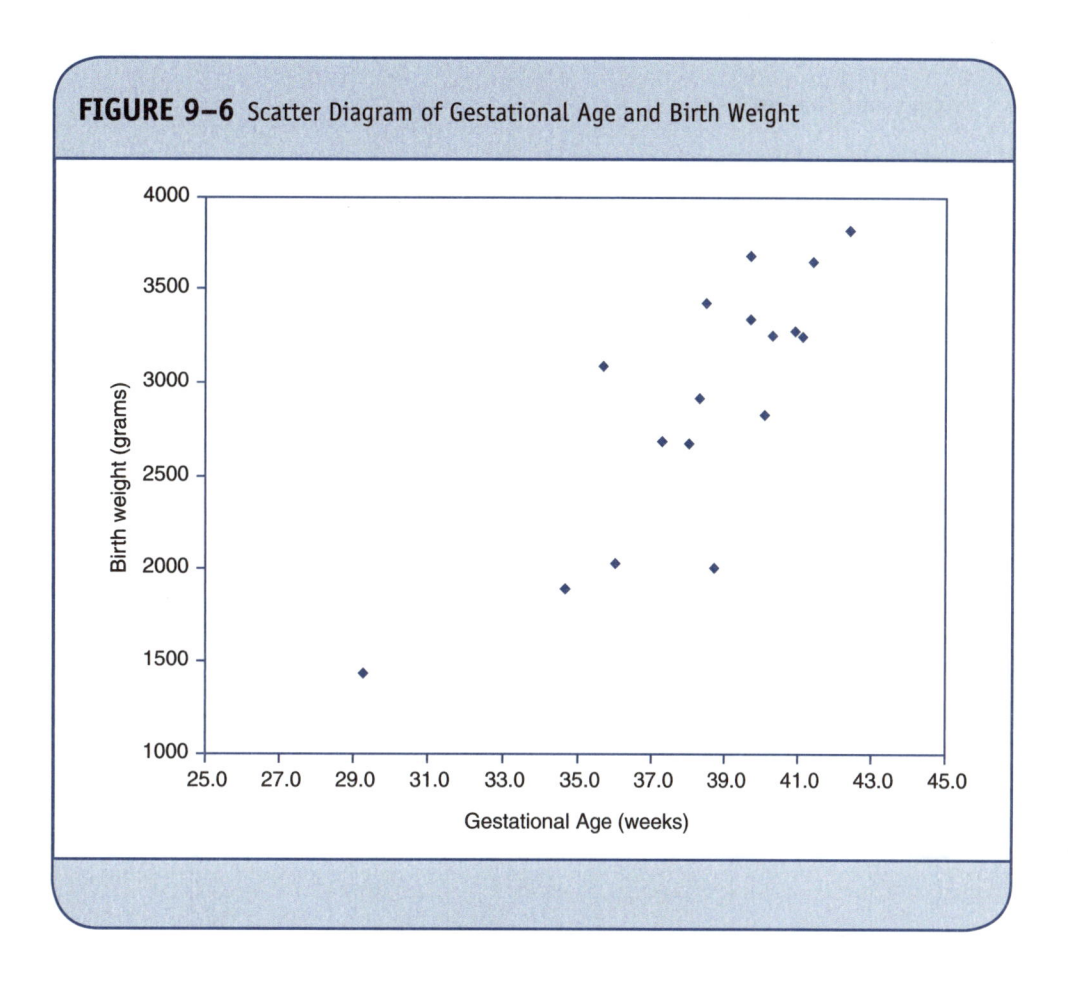

The formula for the sample correlation coefficient is $r = \dfrac{\text{cov}(x,y)}{\sqrt{s_x^2 s_y^2}}$, where cov(x, y) is the *covariance* of x and y, defined as $\text{cov}(x,y) = \dfrac{\Sigma(X-\overline{X})(Y-\overline{Y})}{n-1}$, and s_x^2 and s_y^2 are the sample variances of x and y, defined as $s_x^2 = \dfrac{\Sigma(X-\overline{X})^2}{n-1}$, $s_y^2 = \dfrac{\Sigma(Y-\overline{Y})^2}{n-1}$ (see Chapter 4). The variances of x and y measure the variability of the x scores and y scores around their respective sample means $\overline{X}$ and $\overline{Y}$, considered separately. The covariance measures the variability of the (x, y) pairs around the mean of x and the mean of y, considered simultaneously. To compute the sample correlation coefficient, we need to compute the variance of gestational age, the variance of birth weight, and also the covariance of gestational age and birth weight. We first summarize the gestational age data. The mean gestational age is $\overline{X} = \dfrac{\Sigma X}{n} = \dfrac{652.1}{17} = 38.4$ weeks. To compute the variance of gestational age, we need to sum the squared deviations (or differences) between each observed gestational age and the mean gestational age. The computations are summarized in **Table 9-14**. The variance of gestational age is

$$s_x^2 = \dfrac{\Sigma(X-\overline{X})^2}{n-1} = \dfrac{159.45}{16} = 10.0.$$

Next, we summarize the birth weight data. The mean birth weight is $\overline{Y} = \dfrac{\Sigma Y}{n} = \dfrac{49,334}{17} = 2,902$ grams. To compute the variance of birth weight, we need to sum the squared deviations between each observed birth weight and the mean birth weight. The computations are summarized in **Table 9-15**. The variance of birth weight is $s_y^2 = \dfrac{\Sigma(Y-\overline{Y})^2}{n-1} = \dfrac{7,767,660}{16} = 485,478.8.$

Next we compute the covariance, $\text{cov}(x,y) = \dfrac{\Sigma(X-\overline{X})(Y-\overline{Y})}{n-1}$. To compute the covariance of gestational age and birth weight, we need to multiply the deviation from the mean gestational age by the deviation from the mean birth weight (i.e., $(X-\overline{X})(Y-\overline{Y})$) for each participant. The computations are summarized in **Table 9-16**. Notice that we simply

TABLE 9–14 Variance of Gestational Age

Infant Identification Number	Gestational Age (weeks)	$(X - \bar{X})$	$(X - \bar{X})^2$
1	34.7	−3.7	13.69
2	36.0	−2.4	5.76
3	29.3	−9.1	82.81
4	40.1	1.7	2.89
5	35.7	−2.7	7.29
6	42.4	4.0	16.00
7	40.3	1.9	3.61
8	37.3	−1.1	1.21
9	40.9	2.5	6.25
10	38.3	−0.1	0.01
11	38.5	0.1	0.01
12	41.4	3.0	9.00
13	39.7	1.3	1.69
14	39.7	1.3	1.69
15	41.1	2.7	7.29
16	38.0	−0.4	0.16
17	38.7	0.3	0.09
	$\Sigma X = 652.1$	$\Sigma(X - \bar{X}) = 0$	$\Sigma(X - \bar{X})^2 = 159.45$

TABLE 9–15 Variance of Birth Weight

Infant Identification Number	Birth Weight (grams)	$(Y - \bar{Y})$	$(Y - \bar{Y})^2$
1	1895	−1007	1,014,049
2	2030	−872	760,384
3	1440	−1462	2,137,444
4	2835	−67	4489
5	3090	188	35,344
6	3827	925	855,625
7	3260	358	128,164
8	2690	−212	44,944
9	3285	383	146,689
10	2920	18	324
11	3430	528	278,784
12	3657	755	570,025
13	3685	783	613,089
14	3345	443	196,249
15	3260	358	128,164
16	2680	−222	49,284
17	2005	−897	804,609
	$\Sigma Y = 49,334$	$\Sigma(Y - \bar{Y}) = 0$	$\Sigma(Y - \bar{Y})^2 = 7,767,660$

TABLE 9–16 Covariance of Gestational Age and Birth Weight

Infant Identification Number	$(X - \bar{X})$	$(Y - \bar{Y})$	$(X - \bar{X})(Y - \bar{Y})$
1	−3.7	−1007	3725.9
2	−2.4	−872	2092.8
3	−9.1	−1462	13,304.2
4	1.7	−67	−113.9
5	−2.7	188	−507.6
6	4.0	925	3700.0
7	1.9	358	680.2
8	−1.1	−212	233.2
9	2.5	383	957.5
10	−0.1	18	−1.8
11	0.1	528	52.8
12	3.0	755	2265.0
13	1.3	783	1017.9
14	1.3	443	575.9
15	2.7	358	966.6
16	−0.4	−222	88.8
17	0.3	−897	−269.1

$\Sigma(X - \bar{X})(Y - \bar{Y}) = 28{,}768.4$

copy the deviations from the mean gestational age and birth weight in the previous two tables into columns of the table below and multiply corresponding values. The covariance of gestational age and birth weight is $\mathrm{cov}(x,y) = \dfrac{\Sigma(X-\bar{X})(Y-\bar{Y})}{n-1} = \dfrac{28{,}768.4}{16} = 1798.0$.

We now compute the sample correlation coefficient:

$$r = \frac{\mathrm{cov}(x,y)}{\sqrt{s_x^2 s_y^2}} = \frac{1798.0}{\sqrt{10.0 \times 485{,}478.8}} = \frac{1798.0}{2203.4} = 0.82.$$

The sample correlation coefficient is positive, as expected, and strong (i.e., close to 1). Again, the range of the sample correlation coefficient, r, is between −1 and +1. In practice, meaningful correlations (i.e., correlations that are clinically or practically important) can be as small as 0.4 (or −0.4) for positive (or negative) associations. There are statistical tests to determine whether an observed correlation is statistically significant or not (i.e., statistically significantly different from zero). Procedures to test whether an observed sample correlation is suggestive of a statistically significant correlation are described in detail in Kleinbaum, Kupper, and Muller.[1]

Regression analysis is a very general and widely applied technique. We again focus on the general conceptual framework and somewhat specifically on the application of regression analysis to assess and account for confounding and to assess effect modification. Interested readers should see Kleinbaum, Kupper, and Muller for more details on regression analysis and its many applications.[1]

We first present a simple scenario to establish notation and general principles and then revisit the previous examples. Suppose we want to assess the association between total cholesterol and body mass index (BMI). In this application, total cholesterol is the dependent variable and BMI is the independent variable. We could expand this analysis and consider other potential predictors or independent variables associated with total cholesterol such as age, sex, and smoking. When there are more independent variables, they are denoted x_1, x_2, ..., x_p. Considering only BMI, y = total cholesterol and x = BMI.

When there is a single continuous dependent variable and a single independent variable, the analysis is called a

simple linear regression analysis. This analysis assumes that there is a linear association between the two variables. (If a different relationship is hypothesized, such as a curvilinear or exponential relationship, alternative regression analyses are performed.) **Figure 9–7** displays data on BMI and total cholesterol measured in a sample of $n = 20$ participants. The mean BMI is 27.4 kg/m² with a standard deviation of 3.7 (i.e., $s_x = 3.7$). The mean total cholesterol level is 205.9 mg/dL with a standard deviation of 30.8 (i.e., $s_y = 30.8$).

Figure 9–7 shows a positive or direct association between BMI and total cholesterol. Participants with lower BMI are more likely to have lower total cholesterol levels, and participants with higher BMI are more likely to have higher total cholesterol levels. The correlation between BMI and total cholesterol is $r = 0.78$. In contrast, suppose we examine the association between BMI and HDL cholesterol. **Figure 9–8** is a scatter diagram of BMI and HDL cholesterol measured in the same sample of $n = 20$ participants. The correlation between BMI and HDL cholesterol is $r = -0.72$. The mean HDL cholesterol level is 47.4 mg/dL and the standard deviation is 12.1 (i.e., $s_y = 12.1$).

Figure 9–8 shows a negative or inverse association between BMI and HDL cholesterol. Participants with lower BMI are more likely to have higher HDL cholesterol levels, and participants with higher BMI are more likely to have lower HDL cholesterol levels.

In simple linear regression analysis, we estimate the equation of the line that best describes the association between the independent variable and the dependent variable. The simple linear regression equation is

$$\hat{y} = b_0 + b_1 x,$$

where $\hat{y}$ is the predicted or expected value of the outcome, x is the independent variable or predictor, b_0 is the estimated y-intercept, and b_1 is the estimated slope. The y-intercept and slope are estimated from the sample data and minimize the sum of the squared differences between the observed (y) and the predicted ($\hat{y}$) values of the outcome—i.e., the estimates minimize $\Sigma(\hat{y} - y)^2$. These differences are called residuals. The estimates of the y-intercept and slope minimize the sum of the squared residuals and are called the least squares estimates.[1] The y-intercept is the value of the dependent variable (y)

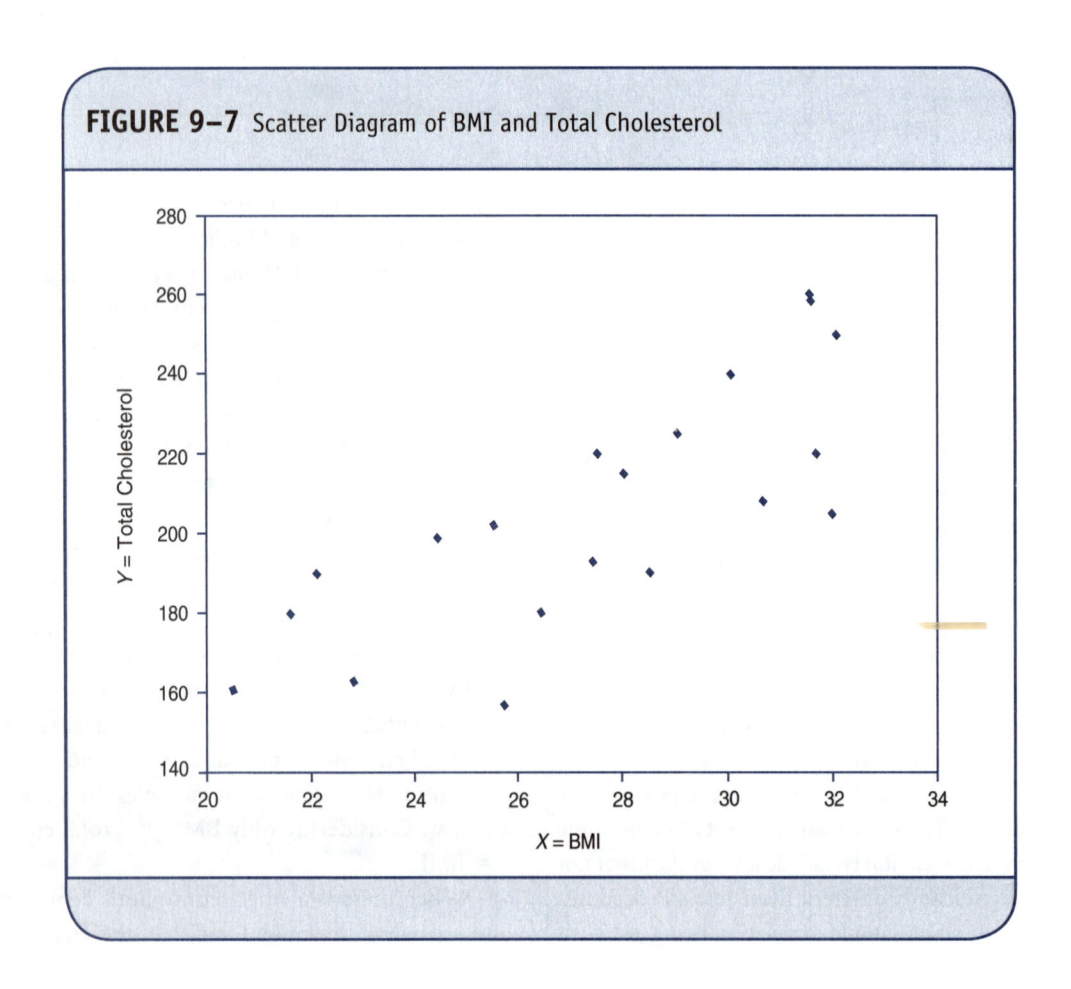

FIGURE 9–7 Scatter Diagram of BMI and Total Cholesterol

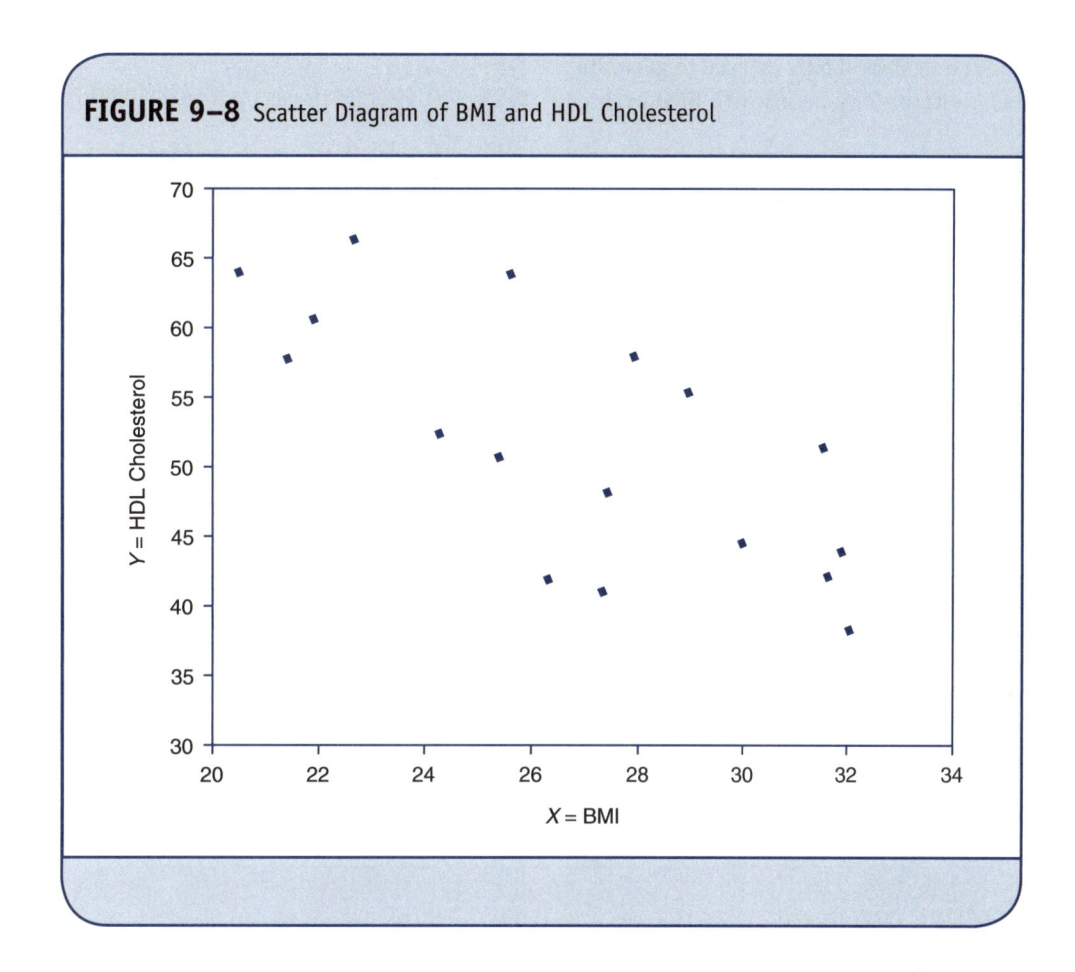

FIGURE 9–8 Scatter Diagram of BMI and HDL Cholesterol

when the independent variable (x) is 0. The slope is the expected change in the dependent variable (y) relative to a one-unit change in the independent variable (x).

The least squares estimations of the y-intercept and the slope are computed as follows:

$$b_1 = r\frac{s_y}{s_x} \text{ and } b_0 = \overline{Y} - b_1\overline{X},$$

where r is the sample correlation coefficient, $\overline{X}$ and $\overline{Y}$ are the means, and s_x and s_y are the standard deviations of the independent variable x and the dependent variable y, respectively.

The least squares estimations of the regression coefficients, b_0 and b_1, describing the relationship between BMI and total cholesterol are $b_0 = 28.07$ and $b_1 = 6.49$. These are computed as follows:

$$b_1 = r\frac{s_y}{s_x} = 0.78\frac{30.8}{3.7} = 6.49 \text{ and}$$

$$b_0 = \overline{Y} - b_1\overline{X} = 205.9 - 6.49(27.4) = 28.07.$$

The estimate of the y-intercept ($b_0 = 28.07$) represents the estimated total cholesterol level when BMI is 0. Because a BMI of 0 is meaningless, the y-intercept is not informative. The estimate of the slope ($b_1 = 6.49$) represents the change in total cholesterol relative to a one-unit change in BMI. For example, if we compare two participants whose BMIs differ by one unit, we would expect their total cholesterol levels to differ by approximately 6.49 mg/dL (with the person with the higher BMI having the higher total cholesterol). The equation of the regression line is

$$\hat{y} = 28.07 + 6.49 \text{ BMI}.$$

The regression equation can be used to estimate a participant's total cholesterol as a function of their BMI. For example, suppose a participant has a BMI of 25. We would estimate their total cholesterol to be 28.07 + 6.49(25) = 190.32 mg/dL. The equation can also be used to estimate total cholesterol for other values of BMI. However, the equation should only be used to estimate cholesterol levels for persons whose BMIs are in the range of the data used to generate the regression equation. In our sample, BMI ranges from

20 to 32; thus the equation should only be used to generate estimates of total cholesterol for persons with BMI in that range.

There are statistical tests that can be performed to assess whether the estimated regression coefficients (b_0 and b_1) provide evidence that the respective coefficients in the population are statistically significantly different from 0. The test of most interest is usually $H_0: \beta_1 = 0$ versus $H_1: \beta_1 \neq 0$, where β_1 is the population slope. If the population slope is significantly different from 0, we conclude that there is a statistically significant association between the independent and dependent variables. The test of significance for the slope is equivalent to the test of significance for the correlation. These tests are conducted in most statistical computing packages (we omit the details here). **Figure 9–9** shows the estimated regression line superimposed on the scatter diagram of BMI and total cholesterol.

The least squares estimations of the regression coefficients, b_0 and b_1, describing the relationship between BMI and HDL cholesterol are $b_0 = 111.79$ and $b_1 = -2.35$. These are computed as follows:

$$b_1 = r\frac{s_y}{s_x} = -0.72\frac{12.1}{3.7} = -2.35 \text{ and}$$

$$b_0 = \overline{Y} - b_1\overline{X} = 47.4 - (-2.35)(27.4) = 111.79.$$

Again, the y-intercept is uninformative because a BMI of 0 is meaningless. The estimate of the slope ($b_1 = -2.35$) represents the expected change in HDL cholesterol relative to a one-unit change in BMI. If we compare two participants whose BMIs differ by one unit, we would expect their HDL cholesterols to differ by approximately 2.35 mg/dL (with the person with the higher BMI having the lower HDL cholesterol). **Figure 9–10** shows the regression line superimposed on the scatter diagram of BMI and HDL cholesterol.

In linear regression analysis, the dependent variable is continuous. There is an assumption that the distribution of the dependent variable (y) at each value of the independent variable (x) is approximately normally distributed. The independent variable can be a continuous variable (e.g., BMI) or a dichotomous variable (also called an **indicator variable**). Dichotomous (or indicator) variables are usually coded as 0 or 1, where 0 is

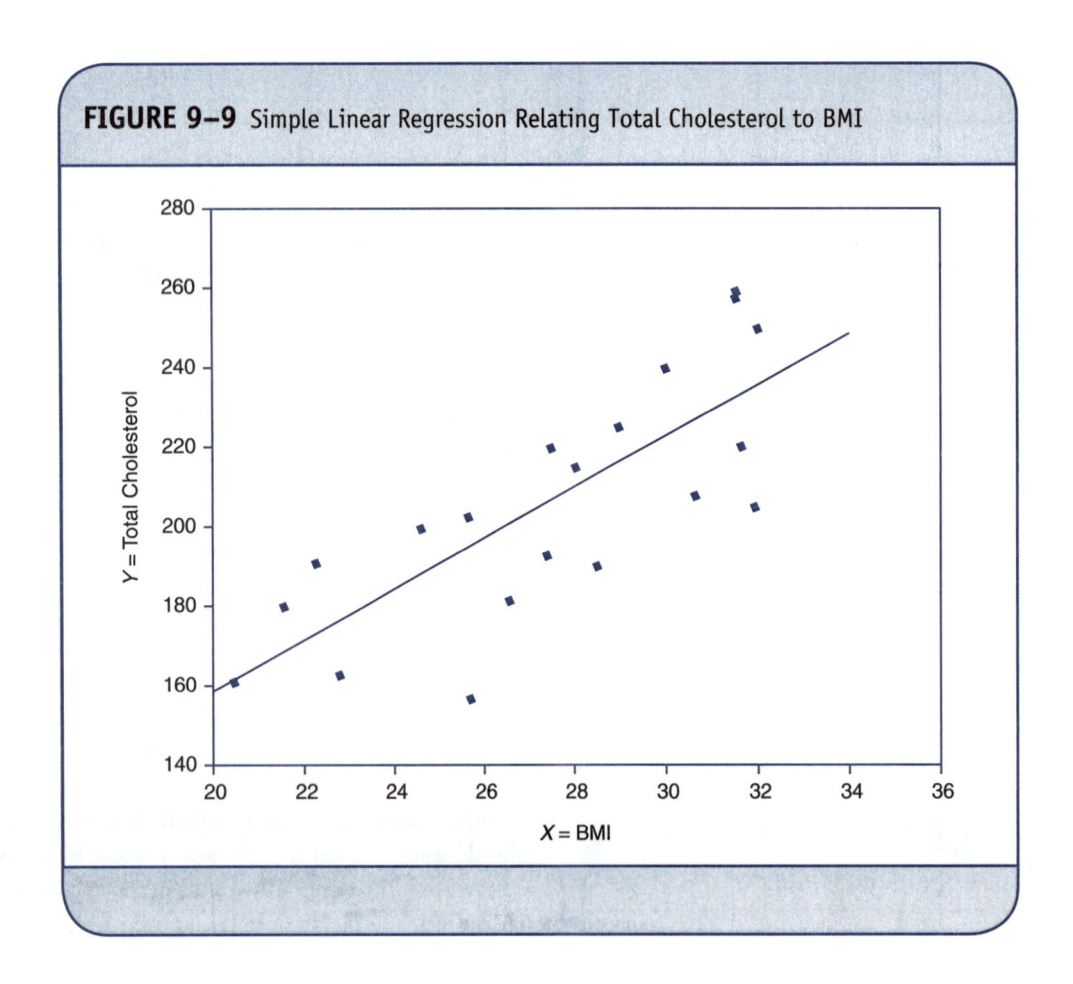

FIGURE 9–9 Simple Linear Regression Relating Total Cholesterol to BMI

FIGURE 9–10 Simple Linear Regression Relating HDL Cholesterol to BMI

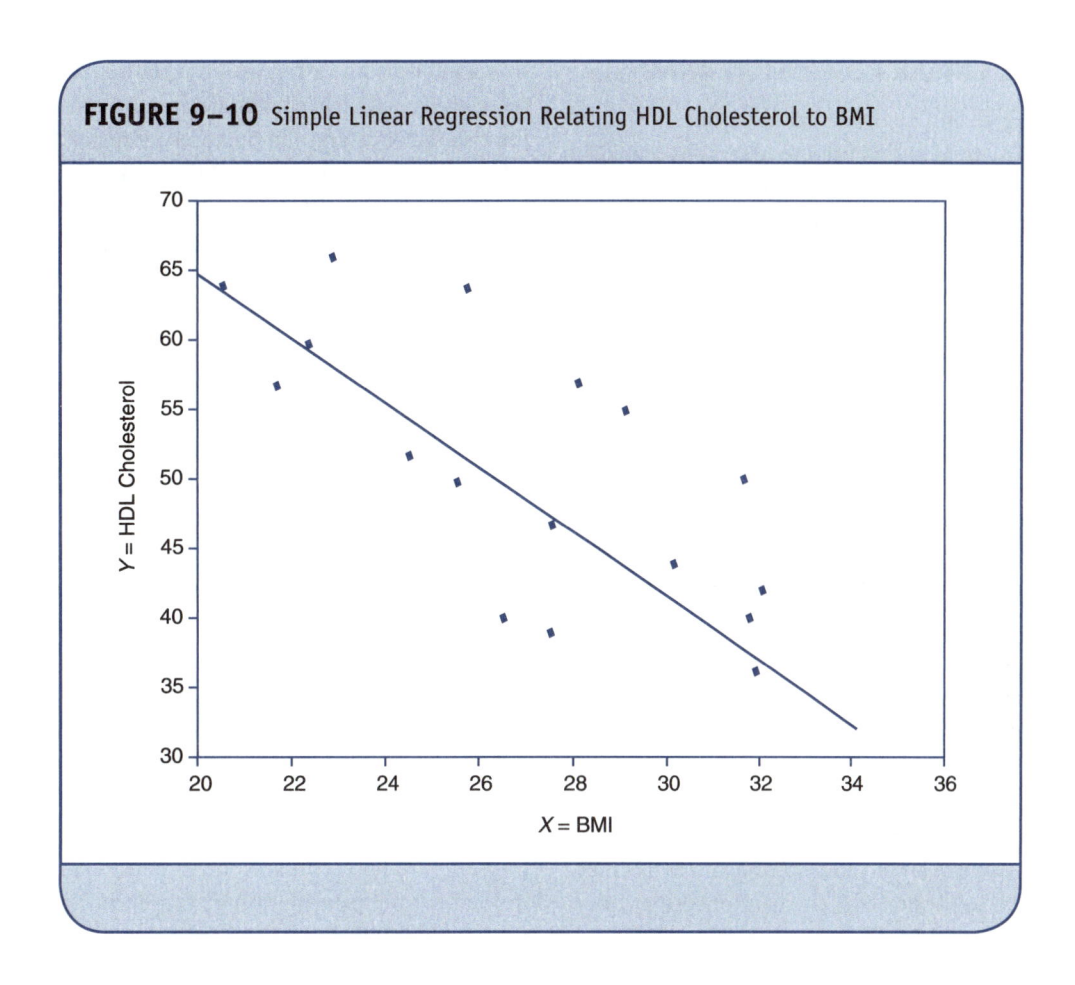

assigned to participants who do not have a particular risk factor, exposure, or characteristic and 1 is assigned to participants who have the particular risk factor, exposure, or characteristic.

In Example 9.2, we considered data from a clinical trial designed to evaluate the efficacy of a new drug to increase HDL cholesterol. One hundred patients enrolled in the study and were randomized to receive either the new drug or a placebo. Summary statistics on the primary outcome, HDL cholesterol, were shown in Table 9–7. We compared the mean HDL levels between treatment groups using a two independent samples t test. Regression analysis can also be used to compare mean HDL levels between treatments. HDL cholesterol is the continuous dependent variable and treatment (new drug versus placebo) is the independent variable. A simple linear regression equation is estimated as

$$\hat{y} = 39.21 + 0.95x,$$

where $\hat{y}$ is the estimated HDL level and x is a dichotomous (or indicator) variable reflecting the assigned treatment. In this example, x is coded as 1 for participants who receive the new drug and as 0 for participants who receive the placebo.

The estimate of the y-intercept is $b_0 = 39.21$. The y-intercept is the expected value of y (HDL cholesterol) when x is 0. In this example, $x = 0$ indicates the placebo group. Thus, the y-intercept is exactly equal to the mean HDL level in the placebo group. The slope is $b_1 = 0.95$. The slope represents the change in y (HDL cholesterol) relative to a one-unit change in x. A one-unit change in x represents a difference in treatment assignment (placebo versus new drug). The slope represents the difference in mean HDL levels between the treatment groups (see Table 9–7; the mean HDL in the placebo group is 39.21 and the difference in means between the placebo and new drug groups is 0.95 mg/dL).

In Section 9.4 and Section 9.5, we extend the simple regression concept to include additional independent variables. In Section 9.4, we introduce multiple linear regression analysis, which applies in situations where the outcome is continuous and there is more than one independent variable. The independent variables can be continuous or dichotomous (indicators). In Section 9.5, we introduce multiple logistic regression analysis, which applies in situations where the outcome is dichotomous (e.g., incident CVD).

9.4 MULTIPLE LINEAR REGRESSION ANALYSIS

Multiple linear regression analysis is an extension of simple linear regression analysis and is used to assess the association between two or more independent variables and a single continuous dependent variable. The multiple linear regression equation is

$$\hat{y} = b_0 + b_1 x_1 + b_2 x_2 + \cdots + b_p x_p,$$

where $\hat{y}$ is the predicted or expected value of the dependent variable, x_1 through x_p are p distinct independent or predictor variables, b_0 is the value of y when all of the independent variables (x_1 through x_p) are equal to 0, and b_1 through b_p are the estimated regression coefficients. Each regression coefficient represents the expected change in y relative to a one-unit change in the respective independent variable holding the remaining independent variables constant. For example, in the multiple regression situation, b_1 is the expected change in y relative to a one-unit change in x_1, holding all other independent variables constant (i.e., when the remaining independent variables are held at the same value or are fixed). Again, statistical tests can be performed to assess whether each regression coefficient is statistically different from 0.

Multiple regression analysis can be used to assess whether confounding exists. In Section 9.1, we mentioned that confounding can be assessed by formal statistical testing (i.e., testing whether the potential confounder is statistically significantly associated with the risk factor and statistically significantly associated with the outcome). We also mentioned that confounding can be assessed by examining the associations between the potential confounder and the risk factor and between the potential confounder and the outcome (e.g., we used Table 9–2 through Table 9–4 to make those assessments in Example 9.1) from a practical or clinical standpoint. Multiple regression analysis can be used to identify confounding as follows: Suppose we have a risk factor or an exposure variable, which we denote x_1 (e.g., obesity or treatment) and an outcome or dependent variable which we denote y. We estimate a simple linear regression equation relating the risk factor (the independent variable) to the dependent variable as

$$\hat{y} = b_0 + b_1 x_1,$$

where b_1 is the estimated regression coefficient that quantifies the association between the risk factor and the outcome.

Suppose we now want to assess whether a third variable (e.g., age) is a confounder. We denote the potential confounder x_2 and then estimate a multiple linear regression equation as

$$\hat{y} = b_0 + b_1 x_1 + b_2 x_2.$$

In the multiple linear regression equation, b_1 is the estimated regression coefficient that quantifies the association between the risk factor x_1 and the outcome, adjusted for x_2 (b_2 is the estimated regression coefficient that quantifies the association between the potential confounder and the outcome). Some investigators assess confounding by assessing the extent to which the regression coefficient associated with the risk factor changes after adjusting for the potential confounder. In this case, we compare b_1 from the simple linear regression model to b_1 from the multiple linear regression model. As an informal rule, if the regression coefficient from the simple linear regression model changes by more than 10%, then x_2 is said to be a confounder. We illustrate this approach in the following examples.

Once a variable is identified as a potential confounder, we can then use multiple linear regression analysis to estimate the association between the risk factor and the outcome adjusting for that confounder. The test of significance of the regression coefficient associated with the risk factor can be used to assess whether the association between the risk factor is statistically significant after accounting for one or more confounding variables. This is also illustrated in Example 9.5.

Example 9.5. Suppose we want to assess the association between BMI and systolic blood pressure using data collected in the seventh examination of the Framingham Offspring Study. A total of $n = 3539$ participants attended the exam, and their mean systolic blood pressure is 127.3 mmHg with a standard deviation of 19. The mean BMI in the sample is 28.2 kg/m^2 with a standard deviation of 5.3. A simple linear regression analysis reveals the results in **Table 9–17**. The simple linear regression model is

$$\hat{y} = 108.28 + 0.67\ \text{BMI},$$

where $\hat{y}$ is the predicted or expected systolic blood pressure. The regression coefficient associated with BMI is 0.67, suggesting that each one-unit increase in BMI is associated with a 0.67 mmHg increase in systolic blood pressure. The association between BMI and systolic blood pressure is also statistically significant ($p = 0.0001$).

TABLE 9–17 Simple Linear Regression Analysis

Independent Variable	Regression Coefficient	t	p-value
Intercept	108.28	62.61	0.0001
BMI	0.67	11.06	0.0001

Suppose we now want to assess whether age (a continuous variable, measured in years), sex, and treatment for hypertension (yes/no) are potential confounders and, if so, appropriately account for these using multiple linear regression analysis. For analytic purposes, treatment for hypertension is coded as 1 = yes and 0 = no. Sex is coded as 1 = male and 0 = female. A multiple regression analysis reveals the results in **Table 9–18**. The multiple regression model is

$$\hat{y} = 68.15 + 0.58 \text{ BMI} + 0.65 \text{ Age} + 0.94 \text{ Male sex} + 6.44 \text{ Treatment for hypertension.}$$

Notice that the association between BMI and systolic blood pressure is smaller (0.58 versus 0.67) after adjustment for age, sex, and treatment for hypertension. BMI remains statistically significantly associated with systolic blood pressure ($p = 0.0001$) but the magnitude of the association is lower after adjustment. The regression coefficient decreases by 13%. Using the informal rule (i.e., a change in the coefficient in either direction by 10% or more), we meet the criteria for confounding. Thus, part of the association between BMI and systolic blood pressure is explained by age, sex, and treatment for hypertension.

As noted previously, most statistical computing packages produce test statistics (t) and p-values for each of the regression coefficients, which can be used to assess the statistical significance of each independent variable. Notice that the p-values for BMI, age, and treatment for hypertension are given as $p = 0.0001$. Assessing only the p-values suggests that these three independent variables are equally statistically significant. The magnitude of the t statistics provides another means to judge relative importance of the independent variables. In this example, age is the most significant independent variable, followed by BMI, treatment for hypertension, and then male sex. In fact, male sex does not reach statistical significance ($p = 0.1133$) in the multiple regression model. Some investigators argue that regardless of whether an important variable such as sex reaches statistical significance, it should be retained in the model. Other investigators only retain variables that are statistically significant. This is yet another example of the complexity involved in multivariable modeling.

The multiple regression model produces an estimate of the association between BMI and systolic blood pressure that accounts for differences in systolic blood pressure due to age, sex, and treatment for hypertension. A one-unit increase in BMI is associated with a 0.58 mmHg increase in systolic blood pressure, holding age, sex, and treatment for hypertension constant. Each additional year of age is associated with a 0.65 mmHg increase in systolic blood pressure, holding BMI, sex, and treatment for hypertension constant. Men have higher systolic blood pressures, by approximately 0.94 mmHg, holding BMI, age, and treatment for hypertension constant; and persons on treatment for hypertension have higher systolic blood pressures, by approximately 6.44 mmHg, holding BMI, age, and sex constant.

The multiple regression equation can be used to estimate systolic blood pressures as a function of a participant's BMI, age, sex, and treatment for hypertension status. For example, we can estimate the blood pressure of a 50-year-old male, with a BMI of 25, who is not on treatment for hypertension as follows:

$$\hat{y} = 68.15 + 0.58 \,(25) + 0.65 \,(50) + 0.94 \,(1) + 6.44 \,(0) = 116.09 \text{ mmHg}$$

We can estimate the blood pressure of a 50-year-old female, with a BMI of 25, who is on treatment for hypertension as follows:

$$\hat{y} = 68.15 + 0.58 \,(25) + 0.65 \,(50) + 0.94 \,(0) + 6.44 \,(1) = 121.59 \text{ mmHg}$$

Example 9.6. In Example 9.2, we considered data from a clinical trial designed to evaluate the efficacy of a new drug to increase HDL cholesterol. One hundred patients enrolled in the study and were randomized to receive either the new drug or a placebo, and summary statistics on the primary outcome, HDL cholesterol, are shown in Table 9–7. In Section 9.3, we compared mean HDL levels between treatments using simple linear regression analysis, and the simple linear regression model was

$$\hat{y} = 39.21 + 0.95x,$$

TABLE 9–18 Multiple Regression Analysis

Independent Variable	Regression Coefficient	t	p-value
Intercept	68.15	26.33	0.0001
BMI	0.58	10.30	0.0001
Age	0.65	20.22	0.0001
Male sex	0.94	1.58	0.1133
Treatment for hypertension	6.44	9.74	0.0001

where $\hat{y}$ is the estimated HDL level and x is a dichotomous (or indicator) variable reflecting the assigned treatment (1 = new drug and 0 = placebo). We know that there is no significant difference in mean HDL levels by treatment, and we fail to reject $H_0: \beta_1 = 0$ ($p = 0.2620$). Recall that investigators were concerned about the small—and statistically insignificant—difference in mean HDL levels by treatment and questioned whether the effect of the treatment might be related to participant sex.

In Table 9-9, we showed that there was an important difference in the efficacy of the new drug depending on the participant's sex. We described this as effect modification by sex. Multiple regression analysis can be used to assess effect modification. This is done by estimating a multiple regression equation relating the outcome of interest (y) to independent variables representing the treatment assignment, sex, and the product of the two (called the treatment by sex interaction variable). For the analysis, we let T = the treatment assignment (1 = new drug and 0 = placebo), M = male sex (1 = male and 0 = female), and TM = T × M, the product of sex and male sex. The multiple regression analysis reveals the results in **Table 9-19**. The multiple regression model is:

$$\hat{y} = 39.24 - 0.36\,T - 0.18\,M + 6.55\,TM.$$

In this model, the regression coefficient associated with the interaction term, b_3, is statistically significant (i.e., $H_0: \beta_3 = 0$ versus $H_1: \beta_3 \neq 0$, $p = 0.0011$). The significance of the interaction term indicates that there is a different relationship between treatment and outcome by sex.

The model shown here can be used to estimate the mean HDL levels for men and women who are assigned to the new drug and to the placebo. To use the model to generate these estimates, we must recall the coding scheme (i.e., T = 1 indicates new drug, T = 0 indicates placebo, M = 1 indicates male sex, and M = 0 indicates female sex).

TABLE 9-19 Multiple Regression Analysis

Independent Variable	Regression Coefficient	t	p-value
Intercept	39.24	65.89	0.0001
T (Treatment)	−0.36	−0.43	0.6711
M (Male sex)	−0.18	−0.13	0.8991
TM (Treatment × Male sex)	6.55	3.37	0.0011

The expected or predicted HDL for men (M = 1) assigned to the new drug (T = 1) is

$$\hat{y} = 39.24 - 0.36(1) - 0.18(1) + 6.55(1)(1) = 45.25 \text{ mg/dL}.$$

The expected HDL for women (M = 0) assigned to the new drug (T = 1) is

$$\hat{y} = 39.24 - 0.36(1) - 0.18(0) + 6.55(1)(0) = 38.88 \text{ mg/dL}.$$

The expected HDL for men (M = 1) assigned to the placebo (T = 0) is

$$\hat{y} = 39.24 - 0.36(0) - 0.18(1) + 6.55(0)(1) = 39.06 \text{ mg/dL}.$$

The expected HDL for women (M = 0) assigned to the placebo (T = 0) is

$$\hat{y} = 39.24 - 0.36(0) - 0.18(0) + 6.55(0)(0) = 39.24 \text{ mg/dL}.$$

Notice that the expected HDL levels for men and women on the new drug and on the placebo are identical to the means shown in Table 9-9.

Because there is effect modification, separate simple linear regression models are estimated to assess the treatment effect in men and women (**Table 9-20**). The regression models are

$$\text{Men: } \hat{y} = 39.06 + 6.19T.$$

$$\text{Women: } \hat{y} = 39.24 - 0.36T.$$

In men, the regression coefficient associated with treatment ($b_1 = 6.19$) is statistically significant ($p = 0.0001$) but in women, the regression coefficient associated with treatment ($b_1 = -0.36$) is not statistically significant ($p = 0.6927$).

TABLE 9-20 Separate Simple Linear Regression Analyses

	Regression Coefficient	t	p-value
Men			
Intercept	39.06	57.09	0.0001
T (Treatment)	6.19	6.56	0.0001
Women			
Intercept	39.24	61.36	0.0001
T (Treatment)	−0.36	−0.40	0.6927

Multiple linear regression analysis is a widely applied technique. We show here how multiple regression analysis is used to assess and account for confounding and to assess effect modification. The techniques we described can be extended to adjust for several confounders simultaneously and to investigate more complex effect modification (e.g., three-way statistical interactions). There are many other applications of multiple regression analysis. A popular application is to assess the relationships between several predictor variables simultaneously and a single, continuous outcome. For example, it may be of interest to determine which predictors in a relatively large set of candidate predictors are most important or most strongly associated with an outcome. It is always important in statistical analysis, particularly in the multivariable arena, that statistical modeling is guided by biologically plausible associations.

Independent variables in regression models can be continuous or dichotomous. Regression models can also accommodate categorical independent variables. For example, it might be of interest to assess whether there is a difference in total cholesterol by race/ethnicity. In Chapter 7 we presented analysis of variance as one way of testing for differences in means of a continuous outcome among several comparison groups. Regression analysis can also be used. However, the investigator must create indicator variables to represent the different comparison groups (e.g., different racial/ethnic groups). The set of indicator variables (also called **dummy variables**) is considered in the multiple regression model simultaneously as a set of independent variables. For example, suppose that participants indicate which of the following best represents their race/ethnicity: White, Black or African American, American Indian or Alaskan Native, Asian, Native Hawaiian or Pacific Islander, or Other Race. This categorical variable has six response options. To consider race/ethnicity as a predictor in a regression model, we create five indicator variables (one less than the total number of response options) to represent the six different groups. To create the set of indicators, or set of dummy variables, we first decide on a reference group or category. In this example, the reference group is the racial group that we will compare the other groups against. Indicator variables are created for the remaining groups and coded 1 for participants who are in that group (e.g., are of the specific race/ethnicity of interest), and all others are coded 0. In the multiple regression model, the regression coefficients associated with each of the dummy variables (representing in this example each race/ethnicity group) are interpreted as the expected difference in the mean of the outcome variable for that race/ethnicity as compared to the reference group, holding all other predictors constant. Example 9.7 illustrates the approach as well as the interpretation of the regression coefficients in the model.

Example 9.7. An observational study is conducted to investigate risk factors associated with infant birth weight. The study involves 832 pregnant women. Each woman provides demographic and clinical data and is followed through the outcome of pregnancy. At the time of delivery, the infant's birth weight is measured, in grams, as is the infant's gestational age, in weeks. Birth weights vary widely and range from 404 to 5400 grams. The mean birth weight is 3367.83 grams, with a standard deviation of 537.21 grams. Investigators wish to determine whether there are differences in birth weight by infant sex, gestational age, mother's age, and mother's race. In the study sample, 421/832 (50.6%) of the infants are male, and the mean gestational age at birth is 39.49 weeks with a standard deviation of 1.81 weeks (range: 22–43 weeks). The mean mother's age is 30.83 years with a standard deviation of 5.76 years (range: 17–45 years). Approximately 49% of the mothers are white, 41% are Hispanic, 5% are black, and 5% identify themselves as "other" race. A multiple linear regression analysis is performed relating infant sex (coded 1 = male, 0 = female), gestational age in weeks, mother's age in years, and three dummy or indicator variables reflecting mother's race to birth weight. The results are shown in **Table 9–21**.

Many of the independent variables are statistically significantly associated with birth weight. Male infants are approximately 175 grams heavier than female infants, adjusting for gestational age, mother's age, and mother's race/ethnicity. Gestational age is highly significant ($p = 0.0001$), with each additional gestational week associated with an increase of 179.89 grams in birth weight, holding infant sex, mother's age, and mother's race/ethnicity constant. Mother's age does

TABLE 9–21 Multiple Regression Analysis

Independent Variable	Regression Coefficient	t	p-value
Intercept	−3850.92	−11.56	0.0001
Male infant	174.79	6.06	0.0001
Gestational age (weeks)	179.89	22.35	0.0001
Mother's age (years)	1.38	0.47	0.6361
Black race	−138.46	−1.93	0.0535
Hispanic race	−13.07	−0.37	0.7103
Other race	−68.67	−1.05	0.2918

not reach statistical significance ($p = 0.6361$). Mother's race is modeled as a set of three dummy or indicator variables. In this analysis, white race is the reference group. Infants born to black mothers have lower birth weight by approximately 140 grams (as compared to infants born to white mothers), adjusting for gestational age, infant gender, and mother's age. This difference is marginally significant ($p = 0.0535$). There are no statistically significant differences in birth weight in infants born to Hispanic versus white mothers or to women who identify themselves as "other" race as compared to white, holding infant sex, gestational age, and mother's age constant.

9.5 MULTIPLE LOGISTIC REGRESSION ANALYSIS

Logistic regression analysis is similar to linear regression analysis except that the outcome is dichotomous (e.g., success/failure, yes/no). Logistic regression is very popular in biostatistical analysis. Simple logistic regression analysis refers to the regression application with one dichotomous outcome and one independent variable, and multiple logistic regression analysis applies when there is a single dichotomous outcome and more than one independent variable. Here again we present the general concept. Hosmer and Lemeshow provide a very detailed description of logistic regression analysis and its applications.[3]

The outcome in logistic regression analysis is often coded as 0 or 1, where 1 indicates that the outcome of interest is present and 0 indicates that the outcome of interest is absent. If we define p as the probability that the outcome is 1, the multiple logistic regression model can be written as

$$\hat{p} = \frac{\exp(b_0 + b_1 x_1 + b_2 x_2 + \cdots + b_p x_p)}{1 + \exp(b_0 + b_1 x_1 + b_2 x_2 + \cdots + b_p x_p)},$$

where $\hat{p}$ is the expected probability that the outcome is present, x_1 through x_p are distinct independent variables, and b_0 through b_p are the regression coefficients.

The multiple logistic regression model is sometimes written differently. In the following form, the outcome is the expected log of the odds that the outcome is present:

$$\ln\left(\frac{\hat{p}}{1-\hat{p}}\right) = b_0 + b_1 x_1 + b_2 x_2 + \cdots + b_p x_p.$$

Notice that the right side of the preceding equation looks like the multiple linear regression equation (i.e., a linear combination of the regression coefficients and independent variables). The technique for estimating the regression coefficients in a logistic regression model is different from that used to estimate the regression coefficients in a multiple linear regression model. (Details can be found in Hosmer and Lemeshow.[3]) The interpretation of the regression coefficients in the multiple logistic regression model is as follows. b_1 is the change in the expected logs odds relative to a one-unit change in x_1, holding all other predictors constant. The antilog of an estimated regression coefficient, $\exp(b_i)$, produces an odds ratio. In Example 9.8, we estimate a multiple logistic regression model and discuss the interpretation of the regression coefficients.

Example 9.8. In Example 9.1, we analyzed data from a study designed to assess the association between obesity (defined as a BMI greater than 30) and incident CVD. Data were collected from participants who were between the ages of 35 and 65 and free of CVD at baseline. Each participant was followed for 10 years for the development of cardiovascular disease. A summary of the data is contained in Table 9–1.

In Example 9.1, we estimated the association between obesity and incident CVD. The unadjusted or crude relative risk was $\hat{RR} = 1.78$, and the unadjusted or crude odds ratio was $\hat{OR} = 1.93$. We also determined that age was a confounder, and using the Cochran–Mantel–Haenszel method, we estimated an adjusted relative risk of $\hat{RR}_{CMH} = 1.44$ and an adjusted odds ratio of $\hat{OR}_{CMH} = 1.52$. We now use logistic regression analysis to assess the association between obesity and incident cardiovascular disease adjusting for age.

A simple logistic regression analysis reveals the results in **Table 9–22**. The simple logistic regression model relates obesity to the log odds of incident CVD:

$$\ln\left(\frac{\hat{p}}{1-\hat{p}}\right) = -2.367 + 0.658\, \text{Obesity}.$$

Obesity is an indicator variable in the model, coded as 1 = obese and 0 = not obese. Among obese persons, the log odds of incident CVD are 0.658 times the log odds among persons who are not obese. If we take the antilog of the regression coefficient, $\exp(0.658) = 1.93$, we get the crude or unadjusted odds ratio. Among obese persons, the odds of developing CVD are 1.93 times the odds among non-obese persons. The association between obesity and incident CVD is statistically

TABLE 9–22 Logistic Regression Analysis

Independent Variable	Regression Coefficient	χ^2	p-value
Intercept	−2.367	307.38	0.0001
Obesity	0.658	9.87	0.0017

significant ($p = 0.0017$). Notice that the test statistics to assess the significance of the regression parameters in logistic regression analysis are χ^2 statistics, as opposed to t statistics, as was the case with linear regression analysis. This is because a different estimation technique, called maximum likelihood estimation, is used to estimate the regression parameters (see Hosmer and Lemeshow[3] for technical details).

Many statistical computing packages generate odds ratios as well as 95% confidence intervals for the odds ratios as part of their logistic regression analysis procedure. In this example, the estimate of the odds ratio is 1.93 and the 95% CI is (1.281, 2.913).

In Example 9.1, we determined that age was a confounder. The following multiple logistic regression model estimates the association between obesity and incident CVD, adjusting for age. In the model, we again consider two age groups (less than 50 years of age and 50 years of age and older). For the analysis, age group is coded as $1 = 50$ years of age and older and $0 = $ less than 50 years of age.

$$\ln\left(\frac{\hat{p}}{1-\hat{p}}\right) = -2.592 + 0.415 \text{ Obesity} + 0.655 \text{ Age group}.$$

If we take the antilog of the regression coefficient associated with obesity, $\exp(0.415) = 1.52$, we get the odds ratio adjusted for age. Among obese persons, the odds of developing CVD are 1.52 times the odds for non-obese persons, adjusting for age. In Section 9.2, we used the Cochran–Mantel–Haenszel method to generate an odds ratio adjusted for age and found $\hat{OR}_{CMH} = 1.52$.

Multiple logistic regression analysis is a widely applied technique. We show here how multiple logistic regression analysis can be used to account for confounding. The models can be extended to account for several confounding variables simultaneously. Multiple logistic regression analysis can also be used to assess effect modification, and the approaches are identical to those used in multiple linear regression analysis. Multiple logistic regression analysis can also be used to examine the impact of multiple risk factors (as opposed to focusing on a single risk factor) on a dichotomous outcome. An example of the latter is a risk function predicting incident cardiovascular disease on the basis of age, sex, blood pressure, cholesterol, smoking, and so on. These types of models have been published extensively in the Framingham Heart Study.[4,5] Similar modeling approaches have been used to evaluate risk factors for other conditions such as breast cancer and diabetes.[8–10]

Example 9.9 Consider again the observational study described in Example 9.7 to investigate risk factors associated with infant birth weight. Suppose that investigators are also concerned with adverse pregnancy outcomes including gestational diabetes, preeclampsia (i.e., pregnancy-induced hypertension), and preterm labor. Recall that the study involves 832 pregnant women who provide demographic and clinical data. In the study sample, 22 women (2.6%) develop preeclampsia, 35 (4.2%) develop gestational diabetes, and 40 (4.8%) develop preterm labor. Suppose we wish to assess whether there are differences in each of these adverse pregnancy outcomes by race/ethnicity, adjusted for maternal age. Three separate logistic regression analyses are conducted relating each outcome, considered separately, to the three dummy or indicator variables reflecting mother's race and mother's age, in years. The results are shown in **Table 9–23** through **Table 9–25**.

TABLE 9–23 Logistic Regression Analysis

Outcome = Preeclampsia	Regression Coefficient	χ^2	p-value	Odds Ratio (95% CI)
Intercept	−3.066	4.518	0.0335	—
Black race	2.191	12.640	0.0004	8.948 (2.673, 29.949)
Hispanic race	−0.1053	0.0325	0.8570	0.900 (0.286, 2.829)
Other race	0.0586	0.0021	0.9046	1.060 (0.104, 3.698)
Mother's age (years)	−0.0252	0.3574	0.5500	0.975 (0.898, 1.059)

TABLE 9–24 Logistic Regression Analysis

Outcome = Gestational Diabetes	Regression Coefficient	χ^2	p-value	Odds Ratio (95% CI)
Intercept	−5.823	22.968	0.0001	—
Black race	1.621	6.660	0.0099	5.056 (1.477, 17.312)
Hispanic race	0.581	1.766	0.1839	1.787 (0.759, 4.207)
Other race	1.348	5.917	0.0150	3.848 (1.299, 11.395)
Mother's age (years)	0.071	4.314	0.0378	1.073 (1.004, 1.147)

TABLE 9–25 Logistic Regression Analysis

Outcome = Preterm Labor	Regression Coefficient	χ^2	p-value	Odds Ratio (95% CI)
Intercept	−1.443	1.602	0.2056	—
Black Race	−0.082	0.015	0.9039	0.921 (0.244, 3.483)
Hispanic Race	−1.564	9.497	0.0021	0.209 (0.077, 0.566)
Other Race	0.548	1.124	0.2890	1.730 (0.628, 4.767)
Mother's age (years)	−0.037	1.198	0.2737	0.963 (0.901, 1.030)

The only statistically significant difference in preeclampsia is between black and white mothers. Black mothers have nearly nine times the odds of developing preeclampsia as compared to white mothers, adjusted for maternal age. The 95% confidence interval for the odds ratio comparing black versus white women who develop preeclampsia is very wide (2.673 to 29.949). This is due to the fact that there are a small number of outcome events (only 22 women develop preeclampsia in the total sample) and a small number of women of black race in the study. Thus, this association should be interpreted with caution. While the odds ratio is statistically significant, the confidence interval suggests that the magnitude of the effect could be anywhere from a 2.7-fold increase to a 29.9-fold increase. A larger study is needed to generate a more precise estimate of effect.

With regard to gestational diabetes (Table 9–24), there are statistically significant differences between black and white mothers ($p = 0.0099$) and between mothers who identify themselves as "other" race as compared to white mothers ($p = 0.0150$), adjusted for mother's age. Mother's age is also statistically significant ($p = 0.0378$), with older women more likely to develop gestational diabetes, adjusted for race/ethnicity.

With regard to preterm labor (Table 9–25), the only statistically significant difference is between Hispanic and white mothers ($p = 0.0021$). Hispanic mothers have 80% lower odds of developing preterm labor than white mothers (odds ratio = 0.209), adjusted for mother's age.

9.6 SUMMARY

Multivariable methods include a number of specific procedures to simultaneously assess the relationships between several exposure or risk factor variables and a single outcome. The methods are computationally complex and generally require the use of a statistical computing package. Multivariable methods can be used to assess and adjust for confounding, to determine whether there is effect modification, or to assess the relationships of several exposures or risk factors on an outcome simultaneously. Multivariable analyses are complex and should always be planned to reflect biologically plausible relationships. Whereas it is relatively easy to consider an additional variable in a multiple linear or multiple logistic regression model, only variables that are clinically meaningful should be included.

We described how multivariable methods are used to assess and account for confounding. It is worth repeating that multivariable methods should not be relied upon to "correct" problems in a study. Multivariable models are used for statistical adjustment of confounding, whereas careful study design can provide much more in the way of minimizing confounding. Randomization is the most effective means of minimizing confounding. Unfortunately, randomization is not always possible. Matching participants in the comparison groups is another option. The objective of matching is to distribute potential confounding variables equally among the comparison groups. However, if matching is employed, then analyses must be tailored to appropriately account for matched (or dependent) samples.

It is important to remember that multivariable models can only adjust or account for differences in confounding variables that are measured in the study. In addition, multivariable models should only be used to account for confounding when there is some overlap in the distribution of the confounder in each of the risk factor groups. For example, in Example 9.1, there was an imbalance in the age distribution in obese as compared to non-obese persons. If the imbalance was extreme (e.g., suppose that all of the obese persons were 50 years of age and older and all of the non-obese persons were less than 50 years of age), then it would not be appropriate to use multivariable techniques to account for confounding. The data would simply not be suitable for the analysis.

With regard to effect modification, a major issue is the fact that analyses must be stratified to appropriately estimate associations in the data. This can present issues with regard to lack of precision or low power. Stratified analyses are informative but if the samples in specific strata are too small, the analyses may lack precision. In planning studies, investigators must pay careful attention to potential effect modifiers. If there is a suspicion that an association between an exposure or risk factor is different in specific subgroups or strata, the study must be designed to ensure sufficient numbers of participants in each of those groups. Specifically, the sample size formulas in Chapter 8 must be used to determine the numbers of subjects required in each strata to ensure adequate precision or power in the analysis.

Multivariable methods are very powerful. In this chapter, we discussed only a few specific applications. Practicing biostatisticians must understand multivariable methods broadly to work effectively in investigative teams, and therefore we encourage students to pursue more coursework in multivariable analysis. That being said, it is critically important to appropriately describe data (using the methods we presented in Chapter 4) and to investigate relationships among variables carefully (using the methods and techniques described in Chapter 6 and Chapter 7). Mastery of the principles and techniques presented in previous chapters provides a solid foundation.

In the opening paragraph of the textbook, we stated that implementing and understanding biostatistical applications is a combination of art and science. This is probably most evident in this chapter. Successful biostatisticians must not only understand the theory and principles of statistics; they must also understand the substantive nature of research problems to effectively address them.

9.7 PRACTICE PROBLEMS

1. A study is conducted to estimate the association between exposure to lead paint in childhood and attention-deficit hyperactivity disorder (ADHD). Data on $n = 400$ children are collected, and data on exposure and ADHD diagnosis are shown in **Table 9–26**.

TABLE 9–26 Exposure to Lead Paint and ADHD

Exposure to Lead Paint	ADHD	No ADHD
Yes	34	71
No	29	266

a. Estimate the crude or unadjusted relative risk.
b. Estimate the crude or unadjusted odds ratio.

2. In the study described in Problem 1, data are also available reflecting whether or not the child's father has ADHD. The relationship between the father's diagnosis, the child's exposure to lead paint, and the child's diagnosis are shown in **Table 9–27**.
   a. Estimate the relative risks and odds ratios for a child's diagnosis of ADHD relative to exposure to lead paint, stratified by the father's diagnosis of ADHD (i.e., for fathers with and without the diagnosis).
   b. How do the results in Part (a) of this problem compare to the results in Problem 1?

3. Use the data in Problem 1 and Problem 2 to conduct tests of hypothesis to determine whether the father's diagnosis of ADHD is a confounder. (Hint: Test if there is a relationship between the father's diagnosis and the child's exposure, and between the father's diagnosis and the child's diagnosis.) Is the father's diagnosis of ADHD a confounder? Justify your conclusion.

4. Use the data in Problem 2 to estimate the relative risk and odds ratio for a child's diagnosis of ADHD relative to exposure to lead paint, adjusted for the father's diagnosis using the Cochran–Mantel–Haenszel method.

5. The data presented in Problem 1 and Problem 2 are analyzed using multiple logistic regression analysis and the models are shown here. In these models, the data are coded as p = the proportion of children with a diagnosis of ADHD, and Child Exposed and Father's Diagnosis are coded as 1 = yes and 0 = no.

TABLE 9–27 Exposure to Lead Paint and ADHD by Father's Diagnosis

Father with ADHD

Exposure to Lead Paint	ADHD	No ADHD
Yes	27	39
No	7	37

Father Without ADHD

Exposure to Lead Paint	ADHD	No ADHD
Yes	7	32
No	22	229

$$\ln\left(\frac{\hat{p}}{1-\hat{p}}\right) = -2.216 + 1.480 \text{ Child exposed} \quad (1)$$

Father with diagnosis: $\ln\left(\frac{\hat{p}}{1-\hat{p}}\right) = -1.665 + 1.297 \text{ Child exposed} \quad (2)$

Father without diagnosis: $\ln\left(\frac{\hat{p}}{1-\hat{p}}\right) = -2.343 + 0.823 \text{ Child exposed} \quad (3)$

$$\ln\left(\frac{\hat{p}}{1-\hat{p}}\right) = -2.398 + 1.056 \text{ Child exposed} + 0.906 \text{ Father's diagnosis} \quad (4)$$

a. Which model produces the unadjusted odds ratio? Compute the unadjusted odds ratio using the regression coefficient from the appropriate model.
b. Which model produces the odds ratio in fathers without a diagnosis of ADHD? Compute the unadjusted odds ratio using the regression coefficient from the appropriate model.
c. What is the odds ratio adjusted for the father's diagnosis?

6. A study is conducted in patients with HIV. The primary outcome is CD4 cell count, which is a measure of the stage of the disease. Lower CD4 counts are associated with more advanced disease. The investigators are interested in the association between vitamin and mineral supplements and CD4 count. A multiple regression analysis is performed relating CD4 count to the use of supplements (coded as 1 = yes and 0 = no) and to the duration of HIV in years (i.e., the number of years between the diagnosis of HIV and the study date). For the analysis, y = CD4 count:

$$\hat{y} = 501.41 + 12.67 \text{ Supplements} - 30.23 \text{ Duration of HIV}.$$

a. What is the expected CD4 count for a patient taking supplements who has had HIV for 2.5 years?
b. What is the expected CD4 count for a patient not taking supplements who was diagnosed with HIV at study enrollment?
c. What is the expected CD4 count for a patient not taking supplements who has had HIV for 2.5 years?
d. If we compare two patients and one has had HIV for 5 years longer than the other, what is the expected difference in their CD4 counts?

7. A clinical trial is conducted to evaluate the efficacy of a new medication to relieve pain in patients undergoing total knee replacement surgery. In the trial, patients are randomly assigned to receive either the new medication or the standard medication. After receiving the assigned medication, patients are asked to report their pain on a scale of 0 to 100, with higher scores indicative of more pain. Data on the primary outcome are shown in **Table 9–28**. Because procedures can be more complicated in older patients, the investigators are concerned about confounding by age. For analysis, patients are classified into two age groups, less than 65 years old and 65 years of age and older. The data are shown in **Table 9–29**. Is there a statistically significant difference in mean pain scores between patients assigned to the new medication as compared to the standard medication? Run the appropriate test at $\alpha = 0.05$. (Ignore age in this analysis.)

8. Use the data in Problem 7 to determine whether age is a confounding variable. Run the tests of hypothesis to determine whether age is related to treatment assignment and whether there is a difference in mean pain scores by age group. Is age a confounder? Justify your conclusion.

9. The data presented in Problem 7 are analyzed using multiple linear regression analysis and the models are shown here. In the models, the data are coded as 1 = new medication and 0 = standard medication, and age 65 and older is coded as 1 = yes and 0 = no.

   $\hat{y} = 53.85 - 23.54$ Medication

   $\hat{y} = 45.31 - 19.88$ Medication $+ 14.64$ Age 65+

   $\hat{y} = 45.51 - 20.21$ Medication $+ 14.29$ Age 65+
   $+ 0.75$ Medication $\times$ Age 65+

   Patients <65: $\hat{y} = 45.51 - 20.21$ Medication

   Patients 65+: $\hat{y} = 59.80 - 19.47$ Medication

   Does it appear that there is effect modification by age? Justify your response using the preceding models.

10. Based on your answers to Problem 8 and Problem 9, how should the effect of the treatment be summarized? Should results be reported separately by age group or combined? Should the effect of treatment be adjusted for age? Justify your response using the models presented in Problem 9.

11. Using the data in Table 9–29, generate a plot to display the mean pain scores by treatment and age group. (Hint: Use Figure 9–4 as an example.)

12. An open-label study (where participants are aware of the treatment they are taking) is run to assess the time to pain relief following treatment in patients with arthritis. The following linear regression equations are estimated relating time to pain relief measured in minutes (dependent variable) to participant's age (in years), sex (coded 1 for males and 0 for females), and severity of disease (a score ranging from 0 to 100, with higher scores indicative of more severe arthritis):

Time to Pain Relief = $-24.2 + 0.9$ Age

Time to Pain Relief = $11.8 + 19.3$ Male Sex

Time to Pain Relief = $3.2 + 0.4$ Severity

Time to Pain Relief = $-19.8 + 0.50$ Age + 10.9 Male Sex + 0.2 Severity

TABLE 9–28 Pain Scores by Treatment

	Sample Size	Mean Pain Score	Std Dev of Pain Score
New medication	60	30.31	7.52
Standard medication	60	53.85	7.44

TABLE 9–29 Pain Scores by Treatment and Age Group

Age < 65

	Sample Size	Mean Pain Score	Std Dev of Pain Score
New medication	40	25.30	2.46
Standard medication	25	45.51	1.83
Total	65	33.07	10.16

Age 65+

	Sample Size	Mean Pain Score	Std Dev of Pain Score
New medication	20	40.33	2.16
Standard medication	35	59.80	2.49
Total	55	52.72	9.74

a. What is the expected time to pain relief for a male following treatment?
b. What is the expected time to pain relief for a participant aged 50 following treatment?
c. In assessing the association between gender and time to pain relief, is there evidence of confounding by age or severity? Briefly justify your answer.

13. A study is run to evaluate risk factors for incident hypertension. All participants are free of hypertension at the start of the study and are followed for 4 years, at which time they are reassessed for hypertension. Risk factors are measured in all participants at the start of the study. A total of $n = 3182$ participants enroll and 1123 develop hypertension over 4 years. A multiple logistic regression model is run and the results are shown in **Table 9–30**.
    a. What is the relative importance (statistical significance) of the risk factors? Justify your answer.
    b. Estimate adjusted odds ratios to quantify the effect of sex and current smoking status on incident hypertension.
    c. Who is more likely to develop hypertension, a man or a woman? Justify your answer.

14. **Table 9–31** displays the numbers of participants who develop hypertension by sex and age group in the study described in Problem 13.
    a. What is the relative risk for hypertension in women 50+ years versus women <50 years of age?

TABLE 9–31 Hypertension by Sex and Age

	Develop Hypertension	Do Not Develop Hypertension	Total
Women			
Age 50+ years	255	237	492
Age <50 years	241	916	1157
Total	496	1153	1649
Men			
Age 50+ years	283	188	471
Age <50 years	344	718	1062
Total	627	906	1533

b. What is the relative risk for hypertension in men 50+ years versus men <50 years of age?

15. Use the data in Problem 14 to estimate the relative risk for hypertension in participants 50+ years versus <50 years of age, adjusted for gender using the Cochran–Mantel–Haenszel method.

REFERENCES

1. Kleinbaum, D., Kupper, L.L., and Muller, K.E. *Applied Regression Analysis and Other Multivariable Methods* (2nd ed.). Boston: PWS-Kent, 1988.
2. Jewell, N.P. *Statistics for Epidemiology.* New York: Chapman and Hall/CRC, 2004.
3. Hosmer, D. and Lemeshow, S. *Applied Logistic Regression.* New York: John Wiley & Sons, 1989.
4. Anderson, M., Wilson, P.W., Odell, P.M., and Kannell, W.B. "An updated coronary risk profile: A statement for health professionals." *Circulation* 1991; 83: 356–362.
5. Wilson, P.W.F., D'Agostino, R.B., Levy, D., Belanger, A.M., Silbershatz, H., and Kannel, W.B. "Prediction of coronary heart disease using risk factor categories." *Circulation* 1998; 97: 1837–1847.
6. SAS Version 9.1. © 2002–2003 by SAS Insitute, Cary, NC.
7. Aschengrau, A. and Seage, G.R. *Essentials of Epidemiology for Public Health.* Sudbury, MA: Jones and Bartlett, 2006.
8. Goldberg, J.I. and Borgen, P.I. "Breast cancer susceptibility testing: Past, present and future." *Expert Review of Anticancer Therapy* 2006; 6(8): 1205–1214.
9. Meigs, J.B., Hu, F.B., Rifai, N., and Manson, J. "Biomarkers of endothelial dysfunction and risk of Type 2 diabetes mellitus." *Journal of the American Medical Association* 2004; 291: 1978–1986.
10. Stern, M.P., Williams, K., and Haffner, S. "Identification of persons at high risk for Type 2 diabetes mellitus: Do we need the oral glucose tolerance test?" *Annals of Internal Medicine* 2002; 136(8): 575–581.

TABLE 9–30 Risk Factors for Hypertension

Risk Factor	Regression Coefficient	χ^2	p-value
Intercept	−18.416	746.103	0.0001
Age (years)	0.0533	95.004	0.0001
Sex*	−0.2524	6.189	0.0129
Systolic blood pressure	0.0629	141.417	0.0001
Diastolic blood pressure	0.0752	80.237	0.0001
BMI	0.0637	29.209	0.0001
Current smoker*	0.3270	10.116	0.0015

*Gender is coded 1 = male and 0 = female. Current smoker is coded 1 = yes and 0 = no.

CHAPTER 10

Nonparametric Tests

When and Why

Key Questions

- When should a nonparametric test be used?
- If nonparametric tests require fewer assumptions than parametric tests, why not always use nonparametric tests?
- Are nonparametric tests just for small samples?

In the News

Rare diseases are defined as diseases with prevalence of less than 200,000 in the United States, and those that affect fewer than 5 out of every 10,000 people in Europe. Drugs that treat rare diseases are often called "orphan drugs," as they are often not of interest to drug developers because of their limited potential for profitability. The United States Congress passed the Orphan Drug Act in 1983; this act offers tax incentives and other benefits for marketing exclusivity for sponsors engaging in development of drugs to treat rare diseases.[1] Studies of rare diseases usually involve small samples, as there are limited numbers of patients available for enrollment.

Dig In

- Approximately 50% of rare diseases affect children. Are there unique challenges in studying children with rare diseases?
- What is the appropriate control group in a randomized trial of a new drug for a rare disease? (What is the standard of care? Is a placebo control ethical?)
- Nonparametric tests generally have lower power than parametric tests. Would you interpret the results of a nonparametric test any differently than the results of a parametric test?

---

[1] U.S. Food and Drug Administration. Regulatory information: Orphan Drug Act. Available at *http://www.fda.gov/regulatoryinformation/legislation/significantamendmentstothefdcact/orphandrugact/default.htm*.

Learning Objectives

By the end of this chapter, the reader will be able to

- Compare and contrast parametric and nonparametric tests
- Identify multiple applications where nonparametric approaches are appropriate
- Perform and interpret the Mann–Whitney U test
- Perform and interpret the Sign test and Wilcoxon Signed Rank test
- Compare and contrast the Sign test and Wilcoxon Signed Rank test
- Perform and interpret the Kruskal–Wallis test
- Identify the appropriate nonparametric hypothesis testing procedure based on type of outcome variable and number of samples

---

In Chapter 7 we presented a number of tests of hypothesis for continuous, dichotomous, categorical, and ordinal outcomes. Tests for continuous outcomes focused on comparing means, while tests for dichotomous, categorical, and ordinal outcomes focused on comparing proportions. All of the tests presented in Chapter 7 are called **parametric tests** and are based on certain assumptions for their appropriate use. For example, when running tests of hypothesis for means of continuous outcomes, all parametric tests assume that the outcome is approximately normally distributed in the population. This does not mean that the data in the observed sample follow a normal distribution, but rather that the outcome follows a normal distribution in the full population, which is not observed. For many outcomes, investigators are comfortable with the normality assumption (i.e., most of the observations are in the center of the distribution while fewer are at either extreme). It also turns out that many statistical tests are robust, which means that they maintain their statistical properties even when assumptions are not entirely met. Tests are robust in the presence of violations of the normality assumption when the sample size is large, based on the Central Limit Theorem (presented in Chapter 5).

When the sample size is small and the distribution of the outcome is not known and cannot be assumed to be approximately normally distributed, then alternative tests called **nonparametric tests** are appropriate.

Nonparametric tests are sometimes called distribution-free tests because they are based on fewer assumptions (e.g., they do not assume that the outcome is approximately normally distributed). Parametric tests involve specific probability distributions (e.g., the normal distribution) and the tests involve estimation of the key parameters of that distribution (e.g., the mean or difference in means) from the sample data. The cost of fewer assumptions is that nonparametric tests are generally less powerful than their parametric counterparts (i.e., when the alternative is true, they may be less likely to reject H_0).

It can sometimes be difficult to assess whether a continuous outcome follows a normal distribution and thus whether a parametric or nonparametric test is appropriate. There are several statistical tests that can be used to assess whether data are likely from a normal distribution. The most popular are the Kolmogorov–Smirnov test, the Anderson–Darling test, and the Shapiro–Wilk test.[1] Each test is essentially a goodness-of-fit test and compares observed data to quantiles of the normal (or other specified) distribution. The null hypothesis for each test is H_0: *Data follow a normal distribution* versus H_1: *Data do not follow a normal distribution*. If the test is significant (e.g., $p < 0.05$), then data do not follow a normal distribution, and a nonparametric test is warranted. It should be noted that these tests for normality can be subject to low power. Specifically, the tests may fail to reject H_0: *Data follow a normal distribution* when in fact the data do not follow a normal distribution. Low power is a major issue when the sample size is small, which is unfortunatley often when we wish to employ these tests. The most practical approach to assessing normality involves investigating the distributional form of the outcome in the sample using a histogram and augmenting that with data from other studies (if available) that may inform the assessment of the distribution of the outcome in the population.

There are some situations when it is clear that the outcome does not follow a normal distribution. These include situations when the outcome is an ordinal variable or a rank, when there are definite outliers, or when the outcome has clear limits of detection. Examples of these types of outcomes are given next.

Consider a clinical trial where study participants are asked to rate their symptom severity following 6 weeks on the assigned treatment. Symptom severity might be measured on a 5-point ordinal scale with the following response options: symptoms got much worse, symptoms got slightly worse, no change, symptoms improved slightly, or symptoms are much improved. Suppose there are a total of $n = 20$ participants in the trial, randomized to an experimental treatment or placebo, and the outcome data are distributed as shown in **Figure 10–1**. The distribution of the outcome (symptom severity) does not appear to be normal; more participants report improvement in symptoms as opposed to worsening of symptoms.

In some studies, the outcome is a rank. For example, in obstetrical studies an APGAR score is often used to assess the health of a newborn. The APGAR score is named after Dr. Virginia Apgar and is used to describe the condition of an infant at birth.[2] The APGAR score is based on five criteria: appearance of the skin, pulse rate, grimace (reflex or reaction to stimulation), activity (or muscle tone), and respiration. Each of the five criteria is rated as 0 (very unhealthy), 1, or 2 (healthy) based on specific clinical criteria. The APGAR score is the sum of the five component scores and ranges from 0 to 10. Infants with scores of 7 or higher are considered normal, scores of 4–6 are considered low, and scores of 0–3 are critically low. Sometimes the APGAR scores are repeated, for example, at 1 minute after birth, at 5 minutes after birth, and at 10 minutes after birth and then are analyzed. APGAR scores generally do not follow a normal distribution; most newborns have scores of 7 or higher (normal range).

In some studies, the outcome is continuous but is subject to outliers or extreme values. For example, the number of days in the hospital following a particular surgical procedure is an outcome that is often subject to outliers. Suppose in an observational study investigators wish to assess whether there is a difference in the days patients spend in the hospital following liver transplant in for-profit versus nonprofit hospitals. Suppose days in the hospital following transplant are measured in $n = 100$ participants, 50 from for-profit hospitals and 50 from nonprofit hospitals; the data are shown in **Figure 10–2**.

Note that 75% of the participants stay at most 16 days in the hospital following a transplant, while at least one patient stays 35 days, which would be considered an outlier. Recall from Chapter 4 that we use $Q_1 - 1.5(Q_3 - Q_1)$ as a lower limit and $Q_3 + 1.5(Q_3 - Q_1)$ as an upper limit to detect outliers. In Figure 10–2, $Q_1 = 12$ and $Q_3 = 16$; thus outliers are values below $12 - 1.5(16 - 12) = 6$ or above $16 + 1.5(16 - 12) = 22$.

In some studies, the outcome is a continuous variable that is measured with some imprecision (e.g., with clear limits of detection). Some instruments or assays cannot measure presence of specific quantities above or below certain limits. For example,

FIGURE 10–1 Distribution of Symptom Severity in Total Sample

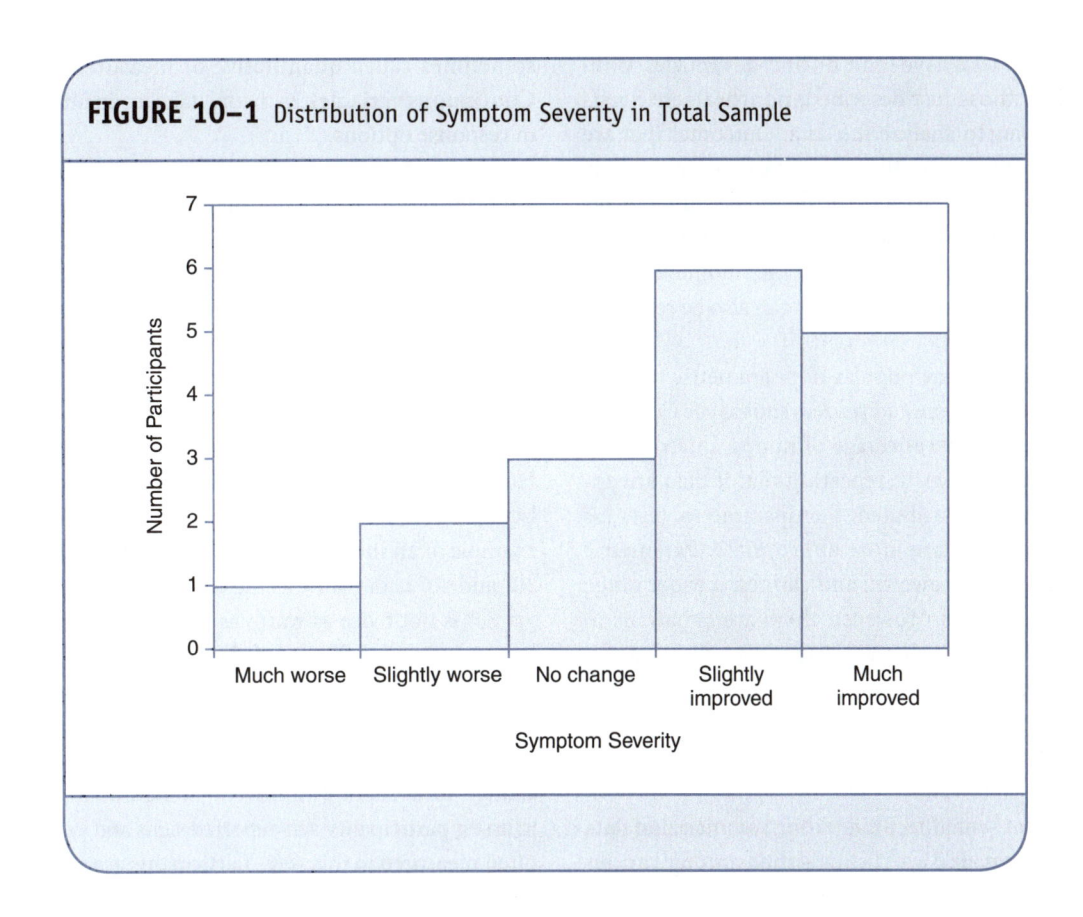

FIGURE 10–2 Distribution of Days in the Hospital Following Transplant

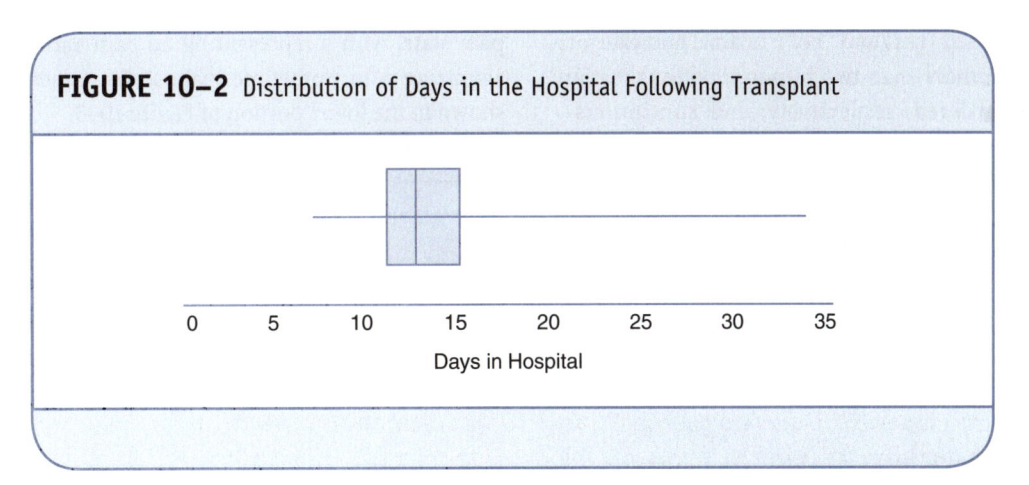

viral load is a measure of the amount of virus in the body and is measured as the number of copies of virus in a certain volume of blood. It can range from "not detected" or "below the limit of detection" to hundreds of millions of copies. Thus, some participants in a sample may have measures like 1,254,000 copies or 874,050 copies, and others are measured as "not detected." If a substantial number of participants have undetectable levels, the distribution of viral load is not normally distributed.

In nonparametric tests, the hypotheses are not about population parameters (e.g., $\mu = 50$ or $\mu_1 = \mu_2$). Instead, the null hypothesis is more general. For example, when comparing two independent groups in terms of a continuous outcome, the null hypothesis in a parametric test is $H_0: \mu_1 = \mu_2$. In a nonparametric test the null hypothesis is that the two populations are equal; often this is interpreted as being equal in terms of their central tendency.

Nonparametric tests have some distinct advantages. With outcomes such as those just described, nonparametric tests may be the only way to analyze the data. Outcomes that are ordinal, ranked, subject to outliers, or measured imprecisely are difficult to analyze with parametric methods without making major assumptions about their distributions, as well as decisions about how to code some values (e.g., "not detected"). As we describe here, nonparametric tests can also be relatively simple to conduct.

Here we describe some popular nonparametric tests for continuous outcomes. Interested readers should see Conover[3] for a more comprehensive coverage of nonparametric tests. Before proceeding, it is worth repeating that if data are approximately normally distributed, then parametric tests (as described in Chapter 7) are more appropriate. Parametric tests are generally more powerful and can test a wider range of alternative hypotheses. However, there are situations in which the assumptions for a parametric test are violated and a nonparametric test is more appropriate.

10.1 INTRODUCTION TO NONPARAMETRIC TESTING

In Chapter 4 we presented techniques for summarizing data collected in a sample, and we distinguished among various types of variables or outcomes. Specifically, we discussed dichotomous outcomes that have only two possible response options (usually coded "yes" and "no"); ordinal and categorical outcomes with more than two response options that are ordered and unordered, respectively; and continuous—sometimes called quantitative or measurement—variables. **Continuous variables**, in theory, take on an unlimited number of response options.

The techniques we describe here apply to outcomes that are ordinal, ranked, or continuous and are not normally distributed. Recall that continuous outcomes are quantitative measures based on a specific measurement scale (e.g., weight in pounds or height in inches). Some investigators make the distinction between continuous and interval- and ordinal-scale data. Interval data are like continuous data in that they are measured on a constant scale (i.e., the same difference exists between adjacent scale scores across the entire spectrum of scores). However, differences between interval scores are interpretable, but ratios are not. Temperature in Celsius or Fahrenheit is an example of an interval-scale outcome. The difference between 30° and 40° is the same as the difference between 70° and 80°, yet 80° is not twice as warm as 40°. Ordinal outcomes can be less specific, and the ordered categories need not be equally spaced. Symptom severity is an example of an ordinal outcome, and it is not clear whether the difference between "much worse" and "slightly worse" is the same as the difference between "no change" and "slightly improved." Some studies use visual scales to assess participants' self-reported signs and symptoms. Pain is often measured in this way. Participants are sometimes shown a visual scale such as that in the upper portion of **Figure 10–3** and are asked to choose the number that best represents their pain state, with 0 representing no pain and 10 representing agonizing pain. Sometimes pain scales use visual anchors, as shown in the lower portion of Figure 10–3.

FIGURE 10–3 Visual Pain Scales

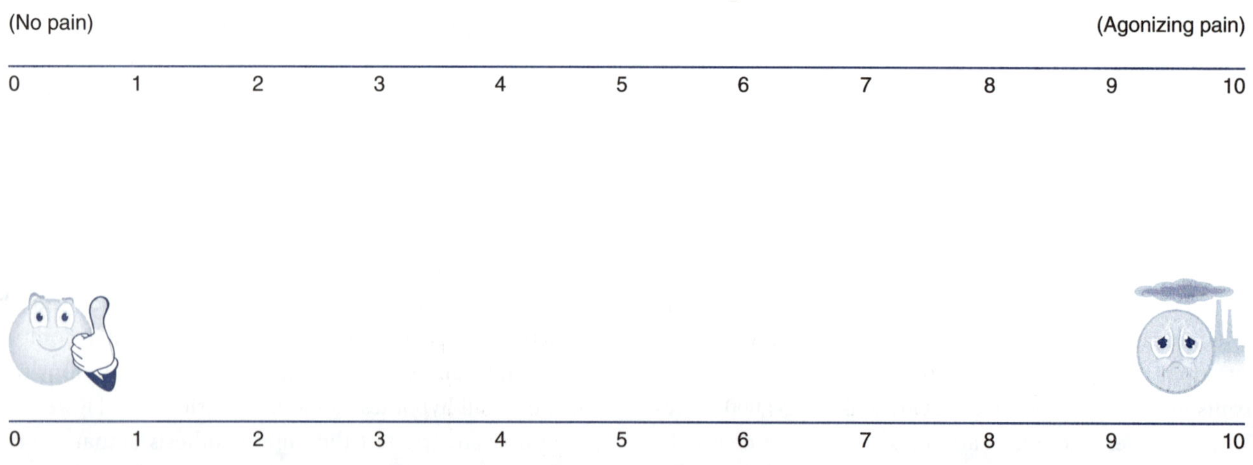

In the upper portion of Figure 10-3, although 10 is worse than 9 which is worse than 8, the difference between adjacent scores may not necessarily be the same. It is important to understand how outcomes are measured to make appropriate inferences based on statistical analysis and, in particular, not to overstate precision.

The nonparametric procedures that we describe here follow the same general procedure. The outcome variable (ordinal, interval, or continuous) is ranked from lowest to highest, and the analysis focuses on the ranks as opposed to the measured or raw values. For example, suppose we measure self-reported pain using a visual analog scale with anchors at 0 (no pain) and 10 (agonizing pain) and record the following responses in a sample of $n = 6$ participants:

$$7 \quad 5 \quad 9 \quad 3 \quad 0 \quad 2$$

The ranks, which are used to perform a nonparametric test, are assigned as follows: First the data are ordered from smallest to largest. The lowest-value score is then assigned a rank of 1, the next lowest a rank of 2, and so on. The largest value is assigned a rank of n (in this example, $n = 6$). The observed data and corresponding ranks are shown below:

Ordered Observed Data: 0 2 3 5 7 9
Ranks: 1 2 3 4 5 6

A complicating issue arises when there are ties in the sample (i.e., the same value is measured in two or more participants). For example, suppose that the following data are observed in our sample of $n = 6$:

Observed Data: 7 7 9 3 0 2

The 4th and 5th ordered values are both equal to 7. When assigning ranks, the recommended procedure is to assign the mean rank of 4.5 to each (i.e., the mean of 4 and 5), as follows:

Ordered Observed Data: 0 2 3 7 7 9
Ranks: 1 2 3 4.5 4.5 6

Using the same approach, suppose that there are three values of 7. In this case, we assign a rank of 5 (the mean of 4, 5, and 6) to the 4th, 5th, and 6th ordered values, as follows:

Ordered Observed Data: 0 2 3 7 7 7
Ranks: 1 2 3 5 5 5

Assigning the mean rank when there are ties ensures that the sum of the ranks is the same in each sample (1 + 2 + 3 + 4 + 5 + 6 = 21, 1 + 2 + 3 + 4.5 + 4.5 + 6 = 21, and 1 + 2 + 3 + 5 + 5 + 5 = 21).

Using this approach, the sum of the ranks will always equal $n(n + 1)/2$. When conducting nonparametric tests, it is useful to check the sum of the ranks before proceeding with the analysis.

To conduct nonparametric tests, we again follow the five-step approach outlined in Chapter 7 for hypothesis testing.

Step 1. Set up hypotheses and select the level of significance α.

Analogous to parametric testing, the research hypothesis can be one- or two-sided (one- or two-tailed), depending on the research question of interest.

Step 2. Select the appropriate test statistic.

The test statistic is a single number that summarizes the sample information. In nonparametric tests, the observed data are converted into ranks and then the ranks are summarized into a test statistic.

Step 3. Set up the decision rule.

The decision rule is a statement that tells under what circumstances to reject the null hypothesis. In some nonparametric tests we reject H_0 if the test statistic is large, while in others we reject H_0 if the test statistic is small. These are distinguished as we describe the different tests.

Step 4. Compute the test statistic.

Here we compute the test statistic by summarizing the ranks using the test statistic identified in Step 2.

Step 5. Conclusion.

The final conclusion is made by comparing the test statistic (which is a summary of the information observed in the sample) to the decision rule. The final conclusion is either to reject the null hypothesis (because it is very unlikely to observe the sample data if the null hypothesis is true) or not to reject the null hypothesis (because the sample data are not very unlikely if the null hypothesis is true).

10.2 TESTS WITH TWO INDEPENDENT SAMPLES

In Chapter 7, Section 7.5, we presented techniques for testing the equality of means in two independent samples. An underlying assumption for appropriate use of the tests described in Section 7.5 was that the continuous outcome was approximately normally distributed or that the samples were sufficiently large (usually $n_1 > 30$ and $n_2 > 30$) to justify their use based on the Central Limit Theorem. When the outcome is not normally distributed and the samples are small, a nonparametric test is appropriate.

A popular nonparametric test to compare outcomes between two independent groups is the Mann–Whitney U test. The Mann–Whitney U test, sometimes called the Mann–Whitney–Wilcoxon test or the Wilcoxon Rank-Sum test, is used to test whether two samples are likely to derive from the same population (i.e., that the two populations have the same shape). Some investigators interpret this test as comparing the medians between the two populations. Recall that the analogous parametric test compares the means (H_0: $\mu_1 = \mu_2$) between independent groups (see Section 7.5).

The null and two-sided research hypotheses for the nonparametric test are stated as follows:

H_0: The two populations are equal.
H_1: The two populations are not equal.

This test is often performed as a two-sided test, and thus the research hypothesis indicates that the populations are not equal as opposed to specifying directionality. A one-sided research hypothesis is used if interest lies in detecting a positive or negative shift in one population as compared to the other. The procedure for the test involves pooling the observations from the two samples into one combined sample, keeping track of which sample each observation comes from, and then ranking lowest to highest from 1 to $n_1 + n_2$, respectively. To illustrate the procedure, consider the following example. We address each of the five steps in the test of hypothesis but include more details to promote understanding. (In subsequent examples, the five-step approach is followed more directly.)

Example 10.1. Consider a Phase II clinical trial designed to investigate the effectiveness of a new drug to reduce symptoms of asthma in children. A total of $n = 10$ participants are randomized to receive either the new drug or a placebo. Participants are asked to record the number of episodes of shortness of breath over a 1-week period following receipt of the assigned treatment. The data are shown below.

Placebo	7	5	6	4	12
New Drug	3	6	4	2	1

The question of interest is whether there is a difference in the number of episodes of shortness of breath over a 1-week period in participants receiving the new drug compared to those receiving the placebo. By inspection, it appears that participants receiving the placebo have more episodes of shortness of breath than those receiving the new drug. The question is whether this observed difference is statistically significant.

In this example, the outcome is a count, and in this sample the data do not follow a normal distribution (see **Figure 10–4**). In addition, the sample size is small ($n_1 = n_2 = 5$).

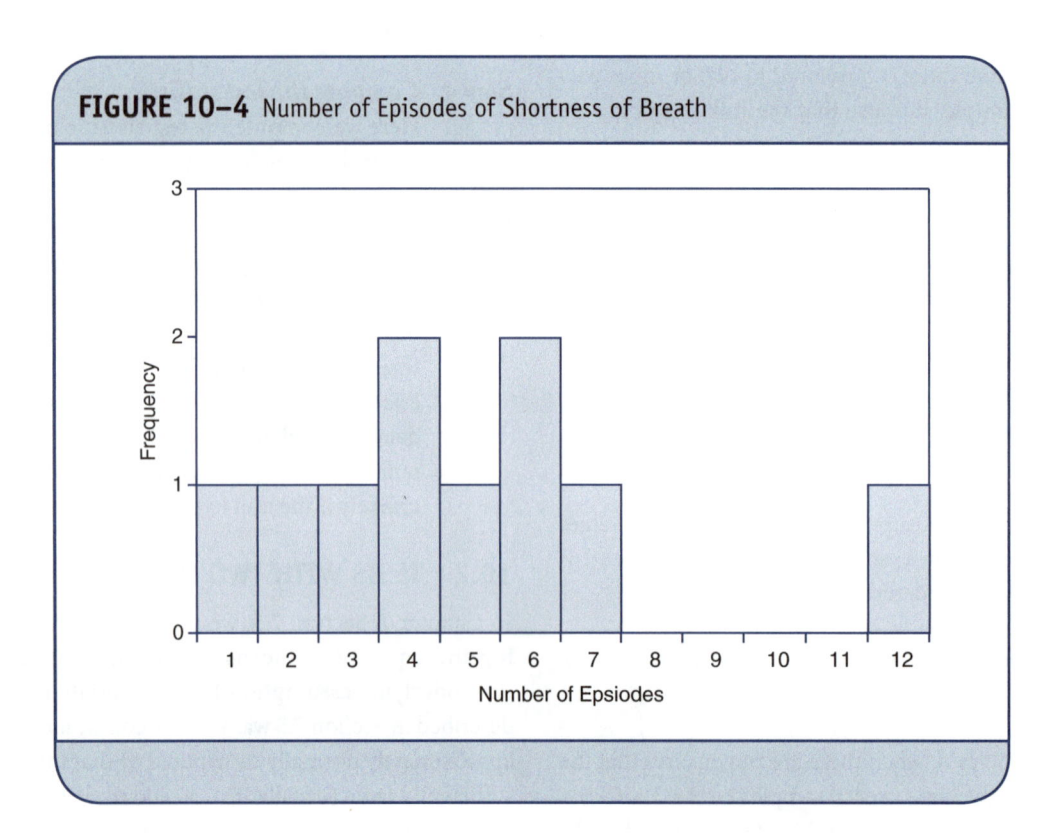

FIGURE 10–4 Number of Episodes of Shortness of Breath

Thus, a nonparametric test is appropriate. The hypotheses to be tested are given below, and we run the test at the 5% level of significance (i.e., $\alpha = 0.05$).

H_0: The two populations are equal.
H_1: The two populations are not equal.

Note that if the null hypothesis is true (i.e., the two populations are equal), we expect to see similar numbers of episodes of shortness of breath in each of the two treatment groups. Specifically, if the numbers of episodes of shortness of breath are equal in participants assigned to the new drug and in those assigned to the placebo, we would expect to see some participants reporting few episodes and some reporting more episodes in each of the comparison groups. This does not appear to be the case with the observed data. A test of hypothesis is needed to determine whether the observed data are evidence of a statistically significant difference in populations.

The first step is to assign ranks, and to do so we order the data from smallest to largest. This is done on the combined or total sample, pooling the data from the two treatment groups and assigning ranks from 1 to 10, as shown in **Table 10–1**. We also need to keep track of the group assignments in the total sample ($n = 10$).

Note that the lower ranks (e.g., 1, 2, and 3) are assigned to responses in the new drug group while the higher ranks (e.g., 9 and 10) are assigned to responses in the placebo group. Again, the goal of the test is to determine whether the observed data support a difference in the populations of responses. Recall that in parametric tests, discussed in Chapter 7, when comparing means between two groups, we analyze the difference in the sample means relative to their variability and summarize the sample information in a test statistic. A similar approach is employed here. Specifically, we produce a test statistic based on the ranks.

First, we sum the ranks in each group. In the placebo group the sum of the ranks is 37, and in the new drug group the sum of the ranks is 18. Recall that the sum of the ranks will always equal $n(n + 1)/2$. As a check on our assignment of ranks, we have $n(n + 1)/2 = 10(11)/2 = 55$, which is equal to $37 + 18 = 55$.

For the test, we call the placebo group 1 and the new drug group 2 (assignment of groups 1 and 2 is arbitrary—we only need to maintain the group assignments throughout the testing procedure). We let R_1 denote the sum of the ranks in group 1 (i.e., $R_1 = 37$) and R_2 denote the sum of the ranks in group 2 (i.e., $R_2 = 18$). If the null hypothesis is true (i.e., if the two populations are equal), we expect R_1 and R_2 to be similar. In this example, the lower values (lower ranks) are clustered in the new drug group (group 2), while the higher values (higher ranks) are clustered in the placebo group (group 1). The question is whether the difference we observe in the sums of the ranks is suggestive of a difference in the populations or is simply due to chance. We now compute a test statistic to summarize the sample information and then compare that test statistic to an appropriate value from a probability distribution corresponding to our selected level of significance.

The test statistic for the Mann–Whitney U test is denoted by U and is the smaller of U_1 and U_2, defined as follows:

$$U_1 = n_1 n_2 + \frac{n_1(n_1 + 1)}{2} - R_1$$

$$U_2 = n_1 n_2 + \frac{n_2(n_2 + 1)}{2} - R_2$$

TABLE 10–1 Assigning Ranks

		Total Sample (Ordered Smallest to Largest)		Ranks	
Placebo	New Drug	Placebo	New Drug	Placebo	New Drug
7	3		1		1
5	6		2		2
6	4		3		3
4	2	4	4	4.5	4.5
12	1	5		6	
		6	6	7.5	7.5
		7		9	
		12		10	

For this example,

$$U_1 = n_1 n_2 + \frac{n_1(n_1+1)}{2} - R_1 = 5(5) + \frac{5(6)}{2} - 37 = 3$$

$$U_2 = n_1 n_2 + \frac{n_2(n_2+1)}{2} - R_2 = 5(5) + \frac{5(6)}{2} - 18 = 22$$

In our example, $U = 3$. Is this evidence in support of the null hypothesis or the research hypothesis?

Before we address this question, we consider the range of the test statistic U in different situations. First, consider the situation where there is complete separation between the groups. This is a situation where the data most clearly support the research hypothesis (i.e., the two populations are not equal). Specifically, suppose that all of the higher numbers of episodes of shortness of breath, and thus all of the higher ranks, are in the placebo group and all of the lower numbers of episodes of shortness of breath, and thus all of the lower ranks, are in the new-drug group and that there are no ties. Then $R_1 = 6 + 7 + 8 + 9 + 10 = 40$ and $R_2 = 1 + 2 + 3 + 4 + 5 = 15$, and $U_1 = 5(5) + \frac{5(6)}{2} - 40 = 0$ and $U_2 = 5(5) + \frac{5(6)}{2} - 15 = 25$. Thus, when there is clearly a difference in the populations, $U = 0$.

Consider a second situation where low and high scores are approximately evenly distributed in the placebo and the new-drug groups. This is a situation where the data most clearly support the null hypothesis (i.e., the two populations are equal). Suppose that ranks of 2, 4, 6, 8, and 10 are assigned to the numbers of episodes of shortness of breath reported in the placebo group and ranks of 1, 3, 5, 7, and 9 are assigned to the numbers of episodes of shortness of breath reported in the new drug group. Then $R_1 = 2 + 4 + 6 + 8 + 10 = 30$ and $R_2 = 1 + 3 + 5 + 7 + 9 = 25$, and $U_1 = 5(5) + \frac{5(6)}{2} - 30 = 10$ and $U_2 = 5(5) + \frac{5(6)}{2} - 25 = 15$. Thus, when there is clearly no difference between populations, $U = 10$. Smaller values of U support the research hypothesis and larger values of U support the null hypothesis. In fact, for any test the theoretical range of U is from 0 (complete separation between groups, H_0 most likely false and H_1 most likely true) to $n_1 \times n_2$ (little evidence in support of H_1). In every test, $U_1 + U_2$ is always equal to $n_1 \times n_2$. In our example, U can range from 0 to 25 and smaller values of U support the research hypothesis (i.e., we reject H_0 if U is small). The procedure for determining exactly when to reject H_0 is described next.

In every test, we must determine whether the observed U supports the null or research hypothesis. This is done following the same approach used in parametric testing. Specifically, we determine a critical value of U such that if the observed value of U is less than or equal to the critical value, we reject H_0 in favor of H_1, and if the observed value of U exceeds the critical value, we do not reject H_0. The critical value of U can be found in Table 5 in the Appendix. To determine the appropriate critical value we need the sample sizes ($n_1 = n_2 = 5$) and our two-sided level of significance ($\alpha = 0.05$). For this example the critical value is 2, and the decision rule is to reject H_0 if $U \leq 2$. We do not reject H_0 because $3 > 2$. We do not have statistically significant evidence at $\alpha = 0.05$ to show that the two populations of numbers of episodes of shortness of breath are not equal. In this example, the failure to reach statistical significance is likely due to low power. The sample data would suggest a difference, but the sample sizes are possibly too small to conclude that there is a statistically significant difference.

Example 10.2. A new approach to prenatal care is proposed for pregnant women living in a rural community. The new program involves in-home visits during the course of pregnancy in addition to the usual or regularly scheduled visits. A pilot randomized trial with 15 pregnant women is designed to evaluate whether women who participate in the program deliver healthier babies than women receiving usual care. The outcome is the APGAR score measured 5 minutes after birth. Recall that APGAR scores range from 0 to 10, where a score of 7 or higher is considered normal (healthy), 4–6 is low, and 0–3 is critically low. The data are shown below.

| Usual Care | 8 | 7 | 6 | 2 | 5 | 8 | 7 | 3 |
| New Program | 9 | 8 | 7 | 8 | 10| 9 | 6 | |

Is there statistical evidence of a difference in APGAR scores in women receiving the new and enhanced care versus usual prenatal care? We run the test using the five-step approach.

Step 1. Set up hypotheses and determine the level of significance.

H_0: The two populations are equal.
H_1: The two populations are not equal.
$\alpha = 0.05$.

Step 2. Select the appropriate test statistic.

Because APGAR scores are not normally distributed and the samples are small ($n_1 = 8$ and $n_2 = 7$), we use the Mann–Whitney U test. The test statistic is U, the smaller of $U_1 = n_1 n_2 + \frac{n_1(n_1+1)}{2} - R_1$ and $U_2 = n_1 n_2 + \frac{n_2(n_2+1)}{2} - R_2$, where R_1 and R_2 are the sums of the ranks in groups 1 and 2, respectively.

Step 3. Set up the decision rule.

The appropriate critical value can be found in Table 5 in the Appendix. To determine the appropriate critical value we need the sample sizes ($n_1 = 8$ and $n_2 = 7$) and our two-sided level of significance ($\alpha = 0.05$). The critical value for this test is 10, and the decision rule is as follows:

$$\text{Reject } H_0 \text{ if } U \leq 10.$$

Step 4. Compute the test statistic.

The first step is to assign ranks of 1 through 15 to the smallest through largest values in the total sample, as shown in **Table 10–2**. Next, we sum the ranks in each group. In the usual-care group the sum of the ranks is $R_1 = 45.5$ and in the new-program group the sum of the ranks is $R_2 = 74.5$. Recall that the sum of the ranks will always equal $n(n + 1)/2$. As a check on our assignment of ranks, we have $n(n + 1)/2 = 15(16)/2 = 120$, which is equal to $45.5 + 74.5 = 120$. We now compute U_1 and U_2, as follows:

$$U_1 = n_1 n_2 + \frac{n_1(n_1+1)}{2} - R_1 = 8(7) + \frac{8(9)}{2} - 45.5 = 46.5$$

$$U_2 = n_1 n_2 + \frac{n_2(n_2+1)}{2} - R_2 = 8(7) + \frac{7(8)}{2} - 74.5 = 9.5$$

Thus, the test statistic is $U = 9.5$.

Step 5. Conclusion.

We reject H_0 because $9.5 < 10$. We have statistically significant evidence at $\alpha = 0.05$ to show that the populations of APGAR scores are not equal in women receiving usual prenatal care as compared to the new program of prenatal care.

Example 10.3. A clinical trial is run to assess the effectiveness of a new antiretroviral therapy for patients with HIV. Patients are randomized to receive a standard antiretroviral therapy (usual care) or the new antiretroviral therapy and are monitored for 3 months. The primary outcome is viral load, which represents the number of HIV copies per milliliter of blood. A total of 30 participants are randomized, and the data are shown below.

Standard Antiretroviral Therapy	7500	8000	2000	550
	1250	1000	2250	6800
	3400	6300	9100	970
	1040	670	400	

New Antiretroviral Therapy	400	250	800	1400	8000
	7400	1020	6000	920	1420
	2700	4200	5200	4100	
	undetectable				

Is there statistical evidence of a difference in viral load in patients receiving the standard therapy versus the new antiretroviral therapy? We run the test using the five-step approach.

TABLE 10–2 Assigning Ranks

		Total Sample (Ordered Smallest to Largest)		Ranks	
Usual Care	New Program	Usual Care	New Program	Usual Care	New Program
8	9	2		1	
7	8	3		2	
6	7	5		3	
2	8	6	6	4.5	4.5
5	10	7	7	7	7
8	9	7		7	
7	6	8	8	10.5	10.5
3		8	8	10.5	10.5
		9			13.5
		9			13.5
		10			15
				$R_1 = 45.5$	$R_2 = 74.5$

Step 1. Set up hypotheses and determine the level of significance.

H_0: The two populations are equal.
H_1: The two populations are not equal.
$\alpha = 0.05$.

Step 2. Select the appropriate test statistic.

Because viral load measures are not normally distributed—generally subject to extreme values or outliers as well as clear limits of detection (e.g., "undetectable")—we use the Mann–Whitney U test. The test statistic is U, the smaller of $U_1 = n_1 n_2 + \frac{n_1(n_1+1)}{2} - R_1$ and $U_2 = n_1 n_2 + \frac{n_2(n_2+1)}{2} - R_2$, where R_1 and R_2 are the sums of the ranks in groups 1 and 2, respectively.

Step 3. Set up the decision rule.

The appropriate critical value can be found in Table 5 in the Appendix. To determine the appropriate critical value we need the sample sizes ($n_1 = n_2 = 15$) and our two-sided level of significance ($\alpha = 0.05$).

The critical value for this test is 64, and the decision rule is as follows:

Reject H_0 if $U \leq 64$.

Step 4. Compute the test statistic.

The first step is to assign ranks of 1 through 30 to the smallest through largest values in the total sample, as shown in **Table 10–3**. Note that the "undetectable" measurement is listed first in the ordered values (smallest) and assigned a rank of 1.

Next, we sum the ranks in each group. In the standard antiretroviral therapy group, the sum of the ranks is $R_1 = 245$, and in the new antiretroviral therapy group, the sum of the ranks is $R_2 = 220$. Recall that the sum of the ranks will always equal $n(n+1)/2$. As a check on our assignment of ranks, we have $n(n+1)/2 = 30(31)/2 = 465$, which is equal to $245 + 220 = 465$.

We now compute U_1 and U_2, as follows:

$$U_1 = n_1 n_2 + \frac{n_1(n_1+1)}{2} - R_1 = 15(15) + \frac{15(16)}{2} - 245 = 100$$

$$U_2 = n_1 n_2 + \frac{n_2(n_2+1)}{2} - R_2 = 15(15) + \frac{15(16)}{2} - 220 = 125$$

Thus, the test statistic is $U = 100$.

Step 5. Conclusion.

We do not reject H_0 because $100 > 64$. We do not have statistically significant evidence at $\alpha = 0.05$ to show that the populations of viral load measures are not equal in patients receiving the standard therapy versus the new antiretroviral therapy.

10.3 TESTS WITH MATCHED SAMPLES

We now describe nonparametric tests to compare two groups with respect to a continuous outcome when the data are collected on matched or paired samples. The parametric procedure was described in detail in Chapter 7, Section 7.6, for the situation in which the continuous outcome followed a normal distribution. Here we consider the situation in which the outcome cannot be assumed to follow a normal distribution.

There are two popular nonparametric tests to compare outcomes between two matched or paired groups. The first is called the Sign test and the second the Wilcoxon Signed Rank test. We describe both tests here.

Recall that when data are matched or paired, we compute a difference score for each individual and analyze the difference scores. The same approach is followed in nonparametric tests. In parametric tests, the null hypothesis is that the mean difference (μ_d) is zero. In nonparametric tests, the null hypothesis is that the median difference is zero.

Example 10.4. Consider a clinical investigation to assess the effectiveness of a new drug designed to reduce repetitive behaviors in children affected with autism. If the drug is effective, children will exhibit fewer repetitive behaviors with treatment as compared with no treatment. A total of 8 children with autism enroll in the study. Each child is observed by the study psychologist for a period of 3 hours both before treatment and then again after treatment, taking the new drug for 1 week. The time that each child is engaged in repetitive behavior during each 3-hour observation period is measured. Repetitive behavior is scored on a scale of 0 to 100, and scores represent the percent of the observation time in which the child is engaged in repetitive behavior. For example, a score of 0 indicates that during the entire observation period the child did not engage in repetitive behavior, while a score of 100 indicates that the child was constantly engaged in repetitive behavior. The data are shown in **Table 10–4**.

The question of interest is whether there is improvement in repetitive behavior after 1 week of treatment as compared to before treatment. Looking at the data, it appears that some children improve (e.g., Child 5 scored 80 before treatment and 20 after treatment), but some get worse (e.g., Child 3 scored 40 before treatment and 50 after treatment).

TABLE 10-3 Assigning Ranks

Standard Antiretroviral	New Antiretroviral	Total Sample (Ordered Smallest to Largest) Standard Antiretroviral	Total Sample (Ordered Smallest to Largest) New Antiretroviral	Ranks Standard Antiretroviral	Ranks New Antiretroviral
7500	400		undetectable		1
8000	250		250		2
2000	800	400	400	3.5	3.5
550	1400	550		5	
1250	8000	670		6	
1000	7400		800		7
2250	1020		920		8
6800	6000	970		9	
3400	920	1000		10	
6300	1420		1020		11
9100	2700	1040		12	
970	4200	1250		13	
1040	5200		1400		14
670	4100		1420		15
400	undetectable	2000		16	
		2250		17	
			2700		18
		3400		19	
			4100		20
			4200		21
			5200		22
			6000		23
		6300		24	
		6800		25	
			7400		26
		7500		27	
		8000	8000	28.5	28.5
		9100		30	
				$R_1 = 245$	$R_2 = 220$

Because the before- and after-treatment measures are paired, we compute difference scores for each child. In this example, we subtract the assessment of repetitive behaviors measured after treatment from that measured before treatment, so that difference scores represent improvement in repetitive behavior. The question of interest is whether there is significant improvement after treatment. Again, notice that some children's repetitive behavior gets worse after treatment as compared to before treatment (and thus they have negative improvement).

The differences are shown in **Table 10-5**. In this small sample, the observed difference (or improvement) scores vary widely and are subject to extremes (e.g., the observed difference of 60 is an outlier). Thus, a nonparametric test is

TABLE 10–4 Repetitive Behavior Before and After Treatment

Child	Before Treatment	After 1 Week of Treatment
1	85	75
2	70	50
3	40	50
4	65	40
5	80	20
6	75	65
7	55	40
8	20	25

TABLE 10–6 Signs of the Differences

Child	Before Treatment	After 1 Week of Treatment	Difference (Before − After)	Sign
1	85	75	10	+
2	70	50	20	+
3	40	50	−10	−
4	65	40	25	+
5	80	20	60	+
6	75	65	10	+
7	55	40	15	+
8	20	25	−5	−

TABLE 10–5 Differences in Repetitive Behavior

Child	Before Treatment	After 1 Week of Treatment	Difference (Before − After)
1	85	75	10
2	70	50	20
3	40	50	−10
4	65	40	25
5	80	20	60
6	75	65	10
7	55	40	15
8	20	25	−5

appropriate to test whether there is significant improvement in repetitive behavior before versus after treatment. The hypotheses are given below and the test is run at a 5% level of significance.

H_0: The median difference is zero.

H_1: The median difference is positive.

$\alpha = 0.05$.

The null hypothesis, as usual, represents the situation where there is no change or no effect. In this example, the null hypothesis represents the situation where there is no difference in scores before versus after treatment, or no effect of treatment. If the null hypothesis is true, we expect to see some positive differences (improvement) and some negative differences (worsening). If the research hypothesis is true, we expect to see more positive differences, or more children with improvement in repetitive behavior after treatment as compared to before.

The Sign test is the simplest nonparametric test for matched or paired data. The approach is to analyze only the signs of the difference scores. The signs here are shown in **Table 10–6**. If the null hypothesis is true (i.e., if the median difference is zero), then we expect to see approximately half of the differences as positive and half of the differences as negative. If the research hypothesis is true, we expect to see more positive differences.

The test statistic for the Sign test is the number of positive signs or the number of negative signs, whichever is smaller. In this example, we observe two negative and six positive signs. Is this evidence of significant improvement or simply due to chance?

We again determine whether the observed test statistic (i.e., the number of negative signs in this example) supports the null or research hypothesis. This is done following the same approach used in parametric testing. Specifically, we determine a critical value such that if the smaller of the number of positive or negative signs is less than or equal to that critical value, then we reject H_0 in favor of H_1, and if the smaller of the number of positive or negative signs is greater than the critical value, then we do not reject H_0. Notice that this is a one-sided decision rule corresponding to our one-sided research hypothesis (the two-sided situation is discussed in the next example).

The critical value for the Sign test can be found in Table 6 in the Appendix. To determine the appropriate critical value we need the sample size, which is equal to the number of matched pairs ($n = 8$), and our one-sided level of significance ($\alpha = 0.05$). For this example the critical value is 1, and the decision rule is to reject H_0 if the smaller of the number of positive or negative signs is ≤ 1. Here we do not reject H_0 because $2 > 1$. We do not have statistically significant evidence at $\alpha = 0.05$ to show that there is improvement in repetitive behavior after treatment as compared to before treatment.

In our usual five-step approach, we set up a decision rule that indicates under what circumstances to reject H_0. For example, with many of our parametric tests, we reject H_0 if the observed test statistic (e.g., Z or t) is more extreme than the critical value. Specifically, we reject H_0 if the test statistic is more extreme than a critical value that comes from a particular probability distribution (e.g., the standard normal distribution) based on a specified level of significance (e.g., $\alpha = 0.05$). We can use the same approach for the Sign test as illustrated above, or we can consider an alternative approach such as the following.

With the Sign test we can readily compute a p-value based on our observed test statistic and compare that to our stated level of significance (e.g., $\alpha = 0.05$). This is analogous to what is done when we run tests of hypothesis with a statistical computing package (e.g., using Excel®). The statistical computing package computes a test statistic and a p-value, and the investigator then compares the p-value to the predetermined level of significance to draw a conclusion about the hypotheses using the following rule: Reject H_0 if $p \leq \alpha$.

The test statistic for the Sign test is the smaller of the number of positive or negative signs, and it follows a binomial distribution with $n =$ the number of subjects in the study and $p = 0.5$ (see Chapter 5, Section 5.6, for more details on the binomial distribution). In this example, $n = 8$ and $p = 0.5$. The following probabilities can be computed with either the binomial distribution formula, $P(x \text{ successes}) = \dfrac{n!}{x!(n-x)!} p^x (1-p)^{n-x}$ or using a statistical computing package such as Excel (see Chapter 5 of the Excel workbook).

Recall that a p-value is the probability of observing a test statistic as or more extreme than that observed. We observe two negative signs. Thus, the p-value for the test is $p = P(x \leq 2)$. Using **Table 10–7**, $P(x \leq 2) = P(0) + P(1) + P(2) = 0.0039 + 0.0313 + 0.1094 = 0.1446$.

Because the p-value 0.1446 exceeds the level of significance ($\alpha = 0.05$), we do not have statistically significant evidence at $\alpha = 0.05$ to show that there is improvement in repetitive behaviors after treatment compared to before treatment. Notice in the

TABLE 10–7 Probabilities Based on the Binomial Distribution Model

$x =$ Number of Successes	$P(x$ successes$)$
0	0.0039
1	0.0313
2	0.1094
3	0.2188
4	0.2734
5	0.2188
6	0.1094
7	0.0313
8	0.0039

table of binomial probabilities (Table 10–7) that we would have had to observe at most one negative sign to declare statistical significance using a 5% level of significance. Recall the critical value for our test was 1, based on Table 6 in the Appendix.

In Example 10.4 we consider a one-sided test (i.e., we hypothesize improvement after treatment). A two-sided test can also be considered, in which we hypothesize a difference in repetitive behavior after treatment compared to before treatment. In Table 6 in the Appendix, we can determine a two-sided critical value and again reject H_0 if the smaller of the number of positive or negative signs is less than or equal to that of the two-sided critical value. Alternatively, we can compute a two-sided p-value. With a two-sided test, the p-value is the probability of observing many or few positive (or negative) signs. If the research hypothesis is a two-sided alternative (i.e., H_1: The median difference is not zero), then the p-value is computed as $p = 2 \times P(x \leq 2)$. Notice that this is equivalent to $p = P(x \leq 2) + P(x \geq 6)$, representing the situation of few or many successes. Recall that in two-sided tests we reject the null hypothesis if the test statistic is extreme in either direction. Thus, in the Sign test, a two-sided p-value is the probability of observing few or many positive (or negative) signs. Here we observe two negative signs (and thus six positive signs). The opposite situation would be six negative signs (and thus two positive signs, as $n = 8$). The two-sided p-value is the probability of observing a test statistic as or more extreme in either direction (i.e., $P(x \leq 2) + P(x \geq 6) = 0.0039 + 0.0313 + 0.1094 + 0.1094 + 0.0313 + 0.0039 = 2 \times (0.1446) = 0.2892$).

There is a special circumstance that needs attention when implementing the Sign test that arises when one or more participants have difference scores of zero (i.e., their paired measurements are identical). If there is just one difference score of zero, some investigators drop that observation and reduce the sample size by 1 (i.e., the sample size for the binomial distribution would be $n - 1$). This is a reasonable approach if there is just one zero. However, if there are two or more zeros, an alternative approach is preferred. If there is an even number of zeros, we randomly assign them positive or negative signs. If there is an odd number of zeros, we randomly drop one (reduce the sample size by 1) and then randomly assign the remaining observations' positive or negative signs. The following example illustrates the approach.

Example 10.5. A new chemotherapy treatment is proposed for patients with breast cancer. Investigators are concerned with patients' ability to tolerate the treatment and, assess their quality of life both before and after receiving the new chemotherapy treatment. Quality of life (QOL) is measured on an ordinal scale and, for analysis purposes, numbers are assigned to each response category as follows: 1 = Poor, 2 = Fair, 3 = Good, 4 = Very Good, 5 = Excellent. The data are shown in **Table 10–8**. The question of interest is whether there is a difference in QOL after chemotherapy treatment as compared to before.

Step 1. **Set up hypotheses and determine the level of significance.**

H_0: The median difference is zero.

H_1: The median difference is not zero.

$\alpha = 0.05$.

TABLE 10–8 Quality of Life Before and After Chemotherapy Treatment

Patient	QOL Before Chemotherapy Treatment	QOL After Chemotherapy Treatment
1	3	2
2	2	3
3	3	4
4	2	4
5	1	1
6	3	4
7	2	4
8	3	3
9	2	1
10	1	3
11	3	4
12	2	3

Step 2. **Select the appropriate test statistic.**

The test statistic for the Sign test is the smaller of the number of positive or negative signs.

Step 3. **Set up the decision rule.**

The appropriate critical value for the Sign test can be found in Table 6 in the Appendix. To determine the appropriate critical value, we need the sample size (or number of matched pairs, $n = 12$) and our two-sided level of significance ($\alpha = 0.05$). The critical value for this two-sided test is 2, and the decision rule is as follows:

Reject H_0 if the smaller of the number of positive or negative signs is ≤ 2.

Step 4. **Compute the test statistic.**

Because the before- and after-treatment measures are paired, we compute difference scores for each patient. In this example, we subtract the QOL measured before treatment from that measured after (see **Table 10–9**).

We now capture the signs of the difference scores. Because there are two zeros, we randomly assign one negative sign (i.e., "−" to patient 5) and one positive sign (i.e., "+" to patient 8), as shown in **Table 10–10**.

The test statistic is the number of negative signs, which is equal to 3.

Step 5. **Conclusion.**

We do not reject H_0 because 3 > 2. We do not have statistically significant evidence at $\alpha = 0.05$ to show that there is a difference in QOL after chemotherapy treatment as compared to before.

We can also compute the p-value directly using the binomial distribution with $n = 12$ and $p = 0.5$. The two-sided p-value for the test is $p = 2 \times P(x \leq 3)$ (which is equivalent to $p = P(x \leq 3) + P(x \geq 9)$). Again, the two-sided p-value is the probability of observing few or many positive (or negative) signs. Here we observe three negative signs (and thus nine positive signs). The opposite situation would be nine negative signs (and thus three positive signs, as $n = 12$). The two-sided p-value is the probability of observing a test statistic as or more extreme in either direction (i.e., $P(x \leq 3) + P(x \geq 9)$). We can compute the p-value using the binomial formula or a statistical computing package, as follows: $p = 2 \times P(x \leq 3) = 2 (P(0) + P(1) + P(2) + P(3)) = 2 \times (0.0002 + 0.0029 + 0.0161 + 0.0537) = 2 \times (0.0729) = 0.1458$.

TABLE 10-9 Differences in Quality of Life Before and After Chemotherapy Treatment

Patient	QOL Before Chemotherapy Treatment	QOL After Chemotherapy Treatment	Difference (After − Before)
1	3	2	−1
2	2	3	1
3	3	4	1
4	2	4	2
5	1	1	0
6	3	4	1
7	2	4	2
8	3	3	0
9	2	1	−1
10	1	3	2
11	3	4	1
12	2	3	1

Because the p-value 0.1458 exceeds the level of significance ($\alpha = 0.05$), we do not have statistically significant evidence at $\alpha = 0.05$ to show that there is a difference in QOL after chemotherapy treatment compared to before treatment.

In each of the two previous examples, we failed to show statistical significance because the p-value was not less than the stated level of significance. While the test statistic for the Sign test is easy to compute, it actually does not take much of the information in the sample data into account. All we measure is the difference in each participant's scores, and we do not account for the magnitude of the difference.

Another popular nonparametric test for matched or paired data is called the Wilcoxon Signed Rank Test. Like the Sign test, it is based on difference scores, but in addition to analyzing the signs of the differences, it also takes into account the magnitudes of the observed differences. We now use the Wilcoxon Signed

TABLE 10-10 Signs of the Differences

Patient	QOL Before Chemotherapy Treatment	QOL After Chemotherapy Treatment	Difference (After − Before)	Sign
1	3	2	−1	−
2	2	3	1	+
3	3	4	1	+
4	2	4	2	+
5	1	1	0	−
6	3	4	1	+
7	2	4	2	+
8	3	3	0	+
9	2	1	−1	−
10	1	3	2	+
11	3	4	1	+
12	2	3	1	+

Rank test to analyze the data in Example 10.4. In the following example, we address each of the five steps in the test of hypothesis but include more details to promote understanding. (In subsequent examples, the five-step approach is followed more directly.)

Example 10.6. Consider again the clinical investigation to assess the effectiveness of a new drug to reduce repetitive behaviors in children affected with autism (described in Example 10.4). A total of 8 children with autism enroll in the study, and the amount of time that each child is engaged in repetitive behavior during three-hour observation periods is measured both before treatment and then again after taking the new drug, for a period of 1 week. The data are shown in **Table 10–11**.

The question of interest is whether there is improvement in repetitive behavior after treatment as compared to before. Again, because the before- and after-treatment measures are paired, we compute difference scores for each child (see **Table 10–12**).

The next step is to rank the difference scores. In the ranking process, we first order the absolute values of the difference scores and assign ranks using the approach outlined in Section 10.1. Specifically, we assign ranks from 1 through n to the smallest through largest absolute values of the difference scores, respectively, and assign the mean rank when there are ties in the absolute values of the difference scores (see **Table 10–13**).

The final step is to attach the signs ("+" or "−") of the observed differences to each rank as shown in **Table 10–14**.

Similar to the Sign test, the hypotheses for the Wilcoxon Signed Rank test concern the population median of the difference scores. The research hypothesis can be one- or two-sided. Here we consider a one-sided test.

H_0: The median difference is zero.
H_1: The median difference is positive.
$\alpha = 0.05$.

The test statistic for the Wilcoxon Signed Rank test is W, defined as the smaller of $W+$ and $W-$, which are the sums of the positive and negative ranks, respectively. If the null hypothesis is true, we expect to see similar numbers of lower and higher ranks that are both positive and negative (i.e., $W+$ and $W-$ would be similar). If the research hypothesis is true, we expect to see more higher and positive ranks (in this example, more children with substantial improvement in repetitive behavior after treatment as compared to before treatment, i.e., $W+$ much larger than $W-$).

In this example, $W+ = 32$ and $W- = 4$. Recall that the sum of the ranks (ignoring the signs) will always equal $n(n + 1)/2$. As a check on our assignment of ranks, we have $n(n + 1)/2 = 8(9)/2 = 36$, which is equal to $32 + 4$. The test statistic is $W = 4$.

TABLE 10–11 Repetitive Behavior Before and After Treatment

Child	Before Treatment	After 1 Week of Treatment
1	85	75
2	70	50
3	40	50
4	65	40
5	80	20
6	75	65
7	55	40
8	20	25

TABLE 10–12 Differences in Repetitive Behavior

Child	Before Treatment	After 1 Week of Treatment	Difference (Before − After)
1	85	75	10
2	70	50	20
3	40	50	−10
4	65	40	25
5	80	20	60
6	75	65	10
7	55	40	15
8	20	25	−5

TABLE 10–13 Ranking the Absolute Values of the Differences

Observed Differences	Ordered Absolute Values of Differences	Ranks
10	−5	1
20	10	3
−10	−10	3
25	10	3
60	15	5
10	20	6
15	25	7
−5	60	8

TABLE 10–14 Attaching Signs to the Ranks

Observed Differences	Ordered Absolute Values of Difference Scores	Ranks	Signed Ranks
10	−5	1	−1
20	10	3	3
−10	−10	3	−3
25	10	3	3
60	15	5	5
10	20	6	6
15	25	7	7
−5	60	8	8

Next we must determine whether the observed test statistic W supports the null or the research hypothesis. This is done following the same approach used in parametric testing. Specifically, we determine a critical value of W such that if the observed value of W is less than or equal to the critical value, we reject H_0 in favor of H_1, and if the observed value of W exceeds the critical value, we do not reject H_0. The critical value of W can be found in Table 7 in the Appendix. To determine the appropriate one-sided critical value from Table 7 in the Appendix, we need the sample size ($n = 8$) and our one-sided level of significance ($\alpha = 0.05$). For this example, the critical value of W is 6, and the decision rule is to reject H_0 if $W \leq 6$. Thus, we reject H_0 because 4 < 6. We have statistically significant evidence at $\alpha = 0.05$ to show that the median difference is positive (i.e., that there is significant improvement in repetitive behavior in children affected with autism after treatment as compared to before treatment).

Notice that when we analyzed the data presented in Example 10.4 using the Sign test, we failed to find statistical significance. However, when we use the Wilcoxon Signed Rank test (Example 10.6), we conclude that there is statistically significant evidence at $\alpha = 0.05$ to show that the median difference is positive (i.e., that there is significant improvement in repetitive behavior in children affected with autism after treatment as compared to before). The discrepant results are due to the fact that the Sign test uses very little information in the data and is a less powerful test (i.e., can fail to reject H_0 when in fact it is false).

Example 10.7. A study is run to evaluate the effectiveness of an exercise program in reducing systolic blood pressure in patients with prehypertension (defined as a systolic blood pressure between 120 mmHg and 139 mmHg or a diastolic blood pressure between 80 mmHg and 89 mmHg). A total of 15 patients with prehypertension enroll in the study, and their systolic blood pressures are measured. Each patient then participates in an exercise training program where they learn proper techniques and execution of a series of exercises. Patients are instructed to do the exercise program 3 times per week for 6 weeks. After 6 weeks, systolic blood pressures are again measured. The data are shown in **Table 10–15**.

TABLE 10–15 Systolic Blood Pressure Before and After Exercise Program

Patient	Systolic Blood Pressure Before Exercise Program	Systolic Blood Pressure After Exercise Program
1	125	118
2	132	134
3	138	130
4	120	124
5	125	105
6	127	130
7	136	130
8	139	132
9	131	123
10	132	128
11	135	126
12	136	140
13	128	135
14	127	126
15	130	132

The question of interest is whether there is a difference in systolic blood pressure after participating in the exercise program compared to before participating in the exercise program.

Step 1. **Set up hypotheses and determine the level of significance.**

H_0: The median difference is zero.
H_1: The median difference is not zero.
$\alpha = 0.05$.

Step 2. **Select the appropriate test statistic.**
The test statistic for the Wilcoxon Signed Rank test is W, defined as the smaller of $W+$ and $W-$, which are the sums of the positive and negative ranks, respectively.

Step 3. **Set up the decision rule.**
The critical value of W can be found in Table 7 in the Appendix. To determine the appropriate critical value from Table 7 in the Appendix, we need the sample size ($n = 15$) and our two-sided level of significance ($\alpha = 0.05$). The critical value for this two-sided test is 25, and the decision rule is as follows:

Reject H_0 if $W \leq 25$.

Step 4. **Compute the test statistic.**
Because the before- and after-systolic blood pressure measures are paired, we compute difference scores for each patient (see **Table 10–16**).

The next step is to rank the ordered absolute values of the difference scores using the approach outlined in Section 10.1. Specifically, we assign ranks from 1 through n to the smallest through largest absolute values of the difference scores, respectively, and assign the mean rank when there are ties in the absolute values of the difference scores (see **Table 10–17**).

The final step is to attach the signs ("+" or "−") of the observed differences to each rank as shown in **Table 10–18**.

In this example, $W+ = 89$ and $W- = 31$. Recall that the sum of the ranks (ignoring the signs) will always equal $n(n + 1)/2$. As a check on our assignment

TABLE 10–16 Differences in Systolic Blood Pressure Before and After Exercise Program

Patient	Systolic Blood Pressure Before Exercise Program	Systolic Blood Pressure After Exercise Program	Difference (Before − After)
1	125	118	7
2	132	134	−2
3	138	130	8
4	120	124	−4
5	125	105	20
6	127	130	−3
7	136	130	6
8	139	132	7
9	131	123	8
10	132	128	4
11	135	126	9
12	136	140	−4
13	128	135	−7
14	127	126	1
15	130	132	−2

TABLE 10–17 Ranking the Absolute Values of the Differences

Observed Differences	Ordered Absolute Values of Differences	Ranks
7	1	1
−2	−2	2.5
8	−2	2.5
−4	−3	4
20	−4	6
−3	−4	6
6	4	6
7	6	8
8	−7	10
4	7	10
9	7	10
−4	8	12.5
−7	8	12.5
1	9	14
−2	20	15

TABLE 10–18 Attaching Signs to the Ranks

Observed Differences	Ordered Absolute Values of Differences	Ranks	Signed Ranks
7	1	1	1
−2	−2	2.5	−2.5
8	−2	2.5	−2.5
−4	−3	4	−4
20	−4	6	−6
−3	−4	6	−6
6	4	6	6
7	6	8	8
8	−7	10	−10
4	7	10	10
9	7	10	10
−4	8	12.5	12.5
−7	8	12.5	12.5
1	9	14	14
−2	20	15	15

of ranks, we have $n(n + 1)/2 = 15(16)/2 = 120$, which is equal to $89 + 31$. The test statistic is $W = 31$.

Step 5. Conclusion.

We do not reject H_0 because $31 > 25$. We do not have statistically significant evidence at $\alpha = 0.05$ to show that the median difference in systolic blood pressures is not zero (i.e., that there is a significant difference in systolic blood pressures after the exercise program as compared to before the exercise program).

10.4 TESTS WITH MORE THAN TWO INDEPENDENT SAMPLES

In Chapter 7, Section 7.8, we presented techniques for testing the equality of means in more than two independent samples using analysis of variance (ANOVA). An underlying assumption for appropriate use of ANOVA was that the continuous outcome was approximately normally distributed or that the samples were sufficiently large (usually $n_j > 30$, where $j = 1, 2, ..., k$ and k denotes the number of independent comparison groups). An additional assumption for appropriate use of ANOVA is equality of variances in the k comparison groups. ANOVA is generally robust when the sample sizes are small but equal. When the outcome is not normally distributed and the samples are small, a nonparametric test is appropriate.

A popular nonparametric test to compare outcomes among more than two independent groups is the Kruskal–Wallis test. The Kruskal–Wallis test is used to compare medians among k comparison groups ($k > 2$) and is sometimes described as an ANOVA with the data replaced by their ranks. The null and research hypotheses for the Kruskal–Wallis nonparametric test are stated as follows:

H_0: The k population medians are equal.

H_1: The k population medians are not all equal.

The procedure for the test involves pooling the observations from the k samples into one combined sample, keeping track of which sample each observation comes from, and then ranking lowest to highest from 1 to N, where $N = n_1 + n_2 + \cdots + n_k$. To illustrate the procedure, we consider the following example, addressing each of the five steps in the test of hypothesis including more details to promote understanding. (In the subsequent example, the five-step approach is followed more directly.)

Example 10.8. A clinical study is designed to assess differences in albumin levels in adults following different low-protein diets. Albumin is measured in grams per deciliter

TABLE 10–19 Albumin Levels in Three Different Diets

5% Protein	10% Protein	15% Protein
3.1	3.8	4.0
2.6	4.1	5.5
2.9	2.9	5.0
	3.4	4.8
	4.2	

(g/dL) of blood, and low-protein diets are often prescribed for patients with kidney disease. Clinically, albumin is also used to assess whether patients get sufficient protein in their diets. Three diets are compared, ranging from 5% to 15% protein, and the 15% protein diet represents a typical American diet. The albumin levels of participants following each diet are shown in **Table 10–19**.

The question of interest is whether there is a difference in albumin levels among the three different diets. For reference, normal albumin levels are generally between 3.4 g/dL and 5.4 g/dL. By inspection, it appears that participants following the 15% protein diet have higher albumin levels than those following the 5% protein diet. The issue is whether this observed difference is statistically significant.

In this example, the outcome is continuous, but the sample sizes are small and not equal across comparison groups ($n_1 = 3$, $n_2 = 5$, $n_3 = 4$). Thus, a nonparametric test is appropriate. The hypotheses to be tested are given below and we run the test at the 5% level of significance (i.e., $\alpha = 0.05$).

H_0: The three population medians are equal.

H_1: The three population medians are not all equal.

To conduct the test we assign ranks using the procedures outlined in Section 10.1. The first step in assigning ranks is to order the data from smallest to largest. This is done on the combined or total sample, pooling the data from the three comparison groups, and assigning ranks from 1 to 12, as shown in **Table 10–20**. We also need to keep track of the group assignments in the total sample ($n = 12$).

Notice that the lower ranks (e.g., 1, 2.5, and 4) are assigned to the 5% protein diet group, while the higher ranks (e.g., 10, 11, and 12) are assigned to the 15% protein diet group. Again, the goal of the test is to determine whether the observed data support a difference in the three population medians. Recall that in the parametric tests, discussed in Chapter 7, when comparing means among more than two groups, we analyze the differences among the sample means (mean square between treatments) relative to their within-group variability and

TABLE 10–20 Assigning Ranks

			Total Sample (Ordered Smallest to Largest)			Ranks		
5% Protein	10% Protein	15% Protein	5% Protein	10% Protein	15% Protein	5% Protein	10% Protein	15% Protein
3.1	3.8	4.0	2.6			1		
2.6	4.1	5.5	2.9	2.9		2.5	2.5	
2.9	2.9	5.0	3.1			4		
	3.4	4.8		3.4			5	
	4.2			3.8			6	
					4.0			7
				4.1			8	
				4.2			9	
					4.8			10
					5.0			11
					5.5			12

summarize the sample information in a test statistic (F statistic). In the Kruskal–Wallis test we again summarize the sample information in a test statistic, one based on the ranks.

The test statistic for the Kruskal–Wallis test is denoted H and is defined as follows:

$$H = \left(\frac{12}{N(N+1)} \sum_{j=1}^{k} \frac{R_j^2}{n_j} \right) - 3(N+1),$$

where k = the number of comparison groups, N = the total sample size, n_j is the sample size in the jth group, and R_j is the sum of the ranks in the jth group.

In this example $R_1 = 7.5$, $R_2 = 30.5$, and $R_3 = 40$. Recall that the sum of the ranks will always equal $n(n+1)/2$. As a check on our assignment of ranks, we have $n(n+1)/2 = 12(13)/2 = 78$, which is equal to $7.5 + 30.5 + 40 = 78$.

In this example,

$$H = \left(\frac{12}{N(N+1)} \sum_{j=1}^{k} \frac{R_j^2}{n_j} \right) - 3(N+1) =$$

$$\frac{12}{12(13)} \left(\frac{7.5^2}{3} + \frac{30.5^2}{5} + \frac{40^2}{4} \right) - 3(13) = 7.52.$$

In every test, we must determine whether the observed test statistic H supports the null or the research hypothesis. This is done following the same approach used in parametric testing. Specifically, we determine a critical value of H such that if the observed value of H is greater than or equal to the critical value, we reject H_0 in favor of H_1, and if the observed value of H is less than the critical value, we do not reject H_0. The critical value of H can be found in Table 8 in the Appendix. To determine the appropriate critical value, we need the sample sizes ($n_1 = 3$, $n_2 = 5$, and $n_3 = 4$) and our level of significance ($\alpha = 0.05$). For this example the critical value is 5.656; thus we reject H_0 because $7.52 > 5.656$. We have statistically significant evidence at $\alpha = 0.05$ to show that there is a difference in median albumin levels among the three different diets.

Notice that Table 8 in the Appendix contains critical values for the Kruskal–Wallis test for tests comparing three, four, or five groups with small sample sizes. If there are three or more comparison groups and five or more observations in each of the comparison groups, it can be shown that the test statistic H approximates a χ^2 distribution with $df = k - 1$.[4] Thus, in a Kruskal–Wallis test with three or more comparison groups and five or more observations in each group, the critical value for the test can be found in Table 3 in the Appendix:

Critical Values of the χ^2 Distribution. The following example illustrates this situation.

Example 10.9. A personal trainer is interested in comparing the anaerobic thresholds of elite athletes. The anaerobic threshold is defined as the point at which the muscles cannot get more oxygen to sustain activity, or the upper limit of aerobic exercise. It is a measure that is also related to maximum heart rate. The data in **Table 10-21** are anaerobic thresholds for distance runners, distance cyclists, distance swimmers, and cross-country skiers. The question of interest is whether there is a difference in anaerobic thresholds among the different groups of elite athletes.

Step 1. Set up hypotheses and determine the level of significance.

H_0: The four population medians are equal.
H_1: The four population medians are not all equal.
$\alpha = 0.05$.

Step 2. Select the appropriate test statistic.

The test statistic for the Kruskal–Wallis test is denoted H and is defined as follows:

$$H = \left(\frac{12}{N(N+1)} \sum_{j=1}^{k} \frac{R_j^2}{n_j} \right) - 3(N+1),$$

where k = the number of comparison groups, N = the total sample size, n_j is the sample size in the jth group, and R_j is the sum of the ranks in the jth group.

Step 3. Set up the decision rule.

Because there are four comparison groups and five observations in each of the comparison groups, we find the critical value in Table 3 in the Appendix for

TABLE 10-21 Anaerobic Thresholds

Distance Runners	Distance Cyclists	Distance Swimmers	Cross-Country Skiers
185	190	166	201
179	209	159	195
192	182	170	180
165	178	183	187
174	181	160	215

$df = k - 1 = 4 - 1 = 3$ and $\alpha = 0.05$. The critical value is 7.81, and the decision rule is to reject H_0 if $H \geq 7.81$.

Step 4. Compute the test statistic.

To conduct the test we assign ranks using the procedures outlined in Section 10.1. The first step in assigning ranks is to order the data from smallest to largest. This is done on the combined or total sample, pooling the data from the four comparison groups, and assigning ranks from 1 to 20. We also need to keep track of the group assignments in the total sample ($n = 20$). **Table 10–22** shows the ordered data.

We next assign the ranks to the ordered values and sum the ranks in each group (**Table 10–23**).

Recall that the sum of the ranks will always equal $n(n + 1)/2$. As a check on our assignment of ranks, we have $n(n + 1)/2 = 20(21)/2 = 210$, which is equal to $46 + 62 + 24 + 78 = 210$.

In this example,

$$H = \left(\frac{12}{N(N+1)} \sum_{j=1}^{k} \frac{R_j^2}{n_j} \right) - 3(N+1) =$$

$$\frac{12}{20(21)} \left(\frac{46^2}{5} + \frac{62^2}{5} + \frac{24^2}{5} + \frac{78^2}{5} \right) - 3(21) = 9.11.$$

Step 5. Conclusion.

Reject H_0 because $9.11 > 7.81$. We have statistically significant evidence at $\alpha = 0.05$ to show that there is a difference in median anaerobic thresholds among the four different groups of elite athletes.

Notice that in this example the anaerobic thresholds of the distance runners, cyclists, and cross-country skiers are comparable (looking only at the raw data). The distance

TABLE 10–22 Ordering the Anaerobic Thresholds

Distance Runners	Distance Cyclists	Distance Swimmers	Cross-Country Skiers	Distance Runners	Distance Cyclists	Distance Swimmers	Cross-Country Skiers
185	190	166	201			159	
179	209	159	195			160	
192	182	170	180	165			
165	178	183	187			166	
174	181	160	215			170	
				174			
					178		
				179			
							180
					181		
					182		
						183	
				185			
							187
					190		
				192			
							195
							201
					209		
							215

Total Sample (Ordered Smallest to Largest)

TABLE 10–23 Assigning Ranks

Total Sample (Ordered Smallest to Largest)				Ranks			
Distance Runners	Distance Cyclists	Distance Swimmers	Cross-Country Skiers	Distance Runners	Distance Cyclists	Distance Swimmers	Cross-Country Skiers
		159				1	
		160				2	
165				3			
		166				4	
		170				5	
174				6			
	178				7		
179				8			
			180				9
	181				10		
	182				11		
		183				12	
185				13			
			187				14
	190				15		
192				16			
			195				17
			201				18
	209				19		
			215				20
				$R_1 = 46$	$R_2 = 62$	$R_3 = 24$	$R_4 = 78$

swimmers appear to be the athletes that differ from the others in terms of anaerobic thresholds. Recall, similar to analysis of variance tests, that we reject the null hypothesis in favor of the alternative hypothesis if any two of the medians are not equal.

10.5 SUMMARY

In this chapter we presented hypothesis testing techniques for situations with small sample sizes and outcomes that are ordinal, ranked, or continuous and cannot be assumed to be normally distributed. Nonparametric tests are based on ranks that are assigned to the ordered data. The tests involve the same five steps as parametric tests, specifying the null hypothesis and the alternative or research hypothesis, selecting and computing an appropriate test statistic, setting up a decision rule, and drawing a conclusion.

Each of the tests discussed here is summarized below.

Mann–Whitney U Test. Used to compare a continuous outcome in two independent samples.

Null Hypothesis H_0: Two populations are equal.

Test Statistic   The test statistic is U, the smaller of

$$U_1 = n_1 n_2 + \frac{n_1(n_1 + 1)}{2} - R_1 \text{ and}$$

$$U_2 = n_1 n_2 + \frac{n_2(n_2 + 1)}{2} - R_2, \text{ where}$$

R_1 and R_2 are the sums of the ranks in groups 1 and 2, respectively.

Decision Rule   Reject H_0 if $U \leq$ critical value from Table 5 in the Appendix.

Sign Test. Used to compare a continuous outcome in two matched or paired samples.

Null Hypothesis H_0: Median difference is zero.

Test Statistic   The test statistic is the smaller of the number of positive or negative signs.

Decision Rule   Reject H_0 if the smaller of the number of positive or negative signs $\leq$ critical value from Table 6 in the Appendix.

Wilcoxon Signed Rank Test. Used to compare a continuous outcome in two matched or paired samples.

Null Hypothesis H_0: Median difference is zero.

Test Statistic   The test statistic is W, defined as the smaller of $W+$ and $W-$, the sums of the positive and negative ranks of the difference scores, respectively.

Decision Rule   Reject H_0 if $W \leq$ critical value from Table 7 in the Appendix.

Kruskal–Wallis Test. Used to compare a continuous outcome in more than two independent samples.

Null Hypothesis H_0: k population medians are equal.

Test Statistic   The test statistic is H:

$$H = \left(\frac{12}{N(N+1)} \sum_{j=1}^{k} \frac{R_j^2}{n_j} \right) - 3(N+1)$$

where k = the number of comparison groups, N = the total sample size, n_j is the sample size in the jth group, and R_j is the sum of the ranks in the jth group.

Decision Rule   Reject H_0 if $H \geq$ critical value from Table 8 in the Appendix.

It is important to note that nonparametric tests are subject to the same errors as parametric tests. A *Type I error* occurs when a test incorrectly rejects the null hypothesis. A **Type II error** occurs when a test fails to reject H_0 when it is false. **Power** is the probability that a test correctly rejects H_0. Nonparametric tests can be subject to low power mainly due to small sample size. Therefore, it is important to consider the possibility of a Type II error when a nonparametric test fails to reject H_0. There may be a true effect or difference, yet the nonparametric test is underpowered to detect it. For more details, interested readers should see Conover[3] and Siegel and Castellan.[4]

10.6   PRACTICE PROBLEMS

1. A company is evaluating the impact of a wellness program offered on-site as a means of reducing employee sick days. A total of 8 employees agree to participate in the evaluation, which lasts 12 weeks. Their sick days in the 12 months prior to the start of the wellness program and again over the 12 months after the completion of the program are recorded and are shown in **Table 10–24**. Is there a significant reduction in the number of sick days taken after completing the wellness program? Use the Sign test at a 5% level of significance.

2. Using the data in Problem 1, assess whether there is a significant reduction in the number of sick days taken after completing the wellness program using the Wilcoxon Signed Rank test at a 5% level of significance.

TABLE 10–24   Numbers of Sick Days Before and After Wellness Program

Employee	Sick Days Taken in 12 Months Prior to Program	Sick Days Taken in 12 Months Following Program
1	8	7
2	6	6
3	4	5
4	12	11
5	10	7
6	8	4
7	6	3
8	2	1

3. A small study (n = 10) is designed to assess whether there is an association between smoking in pregnancy and low birth weight. Low-birth-weight babies are those born weighing <5.5 pounds. The following data represent the birth weights, in pounds, of babies born to mothers who reported smoking in pregnancy and to those who did not.

| Mother smoked during pregnancy | 5.0 | 4.2 | 4.8 | 3.3 | 3.9 |
| Mother did not smoke during pregnancy | 5.1 | 4.9 | 5.3 | 5.4 | 4.6 |

Is there a significant difference in birth weights between mothers who smoked during pregnancy and those who did not? Run the appropriate test at a 5% level of significance.

4. The following data represent the number of playground injuries occurring among children aged 5–9 years over a 3-month period in 12 playgrounds in and around the neighborhoods of Boston. Playground injuries include fractures, internal injuries, lacerations, and dislocations. The question of interest is whether there are differences in the numbers of injuries at playgrounds in various locations. The data below represent the numbers of injuries recorded at four randomly selected playgrounds located on school properties, at day-care centers, and in residential neighborhoods.

School properties	39	51	42	29
Day-care centers	28	25	30	15
Residential neighborhoods	18	16	25	22

Run the appropriate test at a 5% level of significance.

5. The recommended daily allowance of Vitamin A for children between 1–3 years of age is 400 micrograms (mcg). Vitamin A deficiency is linked to a number of adverse health outcomes, including poor eyesight, susceptibility to infection, and dry skin. The following are Vitamin A concentrations in children with and without poor eyesight, a history of infection, and dry skin.

| With poor eyesight, a history of infection, and dry skin | 270 | 420 | 180 | 345 | 390 | 430 |
| Free of poor eyesight, a history of infection, and dry skin | 450 | 500 | 395 | 380 | 430 |

Is there a significant difference in Vitamin A concentrations between children with and without poor eyesight, a history of infection, and dry skin? Run the appropriate test at a 5% level of significance.

6. A study is conducted to assess the potential benefits of an ayurvedic treatment to reduce high cholesterol. Seven patients agree to participate in the study. Each has his or her cholesterol measured at the start of the study and then again after 4 weeks taking a popular herb called arjuna (see **Table 10–25**). Is there a significant difference in total cholesterol after taking the herb? Use the Sign test at a 5% level of significance.

7. Using the data in Problem 6, assess whether there is a difference in total cholesterol after taking the herb using the Wilcoxon Signed Rank test at a 5% level of significance.

8. An investigator wants to test whether there is a difference in endotoxin levels in children who are exposed to endotoxin as a function of their proximity to operating farms. The following are endotoxin levels in units per milligram of dust sampled from children's mattresses, organized by children's proximity to farms.

Within 5 miles	54	62	78	90	70
5–24.9 miles	28	42	39	81	65
25–49.9 miles	37	29	30	50	53
50 miles or more	36	19	22	28	27

Run the appropriate test at a 5% level of significance.

TABLE 10–25 Total Cholesterol Levels Before and After Treatment

Participant	Total Cholesterol Before Treatment	Total Cholesterol After Treatment
1	250	241
2	265	260
3	240	253
4	233	230
5	255	224
6	275	227
7	241	232

9. Using the data in Problem 8, test whether there is a difference in endotoxin levels between children living within 5 miles of an operating farm as compared to those living between 5 miles and 24.9 miles. Run the appropriate test at a 5% level of significance.

10. Using the data in Problem 8, test whether there is a difference in endotoxin levels between children living within 5 miles of an operating farm as compared to those living at a distance of 50 miles or more. Run the appropriate test at a 5% level of significance.

REFERENCES

1. D'Agostino, R.B. and Stevens, M.A. *Goodness of Fit Techniques.* New York: Marcel Dekker, 1986.
2. Apgar, V. "A proposal for a new method of evaluation of the newborn infant." *Current Researches in Anesthesia and Analgesia.* 1953; 32 (4): 260–267.
3. Conover, W.J. *Practical Nonparametric Statistics* (2nd ed.). New York: John Wiley & Sons, 1999.
4. Siegel, S. and Castellan, N.J., Jr. *Nonparametric Statistics for the Behavioral Sciences* (2nd ed.) New York: McGraw–Hill, 1998.

CHAPTER 11

Survival Analysis

When and Why

Key Questions

- What are the specific features of a study that require the use of survival analysis techniques?
- What does it mean to have better survival with a particular treatment choice?
- How do you interpret a 5-year, 10-year, or lifetime risk of developing disease?

In the News

A number of online tools are readily available to the general public to estimate their risks of developing various diseases.

For example, the National Cancer Institute has a breast cancer risk assessment tool that women can use to estimate their risk of developing breast cancer.[1] This tool asks a series of questions that collect data on demographics and medical history. Based on a woman's personal profile, it then produces a 5-year risk and a lifetime risk of developing breast cancer, along with average 5-year and lifetime risks for women of the same age and race/ethnicity.

The Framingham Heart Study offers a number of risk prediction tools based on the data collected in this study that allow users to compute their risks of developing diabetes, hypertension, and various cardiovascular outcomes.[2] Users input demographic and medical history data, and these tools then produce estimates of risk. To assist in interpretation, average risks are also provided for people of the same sex and age (the participants in the Framingham Heart Study are predominantly white, so there is no differentiation by race/ethnicity).

Dig In

- The risk prediction tools mentioned here are readily available to the general public. Do you have any concerns that they might be misused, be misleading, or otherwise cause concern for users?
- Suppose you enter your personal profile into one of the Framingham Heart Study risk prediction tools, and it estimates that

[1] http://www.cancer.gov/bcrisktool/
[2] https://www.framinghamheartstudy.org/risk-functions/

you have a 6% risk of developing cardiovascular disease over the next 10 years. Would knowing about that risk encourage you to change any of your behaviors?
- Is there any other information that you think these risk prediction tools should generate to help users understand their applications and limitations?

Learning Objectives

By the end of this chapter, the reader will be able to

- Identify applications with time-to-event outcomes
- Construct a life table using the actuarial approach
- Construct a life table using the Kaplan–Meier approach
- Perform and interpret the log-rank test
- Compute and interpret a hazard ratio
- Interpret coefficients in Cox proportional hazards regression analysis

In Chapter 4 we presented techniques for summarizing continuous, dichotomous, categorical, and ordinal outcomes. In Chapters 6 and 7 we presented techniques to generate confidence interval estimates and to conduct tests of hypothesis for continuous, dichotomous, categorical, and ordinal outcomes. We now consider a different type of outcome variable, called a **time-to-event variable**. A time-to-event variable reflects the time until a participant has an event of interest (e.g., heart attack, cancer remission, or death). Statistical analysis of time-to-event variables requires different techniques than those described thus far for other types of outcomes because of the unique features of time-to-event variables. Statistical analysis of these variables is called time-to-event analysis or **survival** analysis, even though the outcome is not always death. The questions of interest in survival analysis are questions like these: What is the probability that a participant survives 5 years? Are there differences in survival between groups (e.g., between those assigned to a new drug versus a standard drug in a clinical trial)? How do certain personal, behavioral, or clinical characteristics affect participants' chances of survival?

There are several unique features of time-to-event variables. First, times to event are always positive, and their distributions are often skewed. For example, in a study assessing time to relapse in high-risk patients, the majority of events (relapses) may occur early in the follow-up, with very few occurring later. On the other hand, in a study of time to death in a community-based sample, the majority of events (deaths) may occur later in the follow-up. Standard statistical procedures that assume normality of distributions do not apply. Nonparametric procedures (as discussed in Chapter 10) could be invoked except for the fact that there are additional issues. Specifically, complete data (actual time-to-event data) are not always available on each participant in a study. In many studies, participants are enrolled over a period of time (months or years), and the study ends on a specific calendar date. Thus, participants who enroll later are followed for a shorter period than participants who enroll early. Some participants may drop out of the study before the end of the follow-up period (e.g., move away, become disinterested), and others may die during the follow-up period (assuming the outcome of interest is not death). In each of these instances, we have incomplete follow-up information. True survival time (sometimes called failure time) is not known because the study ends or because a participant drops out of the study before experiencing the event. What we know is that the participant's survival time is greater than his or her last observed follow-up time. These times are called censored times.

There are several different types of censoring. The most common is called **right censoring**, and this occurs when a participant does not have the event of interest during the study, and thus his or her last observed follow-up time is less than the time to event. This can occur when a participant drops out before the study ends or when a participant is event-free at the end of the observation period. In the first instance, the participant's observed time is less than the length of the follow-up, and in the second, the participant's observed time is equal to the length of the follow-up period. These issues are illustrated in the following examples.

Example 11.1. A small prospective study is run and follows 10 participants for the development of myocardial infarction (MI, or heart attack) over a period of 10 years. Participants are recruited into the study over a period of 2 years and are followed for up to 10 years. Participants' experiences are depicted in **Figure 11-1**. Notice that some participants join the study late (some up to 2 years after enrollment begins) and are followed for up to 10 years after enrollment.

During the study period, three participants suffer myocardial infarction (MI), one dies, two drop out of the study (for unknown reasons), and four complete the 10-year follow-up without suffering MI. **Figure 11-2** displays the same data but shows survival time starting at a common time zero (i.e., as if all participants enrolled in the study at the same time).

Based on the data shown in Figure 11-2, what is the likelihood that a participant suffers an MI over 10 years? Three of ten participants suffer MI over the course of follow-up, but 30% is probably an underestimate of the true percentage because two participants dropped out and might have suffered an MI had they been observed for the full 10 years. Their observed times are censored. In addition, one participant dies after 3 years of follow-up. Should these three individuals be included in the analysis, and if so, how? If we exclude all three, the estimate of the likelihood that a participant suffers an MI is 3/7 = 43%, substantially higher than the initial estimate of 30%. The fact that often not all participants are observed over the entire follow-up period makes survival data unique. In this small example, Participant 4 is observed for 4 years and over that period does not have an MI. Participant 7 is observed for 2 years and over that period does not have an MI. While they do not suffer the event of interest, they contribute important information. Survival analysis techniques make use of this information in the estimate of the probability of event. An important assumption is made to make appropriate use of the censored data. Specifically, we assume that censoring is independent or unrelated to the likelihood of developing the event of interest. This is called noninformative censoring and essentially assumes that the participants whose data are censored would have the same distribution of failure times (or times to event) if they were actually observed.

Consider the same study and the experiences of 10 different participants as depicted in **Figure 11-3**. Notice that in Figure 11-3, again, three participants suffer MI, one dies, two drop out of the study, and four complete the 10-year follow-up without suffering MI. However, the events (MIs) occur much earlier, and the drop-outs and death occur later in the course of follow-up. Should the differences in participants' experiences shown in Figure 11-3 as compared to those shown in Figure 11-2 affect the estimate of the likelihood that a participant suffers an MI over 10 years? In survival analysis we analyze not only the numbers of participants who suffer the event of interest (a dichotomous indicator of event status) but also the times at which the events occur.

FIGURE 11–1 Participants' Experiences in 10-Year Study of Myocardial Infarction

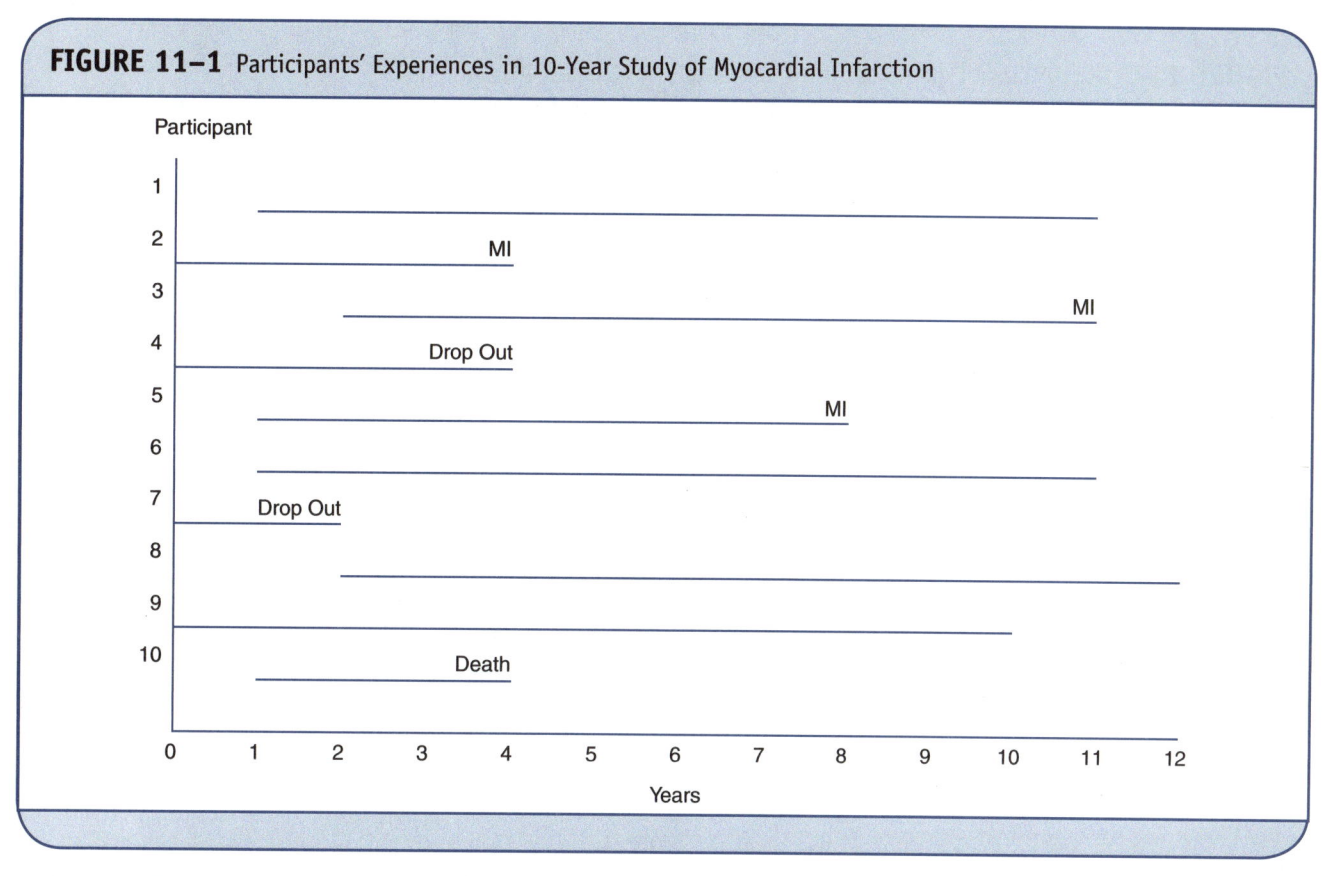

FIGURE 11–2 Participants' Experiences in 10-Year Study of Myocardial Infarction

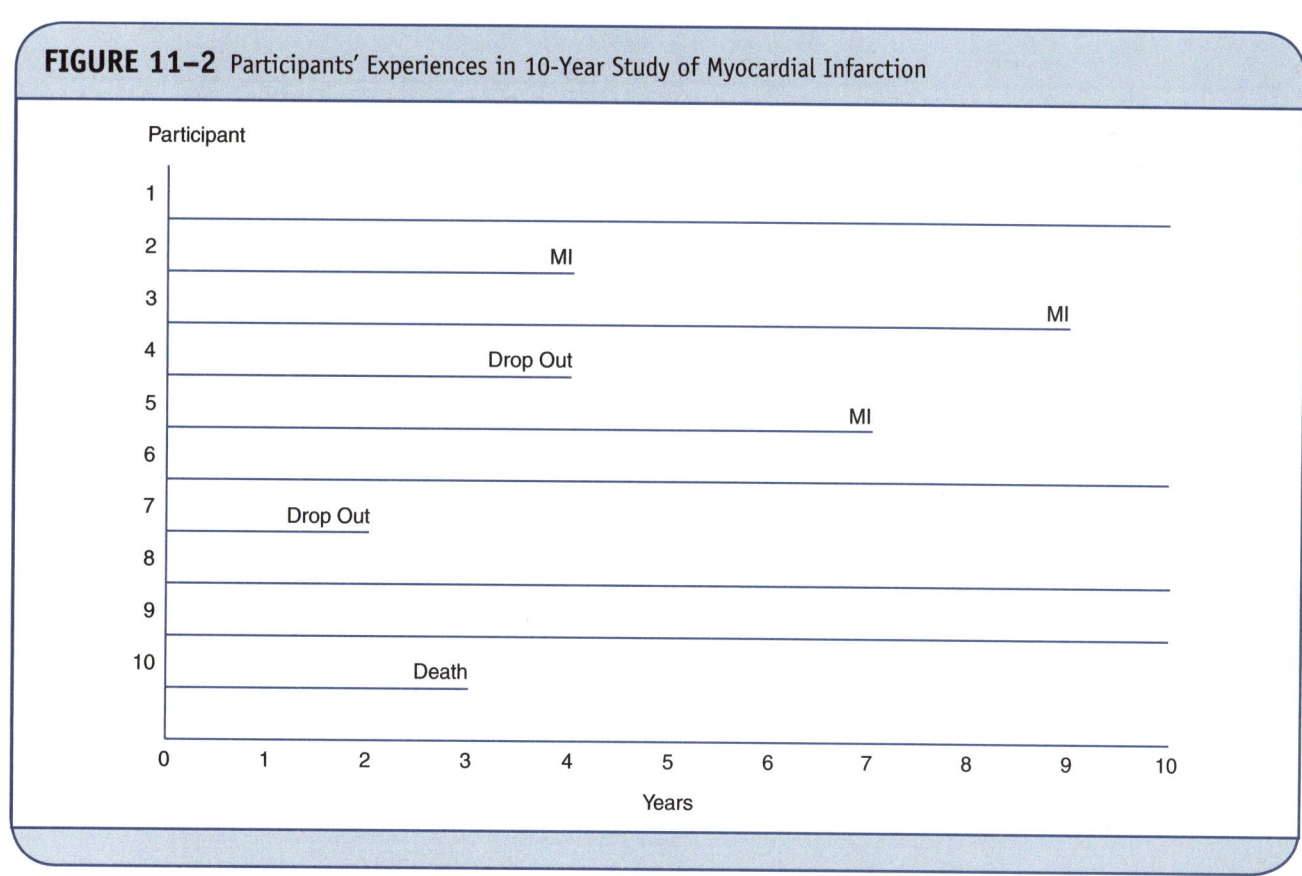

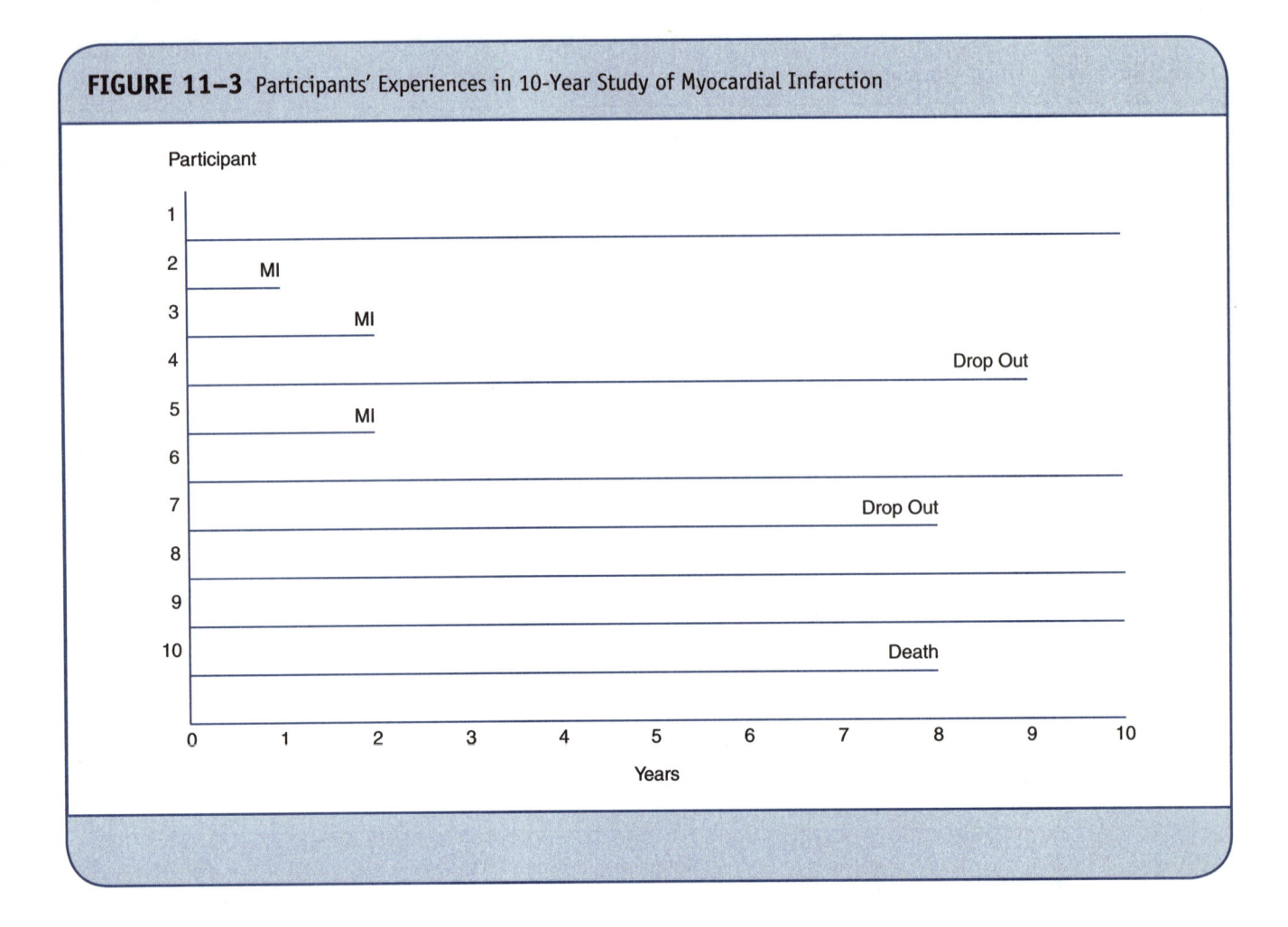

FIGURE 11–3 Participants' Experiences in 10-Year Study of Myocardial Infarction

11.1 INTRODUCTION TO SURVIVAL DATA

In this chapter, we introduce the basic concepts of survival analysis. Interested readers should see Hosmer and Lemeshow[1] and Cox and Oakes[2] for details of more sophisticated applications of survival analysis.

In survival analysis, we measure two important pieces of information. First, we measure whether or not a participant suffers the event of interest during the study period. This is reflected in a dichotomous or indicator variable often coded as 1 = *event occurred* or 0 = *event did not occur during the study observation period*. In addition, for each participant we measure follow-up time. Time zero, or the time origin, is the time at which participants are considered to be at risk for the outcome of interest. In many studies, time at risk is measured from the start of the study (i.e., at enrollment). In a prospective cohort study evaluating time to incident stroke, investigators may recruit participants who are 55 years of age and older because the risk for stroke prior to that age is very low. In a prospective cohort study evaluating time to incident cardiovascular disease, investigators may recruit participants who are 35 years of age and older. In each of these studies, a minimum age might be specified as a criterion for inclusion in the study. Follow-up time is measured from time zero (the start of the study or the point at which the participant is considered to be at risk) until the event occurs, until the study ends, or until the participant is lost, whichever comes first. In a clinical trial, the time origin is usually considered the time of randomization.

Patients often enter or are recruited into cohort studies and clinical trials over a period of several calendar months or years. Thus, it is important to record the entry time so that the follow-up time is accurately measured. Again, our interest lies in the time to event, but for various reasons (e.g., the participant drops out of the study or the study observation period ends) we cannot always measure time to event. For participants who do not suffer the event of interest, we measure follow-up time, which is less than time to event, and these follow-up times are censored.

In survival analysis, we use information on event status and follow-up time to estimate a survival function. The **survival function** is a function of time and represents the probability that a person survives past a certain time point. Consider a 20-year prospective study of survival in patients following a myocardial infarction. In this study, the outcome is all-cause mortality, and the survival function (or survival curve) might look like the one shown in **Figure 11–4**.

The horizontal axis represents time in years, and the vertical axis shows the probability of surviving or the proportion of people surviving. At time zero, the survival probability is 1 (or 100% of the participants are alive). At 2 years, the probability of survival is approximately 0.83 or 83% (see solid lines in Figure 11–4). At 10 years, the probability of survival is approximately 0.47 or 47%. Often it is of interest to estimate the median survival, that is, the time at which 50% of the participants in the study are alive. In Figure 11–4 the median survival is approximately 8.5 years (see dashed lines in Figure 11–4).

A flat survival curve (i.e., one that stays close to 1) suggests very good survival, whereas a survival curve that drops sharply toward 0 suggests poor survival. Figure 11–4 shows the survival function as a smooth curve. In most applications, the survival function is shown as a step function rather than a smooth curve. Popular procedures to estimate survival functions are presented in Section 11.2.

11.2 ESTIMATING THE SURVIVAL FUNCTION

There are several different ways to estimate a survival function or a survival curve. There are a number of popular parametric methods that are used to model survival data, and they differ in terms of the assumptions that are made about the distribution of survival times in the population. Some popular distributions include the exponential, Weibull, Gompertz, and log-normal distributions.[2] Perhaps the most popular is the exponential; it assumes that a participant's likelihood of suffering the event of interest is independent of how long that person has been event-free. Other distributions make different assumptions about the probability that a person develops the event (i.e., it may increase, decrease, or change over time). More details on parametric methods for survival analysis can be found in Hosmer and Lemeshow[1] and Lee and Wang.[3]

We focus here on two nonparametric methods, which make no assumptions about how the probability that a person develops the event changes over time. Using nonparametric

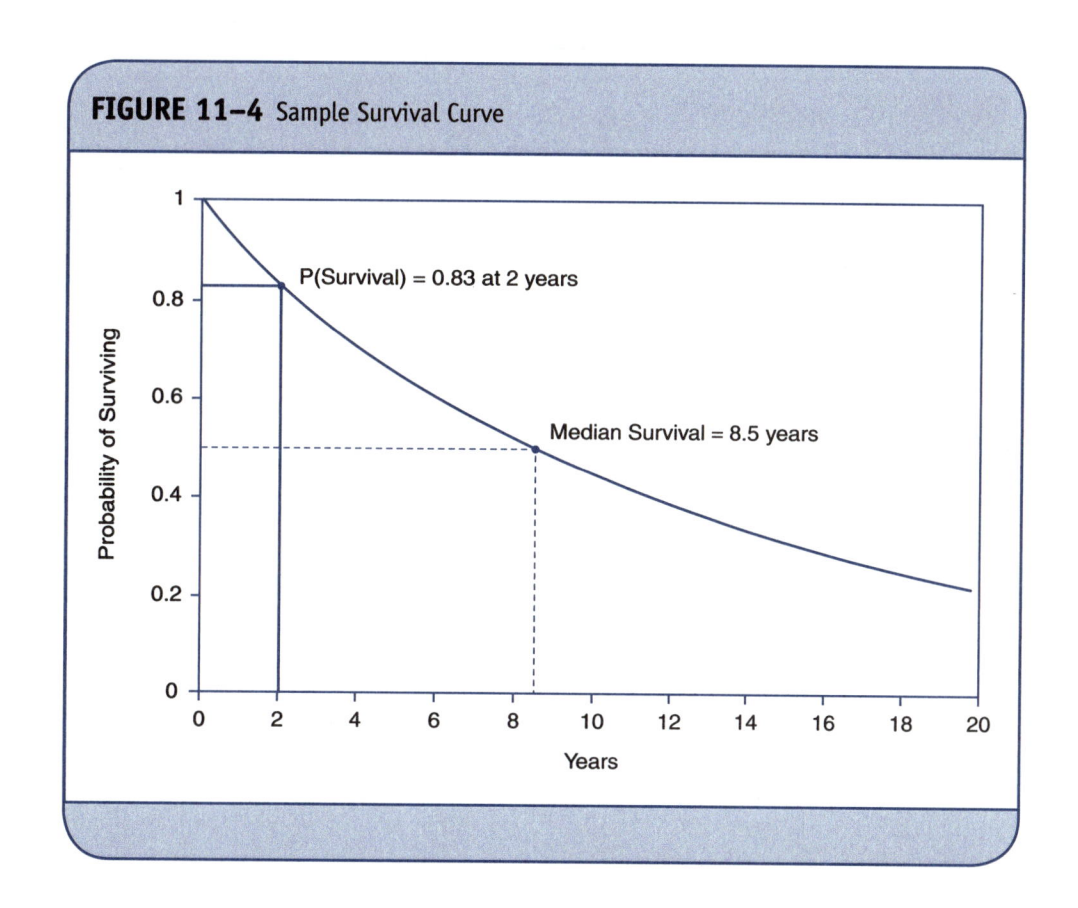

FIGURE 11–4 Sample Survival Curve

methods, we estimate and plot the survival distribution or the survival curve. Survival curves are often plotted as step functions, as shown in **Figure 11–5**. Time is shown on the *x*-axis, and survival (as a proportion or a percentage of people at risk) is shown on the *y*-axis. Note that the percentage of participants "surviving" (as shown on the *y*-axis) does not always represent the percentage of participants who are alive (which assumes that the outcome of interest is death); it can also represent the percentage of participants who are free of another outcome event (e.g., percentage free of MI or cardiovascular disease). It can also represent the percentage of participants who do not experience a healthy outcome (e.g., cancer remission). Notice that the survival probability is 100% for 2 years and then drops to 90%. The median survival is 9 years (i.e., 50% of the population survive past 9 years, see dashed lines).

Example 11.2. Consider a small prospective cohort study designed to study time to death. The study involves participants who are 65 years of age and older who are followed for up to 24 years. The study involves 20 participants who are enrolled over a period of 5 years and are followed until they die, until the study ends, or until they drop out of the study (lost to follow-up). Note that if a participant enrolls after the start of the study, his or her maximum follow-up time is less than 24 years. For example, if a participant enrolls 2 years after the study begins, his or her maximum follow-up time is 22 years. In all biostatistical analyses, we need good data. In survival analysis applications it is critically important that the event of interest, the study start dates, and the study end dates are well defined and accurately measured. The data are shown in **Table 11–1**.

In the study, there are 6 deaths and 3 participants with complete follow-up (i.e., 24 years). The remaining 11 have fewer than 24 years of follow-up due to late enrollment or loss to follow-up.

One way of summarizing the experiences of the participants is with a life table, or **actuarial table**. There are several popular types and uses of life tables; some are used to estimate life expectancy, and some are used in the insurance industry to set premiums. We focus on a particular type of life table used widely in biostatistical analysis called a cohort life table or a follow-up life table. The follow-up life table summarizes the experiences of participants over a predefined follow-up period in a cohort study or in a clinical trial until the time of the event of interest or until the end of the study, whichever comes first.

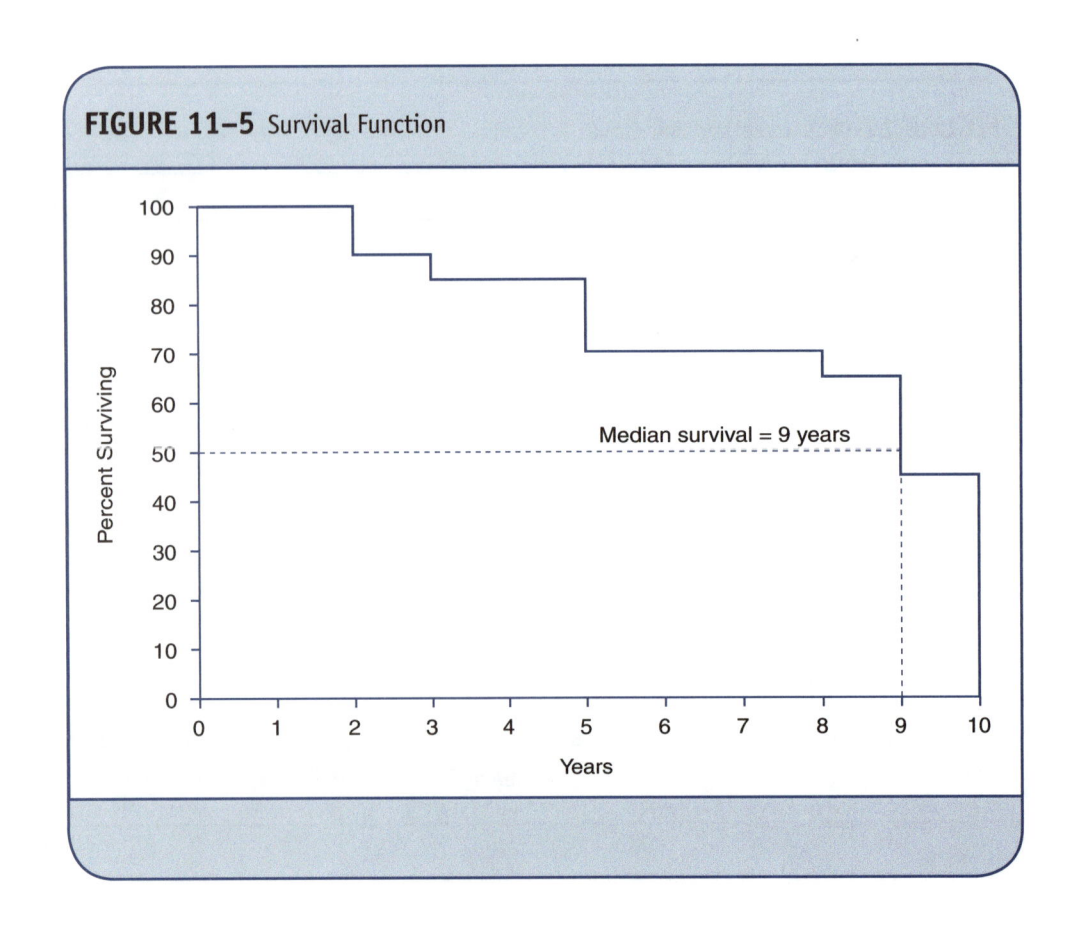

FIGURE 11–5 Survival Function

TABLE 11–1 Year of Death or Year of Last Contact

Participant Identification Number	Year of Death	Year of Last Contact
1		24
2	3	
3		11
4		19
5		24
6		13
7	14	
8		2
9		18
10		17
11		24
12		21
13		12
14	1	
15		10
16	23	
17		6
18	5	
19		9
20	17	

TABLE 11–2 Number Alive, Number of Deaths, and Number Censored in Each Interval

Interval in Years	Number Alive at Beginning of Interval	Number of Deaths During Interval	Number Censored
0–4	20	2	1
5–9	17	1	2
10–14	14	1	4
15–19	9	1	3
20–24	5	1	4

In constructing actuarial life tables, the following assumptions are often made: First, the events of interest (e.g., deaths) are assumed to occur at the end of the interval, and censored events are assumed to occur uniformly (or evenly) throughout the interval. Therefore, an adjustment is often made to N_t to reflect the average number of participants at risk during the interval, N_{t^*}, which is computed as follows: $N_{t^*} = N_t - C_t/2$ (i.e., we subtract half of the censored events).

$q_t =$ proportion dying (or suffering event) during interval t, $q_t = D_t/N_{t^*}$
$p_t =$ proportion surviving (remaining event-free) during interval t, $p_t = 1 - q_t$
$S_t =$ proportion surviving (remaining event-free) past interval t

S_t is the proportion surviving (or remaining event-free) past interval t, sometimes called the cumulative survival probability. It is computed as follows: First, the proportion of participants surviving past time 0 (the starting time) is defined as $S_0 = 1$ (all participants are alive or event-free at time zero or at the start of the study). The proportion surviving past each subsequent interval is computed using principles of conditional probability introduced in Chapter 5.

Specifically, the probability that a participant survives past interval 1 is $S_1 = p_1$. The probability that a participant survives past interval 2 means that they had to survive past interval 1 and through interval 2: $S_2 = $ P(survive past interval 2) = P(survive through interval 2) $\times$ P(survive past interval 1), or $S_2 = p_2 \times S_1$. In general, $S_{t+1} = p_{t+1} \times S_t$. The format of the

To construct a life table, we first organize the follow-up times into equally spaced intervals. In Example 11.2, we have a maximum follow-up of 24 years, and we consider 5-year intervals (0–4 years, 5–9 years, 10–14 years, 15–19 years, and 20–24 years). In each interval, we sum the number of participants who are alive at the beginning of the interval, the number who die, and the number who are censored (see **Table 11–2**).

We use the following notation in our **life-table analysis**. We first define the notation and then use it to construct the life table for the data in Example 11.2.

$N_t =$ number of participants who are event-free and considered at risk during interval t (e.g., in this example, the number alive, as our outcome of interest is death)
$D_t =$ number of participants who die (or suffer the event of interest) during interval t
$C_t =$ number of participants who are censored during interval t
$N_{t^*} =$ the average number of participants at risk during interval t

follow-up life table is shown in **Table 11-3**. The columns of the table include the intervals, the numbers of participants at risk, the number who suffer the event of interest (e.g., die), the number of participants who are lost to follow-up or censored, the proportions who suffer the event of interest and who do not, and the survival probability. At time 0, the start of the first interval (0–4 years), there are 20 participants alive or at risk. Two participants die in the interval, and one participant is censored. We apply the correction for the number of participants censored during that interval to produce $N_{t^*} = N_t - C_t/2 = 20 - (1/2) = 19.5$. The computations of the remaining elements of the life table are outlined in Table 11-3 for the first interval.

We now add the second interval, 5–9 years (see **Table 11-4**). The number at risk is the number at risk in the previous interval (0–4 years) less those who die and are censored (i.e., $N_t = N_{t-1} - D_{t-1} - C_{t-1} = 20 - 2 - 1 = 17$). The probability that a participant survives past 4 years, or past the first interval (using the upper limit of the interval to define the time), is $S_4 = p_4 = 0.897$. The survival probabilities are computed using $S_{t+1} = p_{t+1} \times S_t$. The probability that a participant survives past 9 years is $S_9 = p_9 \times S_4 = 0.937 \times 0.897 = 0.840$.

Table 11-5 is the complete follow-up life table for the data in Example 11.2. Table 11-5 uses the actuarial method to construct the follow-up life table, where the time is divided

TABLE 11-3 Constructing the Life Table—First Interval

Interval in Years	Number at Risk During Interval, N_t	Average Number at Risk During Interval, N_{t^*}	Number of Deaths During Interval, D_t	Lost to Follow-Up, C_t	Proportion Dying During Interval, q_t	Proportion Surviving Interval, p_t	Survival Probability, S_t
0–4	20	20 − (1/2) = 19.5	2	1	2/19.5 = 0.103	1 − 0.103 = 0.897	1(0.897) = 0.897

TABLE 11-4 Constructing the Life Table—Second Interval

Interval in Years	Number at Risk During Interval, N_t	Average Number at Risk During Interval, N_{t^*}	Number of Deaths During Interval, D_t	Lost to Follow-Up, C_t	Proportion Dying During Interval, q_t	Among Those at Risk, Proportion Surviving Interval, p_t	Survival Probability, S_t
0–4	20	20 − (1/2) = 19.5	2	1	2/19.5 = 0.103	1 − 0.103 = 0.897	1(0.897) = 0.897
5–9	17	17 − (2/2) = 16.0	1	2	1/16 = 0.063	1 − 0.063 = 0.937	(0.937)(0.897) = 0.840

TABLE 11-5 Life Table for Data in Example 11.2

Interval in Years	Number at Risk During Interval, N_t	Average Number at Risk During Interval, N_{t^*}	Number of Deaths During Interval, D_t	Lost to Follow-Up, C_t	Proportion Dying During Interval, q_t	Among Those at Risk, Proportion Surviving Interval, p_t	Survival Probability, S_t
0–4	20	19.5	2	1	0.103	0.897	0.897
5–9	17	16.0	1	2	0.063	0.937	0.840
10–14	14	12.0	1	4	0.083	0.917	0.770
15–19	9	7.5	1	3	0.133	0.867	0.688
20–24	5	3.0	1	4	0.333	0.667	0.446

into equally spaced intervals. An issue with this approach is that the survival probabilities can change depending on how the intervals are organized, particularly with small samples. The Kaplan–Meier approach, also called the product-limit approach, is a very popular method that re-estimates the survival probability each time an event occurs (in Example 11.2, each time a death occurs) and addresses this issue. There are several assumptions for appropriate use of the Kaplan–Meier approach. Specifically, we assume that censoring is independent of the likelihood of developing the event of interest and that survival probabilities are comparable in participants who are recruited earlier as well as later into the study. When comparing several groups, it is also important that these assumptions are satisfied in each comparison group so that, for example, censoring is not more likely in one group than in another. This is discussed in more detail in Section 11.3.

Table 11–6 shows the life table for the data in Example 11.2 using the Kaplan–Meier approach. In Table 11–6 we list the times when events or censoring occur, the number of participants at risk at that time (N_t), the number of deaths at that time (D_t), the number censored (C_t), and the survival probability (S_t). Note that we start the table with time = 0 and survival probability = 1. At time = 0 (baseline, or the start of the study), all participants are at risk and the survival probability is 1 (or 100%). With the data layout in Table 11–6, the survival probability is computed using $S_{t+1} = S_t \times ((N_{t+1} - D_{t+1})/N_{t+1})$. The calculations of the survival probabilities are detailed in the first few rows of the table. It is important to note that the calculations using the Kaplan–Meier approach are similar to those using the actuarial life-table approach. The main difference is the time intervals: with the actuarial life-table approach we consider equally spaced intervals, while with the Kaplan–Meier approach we use observed event times and censoring times.

With large datasets, these computations can be tedious. However, many computer programs generate the analyses easily (e.g., SAS®[4]). Microsoft Excel® can also be used to compute the survival probabilities once the data are organized by times and the numbers of events and censored times are summarized (see Chapter 11 in the Excel Workbook).

TABLE 11–6 Life Table for Data in Example 11.2 Using the Kaplan–Meier Approach

Time, years	Number at Risk, N_t	Number of Deaths, D_t	Number Censored, C_t	Survival Probability, $S_{t+1} = S_t \times ((N_{t+1} - D_{t+1})/N_{t+1})$
0	20			1[†]
1	20	1		$1 \times [(20 - 1)/20] = 0.950$
2	19		1	$0.950 \times [(19 - 0)/19] = 0.950$
3	18	1		$0.950 \times [(18 - 1)/18] = 0.897$
5	17	1		$0.897 \times [(17 - 1)/17] = 0.844$
6	16		1	0.844
9	15		1	0.844
10	14		1	0.844
11	13		1	0.844
12	12		1	0.844
13	11		1	0.844
14	10	1		0.760
17	9	1	1	0.676
18	7		1	0.676
19	6		1	0.676
21	5		1	0.676
23	4	1		0.507
24	3		3	0.507

[†]Recall that $S_0 = 1$.

From the life table we can produce a Kaplan–Meier survival curve. The Kaplan–Meier survival curve for the data in Example 11.2 is shown in **Figure 11-6**. In the survival curve shown in Figure 11-6, the symbols represent each event time, either a death or a censored time. From the survival curve, we can also estimate the probability that a participant survives past 10 years by locating 10 years on the x-axis and reading up and over to the y-axis. The proportion of participants surviving past 10 years is 84% (see dashed lines in Figure 11-6). Similarly, the proportion of participants surviving past 20 years is 68%. The median survival is estimated by locating 0.5 on the y-axis and reading over and down to the x-axis. The median survival is approximately 23 years. These estimates of survival probabilities at specific times and of the median survival time are point estimates and should be interpreted as such. There are formulas to produce standard errors and confidence interval estimates of survival probabilities, and they can be generated with many statistical computing packages. A popular formula to estimate the standard error of the survival estimates is called Greenwood's formula[5] and is as follows: $SE(S_t) = S_t \sqrt{\sum \frac{D_t}{N_t(N_t - D_t)}}$. The quantity $\frac{D_t}{N_t(N_t - D_t)}$ is summed for numbers at risk (N_t) and numbers of deaths (D_t) occurring through the time interval of interest (i.e., cumulative across all time intervals before the time interval of interest; see example in **Table 11-7**). Standard errors are computed for the survival estimates for the data in Example 11.2 and are summarized in Table 11-7. Note that the final column shows the quantity $1.96 \times SE(S_t)$, which is the margin of error and is used for computing the 95% CI estimates (i.e., $S_t \pm 1.96 \times SE(S_t)$).

The Kaplan–Meier survival curve (shown as a solid line), along with 95% confidence limits (shown as dotted lines), for the estimates of survival probability is shown in **Figure 11-7**.

Some investigators prefer to generate cumulative incidence curves, as opposed to survival curves that show the cumulative probabilities of experiencing the event of interest. **Cumulative incidence**, or cumulative failure probability, is computed as $1 - S_t$ and can be computed easily from the life table using the Kaplan–Meier approach. **Table 11-8** shows the cumulative failure probabilities for the data in Example 11.2. **Figure 11-8** shows the cumulative incidence of death for participants enrolled in the study described in

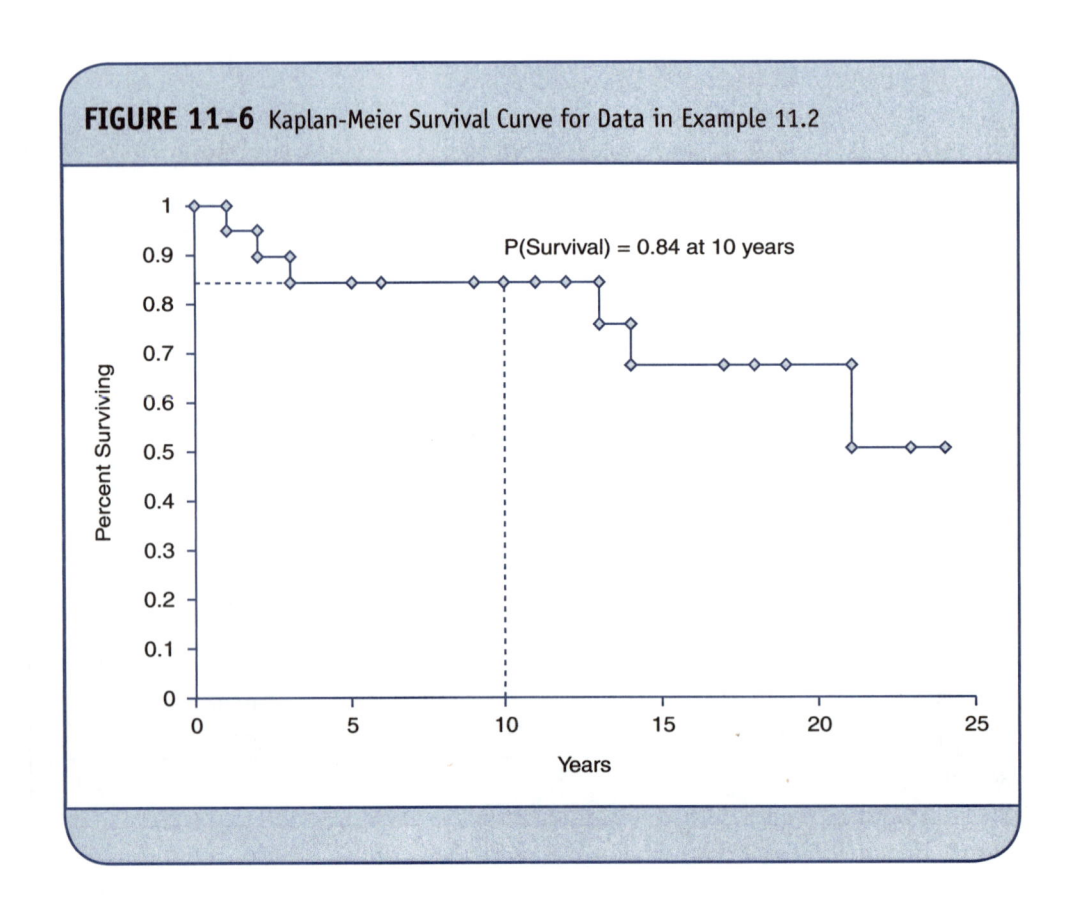

FIGURE 11-6 Kaplan-Meier Survival Curve for Data in Example 11.2

TABLE 11–7 Standard Errors of Survival Estimates for the Data in Example 11.2

Time, years	Number at Risk, N_t	Number of Deaths, D_t	Survival Probability, S_t	$\dfrac{D_t}{N_t(N_t - D_t)}$	$\sum \dfrac{D_t}{N_t(N_t - D_t)}$	$S_t\sqrt{\sum \dfrac{D_t}{N_t(N_t - D_t)}}$	$1.96 \times \text{SE}(S_t)$
0	20		1				
1	20	1	0.950	0.003	0.003	0.049	0.096
2	19		0.950	0.000	0.003	0.049	0.096
3	18	1	0.897	0.003	0.006	0.069	0.135
5	17	1	0.844	0.004	0.010	0.083	0.162
6	16		0.844	0.000	0.010	0.083	0.162
9	15		0.844	0.000	0.010	0.083	0.162
10	14		0.844	0.000	0.010	0.083	0.162
11	13		0.844	0.000	0.010	0.083	0.162
12	12		0.844	0.000	0.010	0.083	0.162
13	11		0.844	0.000	0.010	0.083	0.162
14	10	1	0.760	0.011	0.021	0.109	0.214
17	9	1	0.676	0.014	0.035	0.126	0.246
18	7		0.676	0.000	0.035	0.126	0.246
19	6		0.676	0.000	0.035	0.126	0.246
21	5		0.676	0.000	0.035	0.126	0.246
23	4	1	0.507	0.083	0.118	0.174	0.341
24	3		0.507	0.000	0.118	0.174	0.341

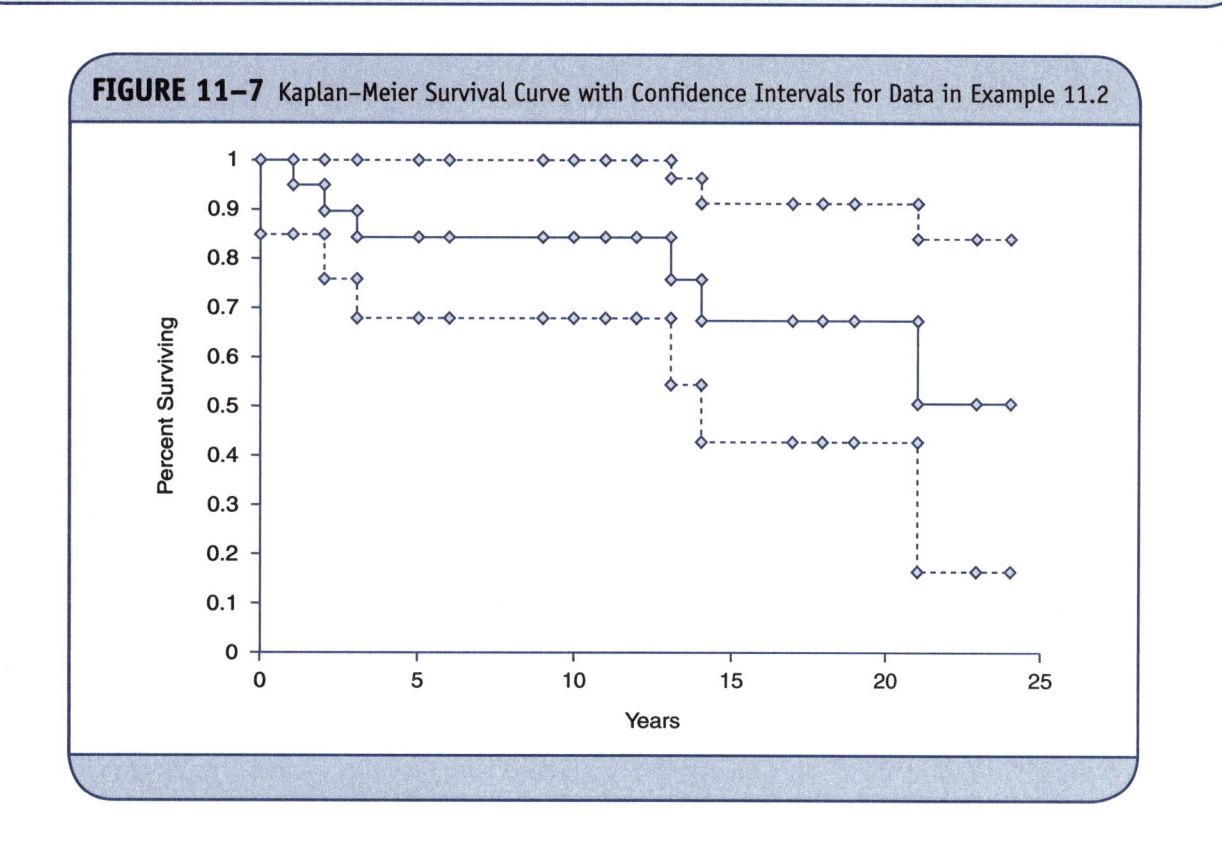

FIGURE 11–7 Kaplan–Meier Survival Curve with Confidence Intervals for Data in Example 11.2

TABLE 11-8 Life Table with Cumulative Failure Probabilities for Data in Example 11.2

Time, years	Number at Risk, N_t	Number of Deaths, D_t	Number Censored, C_t	Survival Probability, S_t	Failure Probability, $1 - S_t$
0	20			1	0
1	20	1		0.950	0.050
2	19		1	0.950	0.050
3	18	1		0.897	0.103
5	17	1		0.844	0.156
6	16		1	0.844	0.156
9	15		1	0.844	0.156
10	14		1	0.844	0.156
11	13		1	0.844	0.156
12	12		1	0.844	0.156
13	11		1	0.844	0.156
14	10	1		0.760	0.240
17	9	1	1	0.676	0.324
18	7		1	0.676	0.324
19	6		1	0.676	0.324
21	5		1	0.676	0.324
23	4	1		0.507	0.493
24	3		3	0.507	0.493

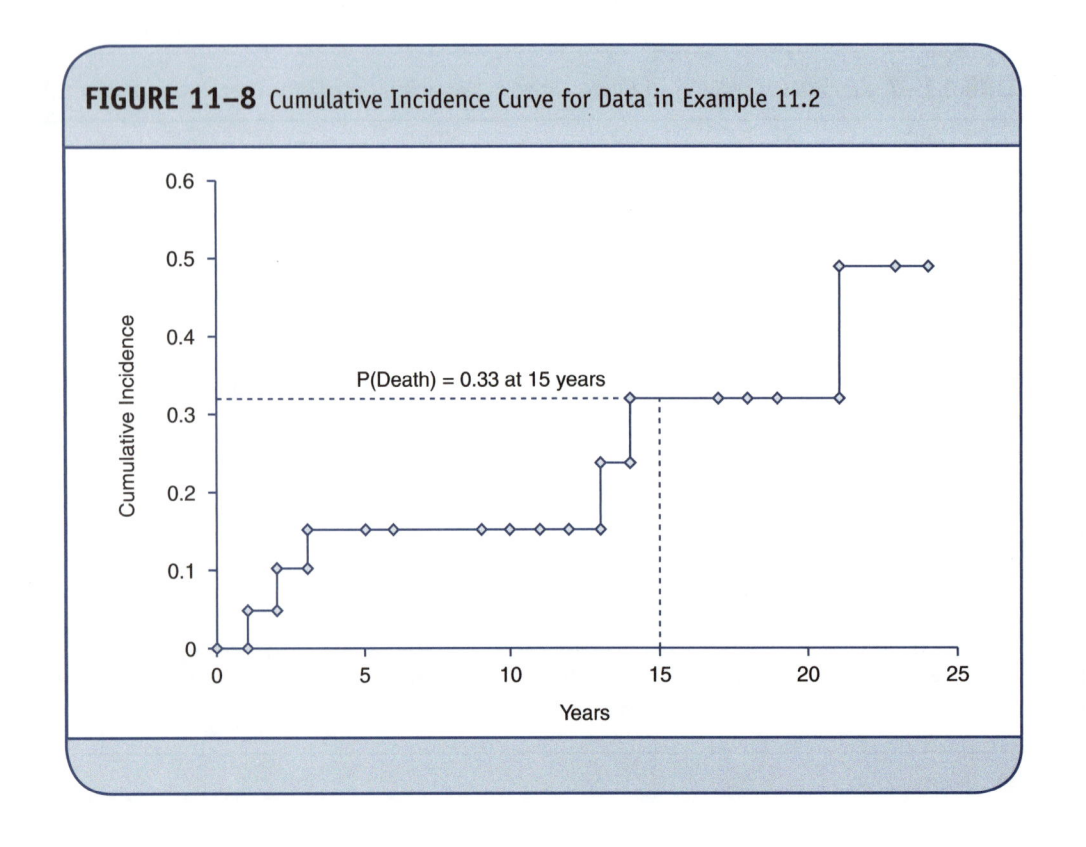

FIGURE 11-8 Cumulative Incidence Curve for Data in Example 11.2

P(Death) = 0.33 at 15 years

Example 11.2. From Figure 11-8, we can estimate the likelihood that a participant dies by a certain time point. For example, the probability of death is approximately 33% at 15 years (see dashed lines).

11.3 COMPARING SURVIVAL CURVES

In many survival analysis applications, we are interested in assessing whether there are differences in survival (or cumulative incidence of event) among different groups of participants. For example, in a clinical trial with a survival outcome, we are interested in comparing survival between participants receiving a new drug as compared to a placebo (or other appropriate comparator). In an observational study, we might be interested in comparing survival between men and women or between participants with and without a particular risk factor (e.g., hypertension or diabetes).

There are several tests available to compare survival among independent groups. The log-rank test is a popular test to test the null hypothesis of no difference in survival between two or more independent groups. The test compares the entire survival experience between groups and can be thought of as a test of whether the survival curves are identical (overlapping) or not. Survival curves are estimated for each group, considered separately, using the Kaplan–Meier method and compared statistically using the log-rank test. To illustrate the procedure, consider the following example. In this example, we address each of the five steps in the test of hypothesis but include more details to promote understanding. (In subsequent examples, the five-step approach for hypothesis testing is followed more directly.)

Several variations of the log-rank test statistic are implemented by various statistical computing packages (e.g., SAS[6] and R[7]). We present one version here that is linked closely to the χ^2 test statistic and compares observed numbers to expected numbers of events at each time point over the follow-up period.

Example 11.3. A small clinical trial is run to compare two combination treatments in patients with advanced gastric cancer. Twenty participants with stage IV gastric cancer who consent to participate in the trial are randomly assigned to receive chemotherapy before surgery or chemotherapy after surgery. The primary outcome is death, and participants are followed for up to 48 months (4 years) following enrollment in the trial. The experiences of participants in each arm of the trial are shown in **Table 11-9**.

Six participants in the chemotherapy-before-surgery group die over the course of follow-up, compared to three participants in the chemotherapy-after-surgery group. Other participants in each group are followed for varying numbers of months, some to the end of the study at 48 months (in the chemotherapy-after-surgery group). We use all available information to compare the two groups with respect to survival.

Using the procedures outlined in Section 11.2, we first construct life tables for each treatment group using the Kaplan–Meier approach. The results are shown in **Table 11-10a** and **Table 11-10b**. The two survival curves are shown in **Figure 11-9**.

Notice that the survival probabilities for the chemotherapy-after-surgery group are higher than the survival probabilities for the chemotherapy-before-surgery group. Figure 11-9 suggests a survival benefit for the chemotherapy-after-surgery group. However, these survival curves are estimated from small samples. Is the difference we see in Figure 11-9 suggestive of a true difference in the population?

TABLE 11-9 Month of Death or Month of Last Contact in Each Treatment Group

Chemotherapy Before Surgery		Chemotherapy After Surgery	
Month of Death	Month of Last Contact	Month of Death	Month of Last Contact
8	8	33	48
12	32	28	48
26	20	41	25
14	40		37
21			48
27			25
			43

TABLE 11–10a Life Tables for Data in Example 11.3 Using the Kaplan–Meier Approach: Chemotherapy Before Surgery Treatment Group

Time, months	Number at Risk, N_t	Number of Deaths, D_t	Number Censored, C_t	Survival Probability, $S_{t+1} = S_t \times ((N_{t+1} - D_{t+1})/N_{t+1})$
0	10			1
8	10	1	1	0.900
12	8	1		0.788
14	7	1		0.675
20	6		1	0.675
21	5	1		0.540
26	4	1		0.405
27	3	1		0.270
32	2		1	0.270
40	1		1	0.270

TABLE 11–10b Life Tables for Data in Example 11.3 Using the Kaplan–Meier Approach: Chemotherapy After Surgery Treatment Group

Time, months	Number at Risk, N_t	Number of Deaths, D_t	Number Censored, C_t	Survival Probability, $S_{t+1} = S_t \times ((N_{t+1} - D_{t+1})/N_{t+1})$
0	10			1
25	10		2	1.000
28	8	1		0.875
33	7	1		0.750
37	6		1	0.750
41	5	1		0.600
43	4		1	0.600
48	3		3	0.600

To compare survival between groups we use the log-rank test. The log-rank test is a nonparametric test that makes no assumptions about the survival distributions. The log-rank test is based on information in the Kaplan–Meier survival curves. The null hypothesis is that there is no difference in survival between the two groups or that there is no difference between the populations in the probability of death at any point. The test essentially compares the observed number of events in each group to what would be expected under the null hypothesis or if the null hypothesis were true (i.e., if the survival curves were identical). The hypotheses to be tested are given below and we run the test at the 5% level of significance (i.e., $\alpha = 0.05$).

H_0: The two survival curves are identical
(or $S_{1t} = S_{2t}$, at all times t).

H_1: The two survival curves are not identical
(or $S_{1t} \neq S_{2t}$, at any time t).

The log-rank statistic follows a χ^2 distribution. There are several forms of the test statistic, and they vary in terms of how they are computed. We use the following:

$$\chi^2 = \sum \frac{\left(\sum O_{jt} - \sum E_{jt}\right)^2}{\sum E_{jt}},$$

FIGURE 11-9 Survival in Each Treatment Group

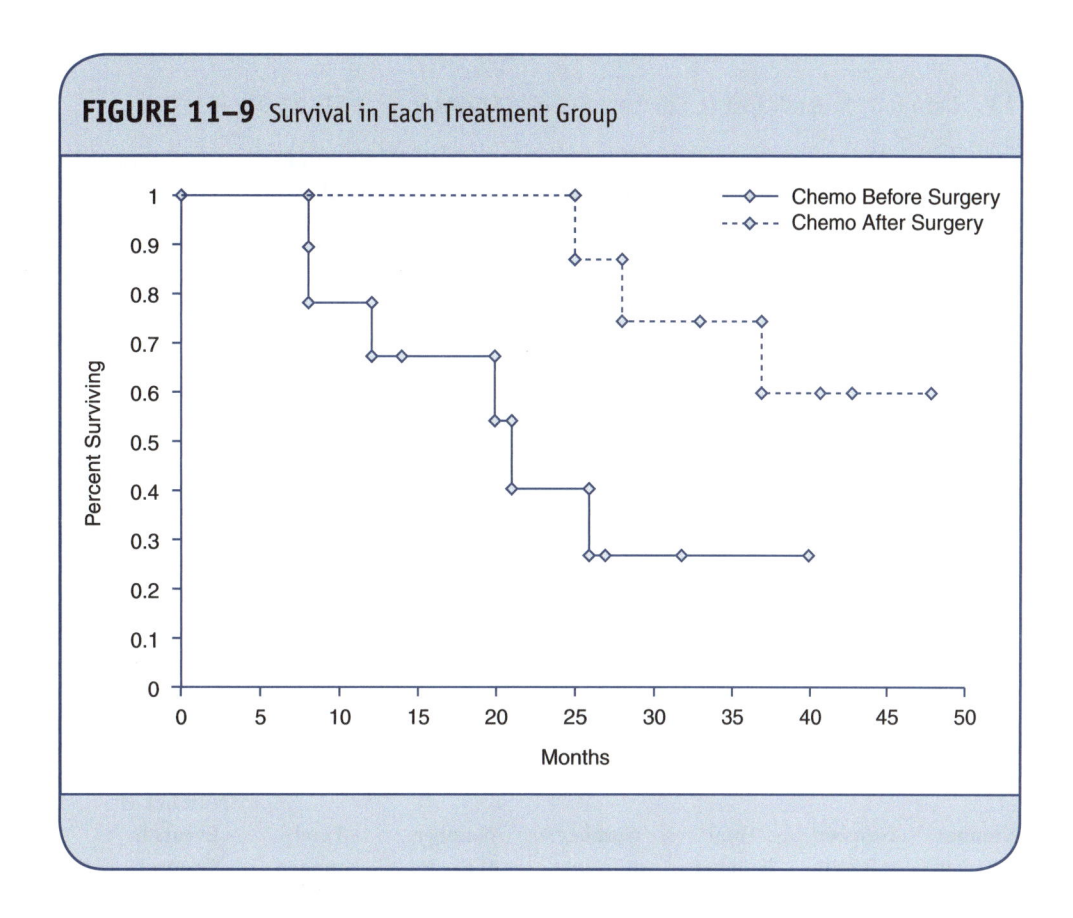

where ΣO_{jt} represents the sum of the observed number of events in the *j*th group over time (e.g., *j* = 1, 2) and ΣE_{jt} represents the sum of the expected number of events in the *j*th group over time. The sums of the observed and expected numbers of events are computed for each event time and are summed for each comparison group. The log-rank statistic has degrees of freedom (*df*) equal to *k* − 1, where *k* represents the number of comparison groups. In this example, *k* = 2, so the test statistic has one degree of freedom (*df* = 1).

To compute the test statistic we need the observed number and the expected number of events at each event time. The observed number of events is from the sample, and the expected number of events is computed assuming that the null hypothesis is true (i.e., that the survival curves are identical). To generate the expected numbers of events, we organize the data into a life table with rows representing each event time, regardless of the group in which the event occurred. We also keep track of group assignment. We then estimate the proportion of events that occur at each time (O_t/N_t) using data from both groups combined under the assumption of no difference in survival (i.e., assuming the null hypothesis is true). We multiply these estimates by the number of participants at risk at that time in each of the comparison groups (N_{1t} and N_{2t} for groups 1 and 2, respectively).

Specifically, we compute, for each event time *t*, the number at risk N_{jt} in each group (where *j* indicates the group, *j* = 1, 2) and the number of events (deaths) O_{jt} in each group. **Table 11-11** contains the information needed to conduct the log-rank test to compare the survival curves for the data in Example 11.3. In Table 11-11 group 1 represents the chemotherapy-before-surgery group and group 2 represents the chemotherapy-after-surgery group.

We next total the number at risk $N_t = N_{1t} + N_{2t}$ at each event time and the number of observed events (deaths) $O_t = O_{1t} + O_{2t}$ at each event time. We then compute the expected number of events in each group. The expected number of events is computed at each event time as follows: $E_{1t} = N_{1t} \times (O_t/N_t)$ for group 1 and $E_{2t} = N_{2t} \times (O_t/N_t)$ for group 2. The calculations for the data in Example 11.3 are shown in **Table 11-12**.

We next sum the observed numbers of events in each group (ΣO_{1t} and ΣO_{2t}) and the expected numbers of events in each group (ΣE_{1t} and ΣE_{2t}) over time. These are shown in the bottom row of **Table 11-13**.

TABLE 11–11 Data for Log-Rank Test to Compare Survival Curves in Example 11.3

Time, months	Number at Risk in Group 1, N_{1t}	Number at Risk in Group 2, N_{2t}	Number of Events (Deaths) in Group 1, O_{1t}	Number of Events (Deaths) in Group 2, O_{2t}
8	10	10	1	0
12	8	10	1	0
14	7	10	1	0
21	5	10	1	0
26	4	8	1	0
27	3	8	1	0
28	2	8	0	1
33	1	7	0	1
41	0	5	0	1

TABLE 11–12 Expected Numbers of Events in Each Group for Data in Example 11.3

Time, months	Number at Risk in Group 1, N_{1t}	Number at Risk in Group 2, N_{2t}	Total Number at Risk, N_t	Number of Events in Group 1, O_{1t}	Number of Events in Group 2, O_{2t}	Total Number of Events, O_t	Expected Number of Events in Group 1, $E_{1t} = N_{1t} \times (O_t/N_t)$	Expected Number of Events in Group 2, $E_{2t} = N_{2t} \times (O_t/N_t)$
8	10	10	20	1	0	1	0.500	0.500
12	8	10	18	1	0	1	0.444	0.556
14	7	10	17	1	0	1	0.412	0.588
21	5	10	15	1	0	1	0.333	0.667
26	4	8	12	1	0	1	0.333	0.667
27	3	8	11	1	0	1	0.273	0.727
28	2	8	10	0	1	1	0.200	0.800
33	1	7	8	0	1	1	0.125	0.875
41	0	5	5	0	1	1	0.000	1.000

We can now compute the test statistic:

$$\chi^2 = \sum \frac{\left(\sum O_{jt} - \sum E_{jt}\right)^2}{\sum E_{jt}} = \frac{(6-2.620)^2}{2.620} + \frac{(3-6.380)^2}{6.380} =$$

$$4.360 + 1.791 = 6.151$$

The test statistic is approximately distributed as χ^2 with one degree of freedom ($df = 1$). Thus, the critical value for the test can be found in Table 3 in the Appendix: Critical values of the χ^2 Distribution. The decision rule for this test is as follows: Reject H_0 if $\chi^2 \geq 3.84$. We observe $\chi^2 = 6.151$, which exceeds the critical value of 3.84, and thus we reject H_0. We have significant evidence at $\alpha = 0.05$ to show that the two survival curves are different.

Example 11.4. An investigator wishes to evaluate the efficacy of a brief intervention to prevent alcohol consumption in pregnancy. Pregnant women with a history of heavy alcohol consumption are recruited into the study and are randomized to receive either the brief intervention focused on abstinence from alcohol or standard prenatal care. The

TABLE 11-13 Total Observed and Expected Numbers of Events in Each Group for Data in Example 11.3

Time, months	Number at Risk in Group 1, N_{1t}	Number at Risk in Group 2, N_{2t}	Total Number at Risk, N_t	Number of Events in Group 1, O_{1t}	Number of Events in Group 2, O_{2t}	Total Number of Events, O_t	Expected Number of Events in Group 1, $E_{1t} = N_{1t} \times (O_t/N_t)$	Expected Number of Events in Group 2, $E_{2t} = N_{2t} \times (O_t/N_t)$
8	10	10	20	1	0	1	0.500	0.500
12	8	10	18	1	0	1	0.444	0.556
14	7	10	17	1	0	1	0.412	0.588
21	5	10	15	1	0	1	0.333	0.667
26	4	8	12	1	0	1	0.333	0.667
27	3	8	11	1	0	1	0.273	0.727
28	2	8	10	0	1	1	0.200	0.800
33	1	7	8	0	1	1	0.125	0.875
41	0	5	5	0	1	1	0.000	1.000
				6	3		2.620	6.380

outcome of interest is relapse to drinking. Women are recruited into the study at approximately 18 weeks' gestation and are followed through the course of pregnancy to delivery (approximately 39 weeks' gestation). The data are shown in **Table 11-14** and indicate whether women relapse to drinking and, if so, the time of their first drink measured in the number of weeks from randomization. For women who do not relapse, we record the number of weeks from randomization that they are alcohol-free.

The question of interest is whether there is a difference in time to relapse between women assigned to standard prenatal care as compared to those assigned to the brief intervention.

TABLE 11-14 Number of Weeks to First Drink or Number of Weeks Alcohol-Free by Treatment Group

Standard Prenatal Care		Brief Intervention	
Relapse	**No Relapse**	**Relapse**	**No Relapse**
19	20	16	21
6	19	21	15
5	17	7	18
4	14		18
			5

Step 1. Set up hypotheses and determine the level of significance.

H_0: Relapse-free time is identical between groups.

H_1: Relapse-free time is not identical between groups.

$\alpha = 0.05$.

Step 2. Select the appropriate test statistic.

The test statistic for the log-rank test is

$$\chi^2 = \sum \frac{\left(\sum O_{jt} - \sum E_{jt}\right)^2}{\sum E_{jt}}$$

Step 3. Set up the decision rule.

The test statistic follows a χ^2 distribution, and so we find the critical value in Table 3 in the Appendix for $df = k - 1 = 2 - 1 = 1$ and $\alpha = 0.05$. The critical value is 3.84, and the decision rule is to reject H_0 if $\chi^2 \geq 3.84$.

Step 4. Compute the test statistic.

To compute the test statistic, we organize the data according to event (relapse) times and determine the numbers of women at risk in each treatment group and the number who relapse at each observed

TABLE 11–15 Number of Relapses in Each Treatment Group

Time, weeks	Number at Risk in Group 1, N_{1t}	Number at Risk in Group 2, N_{2t}	Number of Events Relapses in Group 1, O_{1t}	Number of Events Relapses in Group 2, O_{2t}
4	8	8	1	0
5	7	8	1	0
6	6	7	1	0
7	5	7	0	1
16	4	5	0	1
19	3	2	1	0
21	0	2	0	1

relapse time. In **Table 11–15**, group 1 represents women who receive standard prenatal care, and group 2 represents women who receive the brief intervention.

We next total the number at risk $N_t = N_{1t} + N_{2t}$ at each event time and the number of observed events (relapses) $O_t = O_{1t} + O_{2t}$ at each event time, and determine the expected number of relapses in each group at each event time using $E_{1t} = N_{1t} \times (O_t/N_t)$ and $E_{2t} = N_{2t} \times (O_t/N_t)$. We then sum the observed numbers of events in each group (ΣO_{1t} and ΣO_{2t}) and the expected numbers of events in each group (ΣE_{1t} and ΣE_{2t}) over time. The calculations for the data in Example 11.4 are shown in **Table 11–16**.

We now compute the test statistic:

$$\chi^2 = \Sigma \frac{\left(\Sigma O_{jt} - \Sigma E_{jt}\right)^2}{\Sigma E_{jt}} = \frac{(4-2.890)^2}{2.890} + \frac{(3-4.110)^2}{4.110} = 0.426 + 0.300 = 0.726.$$

TABLE 11–16 Expected Numbers of Relapses in Each Treatment Group

Time, weeks	Number at Risk in Group 1, N_{1t}	Number at Risk in Group 2, N_{2t}	Total Number at Risk, N_t	Number of Relapses in Group 1, O_{1t}	Number of Relapses in Group 2, O_{2t}	Total Number of Relapses, O_t	Expected Number of Relapses in Group 1, $E_{1t} = N_{1t} \times (O_t/N_t)$	Expected Number of Relapses in Group 2, $E_{2t} = N_{2t} \times (O_t/N_t)$
4	8	8	16	1	0	1	0.500	0.500
5	7	8	15	1	0	1	0.467	0.533
6	6	7	13	1	0	1	0.462	0.538
7	5	7	12	0	1	1	0.417	0.583
16	4	5	9	0	1	1	0.444	0.556
19	3	2	5	1	0	1	0.600	0.400
21	0	2	2	0	1	1	0.000	1.000
				4	3		2.890	4.110

Step 5. Conclusion.

Do not reject H_0 because $0.726 < 3.84$. We do not have statistically significant evidence at $\alpha = 0.05$ to show that the time to relapse is different between groups.

Figure 11–10 shows the survival (relapse-free time) in each group. Notice that the survival curves do not show much separation, consistent with the nonsignificant findings in the test of hypothesis.

As noted, there are several variations of the log-rank statistic. Some statistical computing packages use the following test statistic for the log-rank test to compare two independent groups: $\chi^2 = \dfrac{(\Sigma O_{1t} - \Sigma E_{1t})^2}{\Sigma \text{Var}(E_{1t})}$ where ΣO_{1t} is the sum of the observed number of events in group 1 and ΣE_{1t} is the sum of the expected number of events in group 1 taken over all event times. The denominator is the sum of the variances of the expected numbers of events at each event time, which is computed as follows: $\text{Var}(E_{1t}) = \dfrac{N_{1t} \times N_{2t} \times (N_t - O_t)}{N_t^2 \times (N_t - 1)}$. There are other versions of the log-rank statistic as well as other tests to compare survival functions between independent groups.[8-10] For example, a popular test is the modified Wilcoxon test, which is sensitive to larger differences in hazards occurring earlier as opposed to later in follow-up.[11]

11.4 COX PROPORTIONAL HAZARDS REGRESSION ANALYSIS

Survival analysis methods can also be extended to assess several risk factors simultaneously, similar to multiple linear and multiple logistic regression analysis discussed in Chapter 9 for continuous and dichotomous outcomes, respectively. One of the most popular regression techniques for survival analysis is Cox proportional hazards regression. Cox proportional hazards regression is used to relate several risk factors or exposures, considered simultaneously, to survival time.

In a Cox proportional hazards regression model, the measure of effect is the hazard rate, which is the risk of failure (i.e., the risk or probability of suffering the event of interest) conditional on the fact that the participant has survived up to a specific time. The *hazard* (in a single group) is a rate that is

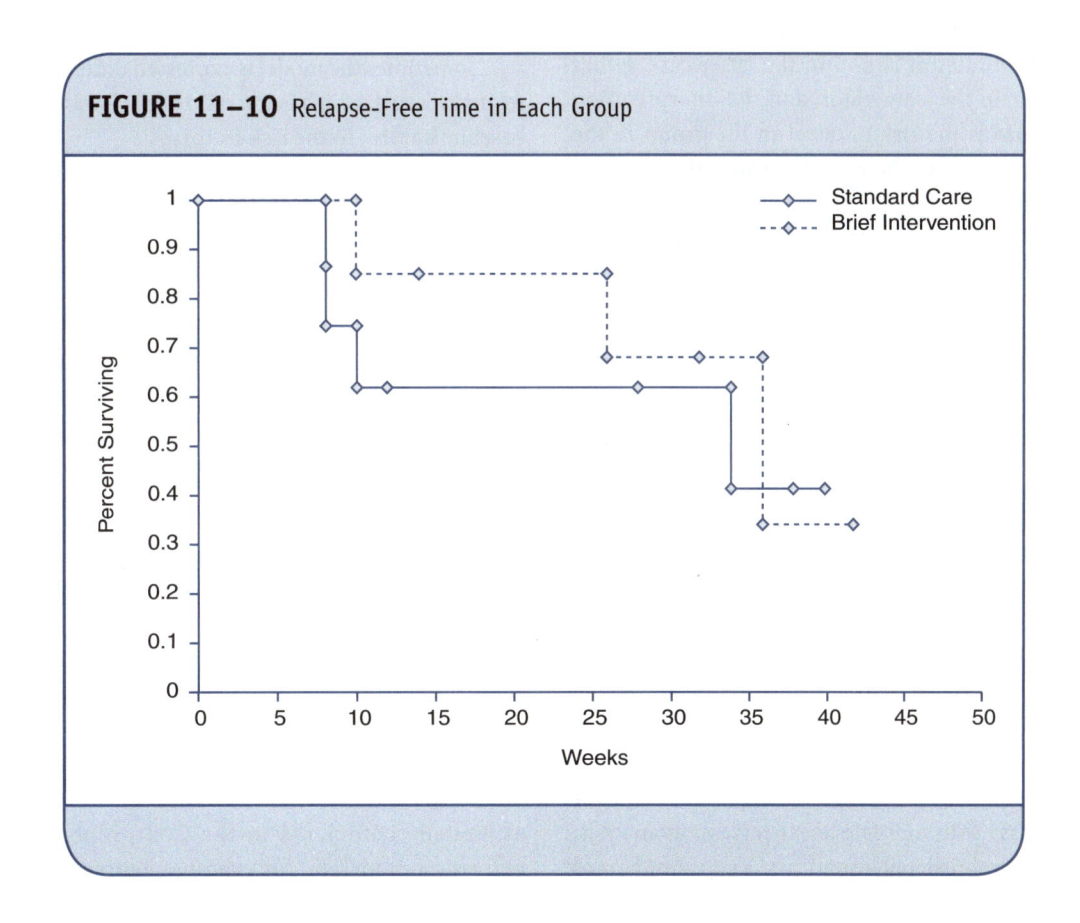

FIGURE 11–10 Relapse-Free Time in Each Group

a positive number that can exceed 1 (as opposed to a probability, which ranges between 0 and 1, inclusive). The hazard represents the expected number of events per unit of time. For example, if the hazard is 0.2 at time t and the time units are months, then, on average, 0.2 event is expected per person at risk per month. Another interpretation is based on the reciprocal of the hazard. For example, $1/0.2 = 5$, which is the expected event-free time (5 months) for each person at risk.

In most situations, we are interested in comparing groups with respect to their hazards, and we use a hazard ratio that is analogous to an odds ratio in the setting of multiple logistic regression analysis. (See Chapter 3 for more details of ratio measures used to compare two independent groups.) The hazard ratio can be estimated from the data we organize to conduct the log-rank test. Specifically, the hazard ratio is the ratio of the total number of observed to expected events in two independent comparison groups:

$$\text{HR} = \frac{\sum O_{\text{Exposed},t} / \sum E_{\text{Exposed},t}}{\sum O_{\text{Unexposed},t} / \sum E_{\text{Unexposed},t}} = \frac{\sum O_{\text{Treatment},t} / \sum E_{\text{Treatment},t}}{\sum O_{\text{Control},t} / \sum E_{\text{Control},t}}$$

In some studies, the distinction between the exposed or treated group as compared to the unexposed or control group is clear. In other studies, it is not. In the latter case, either group can appear in the numerator, and the interpretation of the hazard ratio is the risk of event in the group in the numerator as compared to the risk of event in the group in the denominator.

In Example 11.3 there are two active treatments being compared (chemotherapy before surgery versus chemotherapy after surgery). Thus, it does not matter which appears in the numerator of the hazard ratio. Using the data in Example 11.3, the hazard ratio is estimated as

$$\text{HR} = \frac{\sum O_{\text{Chemo Before Surgery},t} / \sum E_{\text{Chemo Before Surgery},t}}{\sum O_{\text{Chemo After Surgery},t} / \sum E_{\text{Chemo After Surgery},t}} = \frac{6/2.620}{3/6.380} = 4.87$$

Thus, participants in the chemotherapy-before-surgery group have 4.87 times the risk of death as compared to participants in the chemotherapy-after-surgery group.

There are several important assumptions for appropriate use of the Cox proportional hazards regression model, including independence of survival times between distinct individuals in the sample, a multiplicative relationship between the predictors and the hazard (as opposed to a linear one, as was the case with multiple linear regression analysis; this is discussed in more detail below), and a constant hazard ratio over time.

The Cox proportional hazards regression model can be written as follows:

$$h(t) = h_0(t)\exp(b_1 X_1 + b_2 X_2 + \cdots + b_p X_p),$$

where $h(t)$ is the expected hazard at time t, and $h_0(t)$ is the baseline hazard and represents the hazard when all of the predictors $X_1, X_2, \cdots, X_p$ are equal to zero. Notice that the predicted hazard (i.e., $h(t)$), or the rate of suffering the event of interest in the next instant, is the product of the baseline hazard $h_0(t)$ and the exponential function of a linear combination of the predictors. Thus, the predictors have a multiplicative or proportional effect on the predicted hazard.

Consider a simple model with one predictor, X_1. The Cox proportional hazards model is $h(t) = h_0(t)\exp(b_1 X_1)$. Suppose we wish to compare two participants in terms of their expected hazards, and the first has $X_1 = a$ and the second has $X_1 = b$. The expected hazards are $h(t) = h_0(t)\exp(b_1 a)$ and $h(t) = h_0(t)\exp(b_1 b)$, respectively. The hazard ratio is the ratio of these two expected hazards:

$$h_0(t)\exp(b_1 a) / h_0(t)\exp(b_1 b) = \exp(b_1(a - b)),$$

which does not depend on time t. Thus the hazard is proportional over time.

Sometimes the model is expressed differently, relating the relative hazard, which is the ratio of the hazard at time t to the baseline hazard, to the risk factors:

$$h(t) / h_0(t) = \exp(b_1 X_1 + b_2 X_2 + \cdots + b_p X_p)$$

We can take the natural logarithm (ln) of each side of the Cox proportional hazards regression model to produce the following, which relates the log of the relative hazard to a linear function of the predictors. Notice that the right-hand side of the equation looks like the more familiar linear combination of the predictors or risk factors (as seen in the multiple linear regression model).

$$\ln\{h(t) / h_0(t)\} = b_1 X_1 + b_2 X_2 + \cdots + b_p X_p$$

In practice, interest lies in the associations between each of the risk factors or predictors $(X_1, X_2, \ldots, X_p)$ and the outcome. The associations are quantified by the regression coefficients $(b_1, b_2, \ldots, b_p)$. The technique for estimating the regression coefficients in a Cox proportional hazards regression model is beyond the scope of this text and is described in Cox and Oakes.[2] Here we focus on interpretation. The estimated coefficients in the Cox proportional hazards regression model, b_1, for example, represent the change in the expected log of the hazard ratio relative to a one-unit change in X_1, holding all other predictors constant. The antilog of an estimated regression coefficient, $\exp(b_j)$, produces a hazard ratio. If a predictor is dichotomous (e.g., X_1 is an indicator of prevalent cardiovascular disease, or male sex), then $\exp(b_1)$ is the hazard ratio comparing the risk of event for participants with $X_1 = 1$ (e.g., prevalent cardiovascular disease, or male sex) to participants with $X_1 = 0$ (e.g., free of cardiovascular disease, or female sex). If the hazard ratio for a predictor is close to 1, then that predictor does not affect survival. If the hazard ratio is less than 1, then the predictor is protective (i.e., associated with improved survival), and if the hazard ratio is greater than 1, then the predictor is associated with increased risk (or decreased survival). Tests of hypothesis are used to assess whether there are statistically significant associations between predictors and time to event. The examples that follow illustrate these tests and their interpretation.

The Cox proportional hazards model is called a semiparametric model because there are no assumptions about the shape of the baseline hazard function. There are, however, other assumptions as noted above (i.e., independence, changes in predictors produce proportional changes in the hazard regardless of time, and a linear association between the natural logarithm of the relative hazard and the predictors). There are other regression models used in survival analysis that assume specific distributions for the survival times, such as the exponential, Weibull, Gompertz, and lognormal distributions.[1,9] The exponential regression survival model, for example, assumes that the hazard function is constant. Other distributions assume that the hazard is increasing over time, decreasing over time, or increasing initially and then decreasing.

In Example 11.5 we estimate a Cox proportional hazards regression model and discuss the interpretation of the regression coefficients.

Example 11.5. An analysis is conducted to investigate differences in all-cause mortality between men and women participating in the Framingham Heart Study, adjusting for age. A total of 5180 participants aged 45 years and older are followed until time of death or up to 10 years, whichever comes first. Forty-six percent of the sample are male, the mean age of the sample is 56.8 years (standard deviation = 8 years), and the ages range from 45 years to 82 years at the start of the study.

A total of 402 deaths are observed among 5180 participants. Descriptive statistics are shown in **Table 11-17** on the age and sex of participants at the start of the study, classified by whether they die or do not die during the follow-up period.

TABLE 11-17 Description of Study Sample

	Die (*n* = 402)	Do Not Die (*n* = 4778)
Mean (SD) age, years	65.6 (8.7)	56.1 (7.5)
n (%) male	221 (55%)	2145 (45%)

TABLE 11-18 Parameter Estimates and *p*-Values

Risk Factor	Parameter Estimate	*p*
Age, years	0.11149	0.0001
Male sex	0.67958	0.0001

We now estimate a Cox proportional hazards regression model and relate an indicator of male sex and the age, in years, to time to death. The parameter estimates are generated in SAS[4] using the SAS Cox proportional hazards regression procedure[13] and are shown in **Table 11-18** along with their *p*-values.

Note that there is a positive association between age and all-cause mortality and between male sex and all-cause mortality (i.e., there is increased risk of death for older participants and for men). Again, the parameter estimates represent the increase in the expected log of the relative hazard for each unit of increase in the predictor, holding other predictors constant. There is a 0.11149-unit increase in the expected log of the relative hazard for each one-year increase in age, holding sex constant, and there is a 0.67958-unit increase in the expected log of the relative hazard for men as compared to women, holding age constant. For interpretability, we compute hazard ratios by exponentiating the parameter estimates. For age, $\exp(0.11149) = 1.118$. There is an 11.8% increase in the expected hazard relative to a one-year increase in age (or the expected hazard is 1.12 times higher in a person who is one year older than another), holding sex constant. Similarly, $\exp(0.67958) = 1.973$. The expected hazard is 1.973 times higher in men as compared to women, holding age constant.

Suppose we consider additional risk factors for all-cause mortality and estimate a Cox proportional hazards regression model relating an expanded set of risk factors to time to death. The parameter estimates are again generated in SAS using the SAS Cox proportional hazards regression procedure[13] and are

TABLE 11–19 Parameter Estimates and *p*-Values for Risk Factors for All-Cause Mortality

Risk Factor	Parameter Estimate	p	Hazard Ratio (HR) (95% CI for HR)
Age, years	0.11691	0.0001	1.124 (1.111–1.138)
Male gender	0.40359	0.0002	1.497 (1.215–1.845)
Systolic blood pressure	0.01645	0.0001	1.017 (1.012–1.021)
Current smoker	0.76798	0.0001	2.155 (1.758–2.643)
Total serum cholesterol	−0.00209	0.0963	0.998 (0.995–2.643)
Diabetes	−0.20366	0.1585	0.816 (0.615–1.083)

shown in **Table 11-19** along with their *p*-values. Also included in Table 11–19 are the hazard ratios along with their 95% confidence intervals.

All of the parameter estimates are estimated taking the other predictors into account. After accounting for age, sex, blood pressure, and smoking status, there are no statistically significant associations between total serum cholesterol and all-cause mortality or between diabetes and all-cause mortality. This is not to say that these risk factors are not associated with all-cause mortality; their lack of significance is likely due to confounding (interrelationships among the risk factors considered). Notice that for the statistically significant risk factors (i.e., age, sex, systolic blood pressure, and current smoking status), the 95% confidence intervals for the hazard ratios do not include 1 (the null value). In contrast, the 95% confidence intervals for the nonsignificant risk factors (total serum cholesterol and diabetes) include the null value.

Example 11.6. A prospective cohort study is run to assess the association between body mass index and time to incident cardiovascular disease (CVD). At baseline, participants' body mass index is measured along with other known clinical risk factors for cardiovascular disease (e.g., age, sex, and blood pressure). Participants are followed for up to 10 years for the development of CVD. In the study of *n* = 3937 participants, 543 develop CVD during the study observation period. In a Cox proportional hazards regression analysis, we find the association between BMI and time to CVD statistically significant with a parameter estimate of 0.02312 (*p* = 0.0175) relative to a one-unit change in BMI. If we exponentiate the parameter estimate, we have a hazard ratio of 1.023 with a confidence interval of 1.004–1.043. Because we model BMI as a continuous predictor, the interpretation of the hazard ratio for CVD is relative to a one-unit change in BMI (recall BMI is measured as the ratio of weight in kilograms to height in meters squared). A one-unit increase in BMI is associated with a 2.3% increase in the expected hazard. To facilitate interpretation, suppose we create three categories of weight defined by participant's BMI. Normal weight is defined as BMI < 25.0, overweight as BMI between 25 and 29.9, and obese as BMI exceeding 29.9. In the sample, there are 1651 participants (42%) who meet the definition of normal weight, 1648 (42%) who meet the definition of overweight, and 638 (16%) who meet the definition of obese. The numbers of CVD events in each of the three groups are shown in **Table 11-20**.

The incidence of CVD is higher in participants classified as overweight and obese as compared to participants of normal weight. We now use Cox proportional hazards regression analysis to make maximum use of the data on all participants in the study. **Table 11-21** displays the parameter estimates, *p*-values, hazard ratios, and 95% confidence intervals for the hazards ratios when we consider the weight groups alone (unadjusted model), when we adjust for age and sex, and when we adjust for age, sex, and other known clinical risk factors for incident CVD. The latter two models are multivariable models and are performed to assess the association between weight and incident CVD adjusting for confounders. Because we have three weight groups, we need two dummy variables

TABLE 11–20 Incidence of Cardiovascular Disease by Body Mass Index Groups

Group	Number of Participants	Number of CVD Events
Normal Weight	1651	202 (12.2%)
Overweight	1648	241 (14.6%)
Obese	638	100 (15.7%)

TABLE 11–21 Parameter Estimates Relating Body Mass Index to Cardiovascular Disease

	Overweight			Obese		
Model	Parameter Estimate	p	HR (95% CI for HR)	Parameter Estimate	p	HR (95% CI for HR)
Unadjusted or crude model	0.19484	0.0411	1.215 (1.008–1.465)	0.27030	0.0271	1.310 (1.031–1.665)
Age and sex adjusted	0.06525	0.5038	1.067 (0.882–1.292)	0.28960	0.0188	1.336 (1.049–1.701)
Adjusted for clinical risk factors*	0.07503	0.4446	1.078 (0.889–1.307)	0.24944	0.0485	1.283 (1.002–1.644)

* Adjusted for age, sex, systolic blood pressure, treatment for hypertension, current smoking status, and total serum cholesterol.

or indicator variables to represent the three groups. In the models we include the indicators for overweight and obese, and consider normal weight the reference group.

In the unadjusted model, there is an increased risk of CVD in overweight participants as compared to normal weight and in obese participants as compared to normal weight (hazard ratios of 1.215 and 1.310, respectively). However, after adjustment for age and sex, there is no statistically significant difference between overweight and normal-weight participants in terms of CVD risk (hazard ratio = 1.067, $p = 0.5038$). The same is true in the model adjusting for age, sex, and the clinical risk factors. Even after adjustment, the difference in CVD risk between obese and normal-weight participants remains statistically significant, with approximately a 30% increase in risk of CVD among obese participants as compared to participants of normal weight.

11.5 EXTENSIONS

In Section 11.4 we provided a brief outline of the general approach to analyzing the effects of multiple risk factors or predictors, considered simultaneously, on survival time. There are a number of important extensions of the approach that are beyond the scope of this text. For example, we considered the effect of risk factors measured at the beginning of the study period, or at baseline, but there are many applications where the risk factors or predictors change in value over time. Suppose we wish to assess the impact of exposure to nicotine and alcohol during pregnancy on time to preterm delivery. Smoking and alcohol consumption may change during the course of pregnancy. These predictors are called time-dependent covariates and they can be incorporated into survival analysis models. The Cox proportional hazards regression model with time-dependent covariates takes the following form: $\ln\{h(t)/h_0(t)\} = b_1 X_1(t) + b_2 X_2(t) + \cdots + b_p X_p(t)$. Notice that each of the predictors $X_1, X_2, \ldots, X_p$ now has a time component. Many predictors, such as sex and race, are independent of time. Survival analysis models can include both time-dependent and time-independent predictors simultaneously. Many statistical computing packages (e.g., SAS[13]) offer options for the inclusion of time dependent covariates. A difficult aspect of the analysis of time-dependent covariates is the appropriate measurement and management of these data for inclusion in the models.

Second, a very important assumption for the appropriate use of the log-rank test and the Cox proportional hazards regression model is the proportionality assumption. Specifically, we assume that the hazards are proportional over time, which implies that the effect of a risk factor is constant over time. There are several approaches to assess the proportionality assumption; some are based on statistical tests, and others involve graphical assessments. In the statistical testing approach, predictor-by-time interaction effects are included in the model and are tested for statistical significance. If one (or more) of the predictor-by-time interactions reaches statistical significance (e.g., $p < 0.05$), then the assumption of proportionality is violated. An alternative approach to assessing proportionality is through graphical analysis. There are several graphical displays that can be used to assess whether the proportional hazards assumption is reasonable. These are often based on residuals and examination of trends (or lack thereof) over time. More details can be found in Hosmer and Lemeshow.[1] If either a statistical test or a graphical analysis suggests that the hazards are not proportional over time, then the Cox proportional hazards model presented in Section 11.4 is not appropriate, and adjustments must be made to account for nonproportionality. One approach is to stratify the data into groups such that the hazards are proportional within groups and different baseline hazards are estimated in each stratum (as opposed to a single baseline hazard, as was the case for the model presented in Section 11.4). Many statistical computing packages offer this option.

A third issue is related to competing risks. A competing-risks situation is one in which there are several possible outcome events of interest. For example, a prospective study may be conducted to assess risk factors for time-to-incident cardiovascular disease. Cardiovascular disease includes myocardial infarction, coronary heart disease, coronary insufficiency, and many other conditions. The investigator measures whether each of the component outcomes occurs during the observation period as well as the time to each distinct event. The goal of the analysis is to determine the risk factors for each specific outcome, when the outcomes are correlated. Interested readers should see Kalbfleisch and Prentice[11] for more details.

11.6 SUMMARY

Time-to-event data, or survival data, are frequently measured in studies of important medical and public health issues. Because of the unique features of survival data, most specifically the presence of censoring, special statistical procedures are necessary to analyze these data. In survival analysis applications, estimation of the survival function, or survival probabilities over time, is often desired. Several techniques are available; here we presented two popular nonparametric techniques: the life table or actuarial table approach and the Kaplan–Meier approach to constructing cohort life tables or follow-up life tables. Both approaches generate estimates of the survival function that can be used to estimate the probability that a participant survives to a specific time (e.g., 5 years or 10 years). The notation and template for each approach are summarized in **Table 11–22** and **Table 11–23**.

TABLE 11–22 Actuarial, Follow-Up Life Table Approach

Time Intervals	Number at Risk During Interval, N_t	Average Number at Risk During Interval, $N_{t^*} = N_t - C_t/2$	Number of Deaths During Interval, D_t	Lost to Follow-Up, C_t	Proportion Dying, $q_t = D_t/N_{t^*}$	Proportion Surviving, $p_t = 1 - q_t$	Survival Probability, $S_{t+1} = p_{t+1} \times S_t$ ($S_0 = 1$)

TABLE 11–23 Kaplan–Meier Approach

Time	Number at Risk, N_t	Number of Deaths, D_t	Number Censored, C_t	Survival Probability, $S_{t+1} = S_t \times ((N_{t+1} - D_{t+1})/N_{t+1})$ ($S_0 = 1$)

Assessing statistically significant differences in survival between groups is an important element of statistical analysis, whether between competing treatment groups in a clinical trial, between men and women, or between patients with and without a specific risk factor in an observational study. Many statistical tests are available; here we presented the log-rank test, a popular nonparametric test. It makes no assumptions about the survival distributions, and it can be conducted relatively easily using life tables based on the Kaplan–Meier approach.

There are several variations of the log-rank statistic as well as other tests to compare survival curves between independent groups. We use the following test statistic, which follows a χ^2 distribution with degrees of freedom (df) $k - 1$, where k represents the number of independent comparison groups:

$$\chi^2 = \sum \frac{\left(\sum O_{jt} - \sum E_{jt}\right)^2}{\sum E_{jt}}$$

where $\sum O_{jt}$ represents the sum of the observed number of events in the jth group over time and $\sum E_{jt}$ represents the sum of the expected number of events in the jth group over time. The observed and expected numbers of events are computed for each event time and are summed for each comparison group over time.

To compute the log-rank test statistic, we compute, for each event time t, the number at risk in each group N_{jt} (where j indicates the group) and the observed number of events O_{jt} in each group. We then sum the number at risk N_t in each group over time to produce $\sum N_{jt}$ at each event time and sum the number of observed events O_t in each group over time to produce $\sum O_{jt}$ at each event time, and compute the expected number of events in each group using $E_{jt} = N_{jt} \times (O_t/N_t)$ at each time. The expected numbers of events at each time are then summed over time to produce $\sum E_{jt}$ for each group.

Finally, there are many applications in which it is of interest to estimate the effect of several risk factors, considered simultaneously, on survival. Cox proportional hazards regression analysis is a popular multivariable technique for this purpose. The Cox proportional hazards regression model is as follows:

$$h(t) = h_0(t)\exp(b_1 X_1 + b_2 X_2 + \cdots + b_p X_p),$$

where $h(t)$ is the expected hazard at time t and $h_0(t)$ is the baseline hazard and represents the hazard when all of the predictors $X_1, X_2 \ldots, X_p$ are equal to zero. The associations between risk factors and survival time in a Cox proportional hazards model are often summarized by hazard ratios. The hazard ratio for a dichotomous risk factor (e.g., treatment assignment in a clinical trial or prevalent diabetes in an observational study) represents the increase or decrease in the hazard in one group as compared to the other. For example, in a clinical trial with survival time as the outcome, if the hazard ratio is 0.5 comparing participants on a treatment to those on placebo, this suggests that participants in the treatment group have 50% hazard (risk of failure assuming the person survived to a certain point) as compared to participants in the the placebo group. In an observational study with survival time as the outcome, if the hazard ratio is 1.25 comparing participants with prevalent diabetes to those free of diabetes, then participants with diabetes have 1.25 times the risk of death as compared to participants without diabetes. Here we provided only a general description of the Cox proportional hazards regression procedure. There are many more details, and interested readers should consult one of the many very good references on the topic for further details.[1,2,3,12]

11.7 PRACTICE PROBLEMS

1. A study is conducted to estimate survival in patients following kidney transplant. Key factors that adversely affect success of the transplant include advanced age and diabetes. This study involves 25 participants who are 65 years of age and older, and all have diabetes. Following transplant, each participant is followed for up to 10 years. The following are times to death, in years, or the time to last contact (at which time participant was known to be alive).

   Deaths: 1.2, 2.5, 4.3, 5.6, 6.7, 7.3, and 8.1 years
   Alive: 3.4, 4.1, 4.2, 5.7, 5.9, 6.3, 6.4, 6.5, 7.3,
              8.2, 8.6, 8.9, 9.4, 9.5, 10, 10, 10, and 10 years

   a. Use the life-table approach to estimate the survival function.
   b. Use the Kaplan–Meier approach to estimate the survival function.
   c. Sketch the survival function based on the estimates in (b).

2. A clinical trial is run to assess the effectiveness of a new antiarrhythmic drug designed to prevent atrial fibrillation (AF). Thirty participants enroll in the trial and are randomized to receive the new drug or placebo. The primary outcome is AF, and participants are followed for up to 12 months following randomization. The experiences of participants in each arm of the trial are shown in **Table 11–24**.
   a. Estimate the survival functions for each treatment group using the Kaplan–Meier approach.
   b. Test whether there is a significant difference in survival between treatment groups using the log-rank test and a 5% level of significance.

TABLE 11–24 Months to Atrial Fibrillation or Last Contact by Treatment Group

Placebo		New Drug	
Month of AF	Month of Last Contact	Month of AF	Month of Last Contact
4	5	7	6
6	5	9	6
7	6	12	7
8	7		8
9	8		9
11	9		9
	11		10
	12		10
	12		11
			11
			12
			12

TABLE 11–25 Weeks to Joint Failure by Obesity Status

Obese		Not Obese	
Failure	No Failure	Failure	No Failure
28	39	27	37
25	41	31	36
31	37	34	39
32	35		40
	38		36
	36		32
	29		39
			41

3. Using the data in Problem 2, compute and interpret the hazard ratio for atrial fibrillation with the new anti-arrhythmic drug as compared to placebo.

4. Sketch the survival functions for the new anti-arrhythmic drug group and the placebo group using the results from Problem 2.

5. An observational cohort study is conducted to compare time to early failure in patients undergoing joint replacement surgery. Of specific interest is whether there is a difference in time to early failure between patients who are considered obese versus those who are not. The study is run for 40 weeks, and times to early joint failure, measured in weeks, are shown in **Table 11–25** for participants classified as obese or not at the time of surgery.
   a. Estimate the survival functions (time to early joint failure) for each group using the Kaplan–Meier approach.
   b. Test whether there is a significant difference in time to early joint failure between obese and non-obese patients undergoing joint-replacement surgery using the log-rank test and a 5% level of significance.

6. Using the data in Problem 5, compute and interpret the hazard ratio for early joint failure between obese and non-obese patients undergoing joint replacement surgery.

7. Sketch the survival functions for each group (obese and non-obese) using the results from Problem 5.

8. A study of patients with stage I breast cancer is run to assess time to progression to stage II over an observation period of 15 years. Of interest is whether there is a difference in time to progression between women on two different chemotherapy regimens. Times to progression are measured in years from the time at which the chemotherapy regimen was initiated. See **Table 11–26**.

TABLE 11–26 Years to Progression by Chemotherapy Regimen

Regimen 1		Regimen 2	
Progression	No Progression	Progression	No Progression
2	12	9	11
6	14	4	14
7	13	7	13
3	11		9
4	15		14
	10		13
	8		6
	6		9
	9		7
	12		

a. Estimate the survival functions (time to progression) for each chemotherapy regimen using the Kaplan–Meier approach.
b. Test whether there is a significant difference in time to progression between treatment regimens using the log-rank test and a 5% level of significance.

9. A Cox proportional hazards model is estimated relating time to psychiatric hospitalization in patients with severe mental illness. The risk factors include age, sex, prior hospitalization for mental illness, and an indicator of bipolar disorder. The parameter estimates and significance levels for the model are shown in **Table 11–27**.
    a. Which of the predictors are statistically significantly associated with time to psychiatric hospitalization?
    b. Are men or women more likely to be hospitalized? Justify your answer.
    c. Compute hazard ratios for each of the predictors, and provide an interpretation for each.

TABLE 11–27 Risk Factors for Psychiatric Hospitalization

Predictor	Parameter Estimate	p
Age, years	0.0045	0.5647
Sex (0 = female, 1 = male)	–0.4841	0.0341
Prior hospitalization (0 = no, 1 = yes)	0.3726	0.0178
Bipolar disorder (0 = no, 1 = yes)	0.7561	0.0042

10. A clinical trial is conducted to evaluate the efficacy of a new drug for prevention of hypertension in patients with prehypertension (defined as systolic blood pressure between 120 mmHg and 139 mmHg or diastolic blood pressure between 80 mmHg and 89 mmHg). A total of 20 patients are randomized to receive the new drug or a currently available drug for treatment of high blood pressure. Participants are followed for up to 12 months, and time to progression to hypertension is measured. The experiences of participants in each arm of the trial are shown in **Table 11–28**.
    a. Estimate the survival functions (time to progression to hypertension) for each treatment group using the Kaplan–Meier approach.
    b. Test whether there is a significant difference in time to progression between treatment groups using the log-rank test and a 5% level of significance.
    c. Compute and interpret the hazard ratio for time to progression comparing treatment groups.

11. A Cox proportional hazards model is estimated, relating time to development of myocardial infarction (heart attack) over 10 years to high diastolic blood pressure (DBP), defined as DBP $\geq$ 90 mmHg. The unadjusted association is estimated first, then the association is adjusted for age and sex, and then for additional clinical risk factors. The parameter estimates and significance levels for diastolic blood pressure for each model are shown in **Table 11–29**.
    a. Is there a statistically significant association between high DBP and time to MI? Discuss significance in each of the models.

TABLE 11–28 Months to Progression to Hypertension by Treatment Group

New Drug		Currently Available Drug	
Hypertension	Free of Hypertension	Hypertension	Free of Hypertension
7	8	6	8
8	8	7	9
10	8	9	11
	9	10	11
	11	11	12
	12		
	12		

CHAPTER 11 Survival Analysis

TABLE 11–29 Associations Between Diastolic Blood Pressure and Heart Attack

Model	Parameter Estimate	p
Crude effect of high DBP	0.586	0.0001
Effect of high DBP adjusted for age and sex	0.462	0.0012
Effect of high DBP adjusted for age, sex, smoking, total serum cholesterol, and diabetes	0.217	0.0001
Effect of high DBP adjusted for age, sex, smoking, total serum cholesterol, diabetes, and treatment for hypertension	0.059	0.1233

b. What is the unadjusted hazard ratio for MI in participants with high versus low DBP?
c. What is the age- and sex-adjusted hazard ratio for MI in participants with high versus low DBP?
d. Compute the hazard ratios, adjusting for the additional clinical risk factors, and compare them to the unadjusted hazard ratio from (b). Are there differences? If so, what would account for the differences?

12. **Figure 11–11** shows survival probabilities from all-cause mortality over 20 years in participants over 50 years of age who have normal glucose, glucose intolerance, and Type II diabetes mellitus.
    a. Estimate the 5-year survival probability for each group.
    b. Estimate the median survival for each group.

13. The data in **Table 11–30** reflect the time to first surgery in children born with congenital heart disease. Time is measured in years from birth up until the age of 10 years. Construct a life table using the Kaplan–Meier approach. Also include standard errors and 95% confidence limits for the estimates of survival probability.

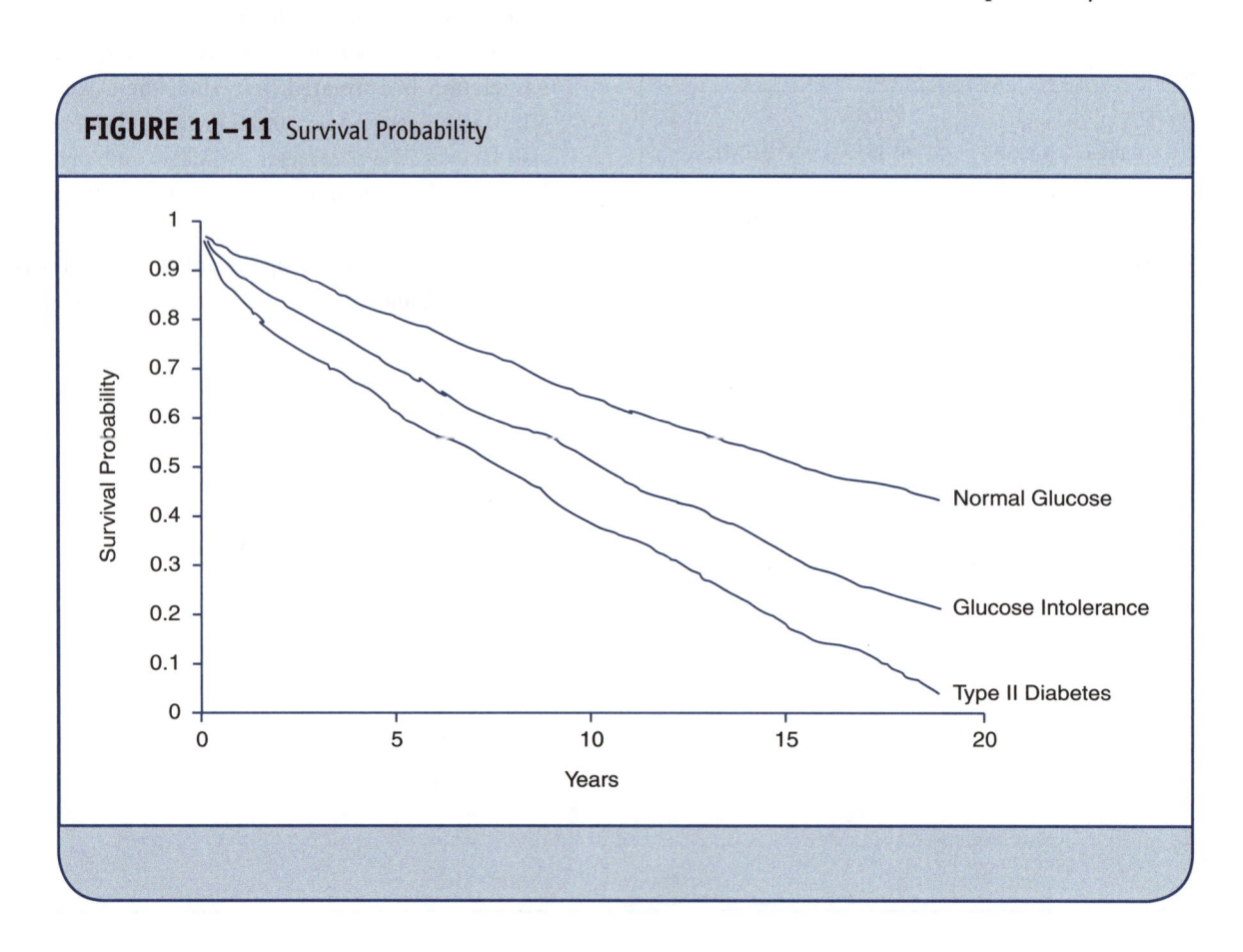

FIGURE 11–11 Survival Probability

TABLE 11–30 Year of First Surgery or Year of Last Contact

Participant Identification Number	Year of First Surgery	Year of Last Contact
1	8	
2		10
3		4
4	4	
5		7
6		6
7		9
8		5
9	3	
10		8
11		9
12		10
13		3
14	2	
15	6	
16		7
17		8
18		9

REFERENCES

1. Hosmer, D.W. and Lemeshow, S. *Applied Survival Analysis: Regression Modeling of Time to Event Data*. New York: John Wiley & Sons, 1999.
2. Cox, D.R. and Oakes, D. *Analysis of Survival Data*. Boca Raton, FL: Chapman and Hall, 1984.
3. Lee, E.T. and Wang, J.W. *Statistical Methods for Survival Data Analysis* (3rd ed.). New York: John Wiley & Sons, 2003.
4. SAS version 9.1 © 2002–2003 by SAS Institute, Inc., Cary, NC.
5. Greenwood, M., Jr. "The natural duration of cancer." *Reports of Public Health and Related Subjects, Vol. 33*. London: HMSO, 1926.
6. SAS version 9.1. © 2002–2003 by SAS Institute, Inc., Cary, NC.
7. Crawley, M.J. *Statistics: An Introduction Using R*. New York: John Wiley & Sons, 2005.
8. Mantel, N. "Evaluation of survival data and two new rank-order statistics arising in its consideration." *Cancer Chemotherapy Reports* 1966; 50 (3): 163–170.
9. Peto, R. and Peto, J. "Asymptotically efficient rank invariant test procedures." *Journal of the Royal Statistical Society, Series A (General)* 1972; 135(2): 185–207.
10. Gehan, E.A. "A generalized Wilcoxon test for comparing arbitrarily singly-censored samples." *Biometrika* 1965; 52: 203–223.
11. Kalbfleisch, J.D. and Prentice, R.L. *The Statistical Analysis of Failure Time Data* (2nd ed.). New York: Wiley, 2002.
12. Kleinbaun, D.G. and Klein, M. *Survival Analysis: A Self-Learning Text*. New York: Springer Science + Business Median, Inc., 2005.
13. Allison, P. *Survival Analysis Using the SAS System*. Cary, NC: SAS Institute, 1995.

CHAPTER 12

Data Visualization

WHEN AND WHY

Key Questions

- How can you tell if someone is trying to hide or exaggerate a message in a graphic?
- Are there certain components that should always be included in a good table of data or statistical results?
- How do you construct a graph that not only grabs attention but also is statistically sound?

In the News

Soda consumption has been shown to increase the risk of developing metabolic syndrome, which is a cluster of risk factors for diabetes and cardiovascular disease that includes waist circumference, blood pressure, triglycerides, high-density lipoprotein cholesterol, and glucose level.[1]

Sugar-sweetened beverages have been shown to increase the risks of obesity, type 2 diabetes, and cardiovascular disease.[2] Consumption of sugar-sweetened beverages has increased dramatically, with teenagers consuming more of these beverages than any other age group. Consumption varies by race/ethnicity and is higher among those groups with low income. Consumers are also more likely to partake of these beverages away from home than when they are at home.[3]

Dig In

You are asked to consult on a campaign for a local school to address consumption of sugar-sweetened beverages.
- How might you frame this problem for students, teachers, and parents?

---

[1] Dhingra, R., Sullivan, L., Jacques, P.F., et al. "Soft drink consumption and risk of developing cardiometabolic risk factors and the metabolic syndrome in middle-aged adults in the community." *Circulation* 2007; 116(5): 480–488.
[2] Malik, V.S., Popkin, B.M., Bray, G.A., Despres, J.P., and Hu, F.B. "Sugar-sweetened beverages, obesity, type 2 diabetes mellitus, and cardiovascular disease risk." *Circulation* 2010; 121(11): 1356–1364.
[3] Centers for Disease Control and Prevention, National Center for Health Statistics. Consumption of sugar drinks in the United States, 2005–2008. Available at *http://www.cdc.gov/nchs/data/databriefs/db71.htm*

- Which data or statistical results would you need?
- Would you use tables and figures differently for the different audiences?

LEARNING OBJECTIVES

By the end of this chapter, the reader will be able to:

- Determine when data and statistical results are best displayed in text, tables, and figures.
- Identify key design features of effective tables and figures.
- Create tables to display data and statistical results.
- Create figures to display data and statistical results.
- Apply sound statistical principles in all manners of communicating data and statistical results.

Strong evidence is needed to guide changes that will improve population health. We must design studies efficiently, collect and organize data carefully, and report findings clearly and accurately. This chapter focuses on the last aspect of this process—the reporting of data and statistical results.

Clearly communicating data and statistical results is essential, and there are many ways in which it can be done. The options available include, but are not limited to, text, tables, and figures. The most appropriate format depends on our understanding of the data and the results, the reasons why they matter, the message we are trying to convey, and the people to whom that message should be addressed (i.e., the audience). Clearly written text and well-designed tables and figures can enhance understanding; conversely, poorly written text and poorly designed tables and figures can create confusion. Text, tables, and figures should be used together to convey results in the clearest possible way, and there must be consistency and synergy across the various formats to best serve the readers and consumers of the information. Some results are best presented in text format, whereas others are best tabulated. Figures and graphical displays are powerful communication tools that can be used to present complex relationships concisely, illustrate patterns and trends, cut down on the length of a report or publication, and enhance the reader's understanding of the results—but only if they are of high quality and carefully constructed.

Effective communication of statistical results is an essential element of any statistical analysis but requires skillful execution on the author's part. To guide researchers in reporting their results, the American Statistical Association has developed "Ethical Guidelines for Statistical Practice" that include eight general topic areas and ethical considerations for each area.[1] The guidelines include, for example, professionalism and responsibilities to human subjects as well as responsibilities to "report sufficient information to give readers, including other practitioners, a clear understanding of the intent of the work, how and by whom it was performed, and any limitations on its validity." The reporting guidelines describe 15 specific points that should be considered when communicating data and statistical results. Thus, ethical statistical practice must include clear, fair, and consistent reporting of data and statistical results.

It is important for statisticians not only to communicate data and statistical results to other statisticians (which is critical to ensure that results are reproducible), but also—and perhaps even more important—to communicate effectively to our non-statistician colleagues and collaborators, as they are often in a position to make changes to improve population health based on empirical evidence.

12.1 DESIGN PRINCIPLES

Before we delve into specific design principles for effective tables and figures, we must first consider one of the several critical questions that guides our process: How will we engage with our audience? If we are presenting results orally to an audience, we must move quickly to conclusions. In an oral presentation, audience members cannot digest every detail or step in an analysis—they need to understand the question of interest, its importance, and our approach to the analysis. The frame in such presentations is wider—that is, the bigger picture. In contrast, if we are distributing a written report of results, we have an opportunity to provide much more detail, as readers can move through the material at their own pace. That said, we must still stay on task and not overwhelm readers with inconsequential data. At the same time, we must provide readers with the necessary and sufficient information to make appropriate comparisons and to draw valid conclusions. Once we have a general approach in mind, we next move to specifics.

A number of excellent resources are available to guide the development of effective tables and figures. Perhaps the most widely recognized of these resources are from Edward E. Tufte, who has published several books on the topic and also offers many online resources.[2,3] Tufte promotes several principles that are generally targeted toward the development of figures or graphical displays, but these principles also apply to the development of tables. He urges the developer to "provoke thought about the subject at hand," to motivate readers or consumers to want to know more, and to become so interested that they want to share the information. It must be clearly evident to readers or consumers that the data or statistical results are important: The reader needs to care about what is being shown and want to engage. When we are developing tables or figures to summarize data and statistical results to tell a story, we need to start our story with a high-level, big-picture view, then move into more detail.

Tufte urges honesty in data visualization and clarity of presentation. One means of achieving this is by minimizing what he calls "chart junk." Chart junk refers to unnecessary features in a table or figure that do not convey any meaning. Tufte argues that chart junk detracts from the main message of the table or figure. In contrast, he favors data presentations that encourage comparison as a specific means of engaging with data and statistical results. To facilitate this type of engagement, tables and figures must be set up in such a way that they provide readers or consumers with the right data and in sufficient detail to make valid comparisons, and to discover meaning as they explore the data or statistical results.

Another set of principles that is aimed at graphical displays, but is also applicable to tabular displays, is the CONVINCE principles for effective data visualization.[4] The CONVINCE principles are consistent with those set forth by Tufte but reinforce key issues and are easy to remember by the clever acronym:

- <u>C</u>onvey meaning, with clear labeling of titles, axes, legends, variables, rows, and columns.
- <u>O</u>bjectivity in presentation, which is maximized with fair scaling of axes so that data and statistical results are not hidden or exaggerated.
- <u>N</u>ecessity of all table and figure elements—that is, avoiding clutter and unnecessary items in tables and figures that detract from the message at hand, consistent with Tufte's recommendation to minimize chart junk.

- Visual honesty—again calling for appropriate use of specific table and chart types, along with scaling and labeling that present data in a fair and balanced way.
- Imagining the audience—reinforcing the notion that we, as the developers of tables and figures, must know our audience, what they need to know, and how we might best meet their needs through our delivery.
- Nimble—a specific principle that perhaps is most salient to graphical displays, through which readers or consumers are able to access more data or additional levels of detail should they need or wish to delve deeper. This is perhaps best achieved through interactive graphics.
- Context—a principle that overlaps with Tufte's principles, urging us to provide readers or consumers with sufficient background to understand the question of interest and its importance, and then appropriate data and statistical results so that they may draw valid conclusions.
- Encourage interaction—engaging readers and consumers to care about the data, and to explore more deeply.

A third set of principles comes from Steven Few, who suggests eight core ideals or principles for data visualization.[5] Again, Few's principles are aimed toward figures but can be generally applied to both tables and figures. He urges us to:

- *Simplify data and statistical results for presentation.* This is always a delicate balance, as we never want to oversimplify data so that important nuances are lost, but rather find the point at which those nuances are clear and the data are digestible.
- *Ask.* Few encourages us to create tables and figures that encourage the reader to want to know more, to ask where the data come from, to consider why we are investigating this particular question or issue, and ideally to think about which actions might be taken based on what we communicate.
- *Compare.* We must develop tables and figures that allow readers to make valid comparisons so that they can interpret differences, effects, and impact.
- *Explore.* We must develop tables and figures that allow readers to explore the data and to discover things on their own.
- *Attend.* This principle refers to our need to understand the readers or consumers of the data or statistical results so that we can most effectively communicate our results. What does our audience need to see to understand the data or statistical results? How best do we communicate our findings to meet their needs? We must highlight what is truly important and ensure that key messages are clearly visible and interpretable, and not lost in cluttered tables or overly complicated figures.
- *Ensure diversity.* Few suggests that data should be presented from different viewpoints to encourage diversity of interpretations—an outcome perhaps most easily achieved through interactive graphical displays as opposed to tabular displays, but nonetheless important to bear in mind in all presentations of data and statistical results. We need to present data and statistical results that are widely accessible to diverse audiences. We must consider the literacy levels, educational backgrounds, and different perspectives of our audience. We might, for example, take a different approach in preparing tables or figures for a research article as compared to a presentation to a community group.
- *Be skeptical.* As both creators and consumers of data and statistical results, we should question what we see. Where did the data come from? Which inferences are appropriate? What are the limitations of the data?
- *Respond.* The notion here relates to sharing data and statistical results. As biostatisticians, we certainly focus on the collection, management, and analysis of data. A key and required additional aspect of our work is the communication of results. In fact, it is our responsibility to share results as outlined in the "Ethical Guidelines for Statistical Practice."[1]

Before we delve even deeper into the development of effective tables and figures, several key messages are worth repeating. No matter which format we choose to report our data and statistical results, we need to make a case as to why the data and statistical results are important—that is, why the reader should care. We must highlight the important elements, be clear in our process (which steps we took and why), to ensure that the reader or consumer can focus on the most critical aspects of the data and statistical results. Our end goal is to make sure our message is accurate, clear, and memorable.

12.2 WHEN AND HOW TO USE TEXT, TABLES, AND FIGURES

Before we decide on the best format to communicate results, we must consider the audience and how we will engage with them. What do they need to know? How will they use the information? What is the best way to reach them? Do we have the opportunity to present and discuss the results, and if so, how much time, space, and flexibility will we have? Or will we provide a stand-alone document or report summarizing results? All of these questions must be considered to determine the best approach for communicating data and statistical results.

Some general guidelines for choosing among text, tables, and figures or graphical displays are offered in this section, based on "Tips on Effective Use of Tables and Figures in Research Papers," by Velany Rodriguez.[6] Regardless of the format used to present data—and often we use multiple formats within the same presentation or report—we must be consistent in the language and terminology that we use.

A text format is suitable for straightforward and limited results—say, three to four numbers. If data, when put into a table, would result in a table with only a few rows and columns, we are best served by reporting those data in text format. Another reason to use text, rather than a table or a figure, is to report results that are not central to the main aims of the investigation. The results of the primary analysis should generally be presented in a table or figure, as readers often turn immediately to tables and figures in a paper or report before reading the text. Ancillary data—that is, data that enhance the reader's understanding or connect the reader to additional resources—should be summarized in the text.

As a rule of thumb, when integers (whole numbers) are presented in the text, numbers less than 10 should be presented in words (e.g., "nine") and numbers 10 and higher should be written as integers (e.g., "14"). The number of digits or decimal places presented should be consistent with the precision of the measure and its purpose. Usually, summary statistics are presented to one more decimal place than the measured values. For example, suppose we measure weight in pounds for each participant in a study of a new weight-loss intervention, and the measurements of weight are to the nearest pound (integer values). To summarize the weights in the study sample, we might present the mean weight and the standard deviation in weight. These data would be presented to the nearest tenths place (e.g., $\overline{X} = 146.2$ pounds and $s = 9.7$ pounds). The number of decimal places should also be consistent, where possible, for all variables shown in the same table. For example, in this same study we might report the mean age of the participants (e.g., $\overline{X} = 53.7$ years) along with some basic clinical measures such as their mean systolic and diastolic blood pressures, and their mean total serum cholesterol levels. All of the summary statistics, in this case, should be reported to the tenths place, or one decimal place. The exception might be a mean of 0.003: We would not report a mean of 0.0 for consistency, but rather 0.003 so that the detail is not lost.

When presenting summary statistics, it is good practice to present a point estimate (e.g., mean, proportion, difference in means, relative risk) along with a measure of variability (e.g., standard deviation, standard error, confidence interval). When summarizing the distribution of a continuous measure, the sample mean is useful along with the standard deviation, assuming that there are no outliers. If there are outliers, then the median and interquartile range are appropriate. When there are no outliers, the standard deviation gives a sense of the variability of the measure in the study sample. For example, if we measure systolic blood pressure (SBP) in millimeters of mercury (mm Hg) in a sample of $n = 75$ patients, we might report a mean SBP of 123.4 mm Hg and a standard deviation of 14.3 mm Hg. If we intend to estimate the mean SBP in a population or generalize our findings, based on our sample of $n = 75$ patients, we might report a point estimate of 123.4 mm Hg and a standard error of $14.3/\sqrt{75} = 1.7$ mm Hg. The standard error is a measure of the variability of the estimate—in this case, the sample mean—sometimes referred to as a measure of sampling variability. Some papers and reports present summary statistics for a continuous variable such as SBP as follows: 123.4 ± 14.3 mm Hg or 123.4 ± 1.7 mm Hg. It is clearer to indicate what is presented in words instead of using the "±" sign, however, as sometimes people show different measures following the "±" sign. The use of the plus/minus sign does not enhance the interpretability of the presentation. However, if this or another symbol is used in text (or in a table), the statistics presented must be indicated (e.g., mean ± std dev, 123.4 ± 14.3 mm Hg).

When presenting data and statistical results, clear and concise writing is essential. Writing is a critical skill that requires planning, drafting, proofing, and editing; it is learned and improved upon through practice and feedback. Writing that communicates statistical results, and other technical writing, requires specific skills including clarity, consistency, and brevity. Technical writing is not reflective or creative writing, but rather focuses on ensuring the clarity of technical details. The technical details must be precisely explained, and any jargon and abbreviations used carefully, thoughtfully, and minimally.

Tables are appropriate when we have many results or statistics to present and their exact values are important. Tables can summarize large amounts of data concisely (e.g., sets of numbers across multiple groups). When organized efficiently, a table often makes it easier for the reader or consumer to make appropriate comparisons and to digest the information, especially data that would be difficult to summarize in text format. As an example, suppose we are interested in monitoring the growth (weight gain) of infants born to mothers of different racial/ethnic backgrounds. Birth weights are measured as are weights at 1, 2, 3, 4, 5, and 6 months of age, all in grams. There are many weights to display, and their exact values are important to appreciate any differences that might exist among racial/ethnic groups. Weights could be presented in a table with mean weights and standard deviations or standard errors (depending on whether the goal is to summarize the distribution of weights in each group or to quantify the variability of the estimates of the mean weights in each group) for each racial/ethnic group at each time point.

Additionally, tables are useful when presenting dichotomous variables (presence or absence) of specific attributes or conditions, some of which are rare. For example, we might wish to report on comorbid conditions in a sample of people participating in a study of a new behavioral intervention for addiction. The sample might include $n = 65$ young adults between the ages of 20 and 39 years. Before evaluating the impact of the intervention, we wish to describe the participants and their medical histories. Given the ages of the participants, we observe few with comorbidities such as high blood pressure, high cholesterol levels, diabetes, heart disease, and cancer. Despite these conditions' low prevalence, it is important to understand the extent to which participants are affected by them. The small percentages would be difficult to present in text, however, and would get lost in a figure. When reporting these data in a table, it is important to indicate the absolute numbers of participants affected along with percentages—not just the percentages by themselves. For example, we might report that 4/65 (6.2%) of participants have high cholesterol levels (e.g., total serum cholesterol exceeding 240 mg/dL).

Each table should have a point to make. When necessary, tables can be combined or split to enhance interpretability. Regardless of what is included, each table must be constructed for a specific purpose (e.g., to describe the study sample, to compare groups on the primary outcome, to summarize risk factors associated with an outcome). The key point or points of every table (and every figure, for that matter) should be referenced in the paper or report in which it is embedded. Otherwise, the table (or figure) should not be included.

As mentioned earlier, readers often turn to tables and figures before reading all of the text in a paper or report. For this reason, tables and figures must be labeled and annotated appropriately to ensure that they are self-explanatory. Clear and concise labeling is essential so that the table can stand alone. The variables included in the table should be clearly defined, along with the units of measurement. The answers to questions about *what*, *where*, *when*, and *how* variables were measured should be addressed in the table. Lastly, if a table (or figure) has been adapted from another source, the original source should be referenced.

Figures are appropriate when we wish to illustrate complex relationships among variables that cannot be shown with tables or text, or to show trends or geographic variations. Many different kinds of figures are possible, ranging from very basic figures discussed in Chapter 4 (e.g., bar charts, histograms, box-whisker plots) to interactive infographics often encountered on websites, social media, news reports, and email communications.

Figures should be constructed very carefully, with the author ensuring that they are self-explanatory. All figures should have clear and descriptive titles, axes should be labeled carefully, and lines, markers, and legends used to strategically draw attention to key points. Figures can be very useful tools for making comparisons among groups, interventions, or both. When designing figures, we must also think about which comparisons are to be made and how the reader or consumer will make those comparisons.

For example, suppose we conduct a clinical trial to evaluate the efficacy of a nonsteroidal anti-inflammatory drug (NSAID) as compared to placebo in reducing pain intensity in $n = 80$ patients with lower back pain. All patients are assessed for pain intensity at baseline (the beginning of the study) on a scale of 0 to 10, with 0 representing no pain and 10 representing unbearable pain. Patients are randomized, in equal numbers, to receive either the NSAID or placebo, and pain intensity is measured again after 4 weeks on the assigned treatment. Suppose that patients receiving the NSAID show a change in pain intensity (computed by subtracting the pain intensity measured at 4 weeks from the baseline pain intensity) of 1.3 units, with a standard error of $3.7/\sqrt{40} = 0.6$ unit. Patients receiving the placebo show a change in pain intensity of 0.2 unit, with a standard error of $4.6/\sqrt{40} = 0.7$ unit. The difference in mean change in pain intensity between patients receiving the NSAID as compared to placebo is 1.1 units, with a 95% confidence interval of (−0.7 to 2.9) units in change in pain intensity. There is no statistical evidence of a difference in change in pain intensity over 4 weeks in patients receiving the NSAID as compared to placebo. Suppose that the investigators hypothesized a priori that there might be a difference in efficacy of the NSAID by patient sex and, consequently, balanced the groups by sex using a stratified randomization scheme. A graphical display showing the changes in pain intensity over 4 weeks in men and women receiving the NSAIDs versus placebo might be a good way to communicate the results.

Figures 12–1 and **12–2** display the information in two different ways. Which is more effective?

Figure 12–1 shows that there is little difference in the change in pain intensity among men who received the NSAID versus placebo (leftmost set of bars in Figure 12–1), but there is a substantial change in pain intensity in women who received the NSAID as compared to placebo (rightmost set of bars in Figure 12–1). Figure 12–2 shows exactly the same data organized differently to promote the comparison of men versus women within treatment. Among patients on the NSAID, the change in pain intensity is greater in women as compared to men, as shown in the leftmost set of bars in Figure 12–2. The rightmost set of bars in Figure 12–2 shows little change in pain intensity among both men and women who received placebo. Figure 12–1 makes it easier to compare the different treatments within the two sexes, whereas Figure 12–2 makes it easier to compare the responses for the two sexes within treatments.

Adjacent bars are easier for readers to compare than bars that are more distant. Thus, in constructing figures we must consider the comparisons of primary interest—they drive the format of the figure. For example, **Figure 12–3** displays the percent response in each of four comparison groups measured at 24 hours and 48 hours post treatment. Which comparison is easiest to make, given the way the figure is designed (i.e., does the display facilitate a comparison in response among groups at specific times or over time within groups)?

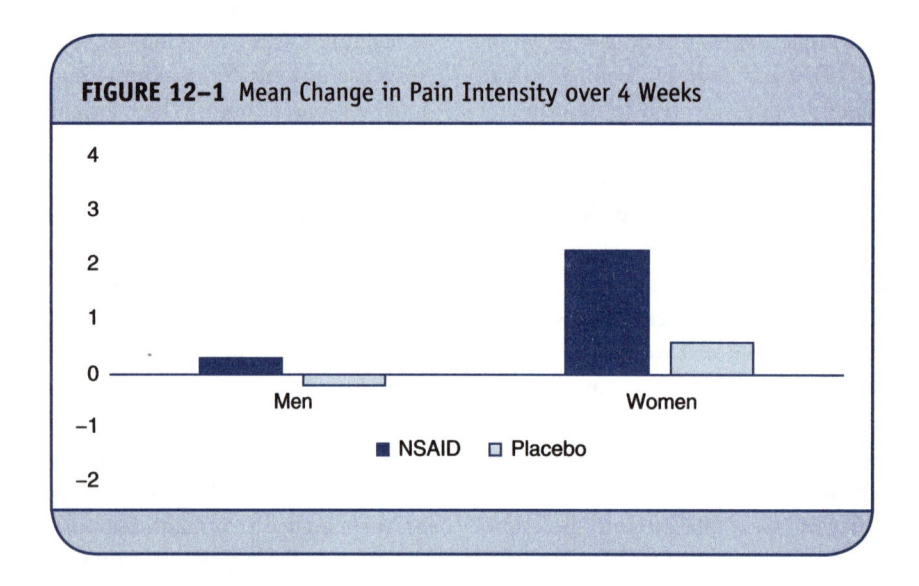

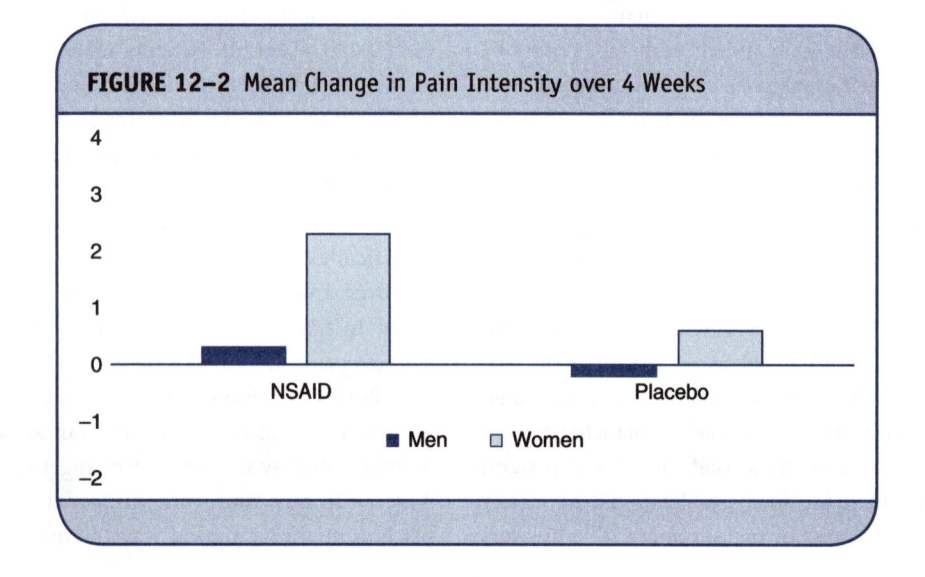

FIGURE 12-1 Mean Change in Pain Intensity over 4 Weeks

FIGURE 12-2 Mean Change in Pain Intensity over 4 Weeks

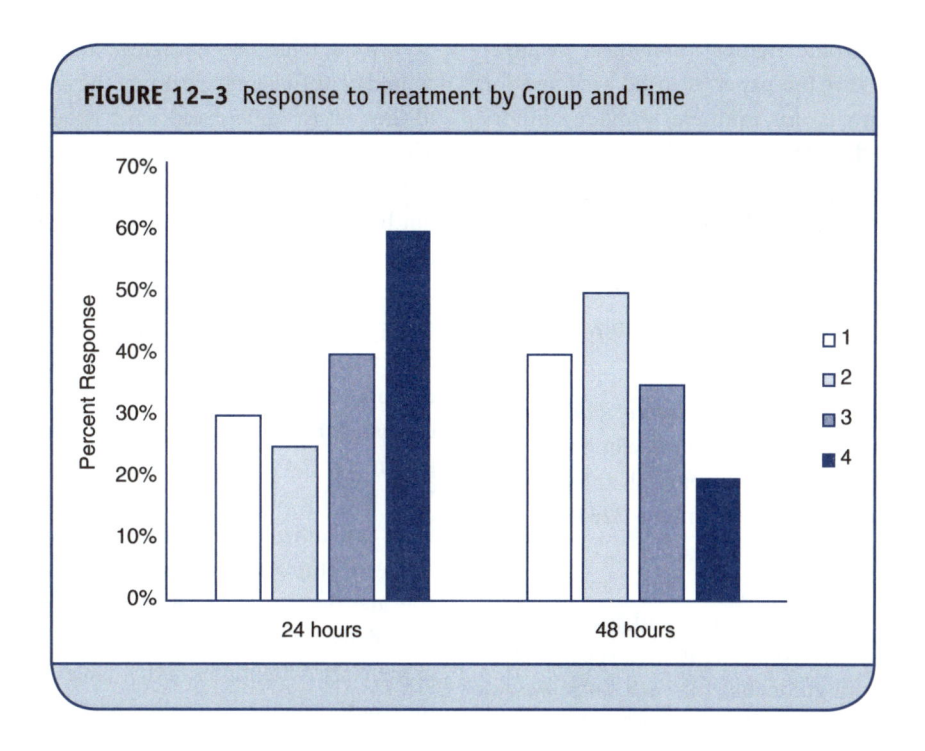

FIGURE 12–3 Response to Treatment by Group and Time

Regardless of which format we use to communicate data and statistical results, we must first decide what we wish to communicate and to whom. We can then choose the format that best communicates to the target audience while always considering the principles of good design.

12.3 PRESENTING DATA AND STATISTICAL RESULTS IN TABLES

When constructing tables, we must first consider the needs of the readers or consumers and how we can best meet their needs as we communicate data and statistical results. A good way to proceed is by thinking about the story we wish to tell. That story begins with the question of interest: Why did we undertake this study or investigation? What is known on the topic and where are the gaps in the knowledge base? The answer to the latter question is usually best presented in text form. Sometimes, however, there may be a substantial background or context to summarize, and we can tabulate key findings from prior studies to provide that context. Once the question or purpose is articulated, along with relevant background information, we can move to more detail on the specific sample we analyzed, what we measured, how we measured it, and what we found. As we map out each table in a presentation, paper, or report, we must think about the message that each table will convey. The presentation, paper, or report should move from the big picture to more granular details in a systematic fashion, and the tables should work together as a cohesive unit. At the same time, each table should stand on its own and have a unique point to make. Tables are effective ways to show data precisely and to facilitate comparisons. The specific comparisons of interest dictate how the tables should be organized and which statistics should be presented.

When presenting data and statistical results in tables to our audience, we must be cognizant of context, time, place, and situation. How will the audience engage with the presentation? Is the presentation designed to be interactive, or will all questions be held until the end? What are the expectations of the audience? Will the presentation be made face-to-face or delivered online? If the presentation is face-to-face, how is the room set up? How close will audience members be to the screen? All of these questions need to be considered as we set up tables for a presentation. There are many challenges in communicating effectively with an audience that we must overcome, so we do not want to introduce additional barriers in the ways we construct our tables.

Journals, websites, publishers, and clients often have specific requirements for abstracts, posters, papers, and reports submitted to them, and we must always conform to their style guides and requirements when preparing tables to summarize data and statistical results. Journals publish papers in different categories including, but not limited to, original research, commentaries, essays, opinions, and systematic reviews. Different types of papers have specific requirements for tables (and figures), and these requirements must be followed if the paper is to be accepted for publication. Most journals, for example, provide authors with access to detailed style guides and instructions for preparing manuscripts for submission. Some of the relevant style guides for the kinds of publications that we might engage in are summarized here. Note that some of these resources are specific to a particular journal, whereas others are endorsed by specific professional organizations and apply across a number of journals.

- The style guide for the *American Journal of Public Health* outlines components of a manuscript along with specific requirements for each: http://ajph.aphapublications.org/userimages/ContentEditor/1432646399120/authorinstructions.pdf.
- The style guide endorsed by the American Statistical Association applies to its family of journals and includes manuscript components and specific style requirements and guidelines for each: http://www.amstat.org/publications/pubdump/ASASTYLE_GUIDE.PDF.
- The American Medical Association endorses a style guide that provides guidance for authors intending to publish in many medical journals,[7] although some medical and biomedical journals have their own specific guidelines. For example, *New England Journal of Medicine* and *Circulation*, the premier journal of the American Heart Association, require authors to conform with "Recommendations for the Conduct, Reporting, Editing and Publication of Scholarly Work in Medical Journals," published by the International Committee of Medical Journal Editors (http://www.icmje.org/).
- The American Psychological Association (APA) style is commonly used in the social sciences, and there are a number of online and for-purchase resources available from APA for potential authors.[8]

These guides lay out some very specific requirements, but also capture some general principles that are consistent with the ideas outlined in the previous section on design principles.

12.3.1 Components of a Table

Before we discuss specific types of tables and their contents, we first outline the components of a table. As a reminder, combining smaller tables or splitting larger tables is not generally recommended, as in the former case the message may get muddled and in the latter case the message may be lost.

Good tables have a clear and concise title. The title should be sufficiently clear that the reader need not refer back to the text to understand what is being shown in the table. The title should describe what is being shown among whom (i.e., key comparison groups) and when. It should generally not be more than two lines and is usually shown at the top of the table. Tables in a paper or report should also be numbered in the order in which they are referred to in the paper or report. Unless the journal, website, or client recommends otherwise, tables should be included in the paper or report in close proximity to where the data or statistical results are discussed, rather than at the end, which might make it challenging for the reader to refer to them while reading the text. Very lengthy tables with more granular detail can be moved to an appendix or, if available, an online supplement.

As we think about structuring a paper or report, we should consider the logical sequence of the tables. The main findings should always come before data or results that are secondary or ancillary. Oftentimes, a first table contains a description of the study participants or other background data to provide the context for the main findings, which soon follow.

The data (sample sizes, means, percentages, results of statistical tests) make up the body of the table. Columns are often organized within the table to establish the most important comparisons of the data. As we set up the structure of a table (sometimes called a table shell), we must consider the comparisons of interest. Data to be compared should be presented, if possible, in adjacent columns. For example, if we want to show improvement in performance on a standardized test before versus after an educational intervention, then the measurements taken before the intervention and after the intervention should be presented in adjacent columns to facilitate that comparison.

Each column in the table should also have a clear heading. The column headings should be sufficiently concise that they do not require multiple lines. One column heading is sufficient per column, as more than one heading can create confusion. A column spanner, which describes the grouping variable (e.g., "Treatment"), appears above the column headings. Sometimes the column headings are sufficient to describe and distinguish the groups, but at other times a column spanner is needed to clarify the meaning. The column spanner is sometimes used to describe the statistics that will be shown in the columns—as an example, see how we tabulate the results of a regression analysis in Example 12.1.

Different variables are often shown in the rows of a table. Variable names and units of measurement should be specified clearly and accurately. The units of measurement are critically important. Some clinical and laboratory measures, for example, might be regularly measured and reported using different scales, so it is critical that the exact units used are reported to ensure that readers and consumers can make valid comparisons. Just as journals have specific style guides for reporting data and statistical results in tables and figures, there are also guidelines for reporting clinical and laboratory measures that should be consulted when reporting such data. Some journals, websites, or clients might prefer metric units (e.g., meters and grams) to U.S. units (e.g., feet, pounds), and their conventions should be followed as appropriate. Data should be rounded as much as is appropriate; usually two decimal places is more than sufficient.

The components of a table are shown in **Figure 12–4**.

Gridlines, which are often seen in spreadsheets, are unnecessary in tables; in fact, they can make it difficult to read data in a table. There should be few, if any, lines in the body of the table. The table should use consistent spacing, and columns should be reasonably close to facilitate comparisons of interest. White space around the table and strategically placed within the table can help to highlight certain aspects of the data and statistical results. The goal is clarity—creating a table that makes it easy for the reader or consumer to hone in on the key messages in the data and statistical results.

Example 12.1. Logistic regression analysis is used to evaluate the associations between body mass index (BMI) categories—normal (BMI ≤ 25.0), overweight (25.0 < BMI < 30.0), and obese (BMI ≥ 30.0)—and three outcomes—incident myocardial infarction, incident cardiovascular disease, and incident stroke, which are considered separately.

Table 12–1 summarizes the associations between BMI categories (normal weight, overweight, and obesity) using odds ratios and 95% confidence intervals for odds ratios. **Table 12–2** shows the same data as in Table 12–1, but without gridlines and a bit more spacing. Which table is easier to read?

Footnotes, such as the notes at the bottoms of Tables 12–1 and 12–2, can be used to provide additional information that is necessary to interpret the data or statistical results presented in the table. Such explanations should not be included in the column headings or in the title, as they might detract from the interpretability of the results. All abbreviations and symbols can be clarified in the footnotes. **Table 12–3** shows the same data as in Tables 12–1 and 12–2, but uses abbreviations for some of the outcomes to increase readability. The abbreviations are then explained in the footnotes. The decision as to whether to use abbreviations depends on the audience. If the audience is familiar with the abbreviations, then they can promote readability. If the audience is unfamiliar with the abbreviations, then it creates more work for them to locate definitions.

It is good practice to draft table shells, or templates, before beginning a statistical analysis. The table shells outline what will be presented in tables and how. The table shells help to organize the author's thinking to develop a clear and logical story of the investigation and results. **Table 12–4** is an example of a table shell designed to summarize background characteristics of participants in a study comparing a group that received an intervention to a control group. Note that in the table shell, variable names are specified along with units (first column). Summary statistics that will be shown are specified, and in some cases the exact number of decimal places that will

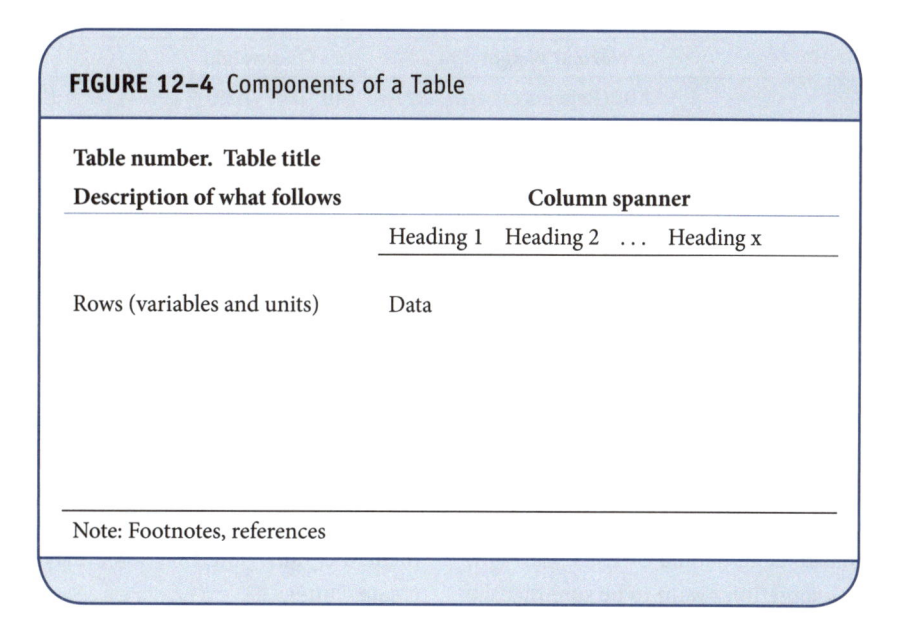

FIGURE 12–4 Components of a Table

Table number. Table title
Description of what follows   Column spanner

                                    Heading 1   Heading 2   ...   Heading x

Rows (variables and units)   Data

Note: Footnotes, references

TABLE 12-1 Association Between BMI Categories and Incident Cardiovascular Disease After Adjustment for Clinical Risk Factors*

	Odds Ratio* (OR) (95% Confidence Interval)		
	Normal Weight	**Overweight**	**Obese**
Incident myocardial infarction	1.00 (Reference)	1.01 (0.69–1.29)	1.14 (1.01–1.50)
Incident cardiovascular disease	1.00 (Reference)	1.21 (0.89–1.37)	1.36 (1.13–2.54)
Incident stroke	1.00 (Reference)	0.99 (0.82–1.08)	1.18 (1.09–1.23)

* Adjusted for age, sex, systolic and diastolic blood pressure, total serum cholesterol, high-density lipoprotein, and smoking; normal weight (body mass index [BMI] ≤ 25.0), overweight (25.0 < BMI < 30.0), and obese (BMI ≥ 30.0).

TABLE 12-2 Association Between BMI Categories and Incident Cardiovascular Disease After Adjustment for Clinical Risk Factors*

	Odds Ratio* (OR) (95% Confidence Interval)		
	Normal Weight	**Overweight**	**Obese**
Incident myocardial infarction	1.00 (Reference)	1.01 (0.69–1.29)	1.14 (1.01–1.50)
Incident cardiovascular disease	1.00 (Reference)	1.21 (0.89–1.37)	1.36 (1.13–2.54)
Incident stroke	1.00 (Reference)	0.99 (0.82–1.08)	1.18 (1.09–1.23)

* Adjusted for age, sex, systolic and diastolic blood pressure, total serum cholesterol, high-density lipoprotein, and smoking; normal weight (body mass index [BMI] ≤ 25.0), overweight (25.0 < BMI < 30.0) and obese (BMI ≥ 30.0).

TABLE 12-3 Association Between BMI Categories and Incident Cardiovascular Disease After Adjustment for Clinical Risk Factors*

	Odds Ratio* (OR) (95% Confidence Interval)		
	Normal Weight	**Overweight**	**Obese**
Incident MI **	1.00 (Reference)	1.01 (0.69–1.29)	1.14 (1.01–1.50)
Incident CVD	1.00 (Reference)	1.21 (0.89–1.37)	1.36 (1.13–2.54)
Incident Stroke	1.00 (Reference)	0.99 (0.82–1.08)	1.18 (1.09–1.23)

* Adjusted for age, sex, systolic and diastolic blood pressure, total serum cholesterol, high-density lipoprotein, and smoking; normal weight (body mass index [BMI] ≤ 25.0), overweight (25.0 < BMI < 30.0) and obese (BMI ≥ 30.0).
** Note: MI = myocardial infarction; CVD = cardiovascular disease.

be shown is indicated (see the last column, where *p*-values will be reported). In Table 12-4, the title is shown at the top of the table, and the columns represent the different intervention groups, which are the primary comparison groups.

Only one typeface should be used in a table, and italics, boldface, and other enhancements should be used sparingly, if at all. Creators of tables sometimes want to be sure that the reader or consumer finds the key data or makes the key comparisons, but if the table is set up appropriately, these should be self-evident. There are exceptions: For example, the use of boldface might be helpful when there are large amounts of data to digest and it might be useful to call out specific aspects of the data. **Table 12-5** and **Table 12-6** show the same data with and without enhancements. Are the enhancements helpful? Note that some may be helpful while others create clutter.

When a paper or report includes a series of tables with the same comparison groups, the organization of the

TABLE 12-4 Example of a Table Shell

Table xx. Demographic Characteristics of Asian American Participants by Study Group

Characteristic	Treatment Intervention Group (n = xx)	Control Group (n = xx)	p-value*
Age, years	$\overline{X}(s)$	$\overline{X}(s)$	x.xx
Years lived in United States, years	$\overline{X}(s)$	$\overline{X}(s)$	x.xx
English proficiency			x.xx
No practical proficiency	n (%)	n (%)	
Elementary proficiency	n (%)	n (%)	
Limited proficiency	n (%)	n (%)	
Full proficiency	n (%)	n (%)	
Native or bilingual proficiency	n (%)	n (%)	

* p-values for continuous measures based on two independent samples' t-test; p-values for categorical characteristics based on chi-square tests of independence.

TABLE 12-5 Impact of High Intake* of Grains and Incident Cardiovascular Disease and Cancer

Grain	Odds Ratio (95% Confidence Interval)
Incident Cardiovascular Disease	
Whole-grain bread	0.85 (0.72, 0.97)
Whole-grain cereal	0.71 (0.63, 0.84)
Refined grain	1.05 (0.87, 1.36)
White bread	1.10 (0.95, 1.30)
Rice (white and brown)	0.99 (0.92, 1.04)
Total grains	1.03 (0.91, 1.24)
Incident Cancer	
Whole-grain bread	0.91 (0.77, 0.99)
Whole-grain cereal	0.91 (0.86, 0.97)
Refined grain	0.97 (0.84, 1.06)
White bread	1.01 (0.95, 1.13)
Rice (white and brown)	0.94 (0.87, 1.02)
Total grains	0.95 (0.86, 1.11)

* High intake is intake above the median.

TABLE 12-6 Impact of High Intake* of Grains and Incident Cardiovascular Disease and Cancer

Grain	Odds Ratio (95% Confidence Interval)
Incident Cardiovascular Disease	
Whole-grain bread	0.85 (0.72, 0.97)
Whole-grain cereal	0.71 (0.63, 0.84)
Refined grain	1.05 (0.87, 1.36)
White bread	1.10 (0.95, 1.30)
Rice (white and brown)	0.99 (0.92, 1.04)
Total grains	1.03 (0.91, 1.24)
Incident Cancer	
Whole-grain bread	0.91 (0.77, 0.99)
Whole-grain cereal	0.91 (0.86, 0.97)
Refined grain	0.97 (0.84, 1.06)
White bread	1.01 (0.95, 1.13)
Rice (white and brown)	0.94 (0.87, 1.02)
Total grains	0.95 (0.86, 1.11)

* High intake is intake above the median.

comparison groups must be consistent across tables. We should never reorder the groups or switch the orientation of the tables (i.e., switch from rows to columns).

12.3.2 Tabulating Summary Statistics

Many scientific papers and professional reports detailing clinical research include a first table describing the participants. It is good practice to provide a summary of the study participants to ensure that readers understand who is involved, so that they can judge the generalizability of the findings as well as any limitations of inferences that might be drawn. When organizing a table to summarize characteristics of the study sample, we carefully define each variable and characteristic, and the units in which each was measured.

When reporting dichotomous, categorical, and ordinal variables, percentages are usually easier to digest than proportions. That said, we should always present the numbers of participants in each response category along with the percentages to allow readers and consumers to organize the data differently should they choose to explore further. Percentages should sum to 100% in the total sample unless there are missing data or other anomalies. The latter possibility should be addressed in the text and in a footnote to the table (again, aiming to ensure that the table is self-explanatory).

Example 12.2. Suppose we investigate the impact of three behavioral interventions (self-help, group therapy, individual therapy) to reduce cigarette smoking among $n = 270$ older adults who reside in assisted living facilities. Before summarizing the main findings, which are based on specific measures of cigarette smoking (e.g., cigarettes smoked per day), we first summarize demographic characteristics of participants assigned to each of the three behavioral intervention groups. The resulting table might look like **Table 12–7**.

Note in Table 12–7 that we present column percentages (percentages that sum to 100% within columns) for each characteristic to facilitate comparisons of variables across intervention groups. When summarizing dichotomous, categorical, and ordinal variables, statistical computing packages often produce frequencies, row percentages, and column percentages as outputs. In putting together tables for presentation, we need to decide which data are most appropriate. Consider the following example, which shows data presented first as frequencies, and then with row and column percentages. Which is the most useful presentation of the data?

Example 12.3. We wish to compare the prevalence of overweight and obesity in men and women participating in the Framingham Heart Study offspring examination 2, which was conducted around 1980. **Table 12–8** shows the numbers of men and women classified as normal weight (BMI ≤ 25.0), overweight (25.0 < BMI < 30.0), and obese (BMI ≥ 30.0).

What is the extent of obesity in the study sample? Is the extent of obesity different in men versus women? There are

TABLE 12–7 Background Characteristics of Study Participants by Intervention Group

	Intervention Group		
Characteristic*	Self-Help ($n = 100$)	Group Therapy ($n = 90$)	Individual Therapy ($n = 80$)
Age, years	78.2 (6.2)	79.6 (5.9)	81.4 (5.7)
Male sex, n (%)	46 (46%)	38 (42%)	28 (35%)
Education, years	9.3 (4.2)	10.7 (3.9)	8.6 (4.1)
Marital status			
Single, never married, n (%)	9 (9%)	11 (12%)	5 (6%)
Married or domestic partnership, n (%)	36 (36%)	36 (40%)	23 (29%)
Widowed, n (%)	43 (43%)	33 (37%)	43 (54%)
Divorced or separated, n (%)	12 (12%)	10 (11%)	9 (11%)

* Means (standard deviations) are shown for continuous measures and n (%) is shown for categorical measures.

TABLE 12–8 Numbers of Men and Women by BMI Category

BMI* Category	Men ($n = 1868$)	Women ($n = 1994$)	Total ($n = 3862$)
Normal weight	631	1289	1920
Overweight	925	452	1377
Obese	312	253	565

* BMI is body mass index = weight$_{kg}$/height$_m^2$; normal weight (BMI ≤ 25.0), overweight (25.0 < BMI < 30.0), and obese (BMI ≥ 30.0).

TABLE 12-9 Distribution of Men and Women in Normal, Overweight, and Obese Categories of BMI

BMI* Category	Men (n = 1868, 48%)	Women (n = 1994, 52%)	Total (n = 3862)
Normal weight	631 (33%)	1289 (67%)	1920
Overweight	925 (67%)	452 (33%)	1377
Obese	312 (55%)	253 (45%)	565

* BMI is body mass index = weight$_{kg}$/height$_m^2$; normal weight (BMI ≤ 25.0), overweight (25.0 < BMI < 30.0), and obese (BMI ≥ 30.0).

565 participants who meet the criteria for obesity: 312 are men and 253 are women. With only the frequencies or numbers of participants in each BMI category, it is difficult to summarize results to make the desired comparison. **Table 12-9** shows the same data with row percentages included.

What is the extent of obesity in the study sample? Is the extent of obesity different in men versus women? The row percentages—which are percentages that sum to 100% in each of the rows of the table—are not helpful in answering the questions of interest. Table 12-9, however, is helpful in answering a different question: What percentage of the participants are women? As shown by the data in the top row of the table, 1994 (52%) participants are women. What percentage of the obese participants are women? Among the 565 participants who meet the criteria for obesity, 253 (45%) are women. **Table 12-10** shows the same data with column percentages—that is, percentages that sum to 100% within each column of the table.

The data in Table 12-10 allow us to answer the questions posed. What is the extent of obesity in the study sample? There are 565 (14%) participants who meet the criteria for obesity. Is the extent of obesity different in men versus women? Similar percentages of men and women meet the criteria for obesity—approximately 17% of men and 13% of women.

When summarizing continuous variables, we usually report either the mean or the median to summarize central tendency and the standard deviation or interquartile range to summarize variability. The choice of specific statistics depends on whether there are outliers in the sample, as discussed Chapter 4. Another option, when describing the study sample in a table of a paper or report, is to organize the continuous variable into categories. In Tables 12-8 to 12-10, we organize BMI, measured as a continuous variable, into three ordinal categories for presentation (normal, overweight, obese). Between 3 and 10 categories is usually considered appropriate when organizing a continuous measure into categories for presentation. Fewer than three categories can result in so much aggregation that the detail in the data is lost. More than 10 categories can make the results difficult to digest when more collapsing is needed. If there are natural or accepted cut-points that define clinically meaningful categories of variables, such as the cut-points for BMI and SBP (see Chapter 4), then those should be used. Even if we choose to summarize a continuous variable using ordinal categories in a table, we can still use the continuous variable in statistical analysis to capture the full extent of variability in the measure unless the clinically relevant categories provide more interpretable results.

When presenting summary statistics (e.g., means, proportions, odds ratios, relative risks, differences in means, regression coefficients) in tables, measures of variability such as standard errors or confidence limits should also be included. Different measures of variability are appropriate in different settings. For example, if we are presenting means of an

TABLE 12-10 Distribution of BMI Categories in Men and Women

BMI* Category	Men (n = 1868)	Women (n = 1994)	Total (n = 3862)
Normal weight	631 (34%)	1289 (65%)	1920 (50%)
Overweight	925 (49%)	452 (23%)	1377 (36%)
Obese	312 (17%)	253 (13%)	565 (14%)

* BMI is body mass index = weight$_{kg}$/height$_m^2$; normal weight (BMI ≤ 25.0), overweight (25.0 < BMI < 30.0), and obese (BMI ≥ 30.0).

outcome measured in each of four comparison groups (e.g., in an analysis where analysis of variance [ANOVA] is used to test for a statistically significant difference in means), means and standard errors might be appropriate. If we are presenting odds ratios quantifying the associations between a set of dichotomous risk factors and a dichotomous outcome, the estimated odds ratios along with 95% confidence intervals might be appropriate. We elaborate in a bit more detail as to how to best present results from regression models in Section 12.3.4.

12.3.3 Tabulating Statistical Significance

When reporting statistical significance levels (p-values) in tables, we should be consistent with the guidelines for the specific journal or report. Some journals, for example, require exact significance levels throughout the text and in tables, and many limit p-values to two or three decimal places (e.g., $p = 0.03$ or $p = 0.032$). Most journals discourage the use of symbols such as "NS" to indicate non-significance when a p-value exceeds a conventional threshold (e.g., $p > 0.05$). Instead, they prefer that authors report the observed p-value, such as $p = 0.27$. Lastly, if a p-value is less than 0.001, it is usually sufficient to report "$p < 0.001$" rather than the exact value. Almost always, p-values are assumed to be two-sided. If a one-sided test is conducted, this choice must be clearly indicated and justified.

When reporting statistical significance to a nontechnical audience, we have to be careful not to overdo the scientific jargon. We need to explain what we mean by statistical significance so that the readers or consumers can understand how to interpret the results being presented.

As an aside, when describing background or baseline characteristics of participants in a parallel-group, randomized, controlled trial (e.g., comparing an experimental treatment to placebo or to an active control), we might construct a table similar to **Table 12–11**, in which the columns represent the two treatments (experimental versus placebo) being compared. While there is some variability in practice, it is not necessary to conduct statistical tests comparing background or baseline characteristics of participants in a randomized controlled trial.[9] It is always important to examine background or baseline characteristics carefully to ensure that groups are balanced, but statistical tests are not needed for this purpose. Consider the data summarized in Table 12–11. Do the treatment groups appear balanced?

12.3.4 Tabulating Regression Results

When tabulating the results of linear regression analysis, the regression coefficients should be reported, along with a measure of variability of the estimated regression coefficients (e.g., standard errors or 95% confidence intervals of the estimated regression coefficients). When tabulating the results of a logistic or Cox proportional hazards regression analysis, the regression coefficients are usually exponentiated to produce odds ratios and hazards ratios, respectively. The 95% confidence intervals are also usually presented along with the estimated odds ratios and hazard ratios to give a sense of the precision (or lack thereof) in the estimates of effect and to judge statistical significance.

Example 12.4. A cohort study is run to evaluate differences in birth weights and incidence of preterm birth in babies born to mothers of different racial/ethnic backgrounds. A total of $n = 453$ mothers participate in the study and identify as white, black, or Hispanic. Mothers report their age, in years, at the start of the study and are followed to delivery, at which time the gestational age and birth weight of the baby are measured. Preterm birth is also recorded, defined as

TABLE 12–11 Characteristics of Participants at Baseline

	Treatment	
Characteristic*	Experimental ($n = 636$)	Placebo ($n = 640$)
Age, years	45.7 (6.4)	45.9 (6.2)
Male sex, n (%)	292 (46%)	289 (45%)
Systolic blood pressure, mm Hg	128.2 (14.7)	127.9 (14.4)
Diastolic blood pressure, mm Hg	78.3 (9.1)	76.9 (8.6)
Total cholesterol, mg/dL	201.5 (42.7)	209.3 (44.5)
Body mass index, kg/m²	26.4 (3.9)	25.9 (4.1)

* Values are means (standard deviations) for continuous characteristics and n (%) for categorical measures.

gestational age at birth less than 37 weeks. **Table 12–12** summarizes the multiple linear regression analysis relating racial/ethnic background, maternal age, and gestational age to birth weight. Note that racial/ethnic groups are modeled as dummy variables, with white race as the hold-out or reference group.

Table 12–13 summarizes a multiple logistic regression analysis relating racial/ethnic background and maternal age to preterm birth. In Table 12–13, it is not necessary to include the regression coefficients in addition to the odds ratios. Odds ratios are generally more interpretable in summarizing the associations.

In reporting regression coefficients, it is critically important to indicate the units of the independent variable, as the interpretation of the regression coefficient depends on these units. As discussed Chapter 9, the regression coefficient (b_i) represents the change in outcome (Y) relative to a one-unit change in the predictor (X). For some predictors, a one-unit change is not clinically meaningful and we might wish to report the change in outcome relative to a larger, more meaningful change.

Consider again Example 12.4. In Table 12–12, the regression coefficient associated with maternal age quantifies the expected change in birth weight associated with a 1-year change in maternal age, adjusted for the racial/ethnic background of the mother and the neonate's gestational age. Similarly, the regression coefficient associated with gestational age quantifies the expected change in birth weight associated with a 1-week change in gestational age, adjusted for the racial/ethnic background of the mother and maternal age. A 1-year difference in maternal age is associated with a lower birth weight by approximately 0.27 gram. It might, for example, be more informative to estimate the expected difference in birth weight associated with a 5-year difference in maternal age, which is −0.27 * 5 = −1.35 grams. These data could be tabulated as shown in **Table 12–14**.

Note that the standard error of the estimate does not change when we modify the effect relative to a 1-year change versus a 5-year change. If we wanted to show a similar effect of maternal age (5 years) on preterm birth, we could make the relevant adjustments in the odds ratio and 95%

TABLE 12–12 Association Between Racial/Ethnic Background, Maternal Age, Gestational Age, and Birth Weight

Characteristic	Regression Coefficient*	Standard Error	*p*-Value
Intercept	−4366.5	188.3	<0.01
Racial/ethnic group			
White	Reference	—	
Black	−46.0	47.0	0.33
Hispanic	46.7	47.6	0.32
Maternal age, years	−0.27*	2.8	0.92
Gestational age, weeks	193.6*	4.7	<0.01

* Regression coefficients are based on multiple linear regression analysis and are relative to a one-year change in maternal age and a one-week change in gestational age.

TABLE 12–13 Association Between Racial/Ethnic Background, Maternal Age, and Preterm Birth

Characteristic	Regression Coefficient*	Odds Ratio (OR)	95% Confidence Interval for OR
Intercept	0.37	—	—
Racial/ethnic group			
White	Reference	—	—
Black	−0.16	0.85	(0.50, 1.47)
Hispanic	0.13	1.14	(0.65, 2.01)
Maternal age, years	0.03*	1.03	(1.00, 1.06)

* Regression coefficients are based on multiple logistic regression analysis and are relative to a one-year change in maternal age.

TABLE 12–14 Association Between Racial/Ethnic Background, Maternal Age, Gestational Age, and Birth Weight

Characteristic	Regression Coefficient*	Standard Error	p-Value
Racial/ethnic group			
White	Reference	—	
Black	−46.0	47.0	0.33
Hispanic	46.7	47.6	0.32
Maternal age, years	−1.4*	2.8	0.92
Gestational age, weeks	193.6*	4.7	<0.01

* Regression coefficients are based on multiple linear regression and are relative to a five-year change in maternal age and a one-week change in gestational age.

confidence interval for the odds ratio. As we discussed in Chapter 9, we exponentiate regression coefficients from a logistic regression analysis to produce odds ratios [e.g., exp(0.03) = 1.03]. Thus, an older mother (by 1 year) has 1.03 times the odds of preterm birth as compared to a younger mother, adjusted for racial/ethnic background. To compute the odds ratio relative to a 5-year difference in maternal age, we use $\exp(c \times b_1)$, where c = the number of units (here, c = 5). The odds of preterm birth relative to a 5-year difference in maternal age, adjusted for the racial/ethnic background of the mother, is exp(5 × 0.03) = 1.16 times higher for the older mother.

The confidence interval for an odds ratio relative to a 1-year difference in a predictor is computed as follows. First, we estimate the logistic regression equation and obtain the regression coefficient (e.g., b_1) and its standard error [e.g., $SE(b_1)$]. Assuming we have a sufficient sample size, the lower 95% confidence limit is computed as $\exp[b_1 - 1.96 \times SE(b_1)]$ and the upper 95% confidence limit is computed as $\exp[b_1 + 1.96 \times SE(b_1)]$. These formulas produce confidence intervals for the odds ratio relative to a one-unit change in the predictor. If we wish to estimate the confidence interval relative to a c-unit change in the predictor, we use the following expression: $\exp[c \times b_1 \pm 1.96 \times c \times SE(b_1)]$. In our example, the confidence interval relative to a 5-year difference in maternal age is exp(5 × 0.03 ± 1.96 × 5 × 0.017), where 0.017 is the standard error of the estimated regression coefficient. This is equivalent to exp(0.15 ± 0.17), or exp(−0.02, 0.32) = (0.98, 1.38). Thus, an older mother (by 5 years) has 1.16 times the odds of preterm birth as compared to a younger mother, adjusted for racial/ethnic background, with a 95% confidence interval for the odds ratio of (0.98, 1.38).

Greek letters (e.g., β) should not be used in the tables (for example, as column headers), as Greek letters generally refer to unknown population parameters and not to sample estimates (which we denoted as b_1, b_2, and so on in Chapter 9). It is often clearer to use text such as "regression coefficient" or "odds ratio" in column headings when tabulating regression results.

When presenting the results of a regression analysis, it is important to provide the complete results. Specifically, it is important to include the estimate of the y-intercept so that the reader has the complete set of results. There are some instances, however, where it might be reasonable to show only partial results in a main table (i.e., a table included in the text) and complete results in an appendix or in an online supplement. For example, if the relationship of interest lies squarely on the association between a particular risk factor and the outcome, the regression coefficient associated with that outcome is primary. So as not to detract from the main message, the y-intercept and regression coefficients associated with other variables included in the model (e.g., confounders) can be shown in full detail in an appendix or in an online supplement.

Example 12.5. Suppose we want to evaluate the association between moderate to heavy drinking (defined as more than one drink per day in women and more than two drinks per day in men) and hypertension. Hypertension is defined as systolic blood pressure of 140 mm Hg or higher or diastolic blood pressure of 90 mm Hg or higher. Data are collected from n = 2500 participants in an observational study. The investigators first run a simple logistic regression model relating the indicator of moderate to heavy drinking to hypertension status. They then estimate a series of models considering additional confounding variables. **Table 12–15** summarizes all of the information for each model, which is more than might be necessary in the body of a paper or report.

Of primary interest is the association between moderate to heavy drinking and hypertension. Because so many other

TABLE 12–15 Association Between Moderate to Heavy Drinking and Hypertension

	Odds Ratio* (95% Confidence Intervals for Odds Ratios)		
Characteristic	**Unadjusted Model**	**Age- and Sex-Adjusted Model**	**Fully Adjusted Model***
Intercept			
Moderate to heavy drinking	1.18 (0.90, 1.53)	1.44 (1.10, 1.90)	1.43 (1.08, 1.89)
Age, years		1.05 (1.04, 1.07)	1.06 (1.04, 1.07)
Male sex		0.92 (0.75, 1.20)	0.92 (0.74, 1.43)
Current smoker			0.93 (0.60, 1.43)
Body mass index, kg/m^2			1.02 (0.99, 1.04)
Moderate to heavy physical activity, hours per week			0.99 (0.94, 1.03)
Lives alone			0.96 (0.63, 1.14)
Diabetes			0.85 (0.63, 1.14)

* Odds ratios are based on multiple logistic regression analysis and are relative to a one-unit change in each predictor, adjusted for other variables in the model.

associations are summarized in the table, however, the main comparison is lost. Sometimes data such as those shown in Table 12–15 are summarized as shown in **Table 12–16**, and then the full detail (Table 12–15) is included in an appendix or in an online supplement.

As a general rule, we want to limit the inclusion of extraneous details or features in tables to allow the reader to focus on what is truly important. Thus, tables should not include multiple horizontal or vertical lines; such lines are generally considered extraneous to the main point of the table—that is, the data or statistics being presented. The goals in visualizing data and statistical results are clarity and accuracy of presentation.

12.4 PRESENTING DATA AND STATISTICAL RESULTS IN FIGURES

We are bombarded with data every day, much of it through the Internet, but also on television and in newspapers, scientific papers, and reports. Websites, blogs, videos, posts, and mobile applications all contain data and graphical information and are available to a far wider audience than could be reached prior to the advent of such mass media. The data presented through these forums are not just for statistics fans. That is, the consumers of information are not all professionals evaluating data and graphical displays in a professional capacity, but also include lay persons who consume the information to make personal decisions based on their understanding of the evidence.

In the field of biostatistics, we have seen increases in the availability of data (e.g., big data) and advances in statistical computing and modeling (e.g., data science) that have accelerated the need for good displays of statistical results. Visualizations are powerful tools for converting large amounts of data into useful information.

Effective visualizations make data come alive. Visual representations, assuming they are developed correctly, are often easier to understand than tables of numbers, particularly

TABLE 12–16 Association Between Moderate to Heavy Drinking and Hypertension

	Odds Ratio for Moderate to Heavy Drinking (95% Confidence Intervals for Odds Ratios)
Unadjusted	1.18 (0.90, 1.53)
Age- and sex-adjusted	1.44 (1.10, 1.90)
Fully adjusted*	1.43 (1.08, 1.89)

* Odds ratios are based on multiple logistic regression analysis; the fully adjusted model includes age, sex, current smoking status, body mass index, moderate to heavy physical activity, living arrangement (alone or with others), and diabetes.

when the tables are lengthy (which makes them difficult to digest). However, there are risks when presenting data in this way. Summarizing complex statistical information into one graphic might result in misleading or confusing information if not presented correctly. Flashy graphical displays might grab our attention, but if not carefully constructed, they can distort the meaning of the data.

Advances in techniques for visualizing data not only increase the flexibility and reach of data presentations, but also elevate the importance of sound statistical principles in ensuring that data are valid. In fact, the field of data visualization has exploded in recent years, with new platforms and systems (e.g., Tableau[10]), making it easy for even novices to create interactive infographics. Data visualization has become highly sophisticated and increasingly important in sales and marketing, for example, bringing together theories of perception and cognitive science with graphic design and visual art.

While new techniques for data visualization continue to emerge, less attention has been paid to graphical displays of statistical information. Gelman and Unwin argue that there is "an unfortunate lack of interaction between the worlds of statistical graphics and information visualization," and they advocate for more collaboration to enhance the outputs of both.[11] Gelman and Unwin recognize that while the goals of statistical graphics and infographics might differ, there should be consistent adherence to sound statistical principles across all types of visualizations. They note that statistical graphics are usually designed to report what was observed in a study sample, whereas infographics might be designed to grab attention, be provocative, or suggest what might be possible in a broader frame. They argue that we do not have to choose between "attractive graphs" and "informational graphs"— it should be possible for a graphic to be both. Optimal results, then, are ideally produced by biostatisticians and designers coming together.

Unfortunately, despite lots of resources outlining strategies for producing good figures, bad graphics continue to proliferate, leaving readers and consumers confused and uninformed in their wake.[12] This presents an opportunity for us to take advantage of new tools to more effectively communicate data and statistical results, and maybe even heighten the appreciation for biostatistics!

12.4.1 Components of a Figure

Figures offer an opportunity to show more information than a table could, in a much smaller space. Just like tables, figures should engage the reader, have a point to make, and be able to stand alone. Here we discuss strategies for developing clear and accurate displays of data and statistical results. Critical in this endeavor is combining sound statistical practice with good design principles to emphasize the important and minimize the trivial.

Before we discuss different types of figures and how best to construct them, we first outline the components of a figure. Good figures have descriptive titles that tell the reader or consumer what to expect, the axes are labeled clearly, the legend describes what is being presented, and key findings are readily accessible. The title should provide readers with enough information to understand what is being presented, and in sufficient detail such that they need not refer back to the text to understand what is being shown in the figure. The x-axis, or horizontal axis, and the y-axis, or vertical axis (usually shown on the left side of the figure—but sometimes on both sides), form the plot area. The axis labels should be concise but indicate what is being presented (e.g., systolic blood pressure, total serum cholesterol, age) along with the units of measure (e.g., mm Hg, mg/dL, years). The x-axis label is usually placed below the axis, and the y-axis label (if it is necessary—sometimes the y-axis is used to show frequencies or relative frequencies, which can be noted in the title) is usually rotated and shown parallel to the y-axis.

Depending on what is being presented, the y-axis should be scaled to be true to the data. Some say that honesty in data visualization requires starting the scale at zero[12]; the y-axis usually starts at zero but there is some variation in practice. Note that the scaling of an axis can distort (exaggerate or mask) a relationship in the data (see Chapter 4 for example figures that commit this sin). If we determine that it is appropriate to rescale the y-axis, we must be careful and clear in disclosing this fact so that the reader is not misled. One way to draw attention to a rescaled y-axis is with the use of a marker or symbol that indicates that the axis has been changed. **Figures 12–5** and **12–6** demonstrate how rescaling is performed.

A related issue involves scaling the y-axis on a log scale— a practice that is sometimes used when we have samples with outliers or extreme values. Another reason to use a log scale is to illustrate a multiplicative effect. In **Figure 12–7**, the y-axis is scaled on a log base 2 scale (i.e., $2^0 = 1$, $2^1 = 2$, $2^2 = 4$, $2^3 = 8$, $2^4 = 16$). The scaling must be completely clear so that readers do not mistakenly interpret the largest increase between years 1 and 2, when in reality there is an even bigger increase between years 3 and 4. The y-axis scaling is on a log base 2 scale, but this is not evident as there is no clear labeling of the axis.

A legend should clarify any markers, symbols, or patterns used to differentiate groups in the figure. Footnotes provide

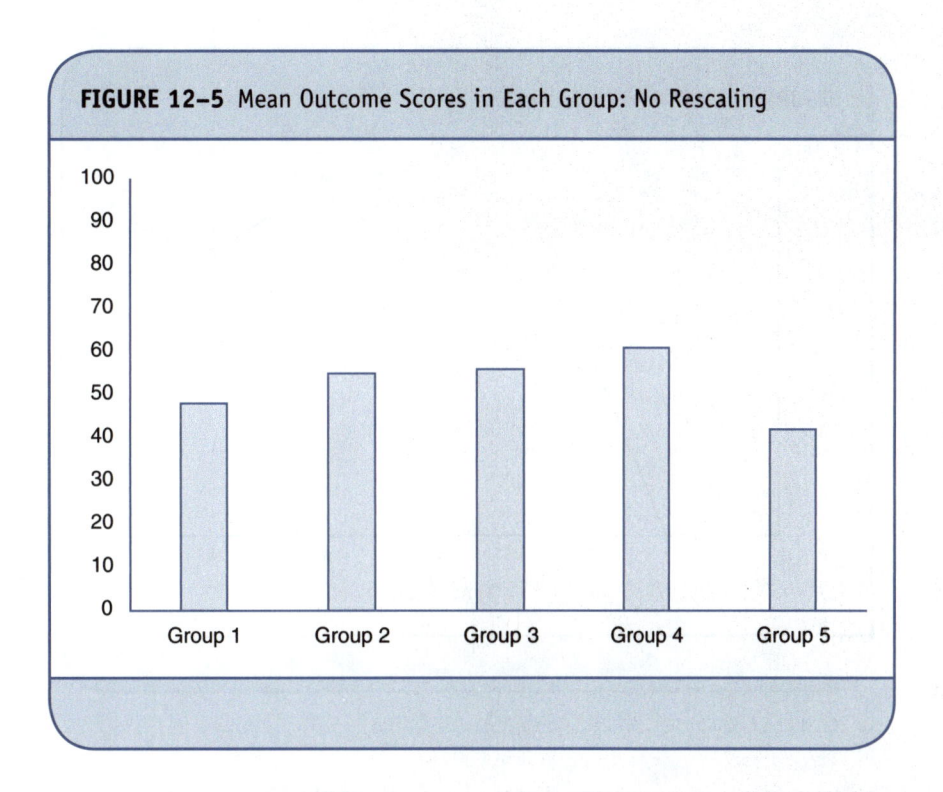

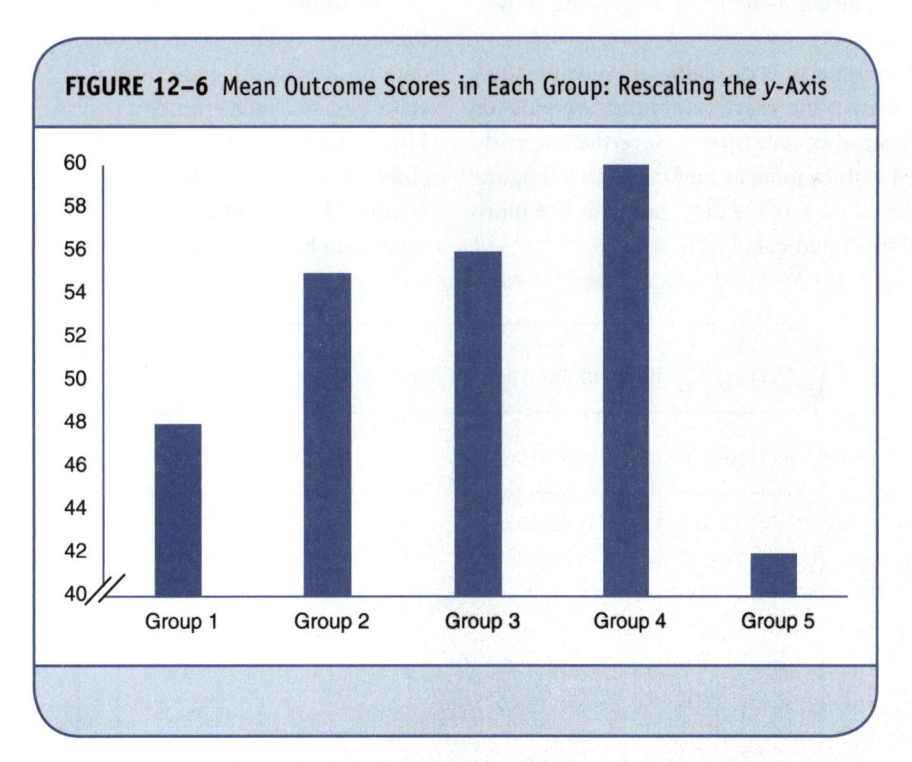

FIGURE 12–5 Mean Outcome Scores in Each Group: No Rescaling

FIGURE 12–6 Mean Outcome Scores in Each Group: Rescaling the y-Axis

additional information that is necessary to interpret the data or statistical results in the figure (e.g., explaining abbreviations or symbols used). Data labels can also be included to show the exact values that are plotted, and are most relevant to bar charts (see Example 12.6). The plot area is the section of the figure that contains the data and is bounded by the x- and y-axes, whereas the chart area includes all of the elements that are associated with the figure (e.g., title, legend,

FIGURE 12-7 y-Axis on a Log$_2$ Scale

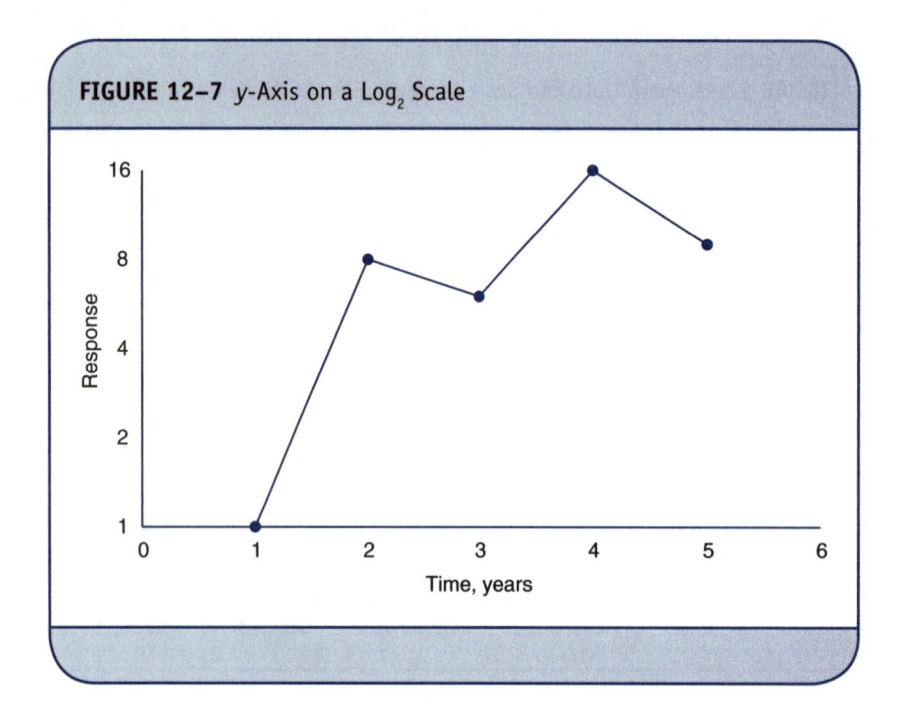

footnotes, plot area). The components of a figure are shown in **Figure 12-8**.

Gridlines are horizontal or vertical lines that are intended to help the reader extract the exact values that are plotted; they can be used instead of data labels. Nevertheless, gridlines should be used with caution, as they can clutter a figure and detract from the message of the data. Example 12.6 illustrates the use of gridlines and data labels.

Example 12.6. The Centers for Disease Control and Prevention (CDC) collects, summarizes, and disseminates data on key health indicators that can be broken down by state, age, sex, race/ethnicity, and annual household income. **Figure 12-9** summarizes the prevalence of asthma among children and adults based on data captured in the 2014 National Health Interview Survey.[13] **Figure 12-10** shows the same data but with gridlines to help the reader decipher the

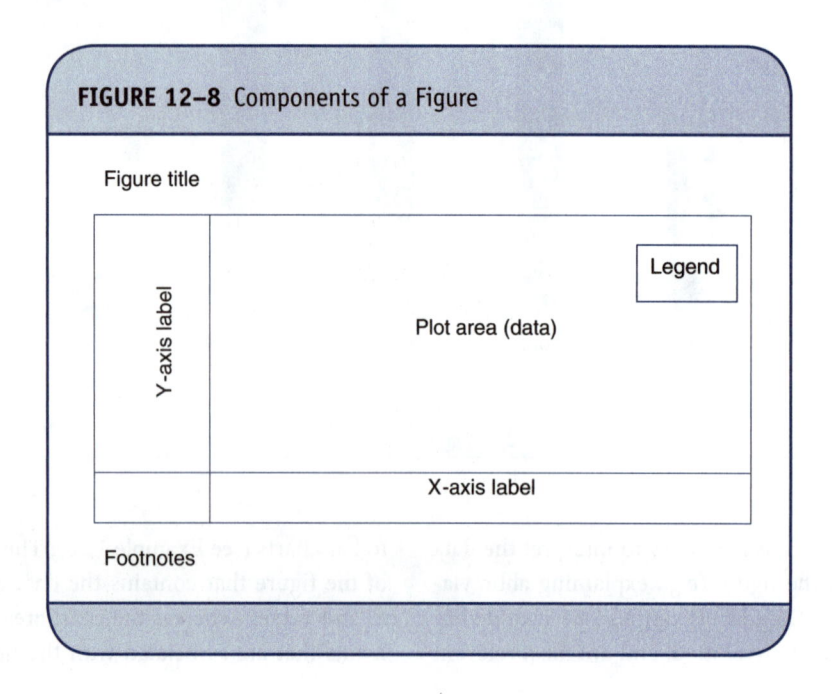

FIGURE 12-8 Components of a Figure

FIGURE 12-9 Prevalence of Asthma Among Children and Adults in the United States, 2014

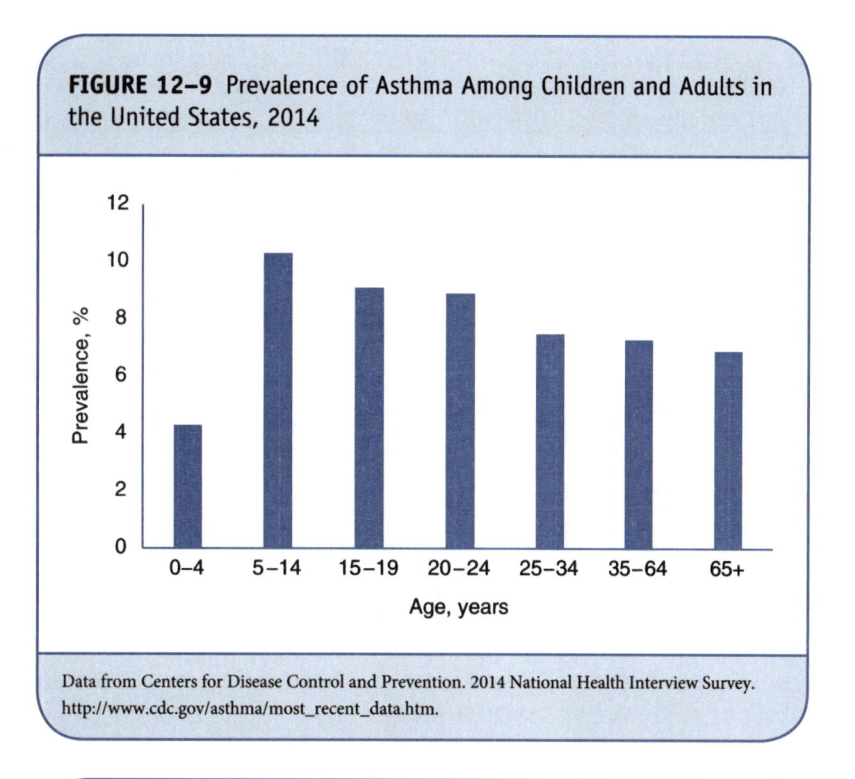

Data from Centers for Disease Control and Prevention. 2014 National Health Interview Survey. http://www.cdc.gov/asthma/most_recent_data.htm.

FIGURE 12-10 Prevalence of Asthma Among Children and Adults in the United States, 2014 (with Gridlines)

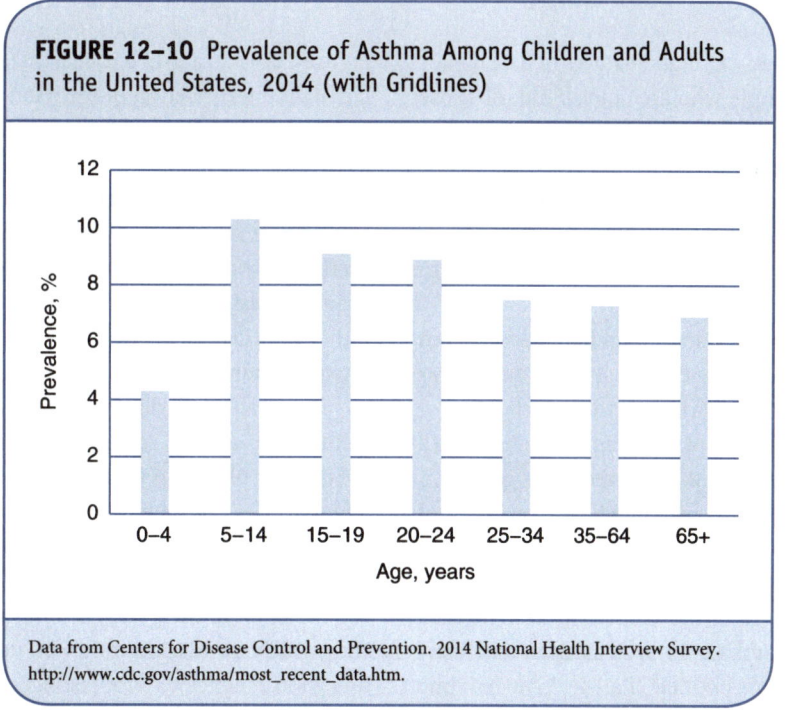

Data from Centers for Disease Control and Prevention. 2014 National Health Interview Survey. http://www.cdc.gov/asthma/most_recent_data.htm.

prevalence percentages in each age group. Do the gridlines help? **Figure 12-11** shows the same data but instead of gridlines, data labels are used to indicate the percentages in each age group. Which figure is easiest to interpret?

As we develop figures, we always want to think about the audience or target group. What do they want and need, and how can we best communicate our data and statistical results? If we decide a figure is the best approach, what is the role of that figure in the presentation, paper, or report? Does the reader or consumer have sufficient context? How do we best generate more interest? In creating good figures, we face an additional challenge in accessibility. Some creators of

FIGURE 12–11 Prevalence of Asthma Among Children and Adults in the United States, 2014 (with Data Labels)

Data from Centers for Disease Control and Prevention. 2014 National Health Interview Survey. http://www.cdc.gov/asthma/most_recent_data.htm.

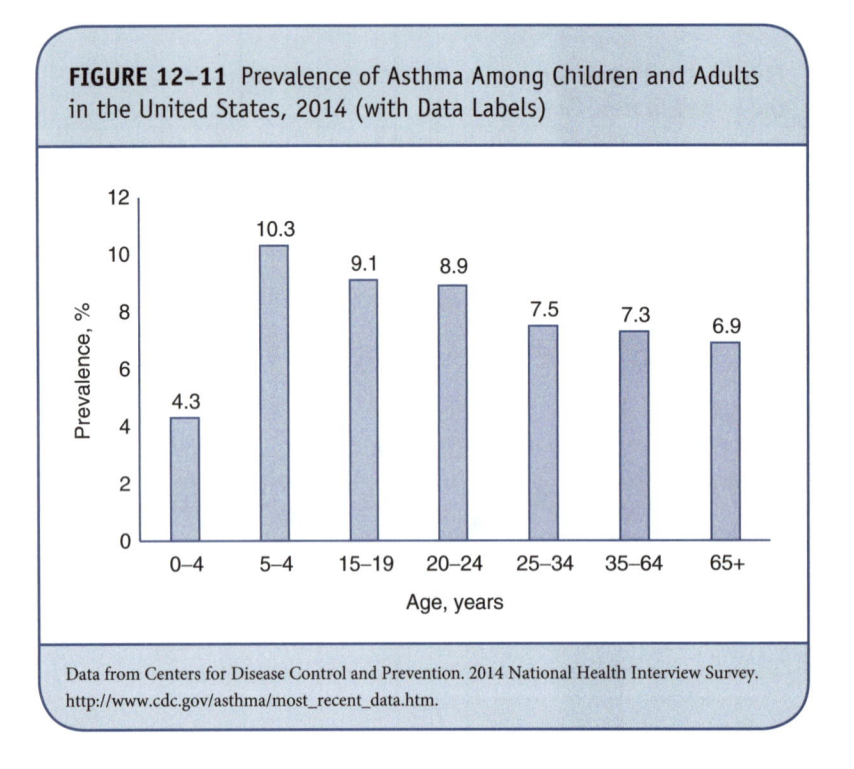

figures use different colors to highlight features of data. Color can be problematic for people who are color blind, or those who might have trouble seeing contrast. For this reason, color should be used carefully, if at all.

Sometimes different symbols are used to represent different groups in a figure. In some instances, the symbols are not sufficiently distinct, making it difficult for the reader to differentiate groups.

Remember that we always want to highlight the data, but not other features or effects that detract from or are extraneous to the data. Gridlines, markers, and axes should not compete with the data. For example, adding a dimension to a figure does not usually enhance its interpretability.

Example 12.7. The U.S. Department of Housing and Urban Development's Office of Community Planning and Development regularly investigates homelessness, with the goal of ending homelessness in the United States. The data shown in **Figures 12–12** and **12–13** are based on data contained in this agency's report entitled "Part I: Point-in-Time Estimates of Homelessness, the 2015 Annual Homeless Assessment Report (HAR) to Congress, November 2015."[14] This report counts homeless persons on a given night (point-in-time) across the United States. In 2015, there were 564,708 persons who were homeless on a given night in the country. Approximately 40% were female and 60% were male. Figures 12–12 and 12–13 display the percent homeless in 2015 by age group. Figure 12–12 is a typical bar chart, whereas Figure 12–13 is a 3-D bar chart. Does the 3-D enhancement facilitate interpretation of the data?

Recall the notion of "chart junk" coined by Tufte; he also advocates maximizing what he calls the data-to-ink ratio.[3] These concepts are actually similar, as both focus on minimizing extraneous elements of a figure to keep the attention on the main player—the data. **Figure 12–14** is an example of a poorly designed graphical display that uses 3-D cylinders rather than bars. This figure also shows income categories of unequal width, making the data difficult to interpret.

To create clear displays, axes can be labeled with lighter colors, such as gray, so that they go to the background while the data are plotted in a bolder, darker color, such as black. Different shades can highlight what is most and least important, as opposed to showing everything with equal value. Be judicious, though: Too much shading can be distracting, so shading needs to be used very cautiously, if at all. Because people vary widely in their comfort with data and graphical displays, it is good practice to pilot test figures with a diverse audience to ensure that the message they receive is consistently clear.

Figure 12–15 is another example of a poorly designed display that attempts to show change over time in control and treatment groups among participants with higher and lower levels of exposure. In this figure, it is very challenging to decipher the magnitude of the changes in each group. For example, what is the change among participants in the treatment group with higher exposure?

FIGURE 12–12 Percent Homeless in the United States by Age Group, 2015

Data from US Department of Housing and Urban Development, Office of Community Planning and Development. Point-in-Time Estimates of Homelessness: The 2015 Annual Homeless Assessment Report (HAR) to Congress. https://www.hudexchange.info/resources/documents/2015-AHAR-Part-1.pdf. November 2015.

FIGURE 12–13 Homelessness in the United States by Age, 2015: 3D Chart Type

Data from US Department of Housing and Urban Development, Office of Community Planning and Development. Point-in-Time Estimates of Homelessness: The 2015 Annual Homeless Assessment Report (HAR) to Congress. https://www.hudexchange.info/resources/documents/2015-AHAR-Part-1.pdf. November 2015.

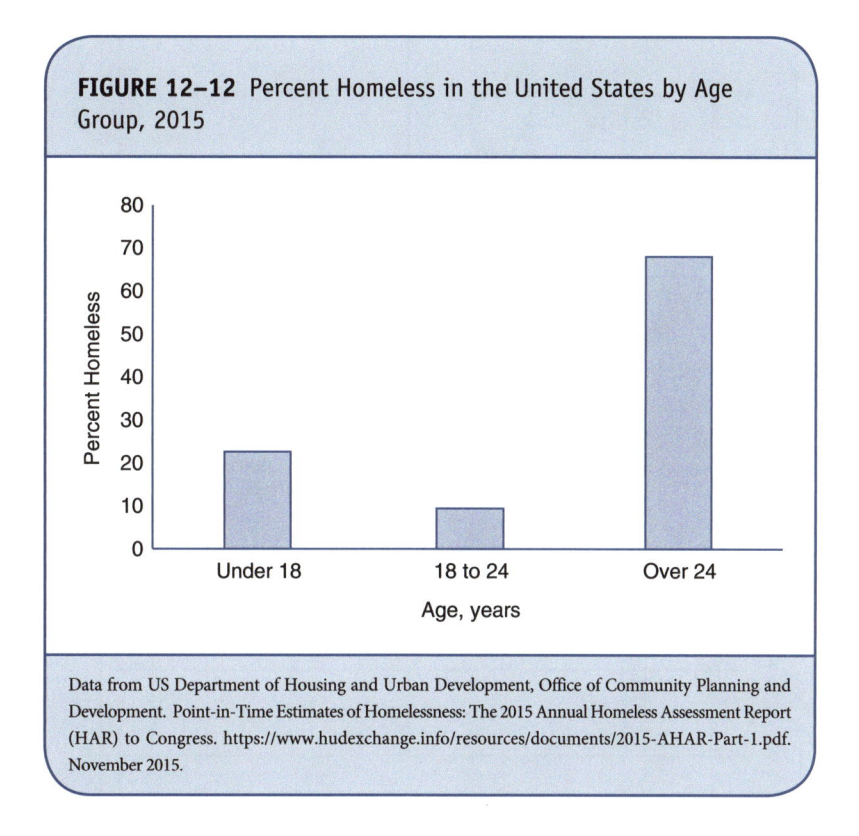

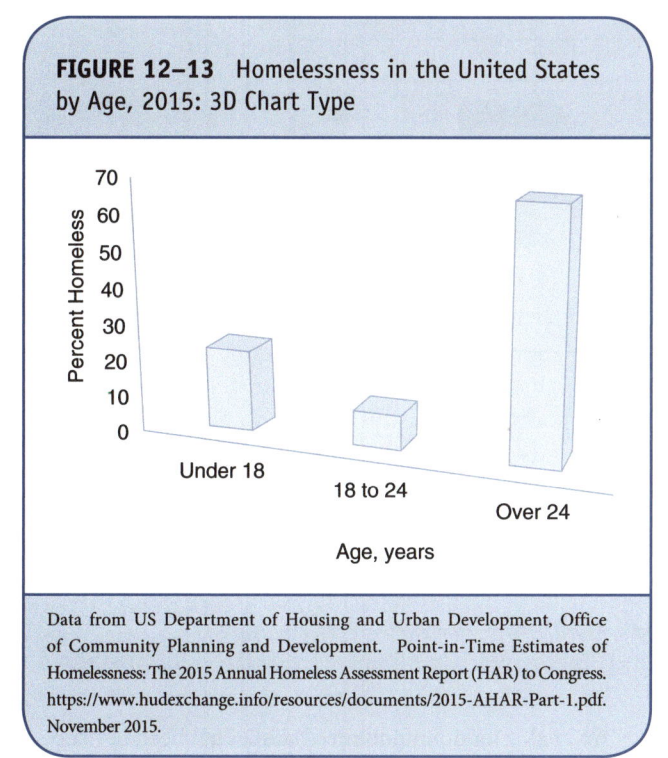

Figures are used for many distinct purposes. In turn, we must select the figure that best matches our purpose or goal. Who is the audience? How much time will they have to interpret the data and statistical results? What is the message? Why is it important? Are we interested in displaying distributions of characteristics in a study sample, or are we trying to make comparisons among groups (who has a higher or lower score)? Are we interested in changes over time? Is the goal to understand correlations or associations between variables? In the remainder of this section, we focus on popular graphical displays. For each, we provide examples of good graphical displays and outline some potential pitfalls.

12.4.2 Participant Flow Charts

Participant flow charts are important graphical displays of study designs and protocols. They can be particularly useful to clarify a complex protocol, as they typically include the number of participants who are eligible, the number excluded and why, the number who withdrew and at which points in the study they withdrew, and so on. Participant flow charts make it easy for the reader to see how the study was run, who was invited to participate, how many consented, and how many ultimately completed the study. These types of graphics are especially important to judge the generalizability of results and to determine whether withdrawals might affect the outcome or any observed associations.

Clinical trials are subject to reporting guidelines set forth by the Consolidates Standards of Reporting Trials (CONSORT), a group of experts who define the reporting standards for clinical trials.[15] These standards include specific

FIGURE 12–14 Income Distribution: Example of a Bad Graphical Display

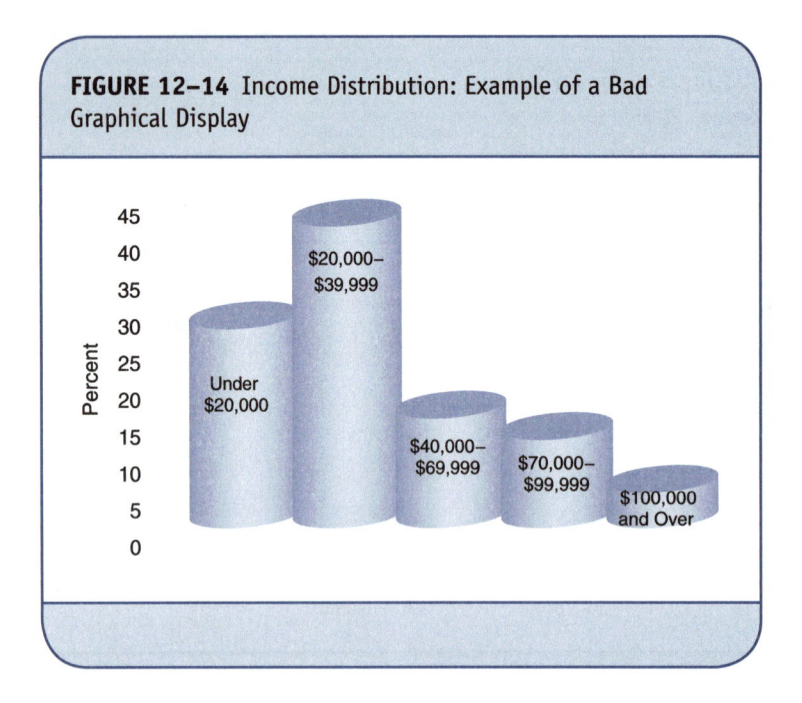

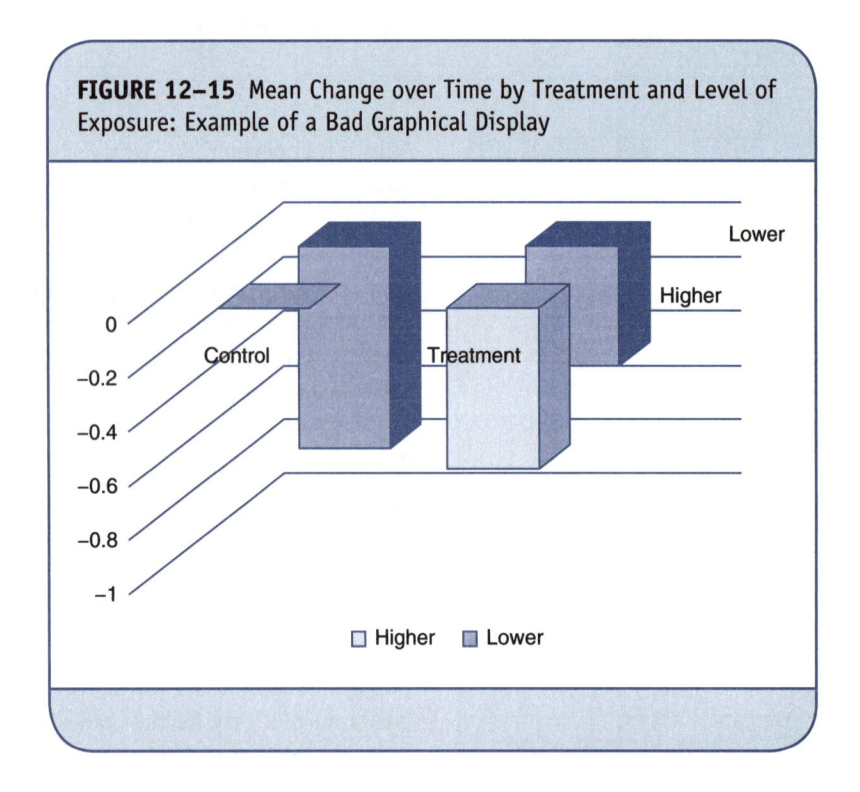

FIGURE 12–15 Mean Change over Time by Treatment and Level of Exposure: Example of a Bad Graphical Display

recommendations for participant flow charts, sometimes called CONSORT charts. **Figure 12-16** shows the format of a CONSORT chart that summarizes participant recruitment and completion; this template is available from the CONSORT website.

An analog for nonrandomized studies has been developed by the Transparent Reporting of Evaluations with Nonrandomized Designs (TREND) group. This template includes a 22-item checklist detailing specific design features for nonrandomized studies and calling for a participant flow diagram.[16]

FIGURE 12–16 Template for Participant Recruitment and Completion

Enrollment

Assessed for eligibility (n=)

Excluded (n=)
- Not meeting inclusion criteria (n=)
- Declined to participate (n=)
- Other reasons (n=)

Randomized (n=)

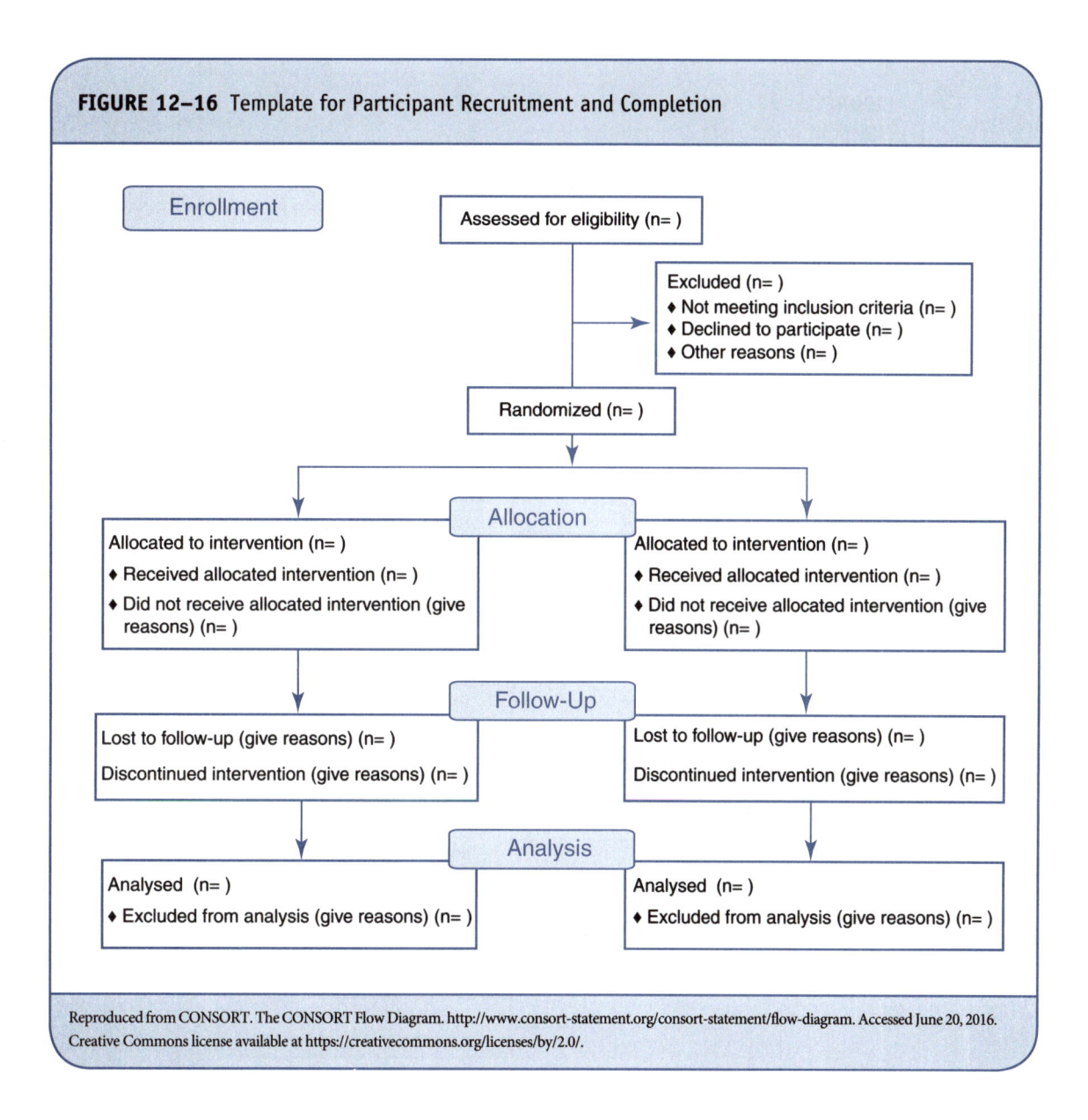

Allocation

Allocated to intervention (n=)
- Received allocated intervention (n=)
- Did not receive allocated intervention (give reasons) (n=)

Allocated to intervention (n=)
- Received allocated intervention (n=)
- Did not receive allocated intervention (give reasons) (n=)

Follow-Up

Lost to follow-up (give reasons) (n=)

Discontinued intervention (give reasons) (n=)

Lost to follow-up (give reasons) (n=)

Discontinued intervention (give reasons) (n=)

Analysis

Analysed (n=)
- Excluded from analysis (give reasons) (n=)

Analysed (n=)
- Excluded from analysis (give reasons) (n=)

Reproduced from CONSORT. The CONSORT Flow Diagram. http://www.consort-statement.org/consort-statement/flow-diagram. Accessed June 20, 2016. Creative Commons license available at https://creativecommons.org/licenses/by/2.0/.

Participant flow charts are also useful to communicate the details of complex study protocols. For example, suppose a study is designed to evaluate the effects of exposure to anesthesia in infancy (defined here as exposure between 0 and 24 months of age) on cognitive development at age 5. Participants for the study are identified through surgical billing records, and cognitive development is measured at 2 and 5 years of age. Participant flow is depicted in **Figure 12–17**, which provides an example of a protocol that is more efficiently depicted in a figure than might be described in the text of a paper or report.

Figures like Figure 12–17 are becoming increasingly popular as investigators use newly available sources of electronic health data to conduct studies. For example, many hospitals now have electronic medical records (EMRs), and these data or records are being used to investigate important clinical issues. Figures like Figure 12–17 describe how samples are developed for analysis.

12.4.3 Displaying Data and Distributions

Some studies have small sample sizes by design (e.g., case series or basic science studies), while others focus on conditions that are rare (e.g., sudden infant death syndrome). If we wish to display data collected in a small study sample, a good option for a continuous variable is a dot plot. **Dot plots** show actual observations, as opposed to summary statistics, so they are appropriate for small samples.

Example 12.8. A study is performed to measure the ages of mothers who experienced the loss of an infant to

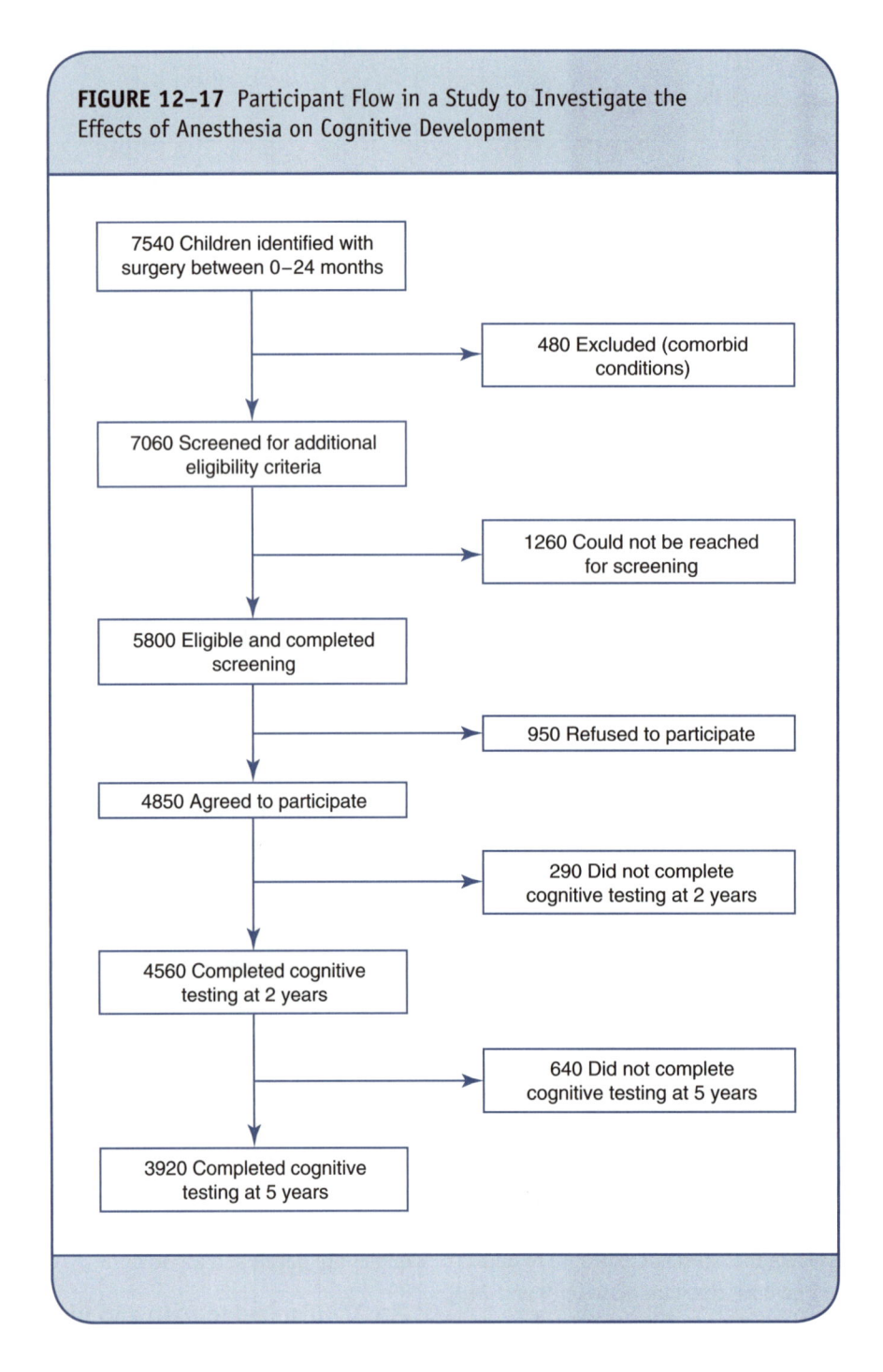

FIGURE 12–17 Participant Flow in a Study to Investigate the Effects of Anesthesia on Cognitive Development

- 7540 Children identified with surgery between 0–24 months
  - 480 Excluded (comorbid conditions)
- 7060 Screened for additional eligibility criteria
  - 1260 Could not be reached for screening
- 5800 Eligible and completed screening
  - 950 Refused to participate
- 4850 Agreed to participate
  - 290 Did not complete cognitive testing at 2 years
- 4560 Completed cognitive testing at 2 years
  - 640 Did not complete cognitive testing at 5 years
- 3920 Completed cognitive testing at 5 years

sudden infant death syndrome (SIDS), defined as death due to unknown cause before 1 year of age. A sample of $n = 9$ infants are identified for whom the cause of death was SIDS. The mother's ages, in years, are as follows and displayed in a dot plot in **Figure 12–18**:

19   21   26   28   30   32   33   39   30

In larger studies, it is important to summarize distributions rather than displaying actual observations. In Chapter, 4 we discussed histograms and bar charts as good ways to summarize the distributions of ordinal and categorical variables, respectively. Here we recap that discussion and add a bit more detail regarding the design of such visualizations.

FIGURE 12–18 Distribution of Mothers' Ages in a Study of Sudden Infant Death Syndrome

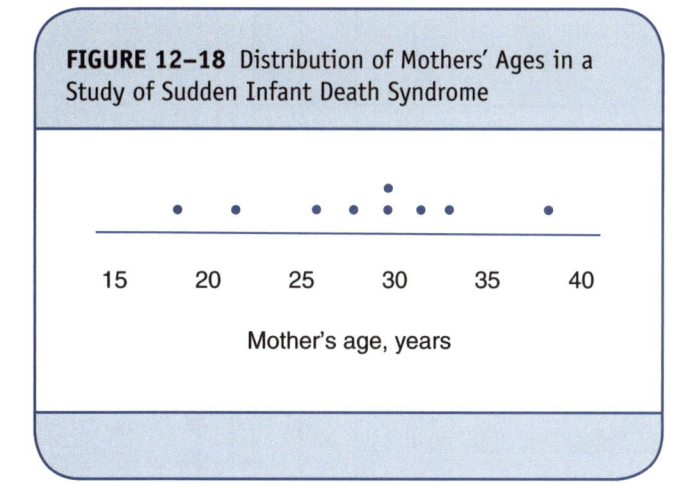

Mother's age, years

Bar charts are used to summarize dichotomous and categorical variables and are considered the easiest figures to construct and understand. In many studies, it is important to show the distribution of a particular characteristic in the study sample or in key subgroups of the sample.

Consider again Example 12.7, which discusses data extracted from the Office of Community Planning and Development's 2015 annual homeless assessment report.[14]

Figure 12–19 shows the racial/ethnic distribution of homeless people (recall from Chapter 4 that race/ethnicity is a categorical variable, so a bar chart is an appropriate graphical display).

The bars in a bar chart can be vertical or horizontal, and we can display frequencies or relative frequencies—whichever makes the most sense to convey the desired message. Figures 12–19 and **12–20** are bar charts displaying the same data, which are presented vertically and then horizontally. Which format is easiest to read and interpret? Do you have a preference?

Figure 12–21 shows the percentages of sheltered and unsheltered homeless people in 2015 by age group.[14] Because homeless persons are classified as either sheltered or unsheltered (a dichotomous variable), we can simplify the presentation and show only one of the responses—for example, the percentages unsheltered (**Figure 12–22**). Which presentation communicates the data more effectively?

Histograms are used to summarize the distributions of ordinal variables. For example, **Figure 12–23** displays the number of hours of physical activity per week (an ordinal variable, with responses ranging from 0 to 15 hours per week) for participants attending the ninth examination of the Framingham Offspring Study.

FIGURE 12–19 Race/Ethnicity of Homeless People in the United States, 2015

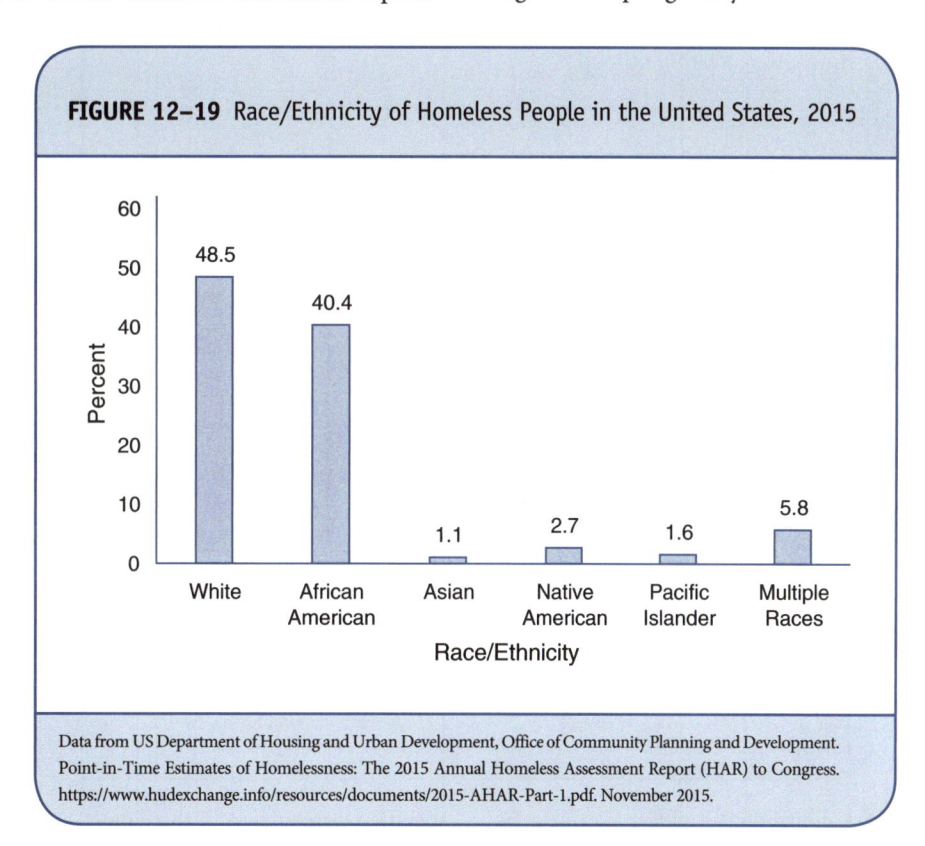

Race/Ethnicity	Percent
White	48.5
African American	40.4
Asian	1.1
Native American	2.7
Pacific Islander	1.6
Multiple Races	5.8

Data from US Department of Housing and Urban Development, Office of Community Planning and Development. Point-in-Time Estimates of Homelessness: The 2015 Annual Homeless Assessment Report (HAR) to Congress. https://www.hudexchange.info/resources/documents/2015-AHAR-Part-1.pdf. November 2015.

FIGURE 12-20 Race of Homeless People in the United States, 2015

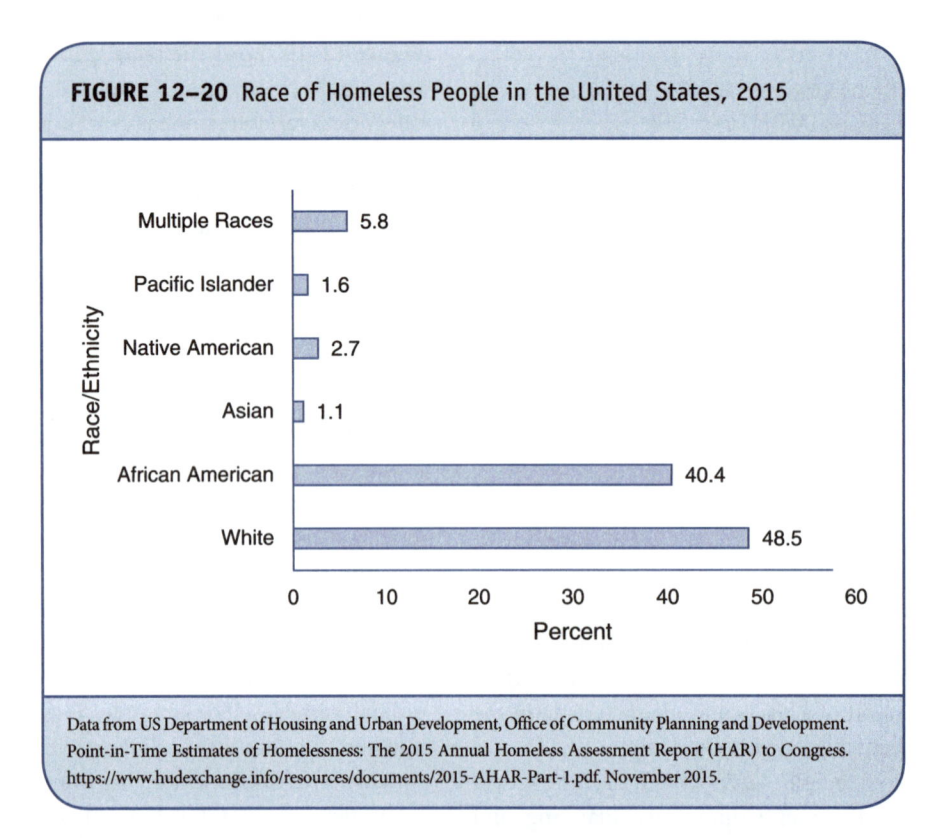

- Multiple Races: 5.8
- Pacific Islander: 1.6
- Native American: 2.7
- Asian: 1.1
- African American: 40.4
- White: 48.5

(Percent on x-axis, Race/Ethnicity on y-axis)

Data from US Department of Housing and Urban Development, Office of Community Planning and Development. Point-in-Time Estimates of Homelessness: The 2015 Annual Homeless Assessment Report (HAR) to Congress. https://www.hudexchange.info/resources/documents/2015-AHAR-Part-1.pdf. November 2015.

FIGURE 12-21 Percentages of Homeless Who Are Sheltered and Unsheltered by Age Group

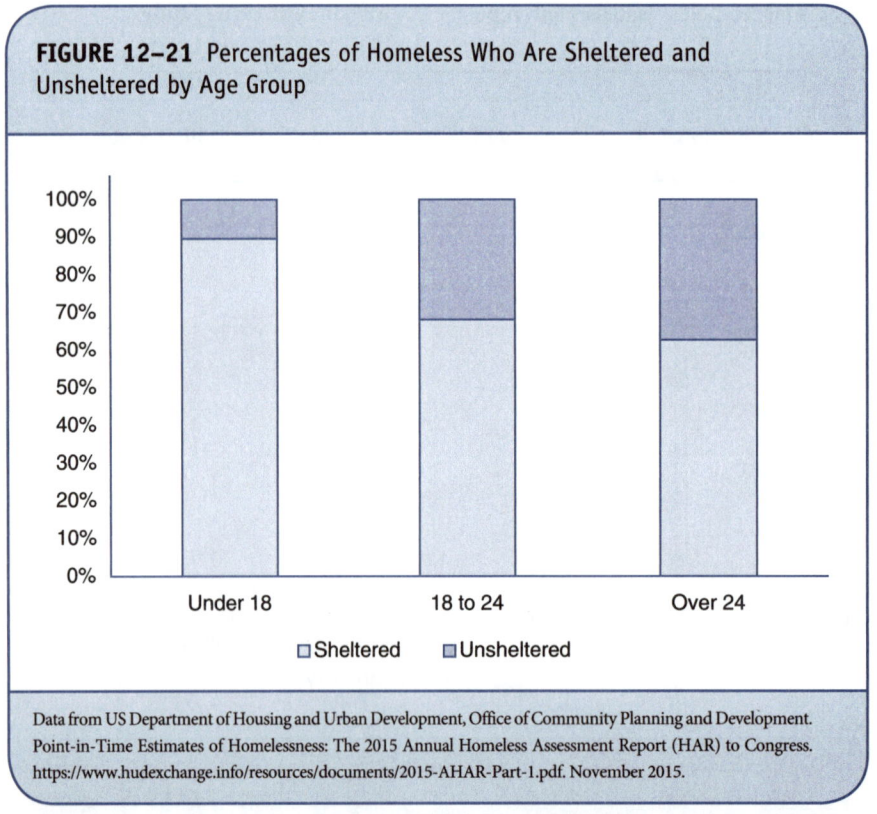

(Stacked bar chart showing Sheltered and Unsheltered percentages for age groups: Under 18, 18 to 24, Over 24)

Data from US Department of Housing and Urban Development, Office of Community Planning and Development. Point-in-Time Estimates of Homelessness: The 2015 Annual Homeless Assessment Report (HAR) to Congress. https://www.hudexchange.info/resources/documents/2015-AHAR-Part-1.pdf. November 2015.

The distribution of a continuous variable can also be displayed in a histogram by creating groups to summarize the data for presentation. The groups can be defined to include specified percentages of the data (e.g., 10 groups, or deciles) or by clinically relevant thresholds (e.g., body mass index categories of normal, overweight, and obese; systolic blood

FIGURE 12–22 Percentages of Homeless Who Are Unsheltered by Age Group

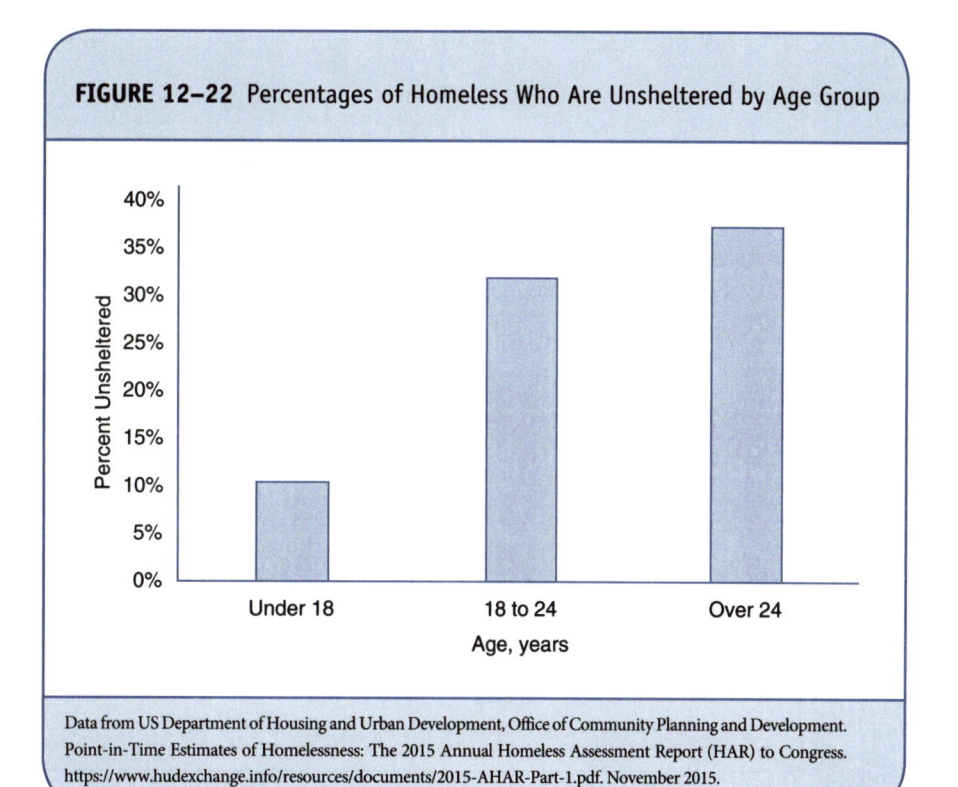

Data from US Department of Housing and Urban Development, Office of Community Planning and Development. Point-in-Time Estimates of Homelessness: The 2015 Annual Homeless Assessment Report (HAR) to Congress. https://www.hudexchange.info/resources/documents/2015-AHAR-Part-1.pdf. November 2015.

FIGURE 12–23 Hours of Physical Activity per Week

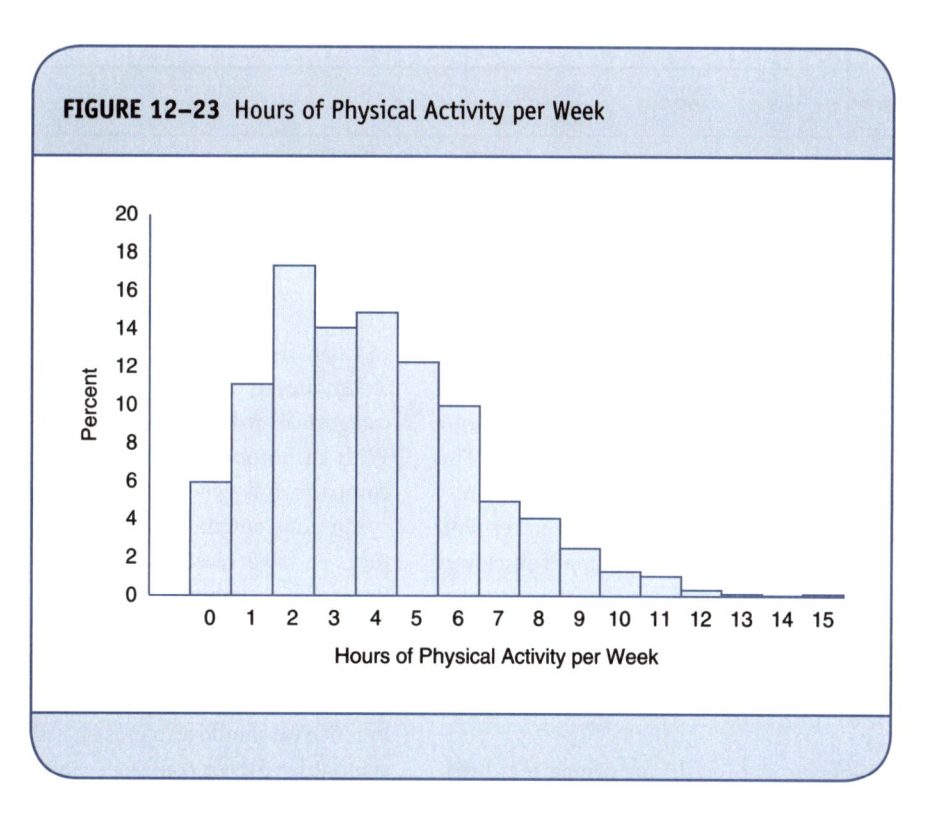

pressure categories of optimal, normal, pre-hypertension, and hypertension). Histograms can be designed to display frequencies or relative frequencies—whichever is best to convey the message. As an example, **Figure 12–24** displays the distribution of total serum cholesterol, measured as a continuous variable in milligrams per deciliter (mg/dL), in

FIGURE 12–24 Distribution of Total Serum Cholesterol (mg/dL)

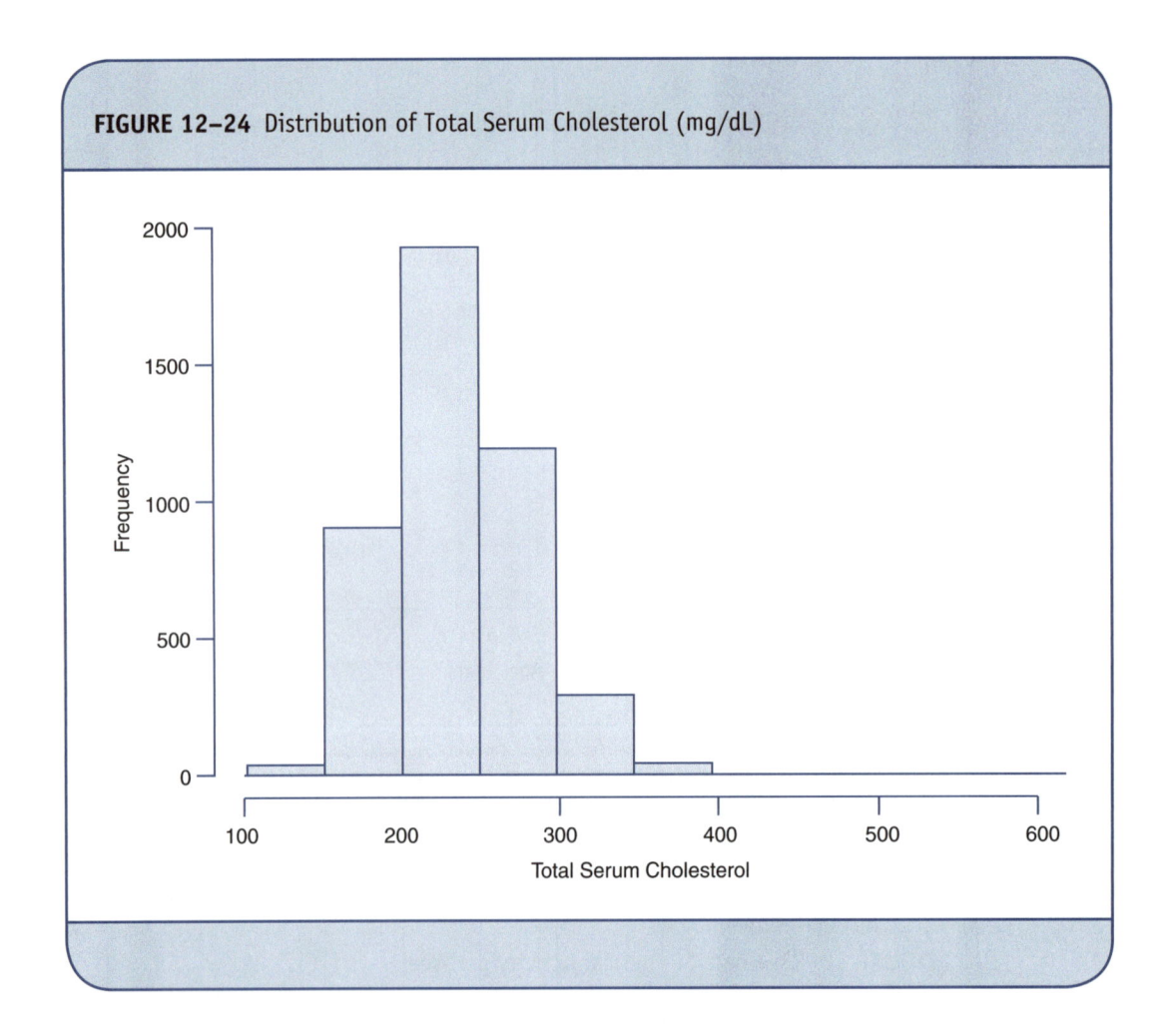

a sample of more than 4000 participants. The continuous measurements are organized into bins of approximately 50 mg/dL units each (e.g., 100–150 mg/dL, 150–200 mg/dL) for presentation.

As discussed in Chapter 4, the distributions of continuous variables can also be displayed using box-whisker plots. The same data shown in Figure 12–24 are depicted in a box-whisker plot in **Figure 12–25**. Recall that in a box-whisker plot, the middle horizontal line is the median, the upper and lower horizontal lines represent the limits for detecting outliers [i.e., $Q_1 - 1.5(Q_3 - Q_1)$ and $Q_3 + 1.5(Q_3 - Q_1)$] and the dots represent outliers. Which figure better conveys to a reader the distribution of total serum cholesterol in the study sample?

Box-whisker plots are also useful for comparing distributions of continuous variables among groups. For example, **Figure 12–26** shows the distributions of total serum cholesterol in men and women. How do total serum cholesterols compare in men and women?

12.4.4 Figures to Compare Estimates Among Groups

In Chapters 6, 7, 10, and 11, we discuss a number of statistical techniques to compare two or more groups on dichotomous, categorical, ordinal, continuous, and time-to-event outcomes. With dichotomous, categorical, and ordinal outcomes, the comparison is generally of proportions or percentages. With continuous outcomes, it generally involves means, assuming that we have reasonably large samples and approximately normally distributed outcomes; otherwise, we focus on comparing distributions using nonparametric tests. With time-to-event outcomes, we usually compare survival curves. Regardless of the statistics that are compared, as a general practice we should also present estimates of variability such as standard errors or confidence intervals for each group so that readers can judge whether groups are significantly different.

Consider again Example 12.4, which described a cohort study that evaluated differences in birth weights and other outcomes of pregnancy in babies born to mothers of different

FIGURE 12–25 Distribution of Total Serum Cholesterol (mg/dL)

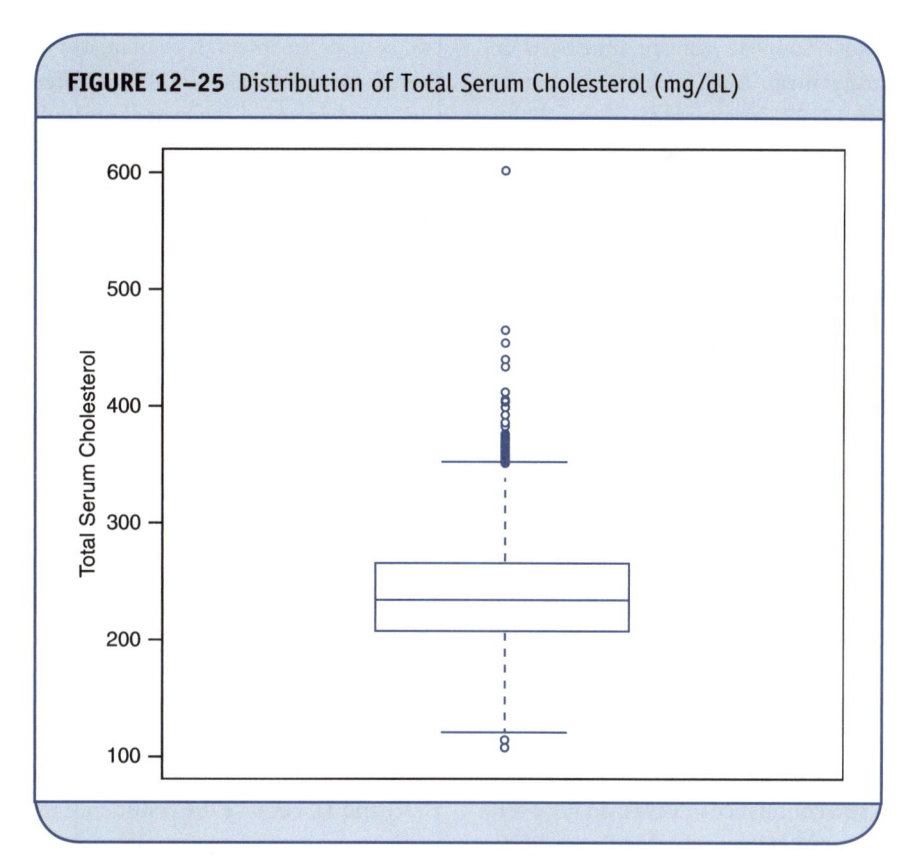

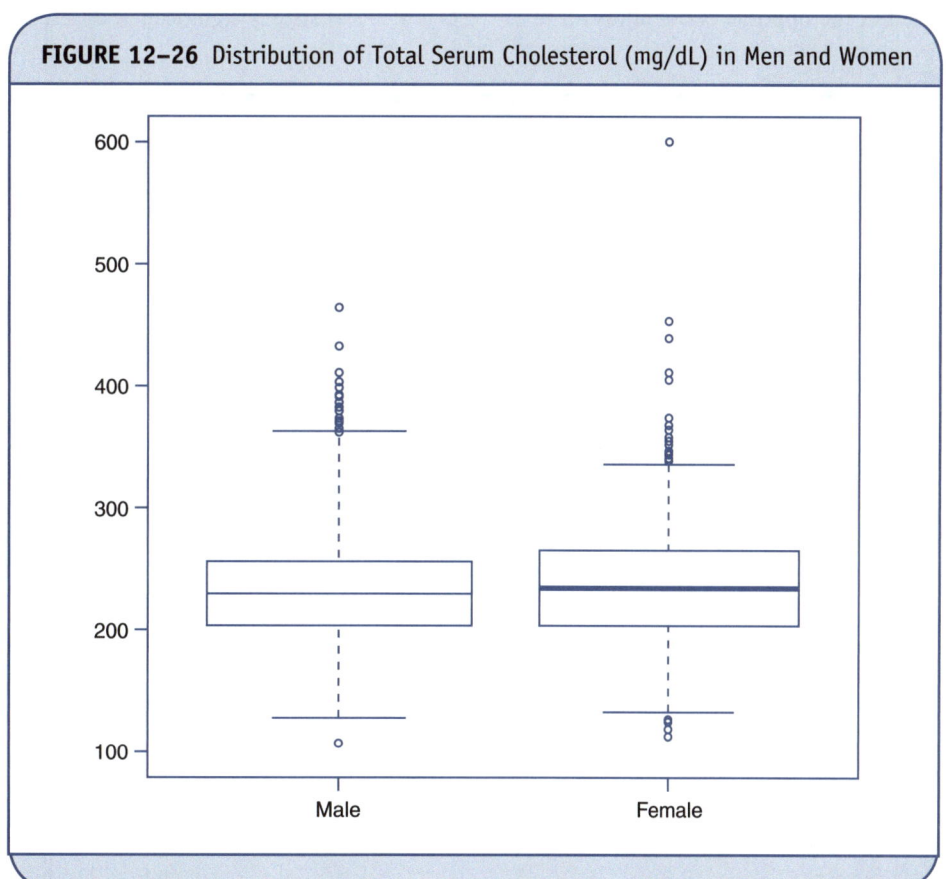

FIGURE 12–26 Distribution of Total Serum Cholesterol (mg/dL) in Men and Women

racial/ethnic backgrounds. Suppose we are interested in estimating and comparing mean birth weights, in grams, by race/ethnicity. **Figure 12–27** depicts mean birth weights along with standard errors of the mean birth weights in each racial/ethnic group.

Standard errors quantify the variability in the estimates of mean birth weight in each group. Suppose an analysis of variance is performed and reveals a statistically significant difference in mean birth weights among the three racial/ethnic groups with $p = 0.04$. It might be of interest to also assess pairwise differences in mean birth weights. We could run three two-independent samples t-tests (comparing mean birth weights between white and Hispanic babies, white and black babies, and Hispanic and black babies), but this increases the overall Type I error rate. A better approach is to conduct a multiple comparison procedure, which allows for multiple pairwise comparisons but controls the overall Type I error rate. Interested readers should see Cabral[17] for more details.

Some readers or consumers of statistical data will compare 95% confidence intervals between groups as a way of judging whether there is statistical evidence of a difference between groups. When two confidence intervals do not overlap, then there is evidence of a statistically significant difference between groups. But the opposite is not always true: If two confidence intervals overlap, it is not always the case that there is no statistically significant difference. We have to bear this point in mind when making judgments based on figures that provide confidence intervals for individual groups. Comparing confidence intervals is not a statistical test.

Figure 12–28 displays mean birth weights along with 95% confidence intervals for mean birth weights in each racial/ethnic group. In this example, all of the confidence intervals overlap, so we cannot draw any conclusions based solely on the within-group confidence intervals about pairwise differences.

An even better display of the data shown in Figure 12–28 appears in **Figure 12–29**, which depicts the means as symbols (boxes), rather than bars. Recall again Tufte's notion of chart junk and the data-to-ink ratio (and the goal of maximizing this ratio).[2] Showing means using bar charts could be viewed as excessive use of ink. In Figure 12–29, we remove the extra "ink" contained in the bars and also make the confidence limits more visible to facilitate the appropriate comparisons.

Consider the data displayed in **Figure 12–30**. Which means are significantly different? The mean in group B is statistically significantly different from the means in groups A, C, and D, because the confidence intervals for the respective comparison groups do not overlap. The mean in group C is also statistically significantly different from the mean in

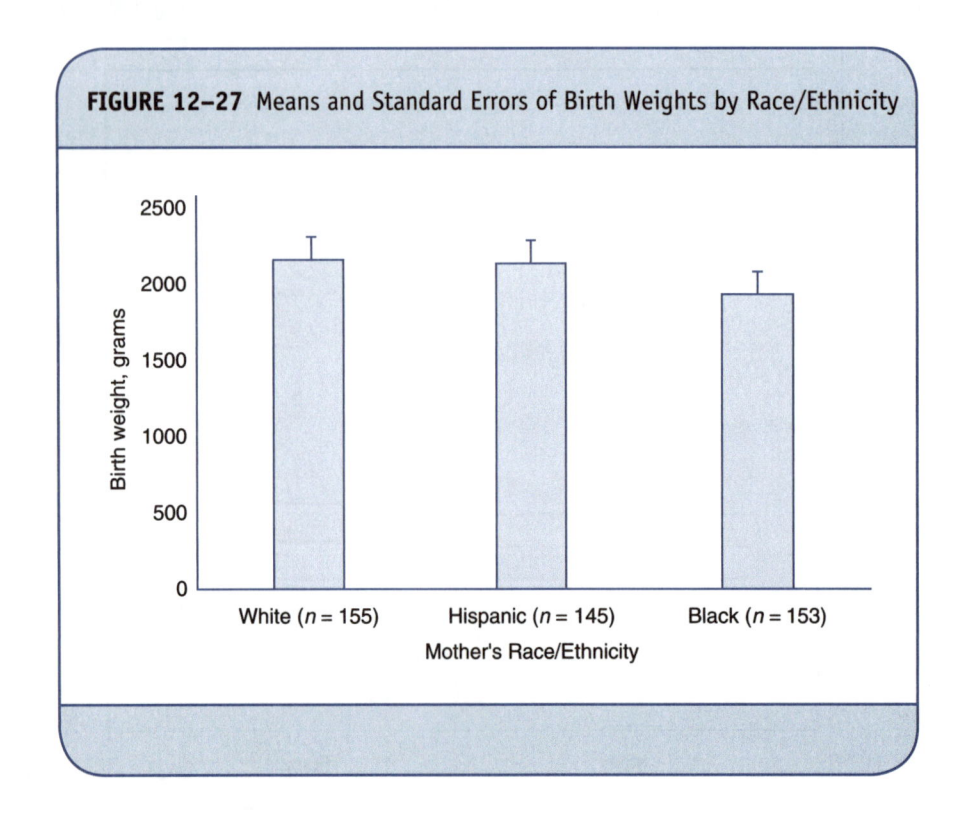

FIGURE 12–27 Means and Standard Errors of Birth Weights by Race/Ethnicity

FIGURE 12-28 Means and 95% Confidence Intervals for Birth Weights by Race/Ethnicity

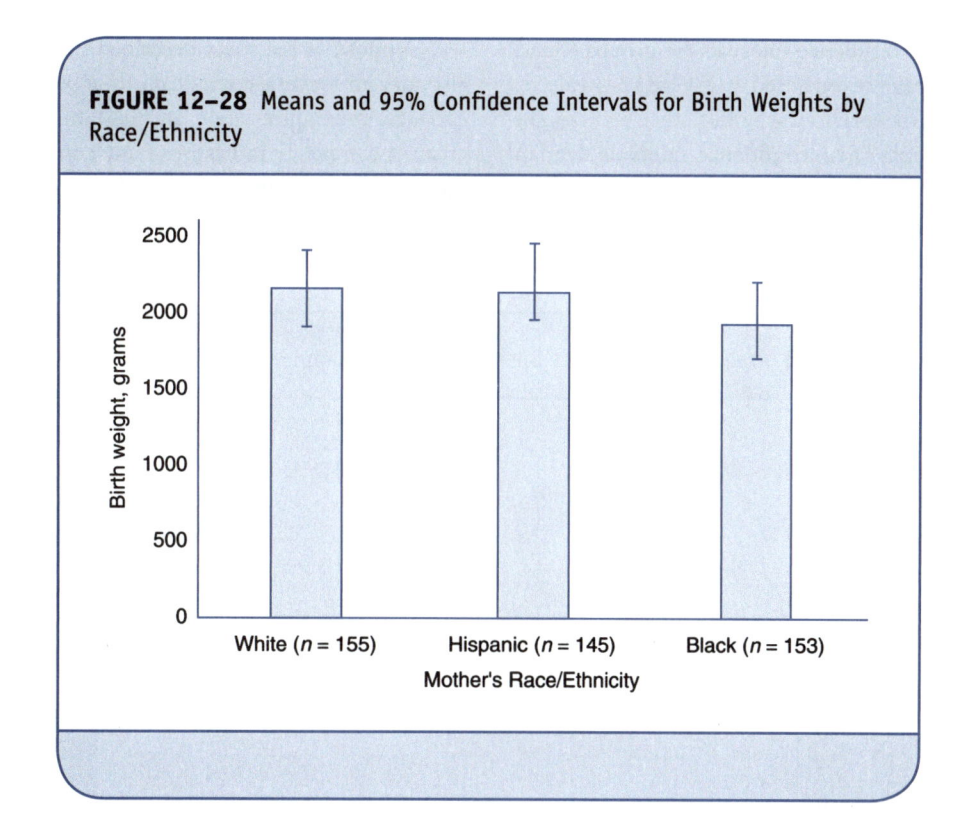

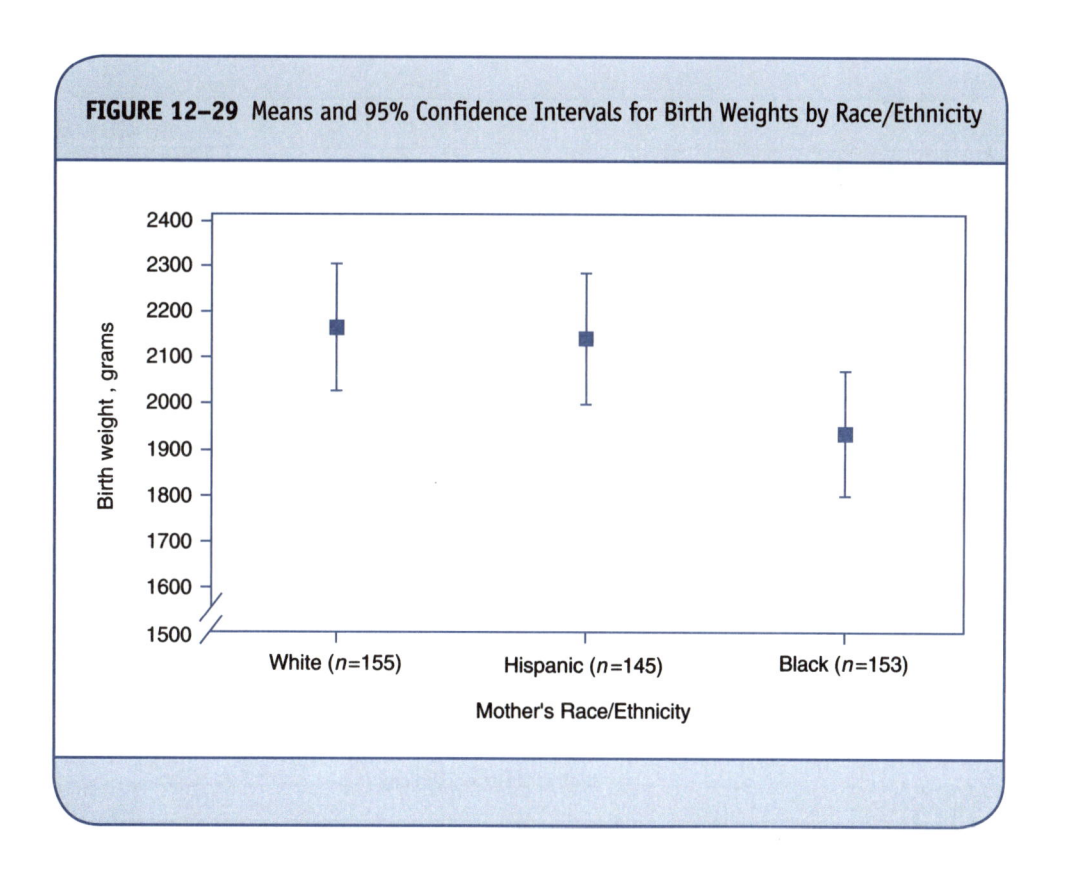

FIGURE 12-29 Means and 95% Confidence Intervals for Birth Weights by Race/Ethnicity

group D, because the confidence intervals for groups C and D also do not overlap. However, based on Figure 12–30 alone, we cannot make judgments about whether other pairs might be statistically different. Their confidence intervals overlap and, therefore, we cannot draw any inferences about statistical significance. If statistical significance is of primary interest, then formal statistical tests should be performed.

Suppose in the study described in Example 12.4 that we also measure several complications of pregnancy, such as miscarriage, premature labor, and gestational diabetes, and we create a composite indicator of any complication (a dichotomous variable). **Figure 12–31** displays the proportions of mothers with complications, along with 95% confidence intervals for the proportions in each racial/ethnic group.

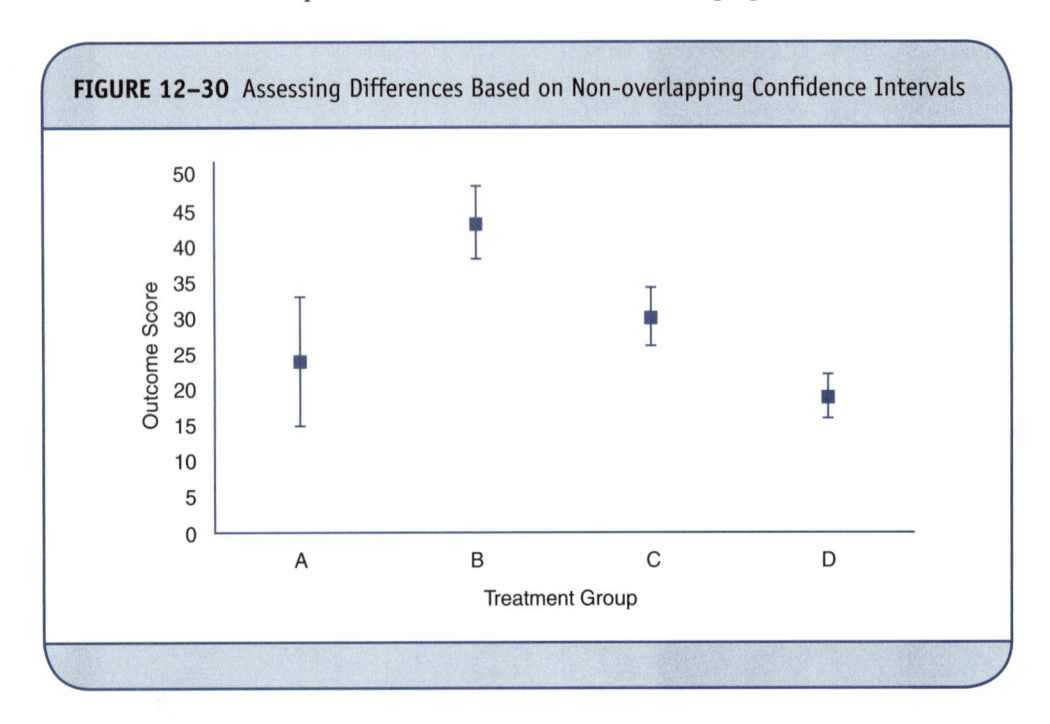

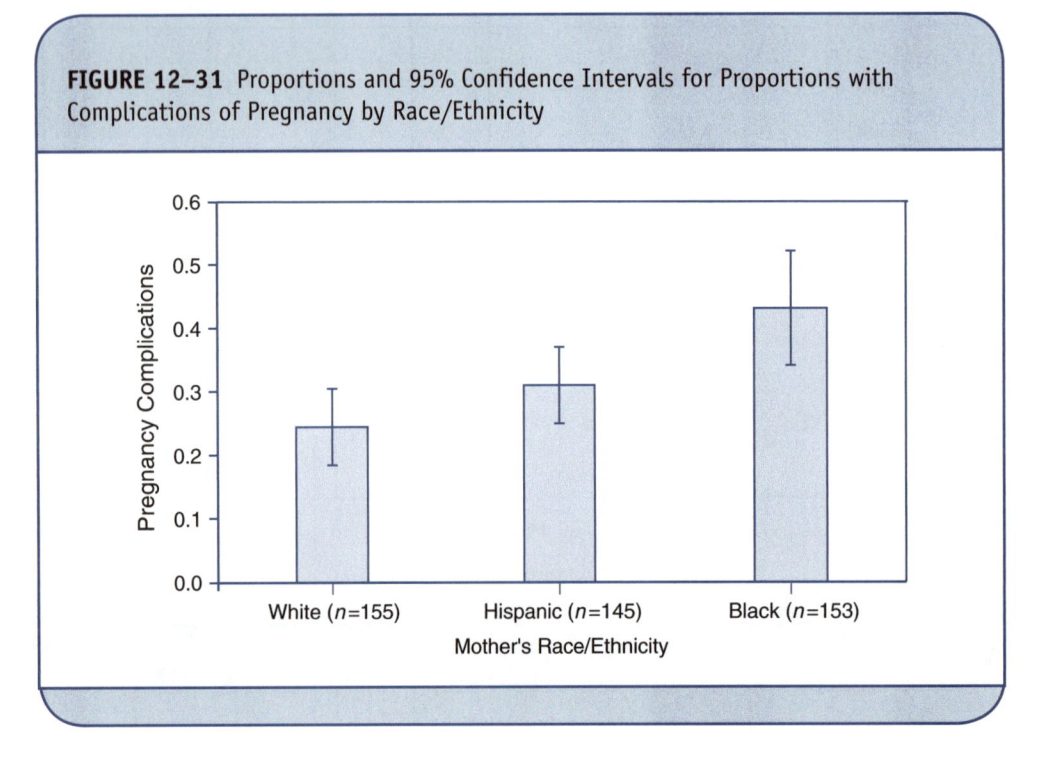

FIGURE 12–30 Assessing Differences Based on Non-overlapping Confidence Intervals

FIGURE 12–31 Proportions and 95% Confidence Intervals for Proportions with Complications of Pregnancy by Race/Ethnicity

A chi-square test of independence is run and finds a statistically significant difference in the proportions of mothers with pregnancy complications by racial/ethnic group with $p = 0.002$. Based on the 95% confidence intervals shown in Figure 12–31, which groups are significantly different? **Figure 12–32** depicts the same data but without the bars. Which presentation is easier to interpret?

As noted earlier, sometimes figures show confidence limits around estimates of means or proportions, and sometimes they show standard error bars. Either choice is acceptable, as long as it is clear to the reader—that is, the choice should be identified in the figure's title or in a footnote. If there are many comparison groups, the confidence limits or error bars can clutter the figure, so it is important to decide whether they are truly important to include. In some instances, the confidence limits or error bars of the comparison groups are approximately equal and they can be shown once—for example, off to the side—as long as this detail is appropriately conveyed to the reader.

Consider again Example 12.4, and suppose we want to compare birth weights both by infant sex and by racial/ethnic group. **Figure 12–33** displays the mean birth weights along with 95% confidence intervals for boys and girls by racial/ethnic group. (Note the rescaling of the y-axis.)

Some figure developers make an error by connecting the bars in a bar chart (**Figure 12–34**). Connecting the means is not appropriate, as lines should be used to illustrate change or trends within the same units. Trend lines should not be included in the display, as no such relationships are depicted in the bar chart. In the next section, we outline the proper use of trend lines in figures.

12.4.5 Trends and Line Charts

Line charts are used to visualize trends or change over time—for example, trends in unemployment rates, percentage uninsured, or incidence of flu over a period of 10 or 20 years. Often the x-axis represents time, and distinct lines can be drawn to illustrate trends in different groups over time.

Example 12.9. Ogden et al. reported on trends in prevalence of obesity among children and adolescents in the United States from 1988 through 2014.[18] These researchers define obesity as BMI at or above the sex-specific 95th percentile in the CDC's BMI-for-age growth charts. **Figure 12–35** displays the trends in prevalence of obesity among 2- to 5-year-olds, 6- to 11-year-olds, and 12- to 19-year-olds from 2000 to 2014.

Figure 12–36 shows the same data but in addition to the estimated prevalence for each age group at each time point, 95% confidence intervals are included. In this figure, the confidence limits are shown with dashed lines. The confidence intervals are useful in providing a sense of the precision in the estimates of prevalence, but they make the line chart difficult to read.

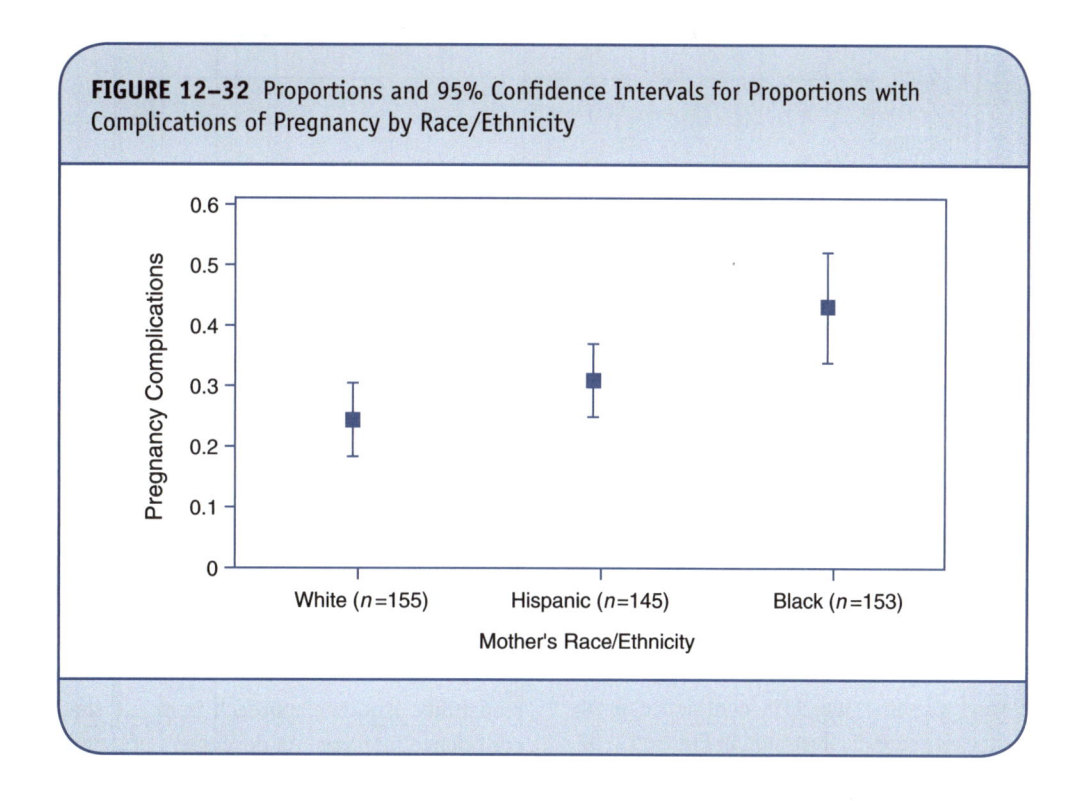

FIGURE 12–32 Proportions and 95% Confidence Intervals for Proportions with Complications of Pregnancy by Race/Ethnicity

FIGURE 12-33 Mean and 95% Confidence Intervals for Mean Birth Weight by Race/Ethnicity and Infant Sex

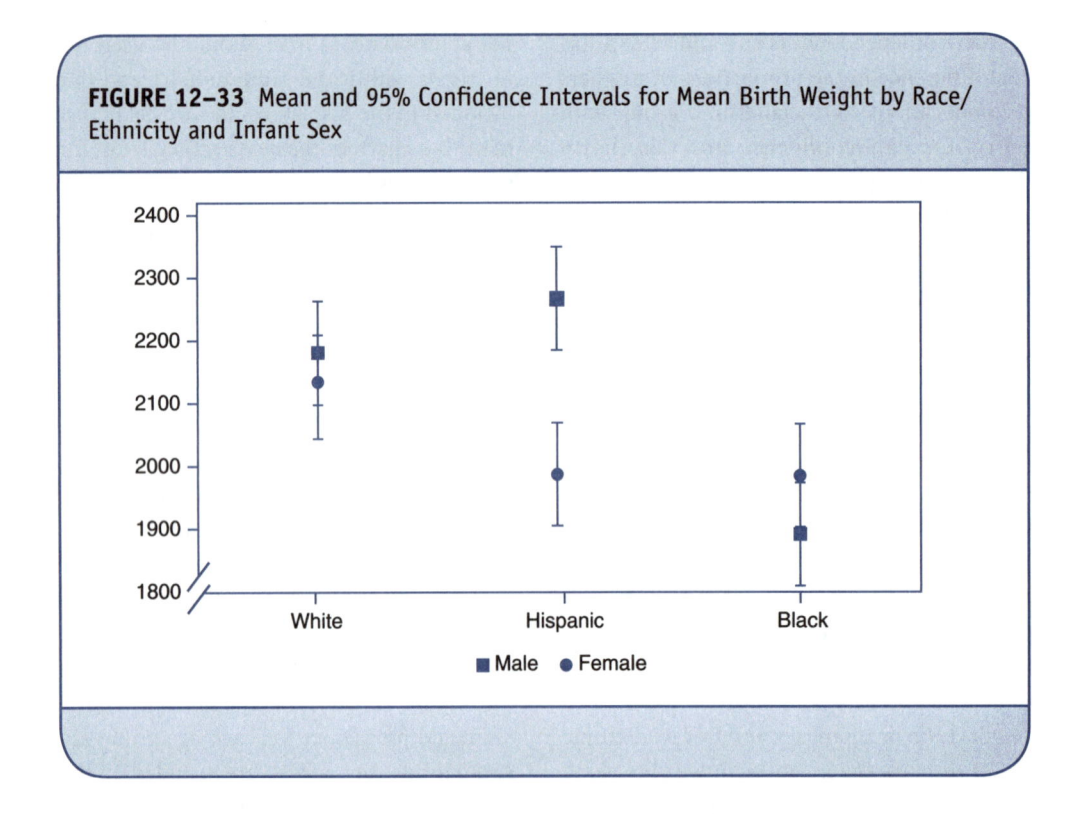

FIGURE 12-34 Mean and 95% Confidence Intervals for Birth Weights by Race/Ethnicity and Infant Sex: Incorrect Use of Trend Lines

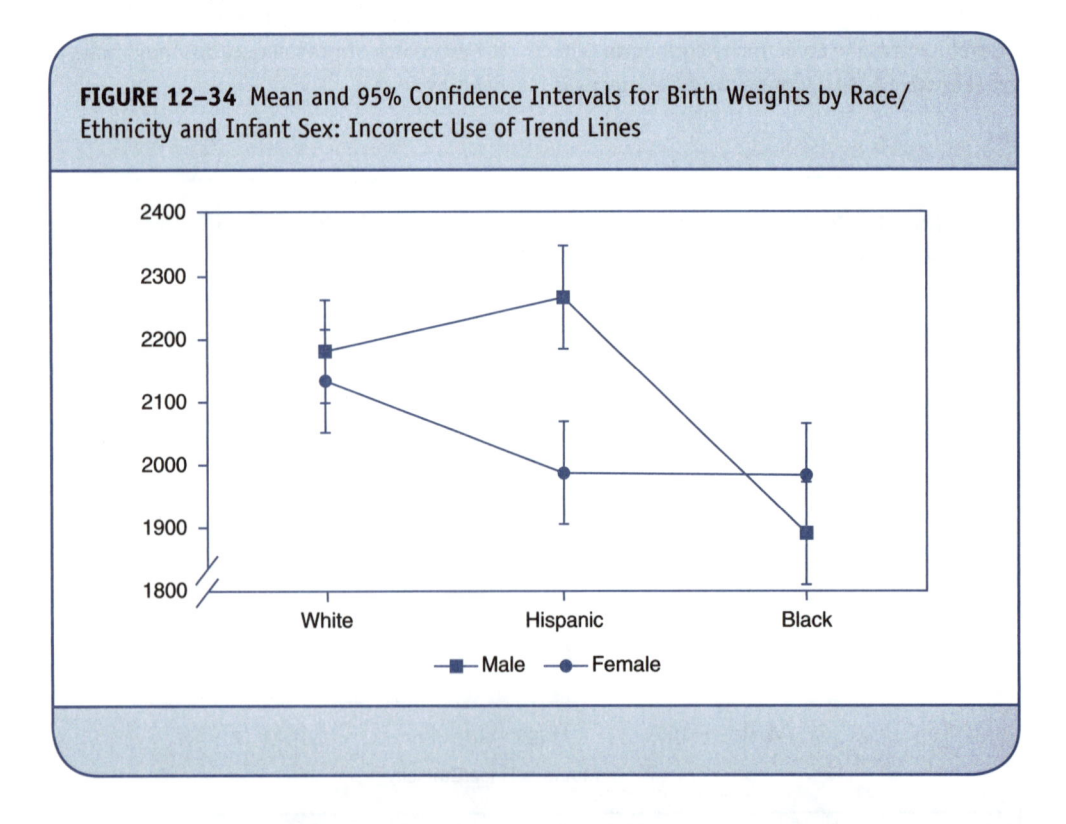

Yet another way to show the 95% confidence limits around the prevalence estimates is illustrated in **Figure 12-37**. The additional white space in this figure enhances its readability. Another approach is to use shading to illustrate confidence intervals. As developers of data displays, we need to experiment with different ways to convey the data accurately

Presenting Data and Statistical Results in Figures

FIGURE 12-35 Prevalence of Obesity Among Children and Adolescents in the United States, 2000–2014

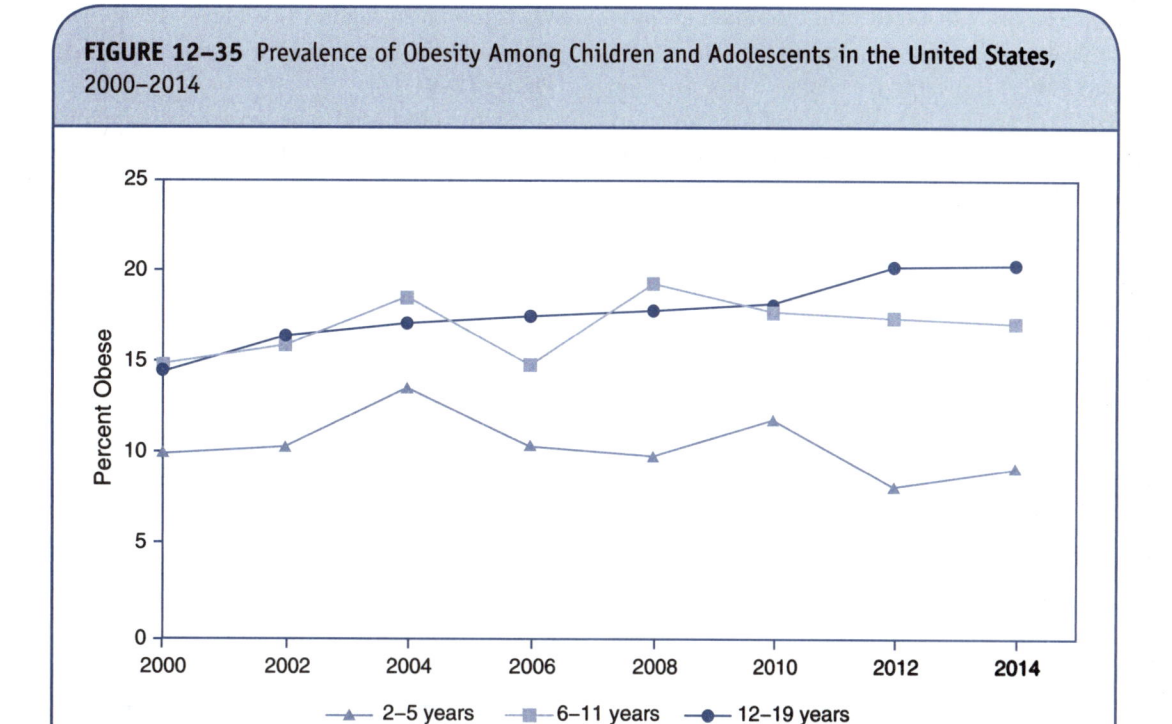

Data from Ogden, C.L., Carroll, M.D., Lawman, H.G., Fryar, C.D., Kruszon-Moran, D., Kit, B.K., and Flegal, K.M. Trends in obesity prevalence among children and adolescents in the United States, 1988–1994 through 2013–2014. JAMA 2016; 315(21): 2292–2299.

FIGURE 12-36 Prevalence and 95% Confidence Intervals of Obesity Among Children and Adolescents in the United States, 2000–2014

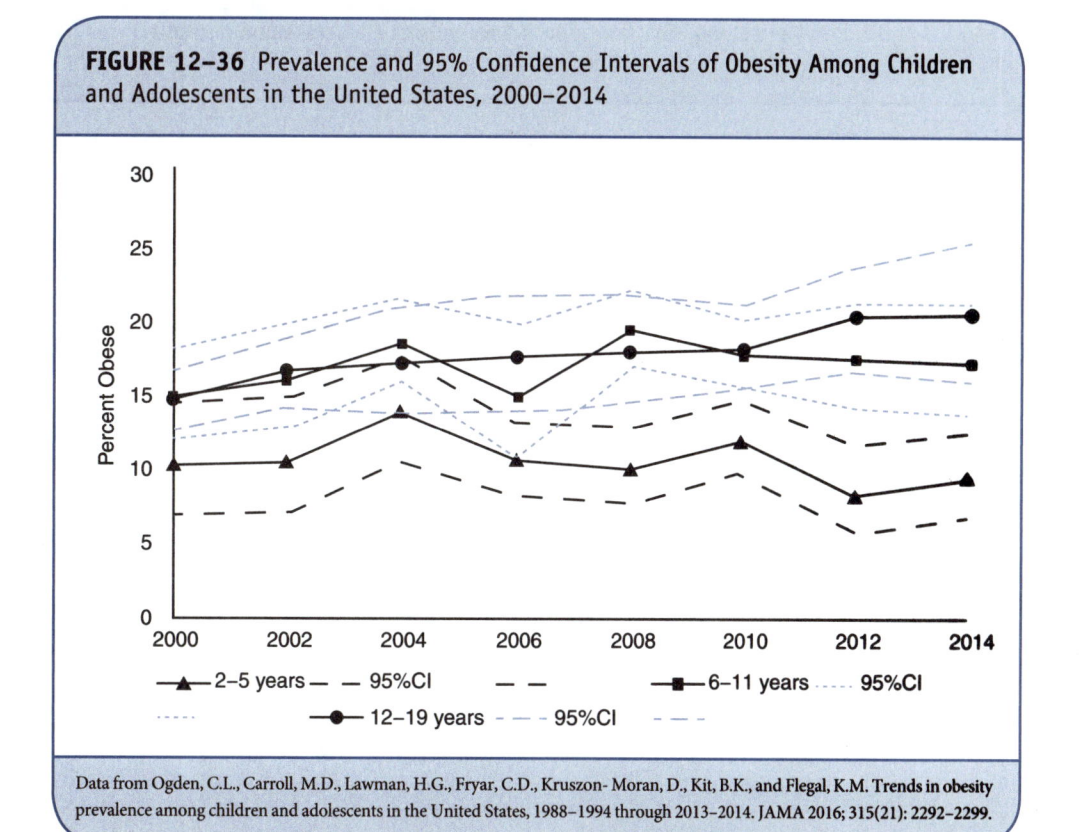

Data from Ogden, C.L., Carroll, M.D., Lawman, H.G., Fryar, C.D., Kruszon-Moran, D., Kit, B.K., and Flegal, K.M. Trends in obesity prevalence among children and adolescents in the United States, 1988–1994 through 2013–2014. JAMA 2016; 315(21): 2292–2299.

and clearly, while ensuring that readers or consumers are provided with sufficient information to interpret the data.

Example 12.10. A clinical trial evaluates a new drug for its effectiveness in lowering LDL cholesterol over a period of 10 years. Participants are randomized to receive either the new drug or placebo; they are followed for 10 years. The participants' LDL cholesterol levels are measured at the start of the study (time 0) and then again after 1, 5, and 10 years of the assigned treatment. **Figure 12-38** summarizes the main findings. Does it appear that the two groups are comparable with respect to LDL cholesterol levels at the start of the study? Is the drug effective? (Notice the differences in effect over time.)

12.4.6 Displaying Associations Among Variables

Scatter plots are useful figures to display associations between two continuous variables—often a continuous outcome or dependent variable (Y) and a continuous predictor (X). **Figure 12-39** is a scatter plot showing the association between systolic blood pressure (in mm Hg) and total serum cholesterol (in mg/dL). **Figure 12-40** shows the same data but with the estimated simple linear regression line ($\hat{y} = -24.36 + 2.0x$, where $b_0 = -24.36$ and $b_1 = 2.0$) added to the display. When producing figures like Figure 12-40, we should not extrapolate the regression line beyond the range of the data (**Figure 12-41**).

Scatter plots need to be constructed carefully, as they can be misleading. Consider the data shown in **Figure 12-42**, which are based on $n = 54$ participants and show the association between days on treatment and response; response is a continuous measure ranging between zero and 50, with higher scores indicative of a more favorable response. The scatter plot suggests a weak, positive association between days on treatment and response. **Figure 12-43** includes the estimated regression line, which has a positive slope ($\hat{y} = 14.59 + 0.17x$, where $b_0 = 14.59$ and $b_1 = 0.17$).

Suppose that there are $n = 34$ females and $n = 20$ males in the sample. **Figure 12-44** is a scatter plot that differentiates responses in males and females by different symbols and shades. Does there appear to be a positive association between days on treatment and response in men or in women? Notice that women are on treatment anywhere from zero to just over 60 days, whereas men receive treatment anywhere between 60 and 100 days. There is also a shift in response between sexes (the mean response among men is 30 as compared to 18 among women).

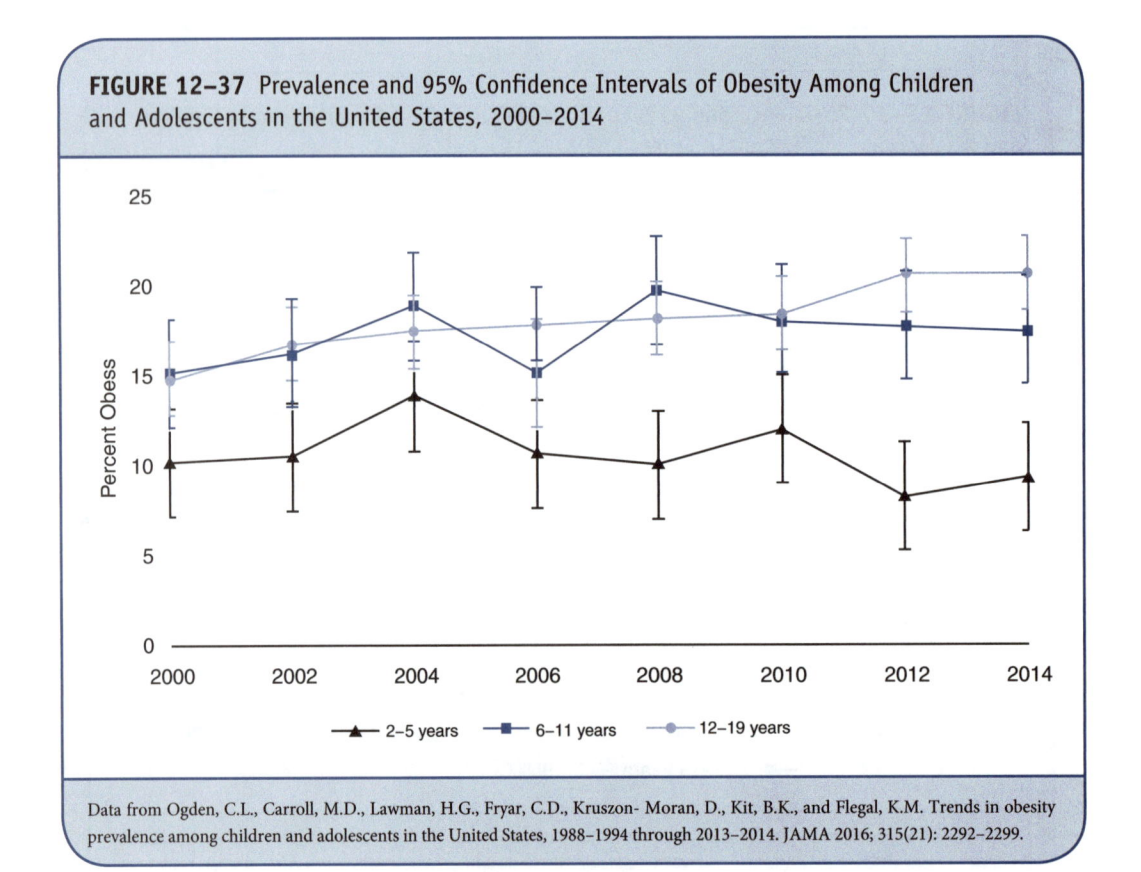

FIGURE 12-37 Prevalence and 95% Confidence Intervals of Obesity Among Children and Adolescents in the United States, 2000–2014

Data from Ogden, C.L., Carroll, M.D., Lawman, H.G., Fryar, C.D., Kruszon-Moran, D., Kit, B.K., and Flegal, K.M. Trends in obesity prevalence among children and adolescents in the United States, 1988–1994 through 2013–2014. JAMA 2016; 315(21): 2292–2299.

FIGURE 12–38 Mean and 95% Confidence Intervals for LDL Cholesterol Levels by Group over Time

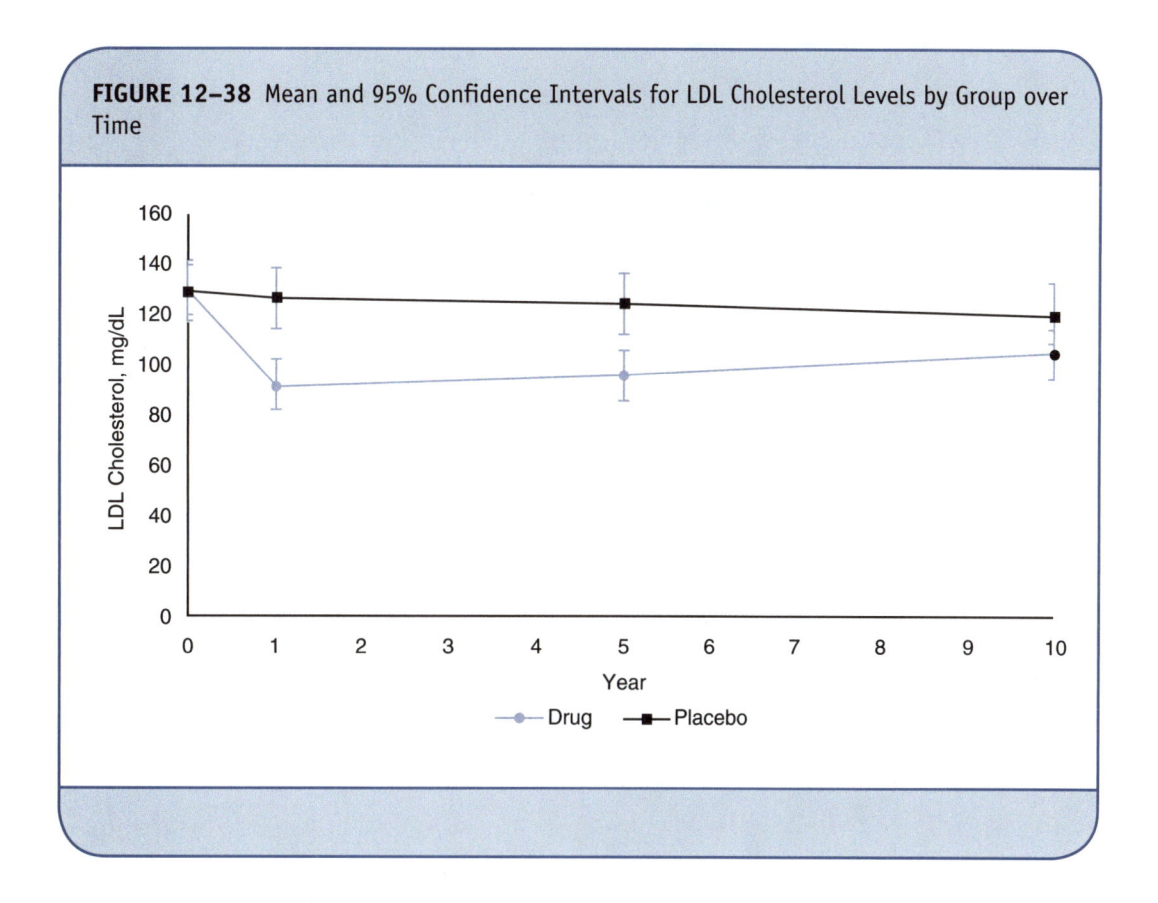

FIGURE 12–39 Association Between Systolic Blood Pressure and Total Serum Cholesterol

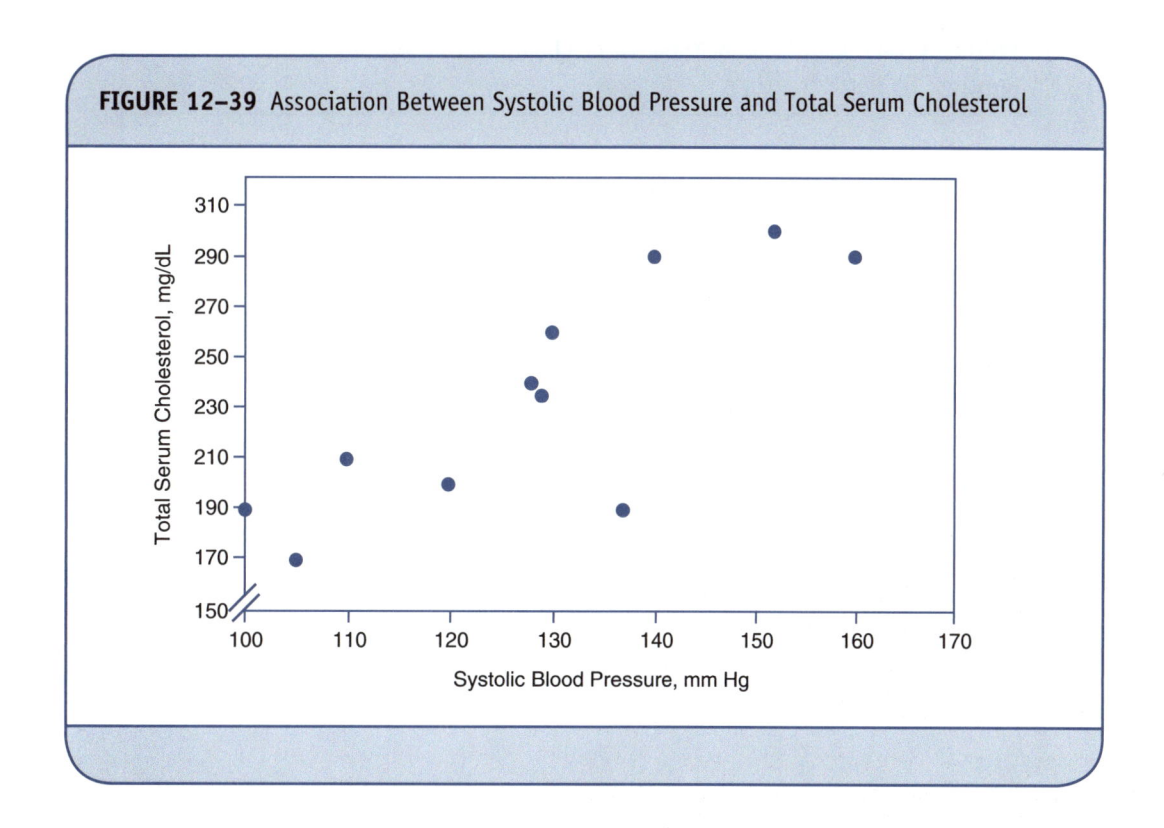

FIGURE 12-40 Regression of Total Serum Cholesterol on Systolic Blood Pressure

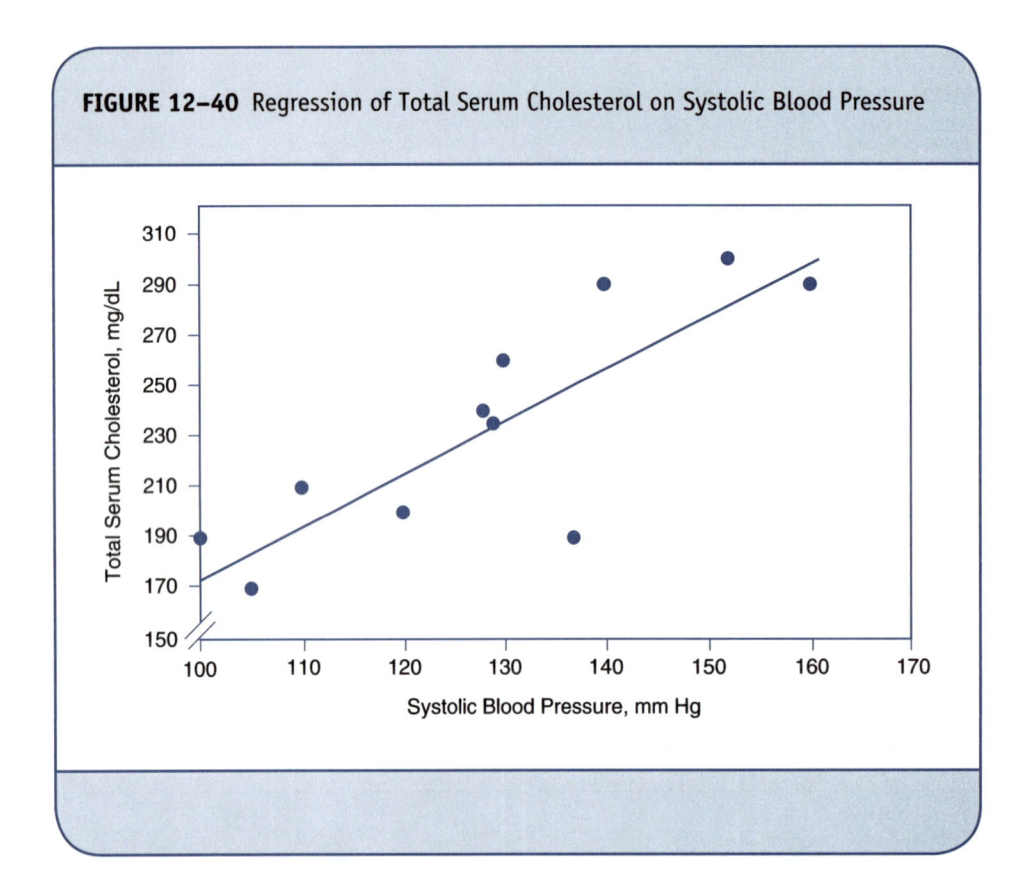

FIGURE 12-41 Regression of Total Serum Cholesterol on Systolic Blood Pressure: Do Not Extrapolate

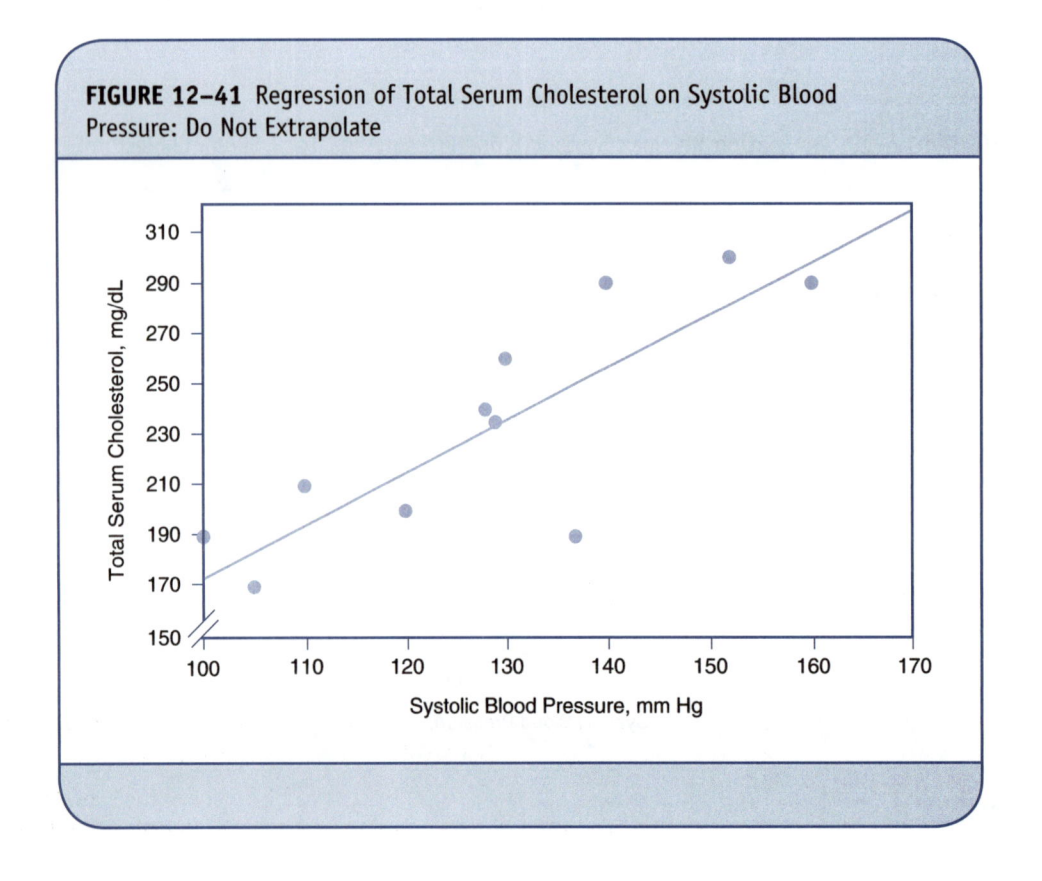

FIGURE 12-42 Association Between Days on Treatment and Response (Total Sample)

FIGURE 12-43 Regression of Response on Days on Treatment (Total Sample)

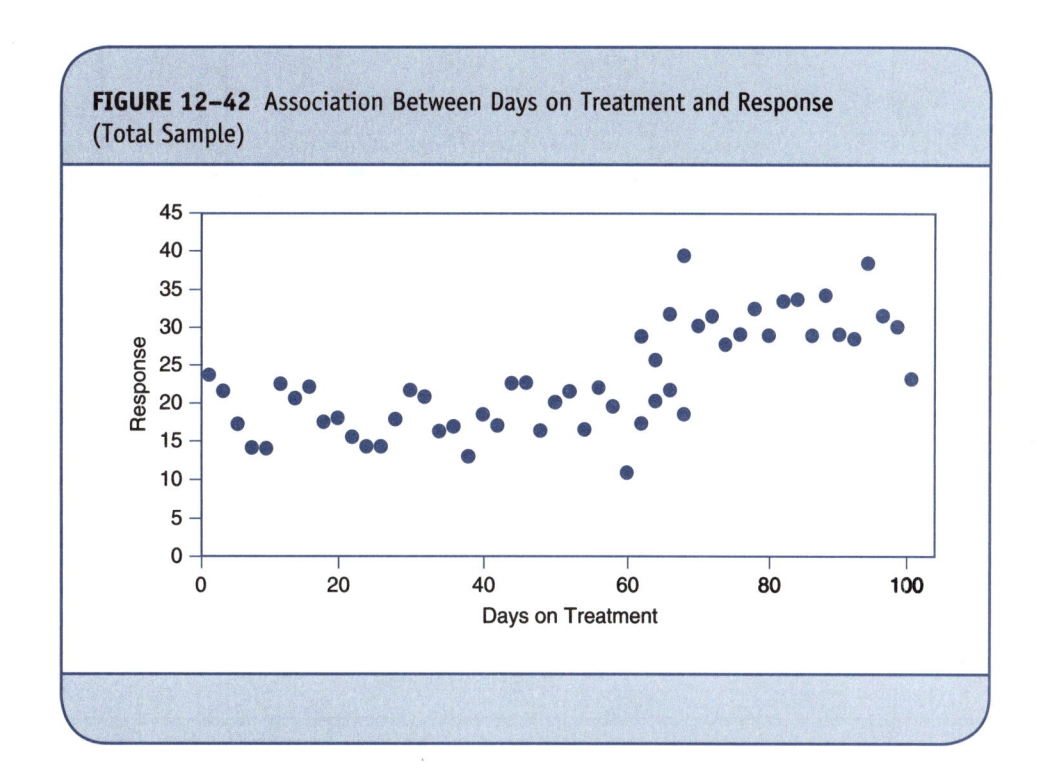

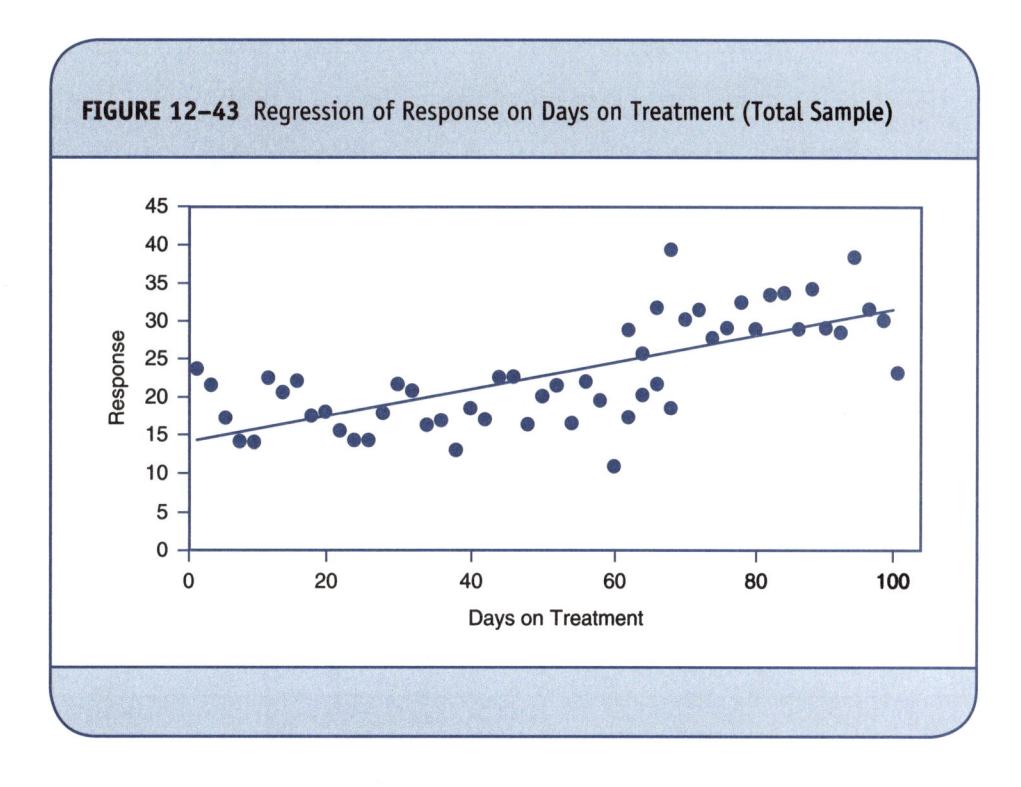

Figure 12-45 includes the regression lines, estimated separately in men and women. Each regression line is essentially flat (the estimated slopes are $b_{1|men} = -0.013$ and $b_{1|women} = 0.0007$). Thus, there is little evidence of an association between days on treatment and response in **either men or women, but pooling the data creates a positive association. This is an example of confounding by sex. As careful analysts, we must always explore the data to investigate potential confounding variables.**

FIGURE 12–44 Association Between Days on Treatment and Responses in Men and Women

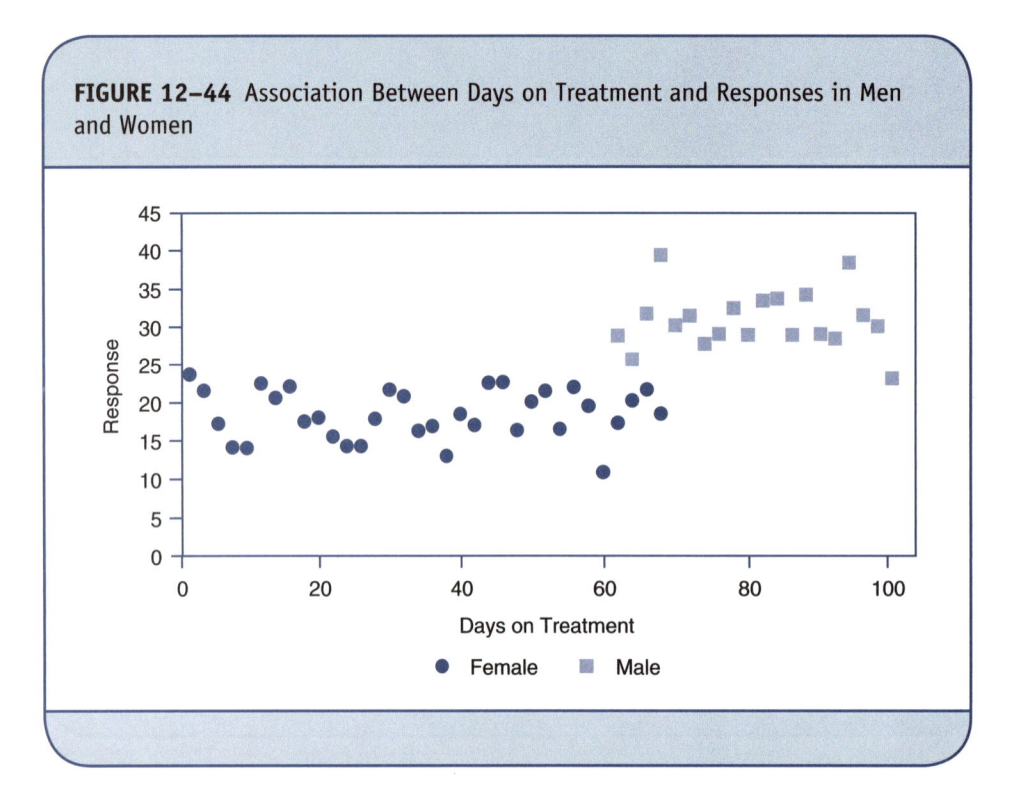

FIGURE 12–45 Regressions of Response on Days on Treatment by Sex

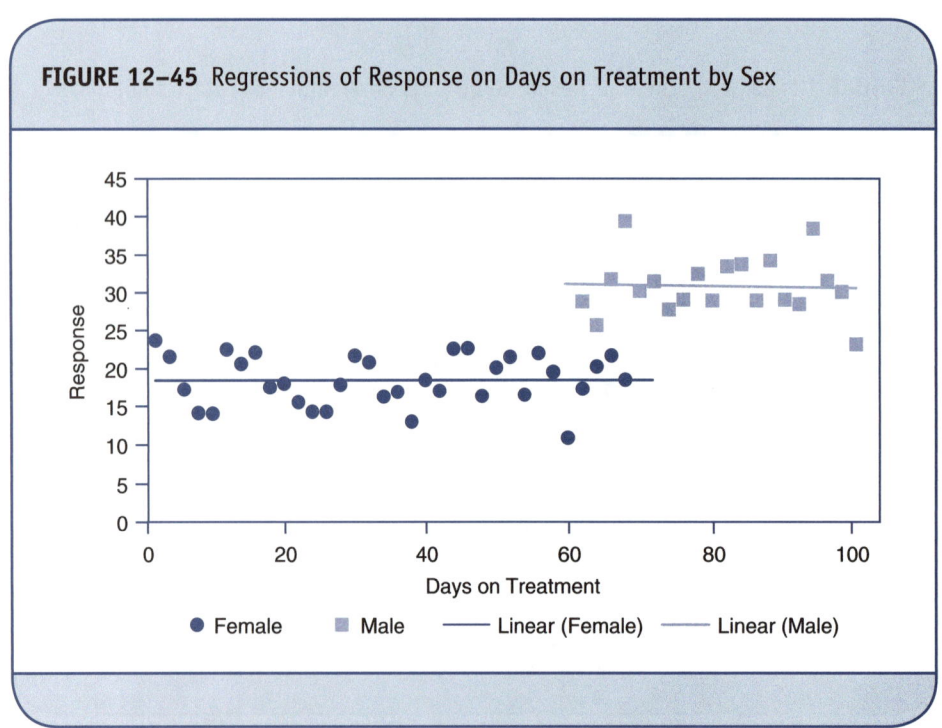

Consider the data shown in **Figure 12–46** based on a small sample of $n = 9$ participants. Assuming that all of the measurements are correct (and none is a data entry error), the observation in the lower-right corner appears to be an outlier—a so-called influential point. It deviates from the remaining observations such that when we fit a regression line, that point pulls the regression line downward. The estimated regression equation is $\hat{y} = 26.83 - 0.66x$, as indicated in **Figure 12–47**. Without that observation (and we should never remove an observation!), the regression equation is $\hat{y} = 23.50 - 0.27x$, based on $n = 8$ observations.

A scatter diagram is a useful tool for examining associations and sometimes identifying extreme observations, relative to the others. We need to be aware of influential observations, particularly when the sample size is small. Figures 12–46 and 12–47 suggest that we might need more data, or a larger sample to explore the entire distribution of values.

We are often involved in studies in which we are investigating associations between a number of potential independent variables and a dependent variable in a multivariable

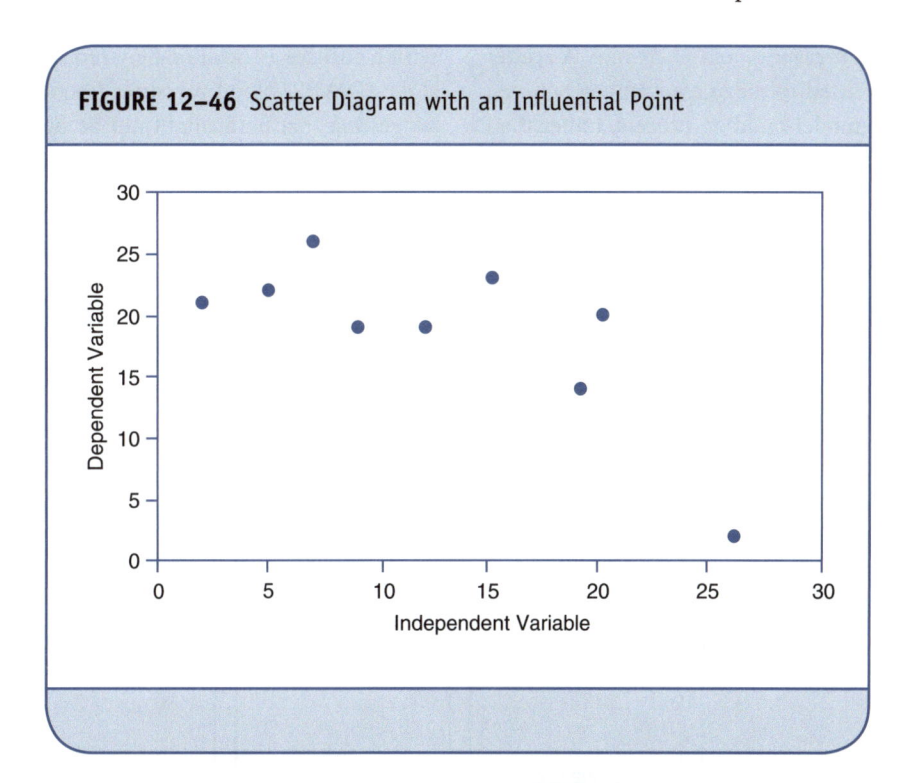

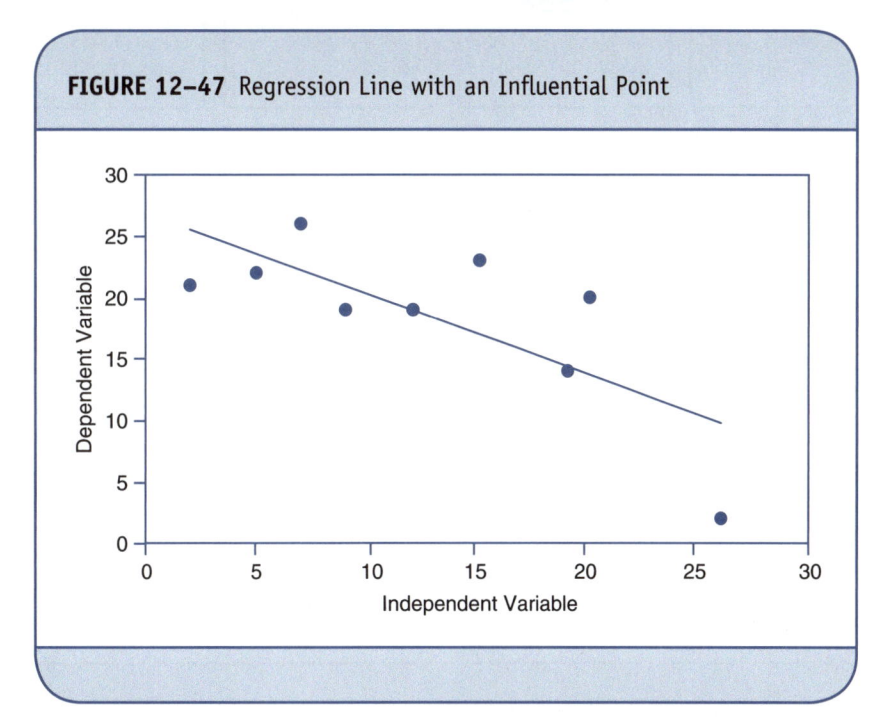

FIGURE 12–46 Scatter Diagram with an Influential Point

FIGURE 12–47 Regression Line with an Influential Point

analysis. With a continuous dependent variable, we might generate a series of scatter diagrams to get a sense of the associations between each independent variable and the dependent variable, considered separately, or between independent variables taken two at a time. **Figure 12–48** is an example of a scatter plot matrix that shows correlations among pairs of independent variables, including total serum cholesterol, systolic and diastolic blood pressures, and heart rate. A scatter plot matrix is generally used to judge associations between variables as part of the model-building process, rather than for presentation purposes. In Figure 12–48, the variables displayed in the rows and columns of the matrix are indicated along the diagonal. Note that the matrix is also symmetric, showing correlations between variables both above and below the diagonal. For example, the correlation shown in the first column, second row (and also in the second column, first row) is the correlation between total serum cholesterol (TOTCHOL) and systolic blood pressure (SYSBP). Note the strong positive association between systolic (SYSBP) and diastolic (DIABP) blood pressures (second column, third row), suggesting that both might not be needed in the multivariable model.

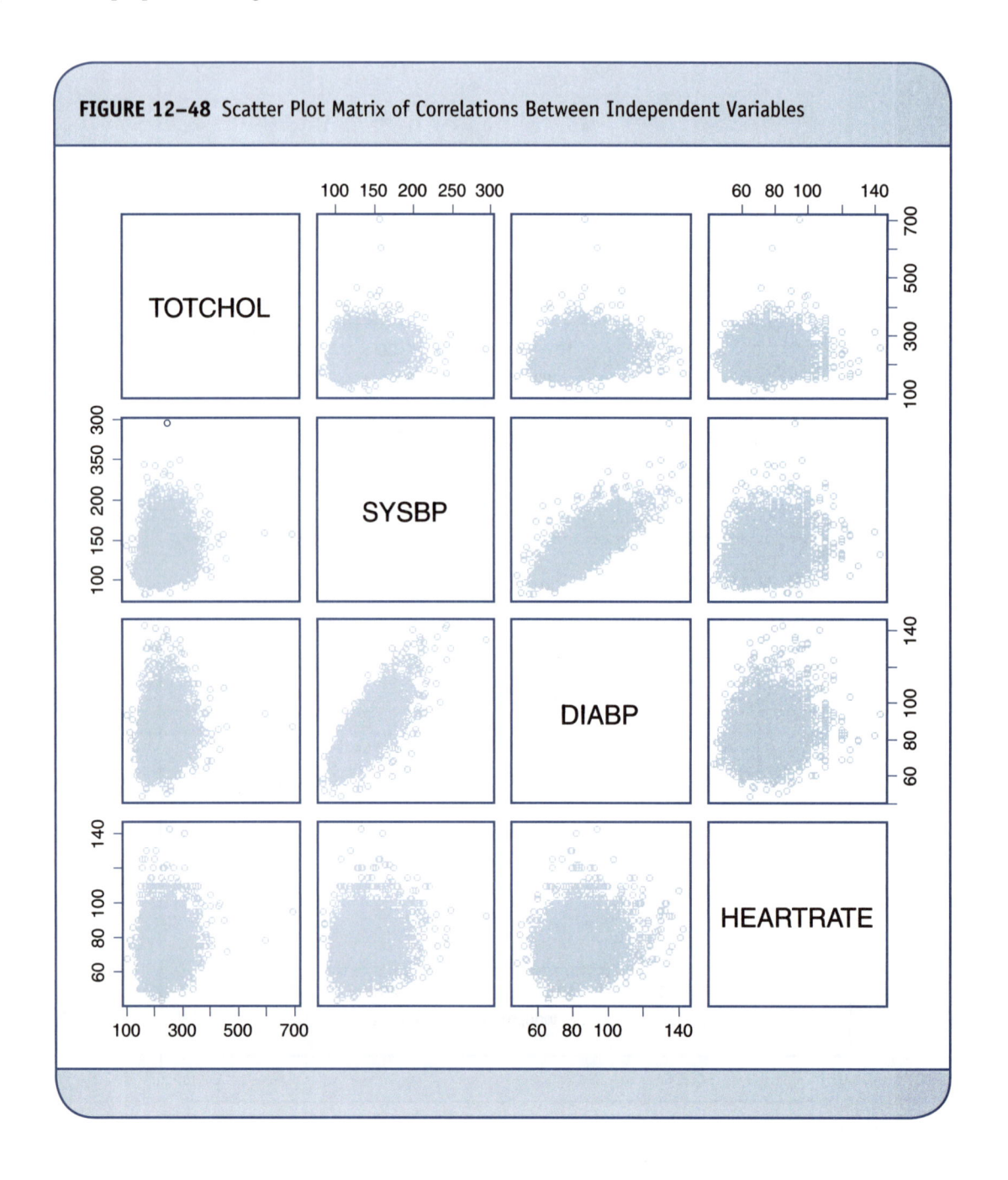

FIGURE 12–48 Scatter Plot Matrix of Correlations Between Independent Variables

There are many ways to display associations between variables. For example, the U.S. Bureau of Labor Statistics regularly reports data on earnings and employment rates as they relate to one another and to other characteristics such as educational attainment.[19] **Figure 12–49** shows median weekly earnings (a continuous measure) and unemployment rates (dichotomous variable) organized by levels of education (an ordinal variable). The data are based on adults age 25 years and older who are working full-time. **Is there an association between educational attainment and earnings? If so, what is the nature of this association (i.e., positive or negative correlation)? Is there an association between educational attainment and unemployment rate?**

Sometimes line plots are used to explore associations between variables. Often there are two distinct y-axes shown on both the right and left sides of the plot area (see **Figure 12–50**,

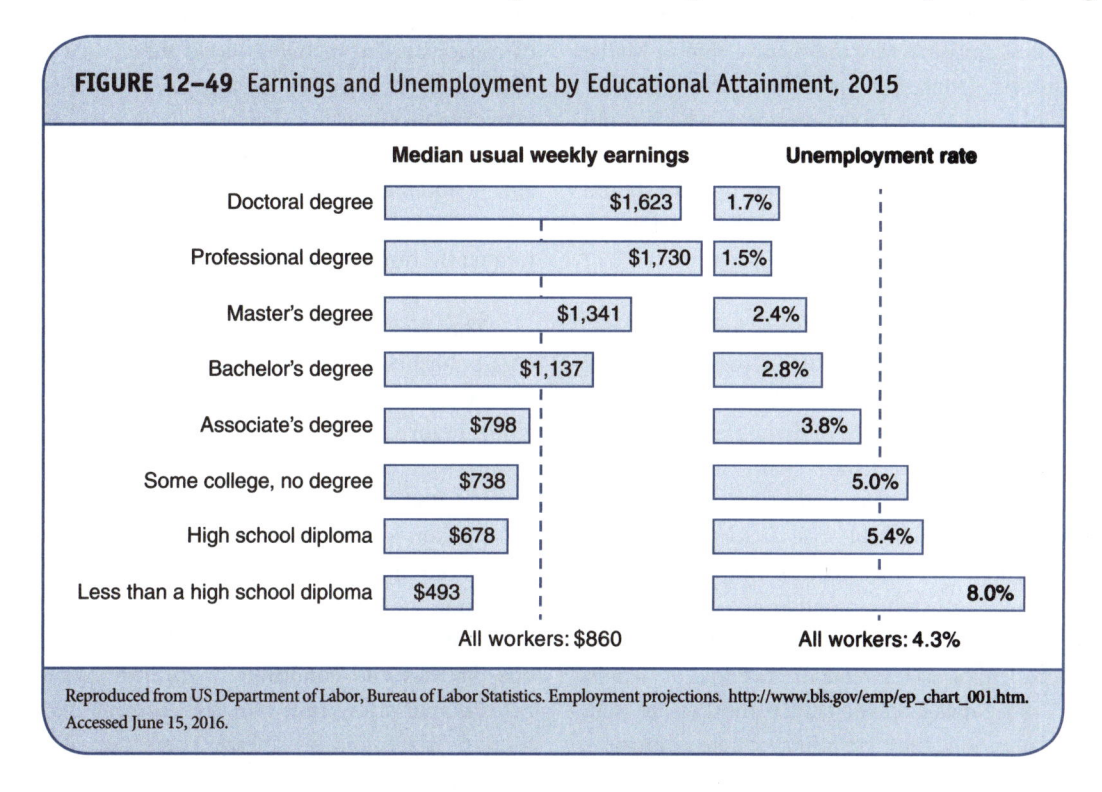

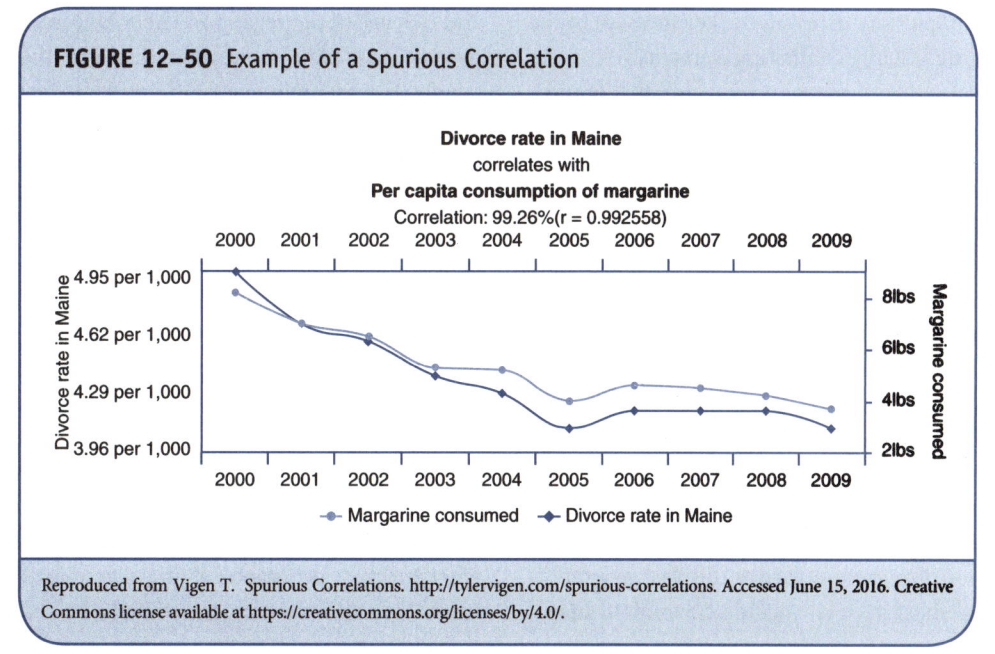

FIGURE 12–49 Earnings and Unemployment by Educational Attainment, 2015

Education	Median usual weekly earnings	Unemployment rate
Doctoral degree	$1,623	1.7%
Professional degree	$1,730	1.5%
Master's degree	$1,341	2.4%
Bachelor's degree	$1,137	2.8%
Associate's degree	$798	3.8%
Some college, no degree	$738	5.0%
High school diploma	$678	5.4%
Less than a high school diploma	$493	8.0%

All workers: $860   All workers: 4.3%

Reproduced from US Department of Labor, Bureau of Labor Statistics. Employment projections. http://www.bls.gov/emp/ep_chart_001.htm. Accessed June 15, 2016.

FIGURE 12–50 Example of a Spurious Correlation

Divorce rate in Maine
correlates with
Per capita consumption of margarine
Correlation: 99.26% (r = 0.992558)

Reproduced from Vigen T. Spurious Correlations. http://tylervigen.com/spurious-correlations. Accessed June 15, 2016. Creative Commons license available at https://creativecommons.org/licenses/by/4.0/.

which was reported in a blog by Tyler Vigen entitled "Spurious Correlations," http://tylervigen.com/spurious-correlations).[20] Line plots with multiple variables can be challenging to develop and interpret, and they need very careful preparation, labeling, and sometimes use of shading or colors to ensure that the meaning of the data is clear. Figure 12–50 illustrates a different point (not the design aspects of the visualization)—namely, an association called a **spurious correlation**. A spurious correlation is an apparent relationship that is likely due to a confounding variable.

It is difficult to imagine why the divorce rate in Maine might correlate with consumption of margarine. A possible explanation is the effect of the economy, which could increase marital stress and also affect people's buying power. Figure 12–50 highlights a point worth repeating: We should conduct statistical analyses only on plausible hypotheses.

12.4.7 Displaying Geographical Variations

Geographical displays are very useful tools for understanding differences in key health indicators across the natural geographical range. For example, we might be interested in comparing prevalence or incidence of disease among neighborhoods, across states in the United States, and or across countries around the world. As we aim to reduce health disparities and identify modifiable risk factors for disease, geographical variations can shed light on risk factors at the country, state, county, or neighborhood level.

Creating maps to display data or statistical results brings some unique challenges. Maps tend to have lots of details, and sometimes those details create clutter; thus, rather than grab a reader's attention, the details may create confusion. Insofar as possible, maps that present data or statistical results should be simple—focusing on the key message in the data and minimizing visually conflicting elements.

The key components of maps include a clear title indicating what is being displayed, how it was measured, and a time stamp, if appropriate. The geographic subunits of interest should also be indicated in the title (e.g., by country, state, county, or neighborhood). The legend should identify symbols and colors used to differentiate key aspects of the data or statistical results. A map scale should be included, if appropriate, to indicate distances between important subunits. Footnotes should be used to explain any abbreviations used or to provide additional context. Likewise, the data source should always be indicated, usually at the bottom of the display. Just as with other figures, a map should be self-explanatory so that the reader need not refer back to the text to understand what is being presented.

Different types of maps are used to display geographical variations. The most popular include **choropleth maps**, which use shading to convey the value of the measure being displayed (e.g., darker colors indicate higher proportions and lighter colors indicate lower proportions), and dot or point maps, which indicate locations of a specific occurrence (e.g., locations of patients who contract a particular disease), with dots being sized proportionally to reflect the density or extent of such occurrences.

The National Cancer Institute (http://statecancerprofiles.cancer.gov/) publishes interactive geographical displays of cancer burden in states across the United States (**Figure 12–51**). Users can click on any state in the map to see an extensive profile of incidence of all cancers, as well as incidence of many distinct cancer types in that state. The profiles also include tabulations of demographic information on the selected state and screening and risk factor data (e.g., how frequently residents of that state undergo regular screening tests; what percentages smoke, exercise regularly, and so on).

The World Health Organization (http://www.who.int/en/) publishes a "Global Health Observatory Map Gallery" on a variety of health indicators and exposures across the globe. **Figure 12–52** displays data on ambient air pollution in countries across the globe in 2014, quantified as concentrations of fine particulate matter in urban areas. Again, users can click on a specific country to display more detailed data for that country. Country rankings are also available at the bottom of the display.

The CDC houses a wealth of data and statistics on various diseases and conditions (e.g., http://www.cdc.gov/diabetes/home/index.html) that are aggregated at the national level, by state, and by county. Users can create interactive graphics by selecting specific disease or condition indicators and a level of aggregation. The CDC website allows users to download data tables that they can then explore more deeply. For example, the CDC conducts regular surveillance of young adults and adolescents through its Youth Risk Behavior Surveillance System (YRBSS; http://www.cdc.gov/healthyyouth/data/yrbs/index.htm), which collects data from high school students approximately every 2 years on healthy and risky behaviors. **Figure 12–53** displays geographical variations in obesity among high school students based on the 2015 YRBSS. The display is online and interactive. The user can click on a particular state in the map to highlight trends over time (shown by the lines on the right-hand side of the figure).

Developing geographical displays that accurately convey data and statistical results is challenging. Interpreting geographical displays is also challenging. While patterns may emerge across geographical lines, we must be careful not to infer associations among individuals.

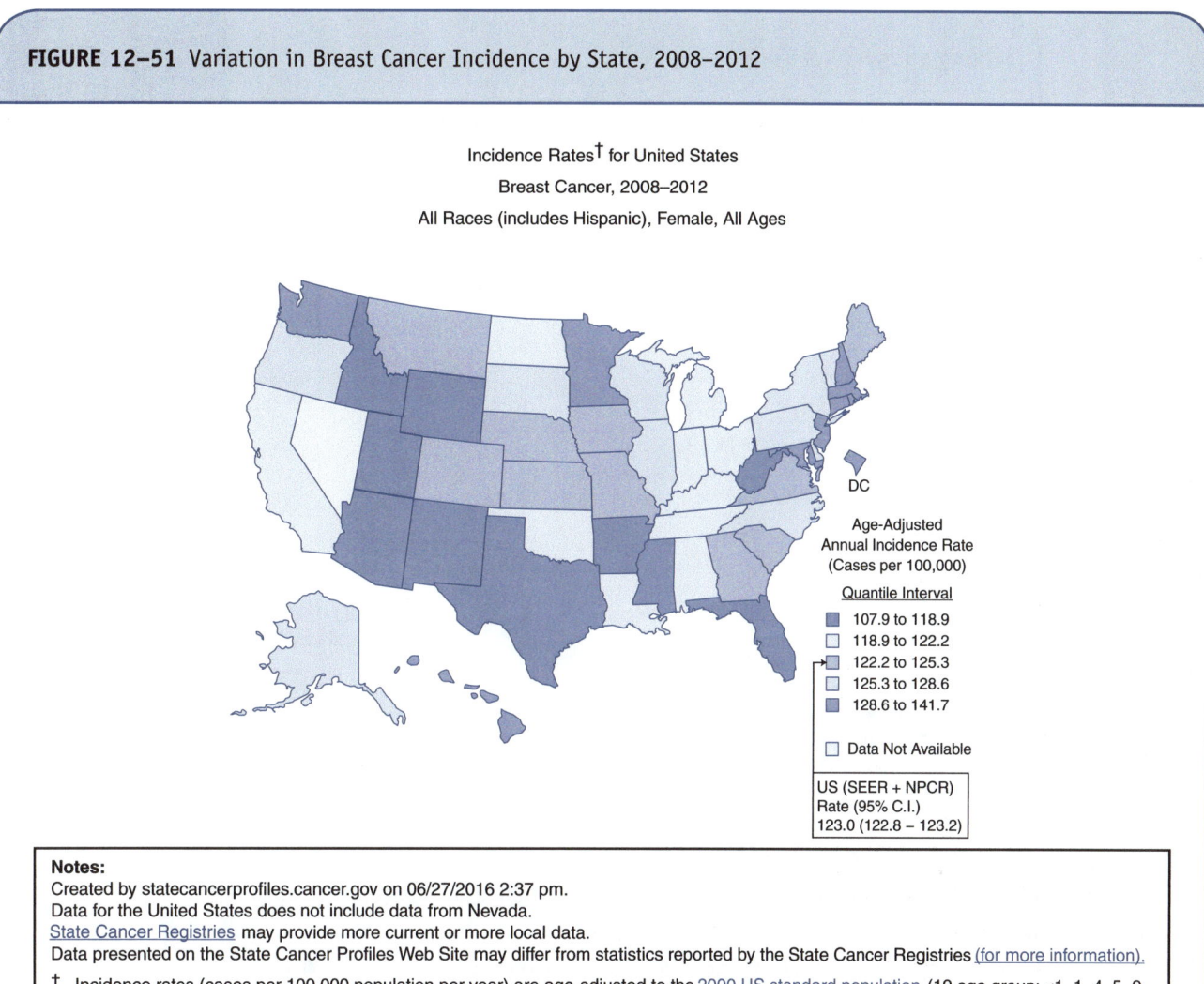

FIGURE 12-51 Variation in Breast Cancer Incidence by State, 2008-2012

Incidence Rates† for United States
Breast Cancer, 2008-2012
All Races (includes Hispanic), Female, All Ages

Age-Adjusted Annual Incidence Rate (Cases per 100,000)

Quantile Interval
- 107.9 to 118.9
- 118.9 to 122.2
- 122.2 to 125.3
- 125.3 to 128.6
- 128.6 to 141.7
- Data Not Available

US (SEER + NPCR) Rate (95% C.I.) 123.0 (122.8 – 123.2)

Notes:
Created by statecancerprofiles.cancer.gov on 06/27/2016 2:37 pm.
Data for the United States does not include data from Nevada.
State Cancer Registries may provide more current or more local data.
Data presented on the State Cancer Profiles Web Site may differ from statistics reported by the State Cancer Registries (for more information).

† Incidence rates (cases per 100,000 population per year) are age-adjusted to the 2000 US standard population (19 age group: <1, 1–4, 5–9,..., 80–84, 85+). Rates are for invasive cancer only (except for bladder which is invasive and in situ) or unless otherwise specified, Rates calculated using SEER*Stat. Population counts for denominators are based on Census populations as modified by NCI.
◊ The 1969–2013 US Populatiuon Data File is used for SEER and NPCR incidence rates.
Data not available for this combination of geography, statistic, age and race/ethnicity.

Reproduced from National Cancer Institute. State Cancer Profiles. http://statecancerprofiles.cancer.gov/map/map.withimage.php?00&001&055&00&2&01&0&1&5&0#results.

12.4.8 Pie Charts

Pie charts are a popular type of graphical display, but we do not generally recommend their use because they can be quite difficult to interpret. Pie charts are designed to represent graphically the various components of a whole (e.g., the percentage distribution of a single categorical or ordinal variable). Unfortunately, it can be difficult for readers and consumers to accurately differentiate the sizes of the segments that make up the pie chart, or to accurately compare segments between pie charts. While techniques to promote interpretability do exist (e.g., organizing the component segments by size), a table or a different graphical display is sometimes more effective.

Consider, for example, the data summarized in **Figure 12–54**, which are available online (http://budget.data.cityof-boston.gov/#/) to interested citizens and detail the city of Boston's fiscal year 2017 capital budget spending (i.e., the city's planned spending on infrastructure and improvement projects). These data are summarized in **Table 12–17** and again in **Figure 12–55**. Which format makes it easiest to compare investments across budget categories?

FIGURE 12-52 Geographical Variation in Ambient Air Pollution, 2014

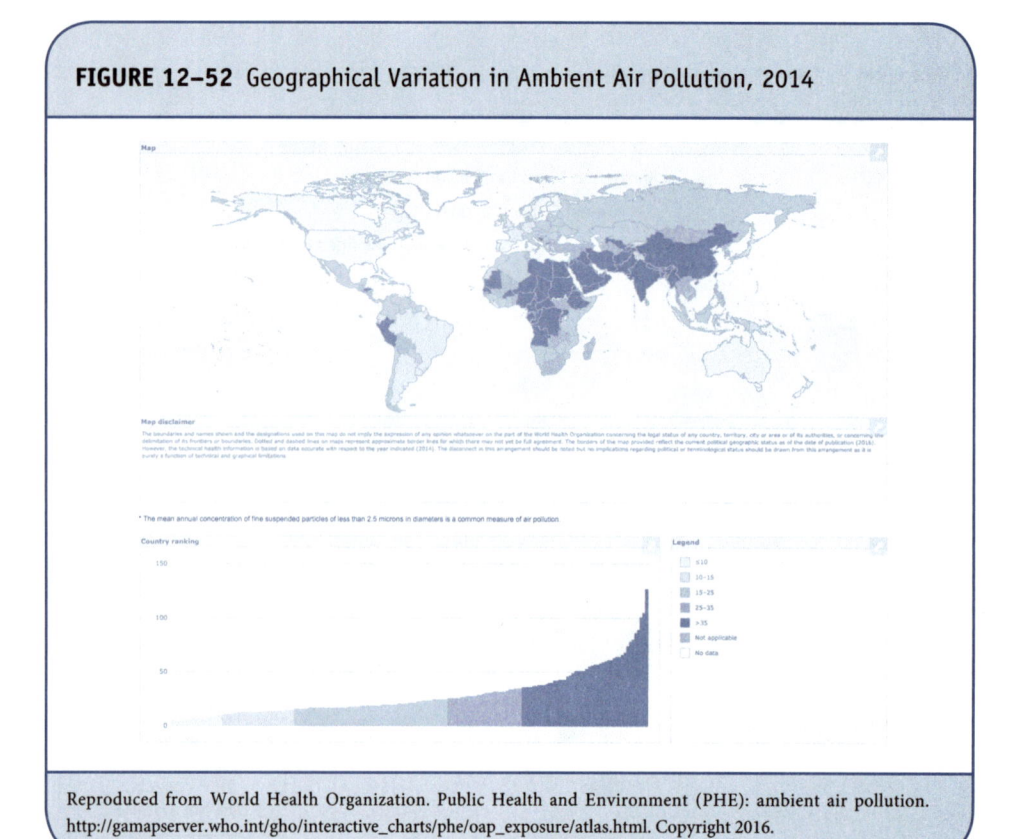

Reproduced from World Health Organization. Public Health and Environment (PHE): ambient air pollution. http://gamapserver.who.int/gho/interactive_charts/phe/oap_exposure/atlas.html. Copyright 2016.

FIGURE 12-53 Variation in Obesity Among U.S. High School Students by State, 2015

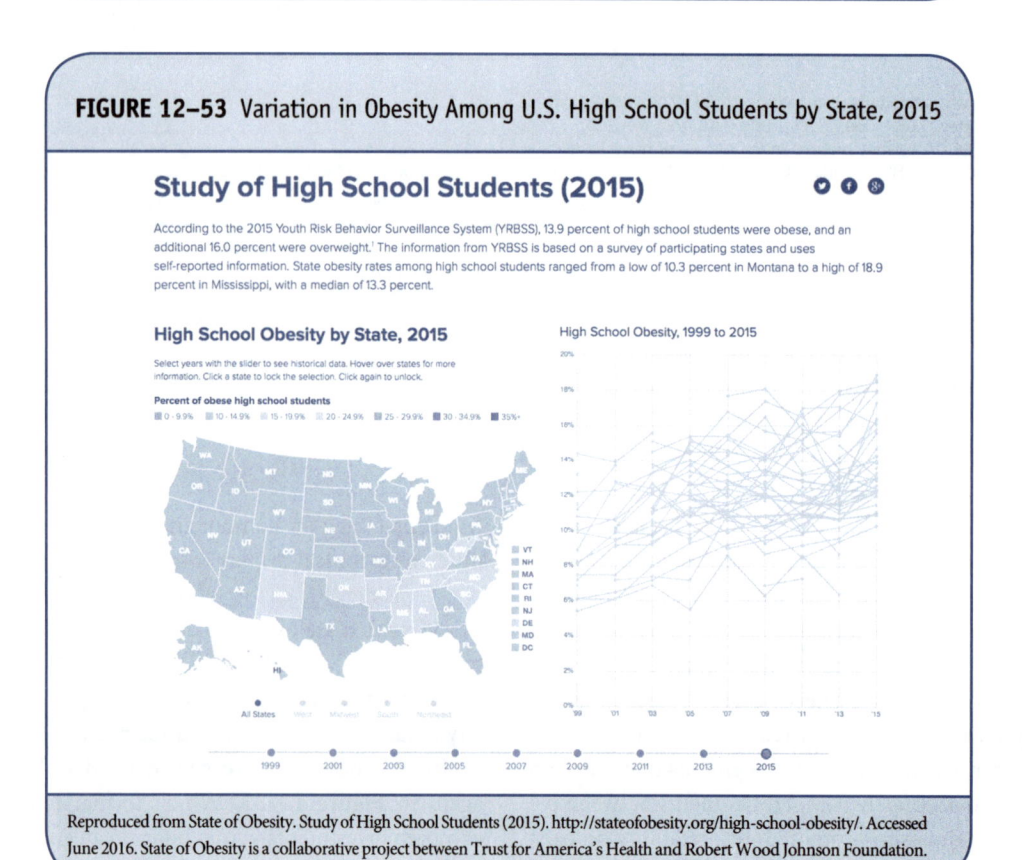

Reproduced from State of Obesity. Study of High School Students (2015). http://stateofobesity.org/high-school-obesity/. Accessed June 2016. State of Obesity is a collaborative project between Trust for America's Health and Robert Wood Johnson Foundation.

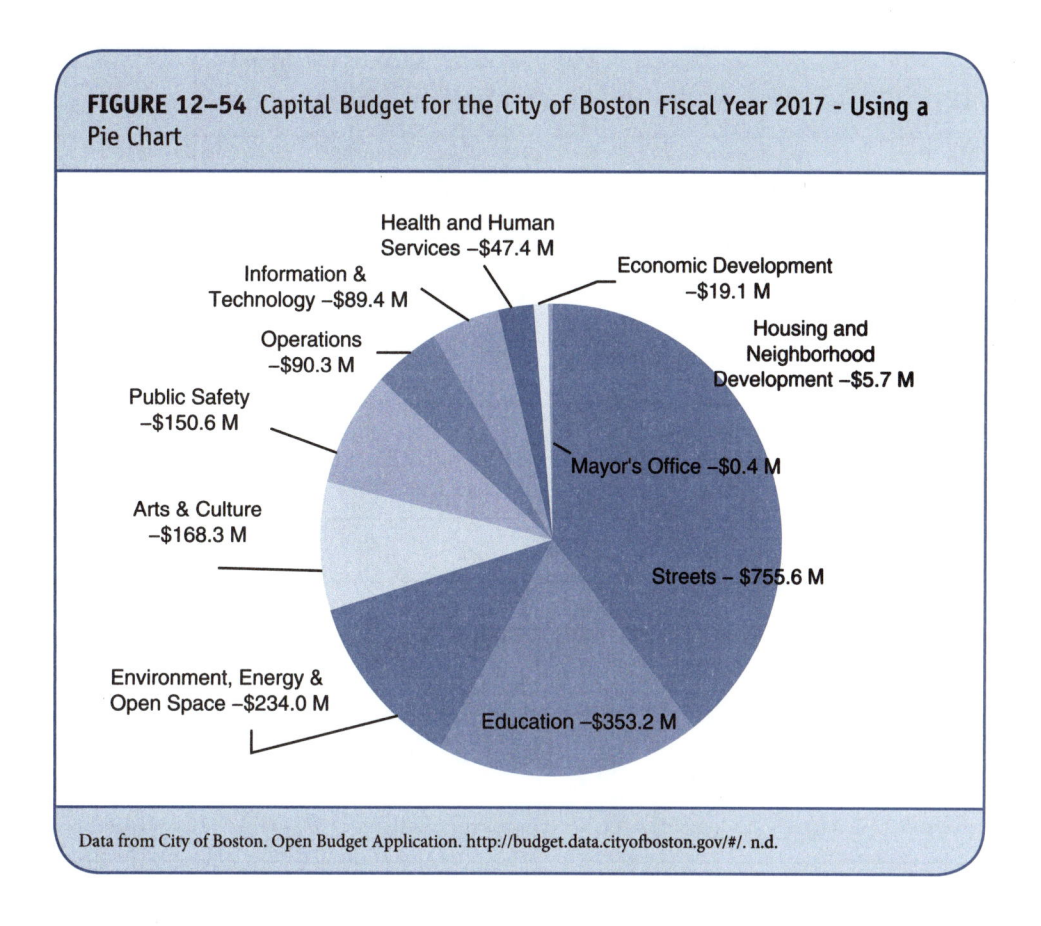

FIGURE 12-54 Capital Budget for the City of Boston Fiscal Year 2017 - Using a Pie Chart

- Health and Human Services –$47.4 M
- Information & Technology –$89.4 M
- Economic Development –$19.1 M
- Operations –$90.3 M
- Housing and Neighborhood Development –$5.7 M
- Public Safety –$150.6 M
- Mayor's Office –$0.4 M
- Arts & Culture –$168.3 M
- Streets – $755.6 M
- Environment, Energy & Open Space –$234.0 M
- Education –$353.2 M

Data from City of Boston. Open Budget Application. http://budget.data.cityofboston.gov/#/. n.d.

TABLE 12-17 Capital Budget for the City of Boston Fiscal Year 2017

Budget Category	Millions	Percentage of Total
Streets	$755.6	39%
Education	$353.2	18%
Environment, Energy & Open Space	$234.0	12%
Arts & Culture	$168.3	9%
Public Safety	$150.6	8%
Operations	$90.3	5%
Information & Technology	$89.4	5%
Health and Human Services	$47.4	2%
Economic Development	$19.1	1%
Housing and Neighborhood Development	$5.7	0%
Mayor's Office	$0.4	0%
Total capital budget	$1,914.0	100%

Data from City of Boston. Open Budget Application. http://budget.data.cityofboston.gov/#/. n.d.

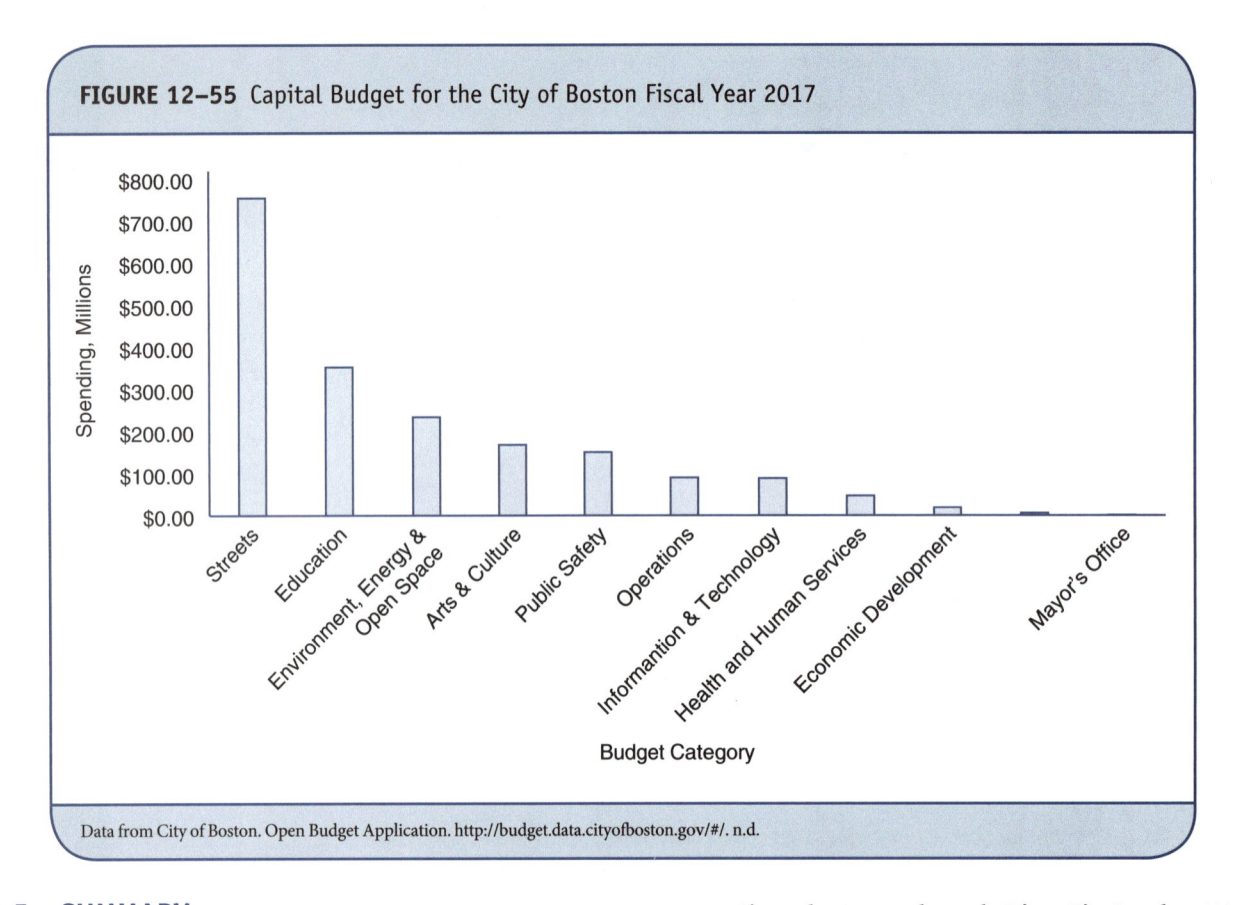

FIGURE 12-55 Capital Budget for the City of Boston Fiscal Year 2017

Data from City of Boston. Open Budget Application. http://budget.data.cityofboston.gov/#/. n.d.

12.5 SUMMARY

Communicating data and statistical results is a critically important step in any analysis. Choosing the right format to present data and statistical results depends largely on the audience, but also on the nature of the data and statistical results to be displayed. Text is most appropriate when we have only a few data points or statistics to report. Tables are best when we have sets of numbers and statistics to present and their exact values are important. Figures are best when we aim to convey complex associations or trends over time in a small amount of space.

Many templates are available to assist in the development of tables and figures; yet there is no shortage of tables and figures that are confusing, incomplete, or even incorrect in the ways in which they present data and statistical results. The primary aims of effective communication of data and statistical results, however, are clarity and accuracy. In presenting data and statistical results, we need to choose the formats that most effectively communicate data and statistical results to our target audience, while always adhering to sound statistical practice and effective design principles.

12.6 PRACTICE PROBLEMS

1. The CDC's 2009 Youth Risk Behavior Survey (YRBS) finds that approximately 31% of students earned mostly A's, 40% earned mostly B's, 19% earned mostly C's, and 6% earned mostly D's or F's. Another 4% did not report their grades. Create a graphical display of the grade distribution among the students. For this display, ignore the 4% of students who did not report grades and indicate this in the footnote to your display.

2. The data in **Table 12–18** are extracted from the CDC's 2009 Youth Risk Behavior Survey and represent the percentages of students who reported each of the indicated behaviors in 2009, organized by the grades they tended to earn (mostly A's, B's, C's, D's, or F's). Create a graphical display to summarize the associations between behaviors (all six together in one graphical display) and grades.

3. Consider the data shown in **Table 12–19**. Identify three modifications, either design or statistical, that would result in a more effective table.

4. Which kind of graphical display would best communicate unemployment percentages in urban and rural areas over the period 2006 to 2015? Sketch out the template of the figure, including a title, axis labels, and any other features that would clarify the data to be presented.

5. We wish to compare the list prices of six popular drugs for HIV infection. The data in **Table 12–20** represent means and standard deviations of the

TABLE 12–18 U.S. Youth Risk Behavior Survey, 2009

Behavior	Grades A's	B's	C's	D's or F's
Carried a weapon at least once in the past 30 days	12%	16%	21%	37%
Smoked at least 1 cigarette per day in past 30 days	10%	19%	27%	45%
Consumed at least 1 alcoholic drink in the past 30 days	32%	43%	51%	62%
Was sexually active at least once in the past three months	24%	34%	43%	54%
Watched more than 3 hours of television per day on an average school day	24%	32%	39%	49%
Was physically active for at least 60 minutes per day on less than 5 days in the past 7 days	56%	63%	68%	76%

Data from Centers for Disease Control and Prevention. Youth risk behavior surveillance system. http://www.cdc.gov/healthyyouth/data/yrbs/index.htm.

TABLE 12–19 Baseline Characteristics of Participants in Randomized Trial*

Characteristic	Treatment Group (*n* = 361)	Placebo Group (*n* = 344)	*p*-Value
Age, years	45.8	47.6	0.5467
Male sex	201	198	0.7561
Systolic blood pressure	128.25 ± 14.2	121.7 ± 13.91	0.645
Race/ethnicity			
Asian	184	180	
White	107	100	
Hispanic	70	64	

prices for each drug based on surveys of 55 pharmacies in the New England area. Prices are for a full course of treatment. Generate a graphical display of the mean prices shown in Table 12–20.

6. Consider again the data shown in Table 12–20. Compute 95% confidence intervals for the mean price of each drug using $\bar{x} \pm t \frac{s}{\sqrt{n}}$. Generate a graphical display showing the means and 95% confidence intervals for the mean drug prices for each drug group. Based only on your figure, which of the drugs have statistically significantly different prices?

7. Consider the data displayed in **Figure 12–56**. Generate a table that summarizes the data.

8. Consider again the data in Figure 12–56. Generate a different graphical display that facilitates the comparison of male and female recent college graduates within employment sectors.

9. Create a table shell using the example in Table 12–24 that would effectively summarize the following characteristics measured in an observational study that is designed to investigate risk factors for new-onset diabetes. Risk factors will be compared between adults with at least one parent with diabetes and adults whose parents are free of diabetes. The risk factors of interest include demographic characteristics such as age (measured in years), sex (male/female), educational achievement (eighth grade or less, some high school, high school graduate, some college, college graduate, postgraduate). Other risk factors include clinical attributes such as systolic and diastolic blood pressures (mm Hg); total, HDL, and LDL cholesterols (mg/dL); and smoking (never, current, former). Lastly, participants were asked to report the number of hours per week they engage in regular physical

TABLE 12-20 Mean and Standard Deviation of Prices of Six HIV Drugs

Drug	Mean Price	Standard Deviation in Price
A	$3,200	$856
B	$6,800	$925
C	$11,352	$754
D	$3,945	$920
E	$2,120	$645
F	$4,856	$540

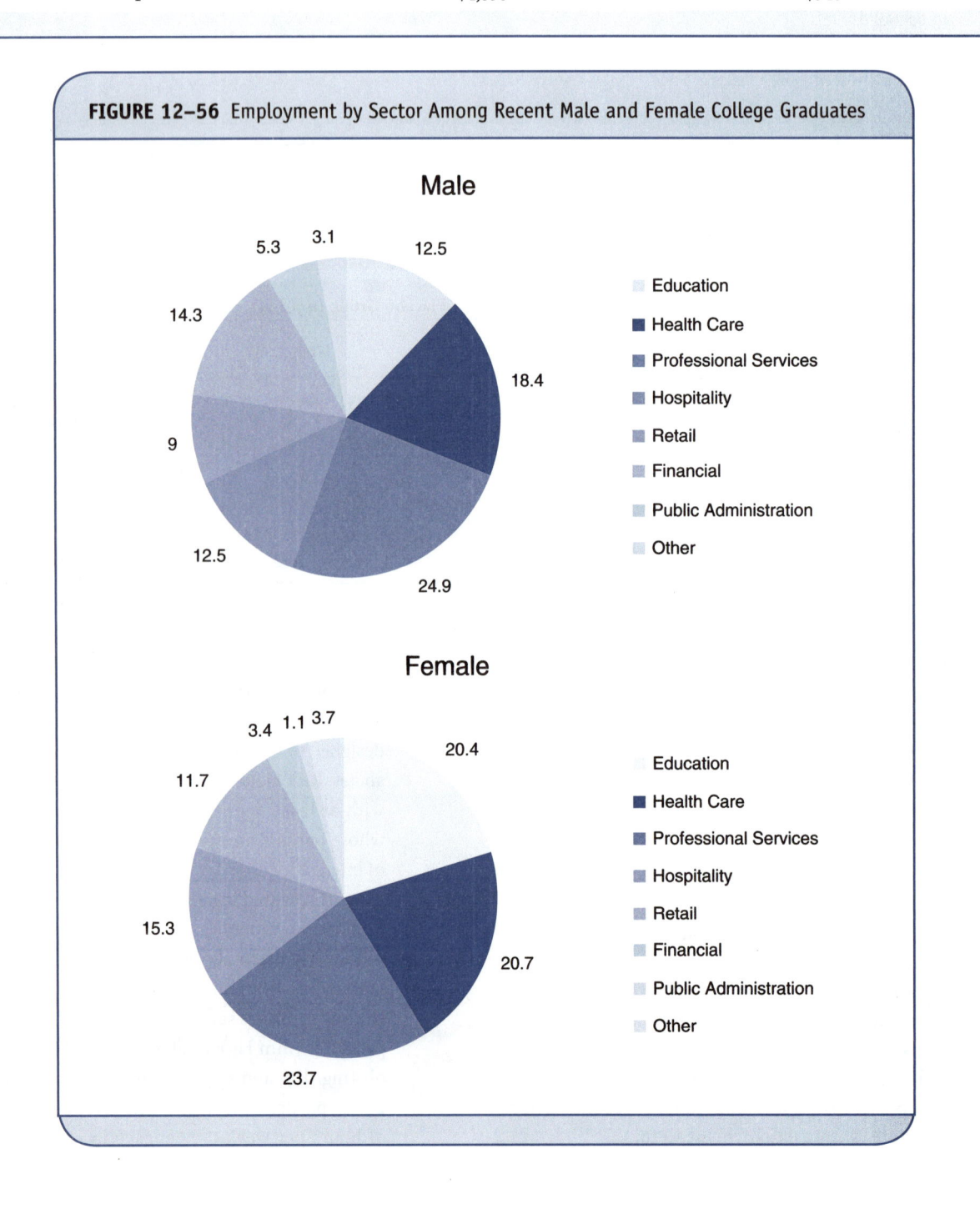

FIGURE 12-56 Employment by Sector Among Recent Male and Female College Graduates

Male

- Education
- Health Care
- Professional Services
- Hospitality
- Retail
- Financial
- Public Administration
- Other

Values: 12.5, 18.4, 24.9, 12.5, 9, 14.3, 5.3, 3.1

Female

- Education
- Health Care
- Professional Services
- Hospitality
- Retail
- Financial
- Public Administration
- Other

Values: 20.4, 20.7, 23.7, 15.3, 11.7, 3.4, 1.1, 3.7

activity. Set up the table to summarize the results. Be sure to include variable names, units of measurement, the statistics that will be presented, and the precision of each (i.e., numbers of decimal places to be reported).

10. The following description summarizes participants in a study comparing emotional well-being for people who commuted to and from work by car with emotional well-being for those who bicycled to and from work. A total of 125 people commuted by car; 74 bicycled to work. The participants who commuted by car were primarily female (75%) and 11% were younger than 18 years, 52% were between 18 and 65 years old, and 37% were older than 65 years. The participants who bicycled were predominantly male (63%) and 37% were younger than 18 years, 49% were between 18 and 65 years old, and 14% were older than 65 years. Fifty percent of the participants who commuted by car were married; 37% were widowed, separated, or divorced; and 13% were never married. Forty-five percent of the participants who bicycled were married; 17% were widowed, separated, or divorced; and 38% were never married. Create a table to summarize the background characteristics of the participants.

REFERENCES

1. Committee on Professional Ethics of the American Statistical Association. *Ethical Guidelines for Statistical Practice.* http://www.amstat.org/about/ethicalguidelines.cfm. Accessed June 15, 2016.
2. Tufte, E.R. *The Visual Display of Quantitative Information* (2nd ed.). Cheshire, CT: Graphics Press, 2001.
3. Tufte, E.R. The work of Edward Tufte and Graphics Press. http://www.edwardtufte.com/tufte/. Accessed June 15, 2016.
4. Isson, J.P. and Harriott, J. *Win with Advanced Business Analytics: Creating Business Value from Your Data.* Hoboken, NJ: John Wiley & Sons, 2012.
5. Hoven, N. "Steven Few on data visualization: 8 core principles." *Tableau.* http://www.tableau.com/blog/stephen-few-data-visualization. Accessed June 15, 2016.
6. Rodrigues, V. "Tips on effective use of tables and figures in research papers." *Editage Insights.* 2015. http://www.editage.com/insights/tips-on-effective-use-of-tables-and-figures-in-research-papers. Accessed June 15, 2016.
7. *American Medical Association Manual of Style*, 10th ed. New York, NY: Oxford University Press, 2007. http://www.amamanualofstyle.com/view/10.1093/jama/9780195176339.001.0001/med-9780195176339.
8. American Psychological Association. APA style. http://www.apastyle.org/index.aspx. Accessed June 15, 2016.
9. Schultz, K.F., Altman, D.G., and Moher, D. "CONSORT 2010 statement: updated guidelines for reporting parallel group randomised trials. *British Medical Journal* 2010; 340(7748): 698–702.
10. Tableau. http://www.tableau.com/. Accessed June 15, 2016.
11. Gelman, A. and Unwin, A. Infovis and statistical graphics: different goals, different looks. http://www.stat.columbia.edu/~gelman/research/published/vis14.pdf. Accessed June 15, 2016.
12. Wainer, H. "How to display data badly." *The American Statistician* 1984; 38(2): 137–147.
13. The 2014 National Health Interview Survey. http://www.cdc.gov/asthma/most_recent_data.htm. Accessed June 15, 2016.
14. Point-in-time estimates of homelessness. *The 2015 Annual Homeless Assessment Report (HAR) to Congress,* November 2015. https://www.hudexchange.info/resources/documents/2015-AHAR-Part-1.pdf. Accessed June 15, 2016.
15. The Consolidated Standards of Reporting Trials (CONSORT). http://www.consort-statement.org/. Accessed June 15, 2016.
16. Des Jarlais, D.C., Lyles, C., Crepaz, N., and TREND Group. "Improving the reporting quality of nonrandomized evaluations of behavioral and public health interventions: the TREND statement." *American Journal of Public Health* 2004; 94: 361–366.
17. Cabral, J. "Statistical primer for cardiovascular research: multiple comparisons procedures." *Circulation* 2008; 117: 698–701.
18. Ogden, C.L., Carroll, M.D., Lawman, H.G., Fryar, C.D., Kruszon-Moran, D., Kit, B.K., and Flegal, K.M. Trends in obesity prevalence among children and adolescents in the United States, 1988–1994 through 2013–2014. *Journal of the American Medical Association* 2016; 315(21): 2292–2299.
19. U.S. Department of Labor, Bureau of Labor Statistics. http://www.bls.gov/. Accessed June 15, 2016.
20. Vigen, T. Spurious correlations. http://tylervigen.com/spurious-orrelations. Accessed June 15, 2016.

Appendix

APPENDIX STATISTICAL TABLES

Table 1. Probabilities of the Standard Normal Distribution Z

Table 1A. Z Values for Percentiles

Table 1B. Z Values for Confidence Intervals

Table 1C. Z Values for Hypothesis Tests

Table 2. Critical Values of the t Distribution

Table 3. Critical Values of the χ^2 Distribution

Table 4. Critical Values of the F Distribution with Upper Tail Area = 0.05

Table 5. Critical Values of the Mann–Whitney U Test

Table 6. Critical Values for the Sign Test

Table 7. Critical Values for the Wilcoxon Signed Rank Test

Table 8. Critical Values for the Kruskal–Wallis Test, $\alpha = .05$ and $\alpha = .01$

TABLE 1. Probabilities of the Standard Normal Distribution Z

Table entries represent $P(Z < Z_i)$
e.g., $P(Z < -1.96) = 0.0250$, $P(Z < 1.96) = 0.9750$

Z_i	.00	.01	.02	.03	.04	.05	.06	.07	.08	.09
−3.0	0.0013	0.0013	0.0013	0.0012	0.0012	0.0011	0.0011	0.0011	0.0010	0.0010
−2.9	0.0019	0.0018	0.0018	0.0017	0.0016	0.0016	0.0015	0.0015	0.0014	0.0014
−2.8	0.0026	0.0025	0.0024	0.0023	0.0023	0.0022	0.0021	0.0021	0.0020	0.0019
−2.7	0.0035	0.0034	0.0033	0.0032	0.0031	0.0030	0.0029	0.0028	0.0027	0.0026
−2.6	0.0047	0.0045	0.0044	0.0043	0.0041	0.0040	0.0039	0.0038	0.0037	0.0036
−2.5	0.0062	0.0060	0.0059	0.0057	0.0055	0.0054	0.0052	0.0051	0.0049	0.0048
−2.4	0.0082	0.0080	0.0078	0.0075	0.0073	0.0071	0.0069	0.0068	0.0066	0.0064
−2.3	0.0107	0.0104	0.0102	0.0099	0.0096	0.0094	0.0091	0.0089	0.0087	0.0084
−2.2	0.0139	0.0136	0.0132	0.0129	0.0125	0.0122	0.0119	0.0116	0.0113	0.0110
−2.1	0.0179	0.0174	0.0170	0.0166	0.0162	0.0158	0.0154	0.0150	0.0146	0.0143
−2.0	0.0228	0.0222	0.0217	0.0212	0.0207	0.0202	0.0197	0.0192	0.0188	0.0183
−1.9	0.0287	0.0281	0.0274	0.0268	0.0262	0.0256	0.0250	0.0244	0.0239	0.0233
−1.8	0.0359	0.0351	0.0344	0.0336	0.0329	0.0322	0.0314	0.0307	0.0301	0.0294
−1.7	0.0446	0.0436	0.0427	0.0418	0.0409	0.0401	0.0392	0.0384	0.0375	0.0367
−1.6	0.0548	0.0537	0.0526	0.0516	0.0505	0.0495	0.0485	0.0475	0.0465	0.0455
−1.5	0.0668	0.0655	0.0643	0.0630	0.0618	0.0606	0.0594	0.0582	0.0571	0.0559
−1.4	0.0808	0.0793	0.0778	0.0764	0.0749	0.0735	0.0721	0.0708	0.0694	0.0681
−1.3	0.0968	0.0951	0.0934	0.0918	0.0901	0.0885	0.0869	0.0853	0.0838	0.0823
−1.2	0.1151	0.1131	0.1112	0.1093	0.1075	0.1056	0.1038	0.1020	0.1003	0.0985
−1.1	0.1357	0.1335	0.1314	0.1292	0.1271	0.1251	0.1230	0.1210	0.1190	0.1170
−1.0	0.1587	0.1562	0.1539	0.1515	0.1492	0.1469	0.1446	0.1423	0.1401	0.1379
−0.9	0.1841	0.1814	0.1788	0.1762	0.1736	0.1711	0.1685	0.1660	0.1635	0.1611
−0.8	0.2119	0.2090	0.2061	0.2033	0.2005	0.1977	0.1949	0.1922	0.1894	0.1867
−0.7	0.2420	0.2389	0.2358	0.2327	0.2296	0.2266	0.2236	0.2206	0.2177	0.2148
−0.6	0.2743	0.2709	0.2676	0.2643	0.2611	0.2578	0.2546	0.2514	0.2483	0.2451
−0.5	0.3085	0.3050	0.3015	0.2981	0.2946	0.2912	0.2877	0.2843	0.2810	0.2776
−0.4	0.3446	0.3409	0.3372	0.3336	0.3300	0.3264	0.3228	0.3192	0.3156	0.3121
−0.3	0.3821	0.3783	0.3745	0.3707	0.3669	0.3632	0.3594	0.3557	0.3520	0.3483
−0.2	0.4207	0.4168	0.4129	0.4090	0.4052	0.4013	0.3974	0.3936	0.3897	0.3859
−0.1	0.4602	0.4562	0.4522	0.4483	0.4443	0.4404	0.4364	0.4325	0.4286	0.4247
−0.0	0.5000	0.4960	0.4920	0.4880	0.4840	0.4801	0.4761	0.4721	0.4681	0.4641

(Continued)

TABLE 1. Probabilities of the Standard Normal Distribution Z (Continued)

Table entries represent $P(Z < Z_i)$

e.g., $P(Z < -1.96) = 0.0250$, $P(Z < 1.96) = 0.9750$

Z_i	.00	.01	.02	.03	.04	.05	.06	.07	.08	.09
0.0	0.5000	0.5040	0.5080	0.5120	0.5160	0.5199	0.5239	0.5279	0.5319	0.5359
0.1	0.5398	0.5438	0.5478	0.5517	0.5557	0.5596	0.5636	0.5675	0.5714	0.5753
0.2	0.5793	0.5832	0.5871	0.5910	0.5948	0.5987	0.6026	0.6064	0.6103	0.6141
0.3	0.6179	0.6217	0.6255	0.6293	0.6331	0.6368	0.6406	0.6443	0.6480	0.6517
0.4	0.6554	0.6591	0.6628	0.6664	0.6700	0.6736	0.6772	0.6808	0.6844	0.6879
0.5	0.6915	0.6950	0.6985	0.7019	0.7054	0.7088	0.7123	0.7157	0.7190	0.7224
0.6	0.7257	0.7291	0.7324	0.7357	0.7389	0.7422	0.7454	0.7486	0.7517	0.7549
0.7	0.7580	0.7611	0.7642	0.7673	0.7704	0.7734	0.7764	0.7794	0.7823	0.7852
0.8	0.7881	0.7910	0.7939	0.7967	0.7995	0.8023	0.8051	0.8078	0.8106	0.8133
0.9	0.8159	0.8186	0.8212	0.8238	0.8264	0.8289	0.8315	0.8340	0.8365	0.8389
1.0	0.8413	0.8438	0.8461	0.8485	0.8508	0.8531	0.8554	0.8577	0.8599	0.8621
1.1	0.8643	0.8665	0.8686	0.8708	0.8729	0.8749	0.8770	0.8790	0.8810	0.8830
1.2	0.8849	0.8869	0.8888	0.8907	0.8925	0.8944	0.8962	0.8980	0.8997	0.9015
1.3	0.9032	0.9049	0.9066	0.9082	0.9099	0.9115	0.9131	0.9147	0.9162	0.9177
1.4	0.9192	0.9207	0.9222	0.9236	0.9251	0.9265	0.9279	0.9292	0.9306	0.9319
1.5	0.9332	0.9345	0.9357	0.9370	0.9382	0.9394	0.9406	0.9418	0.9429	0.9441
1.6	0.9452	0.9463	0.9474	0.9484	0.9495	0.9505	0.9515	0.9525	0.9535	0.9545
1.7	0.9554	0.9564	0.9573	0.9582	0.9591	0.9599	0.9608	0.9616	0.9625	0.9633
1.8	0.9641	0.9649	0.9656	0.9664	0.9671	0.9678	0.9686	0.9693	0.9699	0.9706
1.9	0.9713	0.9719	0.9726	0.9732	0.9738	0.9744	0.9750	0.9756	0.9761	0.9767
2.0	0.9772	0.9778	0.9783	0.9788	0.9793	0.9798	0.9803	0.9808	0.9812	0.9817
2.1	0.9821	0.9826	0.9830	0.9834	0.9838	0.9842	0.9846	0.9850	0.9854	0.9857
2.2	0.9861	0.9864	0.9868	0.9871	0.9875	0.9878	0.9881	0.9884	0.9887	0.9890
2.3	0.9893	0.9896	0.9898	0.9901	0.9904	0.9906	0.9909	0.9911	0.9913	0.9916
2.4	0.9918	0.9920	0.9922	0.9925	0.9927	0.9929	0.9931	0.9932	0.9934	0.9936
2.5	0.9938	0.9940	0.9941	0.9943	0.9945	0.9946	0.9948	0.9949	0.9951	0.9952
2.6	0.9953	0.9955	0.9956	0.9957	0.9959	0.9960	0.9961	0.9962	0.9963	0.9964
2.7	0.9965	0.9966	0.9967	0.9968	0.9969	0.9970	0.9971	0.9972	0.9973	0.9974
2.8	0.9974	0.9975	0.9976	0.9977	0.9977	0.9978	0.9979	0.9979	0.9980	0.9981
2.9	0.9981	0.9982	0.9982	0.9983	0.9984	0.9984	0.9985	0.9985	0.9986	0.9986
3.0	0.9987	0.9987	0.9987	0.9988	0.9988	0.9989	0.9989	0.9989	0.9990	0.9990

Generated in SAS ©. SAS Institute Inc. SAS © *User's Guide: Basics v 9.1.* Cary, NC: SAS Institute Inc.; 2002–2003.

TABLE 1A. Z Values for Percentiles

Percentile	Z
1st	−2.326
2.5th	−1.960
5th	−1.645
10th	−1.282
25th	−0.675
50th	0
75th	0.675
90th	1.282
95th	1.645
97.5th	1.960
99th	2.326

Generated in SAS ©. SAS Institute Inc. SAS © *User's Guide: Basics v 9.1.* Cary, NC: SAS Institute Inc.; 2002–2003.

TABLE 1B. Z Values for Confidence Intervals

Confidence Level	Z
99.99%	3.819
99.9%	3.291
99%	2.576
95%	1.960
90%	1.645
80%	1.282

Generated in SAS ©. SAS Institute Inc. SAS © *User's Guide: Basics v 9.1.* Cary, NC: SAS Institute Inc.; 2002–2003.

TABLE 1C. Z Values for Hypothesis Tests

Lower-Tailed Test α	Z
0.10	−1.282
0.05	−1.645
0.025	−1.960
0.010	−2.326
0.005	−2.576
0.001	−3.090
0.0001	−3.719

Upper-Tailed Test α	Z
0.10	1.282
0.05	1.645
0.025	1.960
0.010	2.326
0.005	2.576
0.001	3.090
0.0001	3.719

Two-Tailed Test α*	Z
0.20	1.282
0.10	1.645
0.05	1.960
0.010	2.576
0.001	3.291
0.0001	3.819

*α is the total tail area, α/2 in each tail.

Generated in SAS ©. SAS Institute Inc. SAS © *User's Guide: Basics v 9.1.* Cary, NC: SAS Institute Inc.; 2002–2003.

TABLE 2. Critical Values of the *t* Distribution

Table entries represent values from *t* distribution with upper-tail area equal to α.
e.g., $P(t_{df} > t) = \alpha$, e.g., $P(t_6 > 1.943) = 0.05$

Condence Level	80%	90%	95%	98%	99%
Two-Sided Test α	.20	.10	.05	.02	.01
One-Sided Test α	.10	.05	.025	.01	.005
df					
1	3.078	6.314	12.71	31.82	63.66
2	1.886	2.920	4.303	6.965	9.925
3	1.638	2.353	3.182	4.541	5.841
4	1.533	2.132	2.776	3.747	4.604
5	1.476	2.015	2.571	3.365	4.032
6	1.440	1.943	2.447	3.143	3.707
7	1.415	1.895	2.365	2.998	3.499
8	1.397	1.860	2.306	2.896	3.355
9	1.383	1.833	2.262	2.821	3.250
10	1.372	1.812	2.228	2.764	3.169
11	1.363	1.796	2.201	2.718	3.106
12	1.356	1.782	2.179	2.681	3.055
13	1.350	1.771	2.160	2.650	3.012
14	1.345	1.761	2.145	2.624	2.977
15	1.341	1.753	2.131	2.602	2.947
16	1.337	1.746	2.120	2.583	2.921
17	1.333	1.740	2.110	2.567	2.898
18	1.330	1.734	2.101	2.552	2.878
19	1.328	1.729	2.093	2.539	2.861
20	1.325	1.725	2.086	2.528	2.845
21	1.323	1.721	2.080	2.518	2.831
22	1.321	1.717	2.074	2.508	2.819
23	1.319	1.714	2.069	2.500	2.807
24	1.318	1.711	2.064	2.492	2.797
25	1.316	1.708	2.060	2.485	2.787
26	1.315	1.706	2.056	2.479	2.779
27	1.314	1.703	2.052	2.473	2.771
28	1.313	1.701	2.048	2.467	2.763
29	1.311	1.699	2.045	2.462	2.756
30	1.310	1.697	2.042	2.457	2.750

(*Continued*)

TABLE 2. Critical Values of the *t* Distribution (Continued)

Table entries represent values from *t* distribution with upper tail area equal to α.
e.g., $P(t_{df} > t) = \alpha$, e.g., $P(t_6 > 1.943) = 0.05$

Condence Level	80%	90%	95%	98%	99%
Two-Sided Test α	.20	.10	.05	.02	.01
One-Sided Test α	.10	.05	.025	.01	.005
df					
31	1.309	1.696	2.040	2.453	2.744
32	1.309	1.694	2.037	2.449	2.738
33	1.308	1.692	2.035	2.445	2.733
34	1.307	1.691	2.032	2.441	2.728
35	1.306	1.690	2.030	2.438	2.724
36	1.306	1.688	2.028	2.434	2.719
37	1.305	1.687	2.026	2.431	2.715
38	1.304	1.686	2.024	2.429	2.712
39	1.304	1.685	2.023	2.426	2.708
40	1.303	1.684	2.021	2.423	2.704
41	1.303	1.683	2.020	2.421	2.701
42	1.302	1.682	2.018	2.418	2.698
43	1.302	1.681	2.017	2.416	2.695
44	1.301	1.680	2.015	2.414	2.692
45	1.301	1.679	2.014	2.412	2.690
46	1.300	1.679	2.013	2.410	2.687
47	1.300	1.678	2.012	2.408	2.685
48	1.299	1.677	2.011	2.407	2.682
49	1.299	1.677	2.010	2.405	2.680
50	1.299	1.676	2.009	2.403	2.678
51	1.298	1.675	2.008	2.402	2.676
52	1.298	1.675	2.007	2.400	2.674
53	1.298	1.674	2.006	2.399	2.672
54	1.297	1.674	2.005	2.397	2.670
55	1.297	1.673	2.004	2.396	2.668
56	1.297	1.673	2.003	2.395	2.667
57	1.297	1.672	2.002	2.394	2.665
58	1.296	1.672	2.002	2.392	2.663
59	1.296	1.671	2.001	2.391	2.662
60	1.296	1.671	2.000	2.390	2.660

(Continued)

TABLE 2. Critical Values of the *t* Distribution (Continued)

Table entries represent values from *t* distribution with upper tail area equal to α.
e.g., $P(t_{df} > t) = \alpha$, e.g., $P(t_6 > 1.943) = 0.05$

Condence Level	80%	90%	95%	98%	99%
Two-Sided Test α	.20	.10	.05	.02	.01
One-Sided Test α	.10	.05	.025	.01	.005
df					
61	1.296	1.670	2.000	2.389	2.659
62	1.295	1.670	1.999	2.388	2.657
63	1.295	1.669	1.998	2.387	2.656
64	1.295	1.669	1.998	2.386	2.655
65	1.295	1.669	1.997	2.385	2.654
66	1.295	1.668	1.997	2.384	2.652
67	1.294	1.668	1.996	2.383	2.651
68	1.294	1.668	1.995	2.382	2.650
69	1.294	1.667	1.995	2.382	2.649
70	1.294	1.667	1.994	2.381	2.648
71	1.294	1.667	1.994	2.380	2.647
72	1.293	1.666	1.993	2.379	2.646
73	1.293	1.666	1.993	2.379	2.645
74	1.293	1.666	1.993	2.378	2.644
75	1.293	1.665	1.992	2.377	2.643
∞	1.282	1.645	1.960	2.326	2.576

Generated in SAS ©. SAS Institute Inc. SAS © *User's Guide: Basics v 9.1*. Cary, NC: SAS Institute Inc.; 2002–2003.

TABLE 3. Critical Values of the χ^2 Distribution

Table entries represent values from χ^2 distribution with upper tail area equal to α.

$P(X^2_{df} > \chi^2) = \alpha$, e.g., $P(X^2_3 > 7.81) = 0.05$

df	α = .10	.05	.025	.01	.005
1	2.71	3.84	5.02	6.63	7.88
2	4.61	5.99	7.38	9.21	10.60
3	6.25	7.81	9.35	11.34	12.84
4	7.78	9.49	11.14	13.28	14.86
5	9.24	11.07	12.83	15.09	16.75
6	10.64	12.59	14.45	16.81	18.55
7	12.02	14.07	16.01	18.48	20.28
8	13.36	15.51	17.53	20.09	21.95
9	14.68	16.92	19.02	21.67	23.59
10	15.99	18.31	20.48	23.21	25.19
11	17.28	19.68	21.92	24.72	26.76
12	18.55	21.03	23.34	26.22	28.30
13	19.81	22.36	24.74	27.69	29.82
14	21.06	23.68	26.12	29.14	31.32
15	22.31	25.00	27.49	30.58	32.80
16	23.54	26.30	28.85	32.00	34.27
17	24.77	27.59	30.19	33.41	35.72
18	25.99	28.87	31.53	34.81	37.16
19	27.20	30.14	32.85	36.19	38.58
20	28.41	31.41	34.17	37.57	40.00
21	29.62	32.67	35.48	38.93	41.40
22	30.81	33.92	36.78	40.29	42.80
23	32.01	35.17	38.08	41.64	44.18
24	33.20	36.42	39.36	42.98	45.56
25	34.38	37.65	40.65	44.31	46.93
26	35.56	38.89	41.92	45.64	48.29
27	36.74	40.11	43.19	46.96	49.64
28	37.92	41.34	44.46	48.28	50.99
29	39.09	42.56	45.72	49.59	52.34
30	40.26	43.77	46.98	50.89	53.67
40	51.81	55.76	59.34	63.69	66.77
50	63.17	67.50	71.42	76.15	79.49
60	74.40	79.08	83.30	88.38	91.95
70	85.53	90.53	95.02	100.4	104.2
80	96.58	101.9	106.6	112.3	116.3
90	107.6	113.1	118.1	124.1	128.3
100	118.5	124.3	129.6	135.8	140.2

Generated in SAS ©. SAS Institute Inc. SAS © *User's Guide: Basics v 9.1.* Cary, NC: SAS Institute Inc.; 2002–2003.

TABLE 4. Critical Values of the F Distribution with Upper Tail Area = 0.05

$$P(F_{df_1, df_2} > F) = 0.05,$$
$$\text{e.g., } P(F_{3,20} > 3.10) = 0.05$$

							df_1							
	1	2	3	4	5	6	7	8	9	10	20	30	40	50
df_2														
1	161.4	199.5	215.7	224.6	230.2	234.0	236.8	238.9	240.5	241.9	248.0	250.1	251.1	251.8
2	18.51	19.00	19.16	19.25	19.30	19.33	19.35	19.37	19.38	19.40	19.45	19.46	19.47	19.48
3	10.13	9.55	9.28	9.12	9.01	8.94	8.89	8.85	8.81	8.79	8.66	8.62	8.59	8.58
4	7.71	6.94	6.59	6.39	6.26	6.16	6.09	6.04	6.00	5.96	5.80	5.75	5.72	5.70
5	6.61	5.79	5.41	5.19	5.05	4.95	4.88	4.82	4.77	4.74	4.56	4.50	4.46	4.44
6	5.99	5.14	4.76	4.53	4.39	4.28	4.21	4.15	4.10	4.06	3.87	3.81	3.77	3.75
7	5.59	4.74	4.35	4.12	3.97	3.87	3.79	3.73	3.68	3.64	3.44	3.38	3.34	3.32
8	5.32	4.46	4.07	3.84	3.69	3.58	3.50	3.44	3.39	3.35	3.15	3.08	3.04	3.02
9	5.12	4.26	3.86	3.63	3.48	3.37	3.29	3.23	3.18	3.14	2.94	2.86	2.83	2.80
10	4.96	4.10	3.71	3.48	3.33	3.22	3.14	3.07	3.02	2.98	2.77	2.70	2.66	2.64
11	4.84	3.98	3.59	3.36	3.20	3.09	3.01	2.95	2.90	2.85	2.65	2.57	2.53	2.51
12	4.75	3.89	3.49	3.26	3.11	3.00	2.91	2.85	2.80	2.75	2.54	2.47	2.43	2.40
13	4.67	3.81	3.41	3.18	3.03	2.92	2.83	2.77	2.71	2.67	2.46	2.38	2.34	2.31
14	4.60	3.74	3.34	3.11	2.96	2.85	2.76	2.70	2.65	2.60	2.39	2.31	2.27	2.24
15	4.54	3.68	3.29	3.06	2.90	2.79	2.71	2.64	2.59	2.54	2.33	2.25	2.20	2.18
16	4.49	3.63	3.24	3.01	2.85	2.74	2.66	2.59	2.54	2.49	2.28	2.19	2.15	2.12
17	4.45	3.59	3.20	2.96	2.81	2.70	2.61	2.55	2.49	2.45	2.23	2.15	2.10	2.08
18	4.41	3.55	3.16	2.93	2.77	2.66	2.58	2.51	2.46	2.41	2.19	2.11	2.06	2.04
19	4.38	3.52	3.13	2.90	2.74	2.63	2.54	2.48	2.42	2.38	2.16	2.07	2.03	2.00
20	4.35	3.49	3.10	2.87	2.71	2.60	2.51	2.45	2.39	2.35	2.12	2.04	1.99	1.97
21	4.32	3.47	3.07	2.84	2.68	2.57	2.49	2.42	2.37	2.32	2.10	2.01	1.96	1.94
22	4.30	3.44	3.05	2.82	2.66	2.55	2.46	2.40	2.34	2.30	2.07	1.98	1.94	1.91
23	4.28	3.42	3.03	2.80	2.64	2.53	2.44	2.37	2.32	2.27	2.05	1.96	1.91	1.88
24	4.26	3.40	3.01	2.78	2.62	2.51	2.42	2.36	2.30	2.25	2.03	1.94	1.89	1.86
25	4.24	3.39	2.99	2.76	2.60	2.49	2.40	2.34	2.28	2.24	2.01	1.92	1.87	1.84
26	4.23	3.37	2.98	2.74	2.59	2.47	2.39	2.32	2.27	2.22	1.99	1.90	1.85	1.82
27	4.21	3.35	2.96	2.73	2.57	2.46	2.37	2.31	2.25	2.20	1.97	1.88	1.84	1.81
28	4.20	3.34	2.95	2.71	2.56	2.45	2.36	2.29	2.24	2.19	1.96	1.87	1.82	1.79
29	4.18	3.33	2.93	2.70	2.55	2.43	2.35	2.28	2.22	2.18	1.94	1.85	1.81	1.77
30	4.17	3.32	2.92	2.69	2.53	2.42	2.33	2.27	2.21	2.16	1.93	1.84	1.79	1.76

(Continued)

TABLE 4. Critical Values of the *F* Distribution with Upper Tail Area = 0.05 (Continued)

$$P(F_{df_1, df_2} > F) = 0.05,$$
$$\text{e.g., } P(F_{3,20} > 3.10) = 0.05$$

						df_1								
	1	2	3	4	5	6	7	8	9	10	20	30	40	50
df_2														
31	4.16	3.30	2.91	2.68	2.52	2.41	2.32	2.25	2.20	2.15	1.92	1.83	1.78	1.75
32	4.15	3.29	2.90	2.67	2.51	2.40	2.31	2.24	2.19	2.14	1.91	1.82	1.77	1.74
33	4.14	3.28	2.89	2.66	2.50	2.39	2.30	2.23	2.18	2.13	1.90	1.81	1.76	1.72
34	4.13	3.28	2.88	2.65	2.49	2.38	2.29	2.23	2.17	2.12	1.89	1.80	1.75	1.71
35	4.12	3.27	2.87	2.64	2.49	2.37	2.29	2.22	2.16	2.11	1.88	1.79	1.74	1.70
36	4.11	3.26	2.87	2.63	2.48	2.36	2.28	2.21	2.15	2.11	1.87	1.78	1.73	1.69
37	4.11	3.25	2.86	2.63	2.47	2.36	2.27	2.20	2.14	2.10	1.86	1.77	1.72	1.68
38	4.10	3.24	2.85	2.62	2.46	2.35	2.26	2.19	2.14	2.09	1.85	1.76	1.71	1.68
39	4.09	3.24	2.85	2.61	2.46	2.34	2.26	2.19	2.13	2.08	1.85	1.75	1.70	1.67
40	4.08	3.23	2.84	2.61	2.45	2.34	2.25	2.18	2.12	2.08	1.84	1.74	1.69	1.66
41	4.08	3.23	2.83	2.60	2.44	2.33	2.24	2.17	2.12	2.07	1.83	1.74	1.69	1.65
42	4.07	3.22	2.83	2.59	2.44	2.32	2.24	2.17	2.11	2.06	1.83	1.73	1.68	1.65
43	4.07	3.21	2.82	2.59	2.43	2.32	2.23	2.16	2.11	2.06	1.82	1.72	1.67	1.64
44	4.06	3.21	2.82	2.58	2.43	2.31	2.23	2.16	2.10	2.05	1.81	1.72	1.67	1.63
45	4.06	3.20	2.81	2.58	2.42	2.31	2.22	2.15	2.10	2.05	1.81	1.71	1.66	1.63
46	4.05	3.20	2.81	2.57	2.42	2.30	2.22	2.15	2.09	2.04	1.80	1.71	1.65	1.62
47	4.05	3.20	2.80	2.57	2.41	2.30	2.21	2.14	2.09	2.04	1.80	1.70	1.65	1.61
48	4.04	3.19	2.80	2.57	2.41	2.29	2.21	2.14	2.08	2.03	1.79	1.70	1.64	1.61
49	4.04	3.19	2.79	2.56	2.40	2.29	2.20	2.13	2.08	2.03	1.79	1.69	1.64	1.60
50	4.03	3.18	2.79	2.56	2.40	2.29	2.20	2.13	2.07	2.03	1.78	1.69	1.63	1.60
75	3.97	3.12	2.73	2.49	2.34	2.22	2.13	2.06	2.01	1.96	1.71	1.61	1.55	1.52
100	3.94	3.09	2.70	2.46	2.31	2.19	2.10	2.03	1.97	1.93	1.68	1.57	1.52	1.48

Generated in SAS ©. SAS Institute Inc. SAS © *User's Guide: Basics v 9.1.* Cary, NC: SAS Institute Inc.; 2002–2003.

TABLE 5. Critical Values of the Mann–Whitney U Test

Two-Sided Test $\alpha = 0.05$

n_2	1	2	3	4	5	6	7	8	9	10	11	12	13	14	15	16	17	18	19	20
2								0	0	0	0	1	1	1	1	1	2	2	2	2
3					0	1	1	2	2	3	3	4	4	5	5	6	6	7	7	8
4				0	1	2	3	4	4	5	6	7	8	9	10	11	11	12	13	13
5			0	1	2	3	5	6	7	8	9	11	12	13	14	15	17	18	19	20
6			1	2	3	5	6	8	10	11	13	14	16	17	19	21	22	24	25	27
7			1	3	5	6	8	10	12	14	16	18	20	22	24	26	28	30	32	34
8		0	2	4	6	8	10	13	15	17	19	22	24	26	29	31	34	36	38	41
9		0	2	4	7	10	12	15	17	20	23	26	28	31	34	37	39	42	45	48
10		0	3	5	8	11	14	17	20	23	26	29	33	36	39	42	45	48	52	55
11		0	3	6	9	13	16	19	23	26	30	33	37	40	44	47	51	55	58	62
12		1	4	7	11	14	18	22	26	29	33	37	41	45	49	53	57	61	65	69
13		1	4	8	12	16	20	24	28	33	37	41	45	50	54	59	63	67	72	76
14		1	5	9	13	17	22	26	31	36	40	45	50	55	59	64	67	74	78	83
15		1	5	10	14	19	24	29	34	39	44	49	54	59	64	70	75	80	85	90
16		1	6	11	15	21	26	31	37	42	47	53	59	64	70	75	81	86	92	98
17		2	6	11	17	22	28	34	39	45	51	57	63	67	75	81	87	93	99	105
18		2	7	12	18	24	30	36	42	48	55	61	67	74	80	86	93	99	106	112
19		2	7	13	19	25	32	38	45	52	58	65	72	78	85	92	99	106	113	119
20		2	8	13	20	27	34	41	48	55	62	69	76	83	90	98	105	112	119	127

$n_1 \leq n_2$

(Continued)

TABLE 5. Critical Values of the Mann–Whitney U Test (Continued)

One-Sided Test $\alpha = 0.05$

n_2	1	2	3	4	5	6	7	8	9	10	11	12	13	14	15	16	17	18	19	20
2					0	0	0	1	1	1	1	2	2	2	3	3	3	4	4	4
3			0	0	1	2	2	3	3	4	5	5	6	7	7	8	9	9	10	11
4			0	1	2	3	4	5	6	7	8	9	10	11	12	14	15	16	17	18
5		0	1	2	4	5	6	8	9	11	12	13	15	16	18	19	20	22	23	25
6		0	2	3	5	7	8	10	12	14	16	17	19	21	23	25	26	28	30	32
7		0	2	4	6	8	11	13	15	17	19	21	24	26	28	30	33	35	37	39
8		1	3	5	8	10	13	15	18	20	23	26	28	31	33	36	39	41	44	47
9		1	3	6	9	12	15	18	21	24	27	30	33	36	39	42	45	48	51	54
10		1	4	7	11	14	17	20	24	27	31	34	37	41	44	48	51	55	58	62
11		1	5	8	12	16	19	23	27	31	34	38	42	46	50	54	57	61	65	69
12		2	5	9	13	17	21	26	30	34	38	42	47	51	55	60	64	68	72	77
13		2	6	10	15	19	24	28	33	37	42	47	51	56	61	65	70	75	80	84
14		2	7	11	16	21	26	31	36	41	46	51	56	61	66	71	77	82	87	92
15		3	7	12	18	23	28	33	39	44	50	55	61	66	72	77	83	88	94	100
16		3	8	14	19	25	30	36	42	48	54	60	65	71	77	83	89	95	101	107
17		3	9	15	20	26	33	39	45	51	57	64	70	77	83	89	96	102	109	115
18		4	9	16	22	28	35	41	48	55	61	68	75	82	88	95	102	109	116	123
19	0	4	10	17	23	30	37	44	51	58	65	72	80	87	94	101	109	116	123	130
20	0	4	11	18	25	32	39	47	54	62	69	77	84	92	100	107	115	123	130	138

$n_1 \leq n_2$

Reproduced from Johnson RR. *Elementary Statistics*. Reproduced with permission of PWS-KENT PUB. CO; 1992.

TABLE 6. Critical Values for the Sign Test

Two-Sided Test α	.10	.05	.02	.01
One-Sided Test α	.05	.025	.01	.005
n				
1				
2				
3				
4				
5	0			
6	0	0		
7	0	0	0	
8	1	0	0	0
9	1	1	0	0
10	1	1	0	0
11	2	1	1	0
12	2	2	1	1
13	3	2	1	1
14	3	2	2	1
15	3	3	2	2
16	4	3	2	2
17	4	4	3	2
18	5	4	3	3
19	5	4	4	3
20	5	5	4	3
21	6	5	4	4
22	6	5	5	4
23	7	6	5	4
24	7	6	5	5
25	7	7	6	5

Reproduced from Triola MF. *Elementary Statistics*. 7th ed. New York, NY: Pearson Education, Inc.; 1998. Reprinted by permission of Pearson Education, Inc. © 1998.

TABLE 7. Critical Values for the Wilcoxon Signed Rank Test

Two-Sided Test α	.10	.05	.02	.01
One-Sided Test α	.05	.025	.01	.005
n				
5	1			
6	2	1		
7	4	2	0	
8	6	4	2	0
9	8	6	3	2
10	11	8	5	3
11	14	11	7	5
12	17	14	10	7
13	21	17	13	10
14	26	21	16	13
15	30	25	20	16
16	36	30	24	19
17	41	35	28	23
18	47	40	33	28
19	54	46	38	32
20	60	52	43	37
21	68	59	49	43
22	75	66	56	49
23	83	73	62	55
24	92	81	69	61
25	101	90	77	68
26	110	98	85	76
27	120	107	93	84
28	130	117	102	92
29	141	127	111	100
30	152	137	120	109

Reproduced from Triola MF. *Elementary Statistics.* 7th ed. New York, NY: Pearson Education, Inc.; 1998. Reprinted by permission of Pearson Education, Inc. © 1998.

TABLE 8. Critical Values for the Kruskal–Wallis Test, $\alpha = .05$ and $\alpha = .01$

\multicolumn{3}{c}{Three groups}			\multicolumn{4}{c}{Four groups}		
n_1	n_2	n_3	$\alpha = .05$	$\alpha = .01$	n_1	n_2	n_3	n_4	$\alpha = .05$	$\alpha = .01$
2	2	2			2	2	1	1		
3	2	1			2	2	2	1	5.679	
3	2	2	4.714		2	2	2	2	6.167	6.667
3	3	1	5.143		3	1	1	1		
3	3	2	5.361		3	2	1	1		
3	3	3	5.600	7.200	3	2	2	1	5.833	
4	2	1			3	2	2	2	6.333	7.133
4	2	2	5.333		3	3	1	1	6.333	
4	3	1	5.208		3	3	2	1	6.244	7.200
4	3	2	5.444	6.444	3	3	2	2	6.527	7.636
4	3	3	5.791	6.745	3	3	3	1	6.600	7.400
4	4	1	4.967	6.667	3	3	3	2	6.727	8.015
4	4	2	5.455	7.036	3	3	3	3	7.000	8.538
4	4	3	5.598	7.144	4	1	1	1		
4	4	4	5.692	7.654	4	2	1	1	5.833	
5	2	1	5.000		4	2	2	1	6.133	7.000
5	2	2	5.160	6.533	4	2	2	2	6.545	7.391
5	3	1	4.960		4	3	1	1	6.178	7.067
5	3	2	5.251	6.909	4	3	2	1	6.309	7.455
5	3	3	5.648	7.079	4	3	2	2	6.621	7.871
5	4	1	4.985	6.955	4	3	3	1	6.545	7.758
5	4	2	5.273	7.205	4	3	3	2	6.795	8.333
5	4	3	5.656	7.445	4	3	3	3	6.984	8.659
5	4	4	5.657	7.760	4	4	1	1	5.945	7.909
5	5	1	5.127	7.309	4	4	2	1	6.386	7.909
5	5	2	5.338	7.338	4	4	2	2	6.731	8.346
5	5	3	5.705	7.578	4	4	3	1	6.635	8.231
5	5	4	5.666	7.823	4	4	3	2	6.874	8.621
5	5	5	5.780	8.000	4	4	3	3	7.038	8.876
6	1	1			4	4	4	1	6.725	8.588
6	2	1	4.822		4	4	4	2	6.957	8.871
6	2	2	5.345	6.655	4	4	4	3	7.142	9.075
6	3	1	4.855	6.873	4	4	4	4	7.235	9.287

(Continued)

TABLE 8. Critical Values for the Kruskal–Wallis Test, $\alpha = .05$ and $\alpha = .01$ (Continued)

\multicolumn{3}{c	}{Three groups}			\multicolumn{5}{c	}{Five groups}		
n_1	n_2	n_3	$\alpha = .05$	$\alpha = .01$	n_1	n_2	n_3	n_4	n_5	$\alpha = .05$	$\alpha = .01$
6	3	2	5.348	6.970	2	2	1	1	1		
6	3	3	5.615	7.410	2	2	2	1	1	6.750	
6	4	1	4.947	7.106	2	2	2	2	1	7.133	7.533
6	4	2	5.340	7.340	2	2	2	2	2	7.418	8.291
6	4	3	5.610	7.500	3	1	1	1	1		
6	4	4	5.681	7.795	3	2	1	1	1	6.583	
6	5	1	4.990	7.182	3	2	2	1	1	6.800	7.600
6	5	2	5.338	7.376	3	2	2	2	1	7.309	8.127
6	5	3	5.602	7.590	3	2	2	2	2	7.682	8.682
6	5	4	5.661	7.936	3	3	1	1	1	7.111	
6	5	5	5.729	8.028	3	3	2	1	1	7.200	8.073
6	6	1	4.945	7.121	3	3	2	2	1	7.591	8.576
6	6	2	5.410	7.467	3	3	2	2	2	7.910	9.115
6	6	3	5.625	7.725	3	3	3	1	1	7.576	8.424
6	6	4	5.725	8.000	3	3	3	2	1	7.769	9.051
6	6	5	5.765	8.124	3	3	3	2	2	8.044	9.505
6	6	6	5.801	8.222	3	3	3	3	1	8.000	9.451
7	7	7	5.819	8.378	3	3	3	3	2	8.200	9.876
8	8	8	5.805	8.465	3	3	3	3	3	8.333	10.200

Reproduced from Neave HR. *Statistics Tables: for mathematicians, engineers, economists and the behavioural and management sciences.* New York, NY: Taylor & Francis; 1978. © 1978 Taylor & Francis, reproduced by permission of Taylor & Francis Books UK.

Glossary

Active-Controlled Trial—a randomized controlled trial with an active treatment or intervention as the comparator (as opposed to a placebo comparator, which is used in a placebo-controlled trial).

Actuarial Table—also called a life table or a follow-up life table. Summarizes the survival experiences of participants over a predefined follow-up period in a cohort study or a clinical trial until the time of the event of interest or the end of the study, whichever comes first.

Adjusted Analysis—an analysis that incorporates important covariates (i.e., variables that are associated with the outcome) or confounding variables (i.e., variables that are associated with the outcome and also with the risk factor(s) or exposure(s) of interest).

Alpha (α)—the level of significance in a test of hypothesis. Defined as the probability of a Type I error, or the probability that we reject H_0 when H_0 is true, P(reject H_0 | H_0 true).

Alternative Hypothesis—also called research hypothesis. Represents the investigator's belief in terms of the association or effect under study.

Analysis of Variance (ANOVA)—a popular procedure for testing the equality of k ($k > 2$) independent group means.

Attrition—the loss of participants from a study.

Bar Chart—a graphical display for a categorical variable where the distinct response options are shown on the horizontal axis. Bars are centered over each response option with spaces in between adjacent responses, and the heights of the bars represent either the frequencies or relative frequencies of each response shown on the vertical axis.

Bayes' Rule—also called Bayes' theorem. A procedure for updating a probability based on new information. The rule can be used to compute a conditional probability based on specific, available information. There are several versions of the theorem, ranging from simple to more involved. The simple statement of the rule is $P(A | B) = \dfrac{P(B | A) P(A)}{P(B)}$.

Beta (β)—the probability of a Type II error in a test of hypothesis, or the probability that we do not reject H_0 when H_0 is false, P(do not reject H_0 | H_0 false).

Bias—a systematic error that introduces uncertainty in estimates of effect or association.

Binomial Distribution—a probability distribution model for an application or process in which each replication of the process results in one of two possible outcomes (success or failure), in which the probability of success is the same for each replication and in which the replications are independent, meaning that success in one replication (e.g., patient) does not influence the probability of success in another. The probability model is as follows: P(x successes) = $\dfrac{n!}{x!(n-x)!} p^x (1-p)^{n-x}$, where n represents the number of replications, x is the number of successes of interest, and p = P(success).

Biostatistics—the application of statistical principles to medical, public health, or biological problems.

Blind—the state whereby a participant or an investigator is unaware of treatment status (e.g., experimental drug or placebo).

Box and Whisker Plot—a graphical display for a continuous variable that shows the range of the data (minimum to maximum values), the first quartile, the median, and the third quartile.

Case—often used to describe a person with the outcome of interest (e.g., disease) in a study.

Case-Control Study—a study where participants are selected on the basis of their outcome status. Specifically, we select a set of cases, or persons with the outcome of interest. We then select a set of controls, who are persons like the cases except for the fact that they are free of the outcome of interest. We then assess exposure or risk factor status retrospectively.

Case Report—a very detailed report of the specific features of a particular patient or case.

Case Series—a systematic review of the interesting and common features of a small collection, or series, of cases.

Categorical Variable—a variable with a fixed number of unordered response options.

Censored Data—occur in survival analysis when true survival time (sometimes called failure time) is not known because the study ends or because a participant drops out of the study before experiencing the event. What we know is that the participant's survival time is greater than his or her last observed follow-up time, and these times are called censored times or censored data.

Central Limit Theorem—a very important theorem in statistics that states that if we take simple random samples of size n from a population with replacement, then for large samples (usually defined as samples with $n \geq 30$), the sampling distribution of the sample means is approximately normally distributed with a mean of $\mu_{\bar{x}} = \mu$ and a standard deviation of $\sigma_{\bar{x}} = \frac{\sigma}{\sqrt{n}}$.

Chi-square Goodness-of-Fit Test—a test of hypothesis used to assess whether the responses to a categorical or ordinal variable follow a prespecified distribution.

Chi-square (χ^2) Statistic—a test statistic that follows the χ^2 distribution, which is positive and skewed. This test statistic is used in the χ^2 goodness-of-fit test as well as the χ^2 test of independence.

Chi-square Test of Independence—a test of hypothesis used to assess whether there is a difference in the distribution of responses to a categorical or ordinal variable among independent comparison groups.

Choropleth Map—a data visualization using different shades and colors to depict varying levels of the characteristic being displayed.

Clinical Trial—a specific type of study involving human participants who are randomized to the comparison groups.

Cochran-Mantel-Haenszel Method—a technique that generates an estimate of an association between a risk factor or exposure and an outcome that accounts for confounding. The method is used with a dichotomous outcome variable and a dichotomous risk factor, and essentially computes a weighted average of the relative risks (or odds ratios, whichever measure is used to quantify association) across the strata (or groups) defined by the confounding variable.

Cohort—a group of participants that usually share some common characteristics, who are monitored or followed over time.

Cohort Study—a study that generally involves a group of individuals that meet a set of inclusion criteria at the study start (e.g., free of the disease of interest). The cohort is followed prospectively, and associations are estimated between risk factor(s) and disease or other outcomes.

Complementary Events—events that are mutually exclusive. A and B are said to be complements of one another if $P(A) + P(B) = 1$.

Concurrent—at the same time. Optimally, comparison treatments are evaluated concurrently or in parallel.

Conditional Probability—the probability that an event A occurs given or assuming that some other event B has occurred, $P(A|B) = P(A \text{ and } B)/P(B)$.

Confidence Interval—a range of plausible values for a population parameter with a level of confidence attached (e.g., 95% confidence that the interval contains the unknown parameter).

Confidence Level—the theoretical probability that a confidence interval contains the true, unknown parameter. In practice, confidence levels of 90%, 95%, and 99% are often used.

Confounding—a distortion of the effect of an exposure or risk factor on an outcome by another characteristic or variable.

Confounding Variable—a variable associated with the outcome of interest and also with the exposure or risk factor of interest that modifies the relationship between the exposure and the outcome.

Continuous Variable—sometimes called a quantitative or measurement variable. Takes on an unlimited number of responses between a defined minimum and maximum value.

Control—often used to describe a person free of the outcome of interest (e.g., free of disease) in a study.

Convenience Sample—a non-probability-based sample where individuals are selected into the sample by any convenient contact.

Correlation Analysis—used to quantify the association between two continuous variables.

Correlation Coefficient—also called the Pearson Product Moment correlation coefficient. Ranges between −1 and +1 and quantifies the direction and strength of the linear association between two continuous variables.

Covariance—measures the variability of (x, y) pairs around the mean of X and mean of Y, considered simultaneously. Defined as $\text{cov}(X,Y) = \dfrac{\sum(X-\overline{X})(Y-\overline{Y})}{n-1}$.

Covariate—a variable that is associated with the outcome of interest.

Cox Proportional Hazards Regression—a technique used in survival analysis to relate several risk factors or exposures, considered simultaneously, to time to event.

Critical Value—a threshold or cut-off point usually determined from a probability distribution that is used to determine when we reject the null hypothesis in a test of hypothesis.

Crossover Trial—a clinical trial where each participant is assigned to two or more treatments sequentially (typically, participants are randomly assigned to the first treatment in period 1).

Cross-sectional—at a point in time.

Cross-sectional Survey—a nonrandomized study that involves a group of participants who are identified at a point in time, and information is collected at that point in time.

Crude Analyses—associations between a risk factor or exposure and an outcome or the difference in outcomes between comparison groups that does not account for covariates or confounding variables.

Cumulative Frequency—the sum of the number of responses up to and including the current response.

Cumulative Incidence—the ratio of the number of new cases of disease to the total number of participants who are at risk, or free of disease at the start of the study. In survival analysis, cumulative incidence, or cumulative failure probability, is computed as the complement of the survival probability, which can be computed from a life table using the Kaplan–Meier approach.

Cumulative Relative Frequency—the ratio of the sum of the number of responses up to and including the current response (cumulative frequency) to the total sample size.

Decision Rule—a statement that indicates the circumstances in which to reject the null hypothesis in a test of hypothesis.

Dependent Samples—also called matched or paired samples, where observations within each matched pair are related. This relationship must be reconciled in statistical analysis.

Dependent Variable—the primary response or outcome variable in an analysis.

Descriptive Statistics—numerical or graphical summaries of data collected in a sample.

Dichotomous Variable—a variable with exactly two possible responses (e.g., yes/no), usually coded 0 = no and 1 = yes for analysis purposes.

Dispersion—variability.

Dot Plot—a graphical display of observed values of a continuous variable, most suitable for small samples.

Double Blind—A study in which both the participant and the outcome assessor are unaware of the treatment status (masking is a term equivalent to blinding).

Dummy Variables—a set of indicator (or dichotomous) variables used to differentiate among more than two levels of a categorical or grouping variable.

Effect, Effect Size—a measure of the strength of the association or difference in the parameter of interest between comparison groups.

Effect Modification—a situation that occurs when there is a different relationship between the exposure or risk factor and the outcome depending on the level of another characteristic or variable. Also called statistical interaction.

Efficacy—the capacity of a drug, treatment, device, or intervention to produce a beneficial change.

Epidemiology—a field of study focused on health and disease in human populations, patterns of health or disease, and the factors that influence these patterns.

Equally Likely—outcomes are said to be equally likely if each has the same probability of occurrence.

Estimation—the process of determining likely values for a population parameter based on a random sample.

Expected Frequency—the number of participants that would be expected in a group under certain assumptions. Expected frequencies are part of χ^2 tests, where they are computed based on the assumption that the null hypothesis is true.

False Negative Fraction—the proportion of diseased participants who test negative, P(test negative | disease).

False Positive Fraction—the proportion of disease-free participants who test positive, P(test positive | disease-free).

First Quartile—denoted Q_1, is the value in the dataset that holds 25% of the values below it. Equivalent to the 25th percentile.

Frequency—the number of participants in a particular group.

Frequency Distribution Table—a table used to summarize responses to a dichotomous, categorical, or ordinal variable, including each of the unique response options, the frequencies, and the relative frequencies of each of the response options. For ordinal variables, the table often includes the cumulative and cumulative relative frequencies.

Gaussian Distribution—normal distribution.

Generalizability—in biostatistics, results observed in a sample are said to be generalizable to the population as long as the sample is truly representative of that population.

Geographical Display—representation of data or statistical results using a map.

Hazard—the expected number of events per unit of time.

Hazard Ratio—the ratio of hazards in two independent comparison groups, interpreted as the risk of event in the group in the numerator as compared to the risk of event in the group in the denominator.

Histogram—a graphical display for an ordinal variable where the distinct response options are shown on the horizontal axis. Bars are centered over each response option, and the heights of the bars represent either the frequencies or relative frequencies of each response shown on the vertical axis. There is no gap between response options on the horizontal axis, suggestive of the underlying continuum in ordered response options.

Historical Control—a group of participants treated in the past that are used as a comparator in a current study.

Hypothesis Testing—a process whereby a specific statement or hypothesis is generated about a population parameter and in which sample statistics are used to assess the likelihood that the hypothesis is true.

Incidence (of disease)—the number of new cases (of disease) over a period of time.

Incidence Rate—also called incidence density. Computed as the ratio of the number of new cases of disease to the total follow-up time (i.e., the sum of all disease-free person-time).

Independent Events—two events are said to be independent if the probability of one is not affected by the occurrence or nonoccurrence of the other. Two events A and B are independent if $P(A \mid B) = P(A)$, if $P(B \mid A) = P(B)$, or if $P(A \text{ and } B) = P(A)P(B)$.

Independent Variable—the predictor, risk factor, or exposure in an analysis that is hypothesized to predict or be associated with an outcome or dependent variable.

Indicator Variable—a dichotomous variable used to differentiate participants with and without a particular characteristic and generally coded as 0 = No, 1 = Yes.

Intention-to-Treat—an analytic strategy whereby participants are analyzed in the treatment they are assigned regardless of whether they follow the study protocol completely (e.g., regardless of whether they take all of the assigned medication).

Interaction—also called statistical interaction or effect modification. Occurs when there is a different relationship between the exposure or risk factor and the outcome depending on the level of another characteristic or variable.

Intercept—also called the y-intercept. In linear regression analysis, the value of the dependent variable (y) when the independent variable(s) are all equal to zero.

Interquartile Range—a measure of variability, computed as the difference between the first and third quartiles, Interquartile Range = $Q_3 - Q_1$.

Interval Data—like continuous data in that they are measured on a constant scale (i.e., the same difference between adjacent scale scores exists across the entire spectrum of scores). Differences between interval scores are interpretable, but ratios are not (e.g., temperature is measured on an interval scale, and the difference between 20 and 40 degrees is the same as the difference between 40 and 60 degrees; however, 40 degrees is not twice as warm as 20 degrees).

Kaplan–Meier Approach—also called the product-limit approach. A very popular approach to summarizing the survival experiences of participants over a predefined follow-up period in a cohort study or in a clinical trial until the time of the event of interest or until the end of the study, whichever comes first. In this approach, we re-estimate the survival probability each time an event occurs (as opposed to using equally spaced intervals, as in the life table or actuarial approach).

Kruskal–Wallis Test—a nonparametric test used to compare medians among k independent comparison groups ($k > 2$), sometimes described as an ANOVA with the data replaced by their ranks.

Level of Significance—denoted α. Used in a test of hypothesis and defined as the probability of a Type I error, or the probability that we reject H_0 when H_0 is true, $P(\text{reject } H_0 \mid H_0 \text{ true})$.

Life Table Analysis—also called actuarial table or follow-up life table analysis. An approach to summarizing the survival experiences of participants over a predefined follow-up period in a cohort study or in a clinical trial until the time of the event of interest or until the end of the study, whichever comes first.

Line Chart—a graphical display of trends in a measure over time.

Linear Regression Analysis—a technique used to estimate the equation that best describes the association between one or more independent variables and a single continuous outcome or dependent variable.

Logistic Regression Analysis—a technique used to estimate the association between one or more independent variables and a single dichotomous outcome or dependent variable.

Logit—also called the log odds. Used in logistic regression analysis and computed by taking the log of the odds of the probability of the outcome or dependent variable.

Log-Rank Test—a nonparametric test to test the null hypothesis of no difference in survival between two or more independent groups.

Lower-tailed Test—a test in which the decision rule has investigators reject H_0 if the test statistic is smaller than the critical value.

Mann–Whitney U Test—sometimes called the Mann–Whitney–Wilcoxon test or the Wilcoxon Rank Sum test. A popular nonparametric test used to compare outcomes between two independent groups, to determine whether two samples are likely to derive from the same population, or that the two populations have the same shape.

Margin of Error—the product of the value that reflects the desired confidence level (e.g., Z = 1.96 for 95% confidence) and the standard error of the point estimate.

Matching—a process of organizing comparison groups on the basis of similar characteristics.

Maximum—the largest value.

McNemar's Test—a test for the equality of two proportions when the samples are matched or paired, sometimes called McNemar's test for dependent proportions.

Mean—a measure of central tendency or location, computed as the ratio of the sum of a set of values to its size (e.g., the population mean is $\mu = \frac{\Sigma X}{N}$ and the sample mean is $\overline{X} = \frac{\Sigma X}{n}$).

Median—a measure of central tendency or location, defined as the value that separates the top 50% of values from the bottom 50%, sometimes called the middle value or the 50th percentile.

Minimum—the smallest value.

Misclassification Bias—bias introduced by incorrect classification of outcome status (e.g., case or control) or incorrect classification of exposure status (e.g., exposed or not).

Mode—the most frequently occurring value in a set of observations.

Multicenter Trial—a trial conducted at several sites or clinical centers (as opposed to a single-center trial).

Multiple Linear Regression Analysis—a technique used to estimate the equation that best describes the association between two or more independent variables and a single continuous outcome or dependent variable.

Multiple Logistic Regression Analysis—a technique used to estimate the equation that best describes the association between two or more independent variables and a single dichotomous outcome or dependent variable.

Multivariable Analysis—analyses designed to assess the interrelationships among several risk factors or exposure variables and a single outcome. Sometimes confused with multivariate analysis, which refers to the situation where we assess interrelationships among several risk factors or exposures and several outcomes, considered simultaneously.

Mutually Exclusive Events—events that have no outcomes in common. A and B are said to be mutually exclusive if P(A and B) = 0.

Negative Predictive Value—the proportion of participants that test negative who are disease-free, P(disease-free | test negative).

Nested Case-Control Study—a specific type of case-control study that is usually designed from a cohort study.

Nonparametric Tests—sometimes called distribution-free tests because they are based on few assumptions (e.g., they do not assume that the outcome follows a particular distribution).

Non-probability Sampling—a procedure for sampling participants from a population where members of the population are selected into the sample without the use of probability.

Normal Distribution—a probability distribution for a continuous outcome that follows what is called the Gaussian distribution or is well described by a bell-shaped curve. In a normal distribution, mean = median = mode, and the distribution is symmetric about mean. In addition, approximately 68% of the values fall between the mean and one standard deviation in either direction, i.e., $\mu - \sigma < X < \mu + \sigma$, where μ is the population mean and σ is the population standard deviation; approximately 95% of the values fall between the mean and two standard deviations in either direction, i.e., $\mu - 2\sigma < X < \mu + 2\sigma$; and approximately 99.9% of the values fall between the mean and three standard deviations in either direction, i.e., $P(\mu - 3\sigma < X < \mu + 3\sigma)$.

Null Hypothesis—statement of no difference, no association, or no effect. Tested against an alternative or research hypothesis, which postulates a difference, an association, or an effect.

Observational Study—sometimes called a descriptive or associational study. A nonrandomized or historical study, where we generally observe whether participants engage in a particular risk behavior or whether they are exposed, in contrast to a randomized study, where we intervene and measure a response.

Odds—the ratio of the number of events to nonevents, or the ratio of the probability of event to its complement.

Odds Ratio—the ratio of two odds. For example, the ratio of the odds of disease in an exposed group as compared to the odds of disease in an unexposed group.

One-tailed or One-sided Test—a test of hypothesis in which the alternative or research hypothesis has investigators reject H_0 if the test statistic is extreme in a particular direction.

Ordinal Variable—a variable with a fixed number of ordered response options.

Outcome—the primary response or dependent variable in an analysis.

Outliers—values outside the range of those generally expected. Popular guidelines to detect outliers include the following: Outliers are values below $Q_1 - 1.5 \times IQR$ or above $Q_3 + 1.5 \times IQR$, where IQR is the interquartile range, or outliers are values below $\bar{X} - 3s$ or above $\bar{X} + 3s$.

p-value—the exact significance of the data, the likelihood of observing the sample data if the null hypothesis is true, or the smallest level of significance where we still reject H_0.

Paired t Test—a test for the equality of means in two matched or paired samples, based on an analysis of difference scores.

Parameter—any summary measure computed on a population.

Parametric Test—a test of hypothesis where the outcome is assumed to follow a particular probability distribution (e.g., the normal distribution) and where the test involves estimation of the key parameters of that distribution (e.g., the mean or difference in means) from the sample data.

Participant Flow Chart—a graphical display of a study design or protocol that details the numbers of participants recruited, consented, enrolled and completing various components of the study.

Per Protocol—an analytic strategy whereby only participants who adhere to the study protocol are analyzed, an analysis of only those assigned to a particular group who follow all procedures for that group.

Percentile—a value in a distribution that holds a specified percentage of the population below it. Percentiles of the normal distribution are computed using $X = \mu + Z\sigma$, where μ is the mean and σ is the standard deviation of the variable X, and Z is the value from the standard normal distribution for the desired percentile.

Phase I Clinical Trial—also called a *first time in humans study*. The main objectives are to assess the toxicology and safety of proposed treatment in humans and to assess the pharmacokinetics of the proposed treatment (how fast the drug is absorbed in, flows through, and is secreted from the body).

Phase II Clinical Trial—also called a *feasibility study* or *dose-finding study* where the focus is on safety of a proposed treatment; of primary interest are side effects and adverse events (which may or may not be directly related to the drug). Another objective in the Phase II study is to assess efficacy, but based on descriptive analyses.

Phase III Clinical Trial—also called a *confirmatory clinical trial*, where the focus is efficacy, although data are also collected to monitor safety.

Phase IV Clinical Trial—also called a *post-marketing trial*, which is undertaken to understand the long-term effects (efficacy and safety) of a drug.

Pie Chart—a graphical display for a single categorical or ordinal variable that displays relative frequencies in each category as slices or components of a pie.

Placebo—an inert substance designed to look, feel, and taste like the active or experimental treatment (e.g., saline solution would be a suitable placebo for a liquid medication).

Placebo-controlled Trial—a randomized controlled trial with a placebo comparator.

Point Estimate—a single-valued estimate of a population parameter derived from a sample.

Point Prevalence—extent of disease at a specific point in time computed as the ratio of the number of persons with disease to the number of persons in the study population.

Population—the collection of all individuals about whom investigators are interested in making generalizations or inferences.

Population Attributable Risk—the ratio of the difference in prevalence or incidence between the total sample and the unexposed group to the prevalence or incidence in the total sample, sometimes interpreted as the percentage of the risk that could be eliminated if the exposure or risk factor were removed.

Population Size—the number of participants in the population, denoted N.

Positive Predictive Value—the proportion of participants that test positive who have the disease, P(disease | test positive).

Power—the probability that a test of hypothesis correctly rejects the null hypothesis when it is false. Power is defined as P(reject H_0|H_0 false) = 1 − P(Type II error) = 1 − β.

Prevalence (of disease)—the proportion of individuals with a disease at a point in time.

Probability—a number that reflects the likelihood that a particular event occurs.

Probability Model—a mathematical equation or formula used to generate probabilities based on certain assumptions about a process.

Probability Sampling—a procedure for sampling participants from a population where each member of the population has a known probability of being selected into the sample.

Prognostic Factor—a characteristic that is strongly associated with an outcome (e.g., disease) such that it could be used to reasonably predict whether a person is likely to experience the outcome or not.

Prospective—looking forward in time.

Prospective Cohort Study—a study where participants are enrolled and followed going forward in time. In some studies the cohort is drawn from the general population, while in other studies a cohort is assembled (e.g., a special exposure cohort).

Protocol—a step-by-step plan for a study that details every aspect of the study design and data collection plan.

Quartiles—values that divide a distribution into four equally sized groups.

Quasi-experimental Design—a design in which subjects are not randomly assigned to comparison groups or competing treatments.

Quota Sample—a procedure whereby individuals are selected from a population into a sample until a prespecified number of individuals have been selected in each of several predefined groups.

Random Sample—a sample selected from a population in such a way that samples of equal size are as likely to be observed.

Randomization—a process by which participants are assigned to receive different treatments; this is usually based on a probability scheme.

Randomized Controlled Trial—sometimes called an analytic or experimental study. Used to test specific hypotheses or to evaluate the effect of an intervention (e.g., a behavioral or pharmacologic intervention) which is based on randomizing participants to the comparison groups.

Range—a measure of variability, computed by taking the difference between the minimum and maximum values, range = maximum − minimum.

Rank—a number that indicates relative order.

Rate—the likelihood that an individual changes status (e.g., develops disease) in a specified unit of time.

Rate Ratio—also called the incidence density ratio. Computed as the ratio of incidence rates in two independent groups.

Recall Bias—a distortion of the association between exposure and outcome due to differential recall of exposure status between comparison groups (e.g., cases versus controls).

Regression Analysis—a technique used to relate an outcome or dependent variable to one or more predictor or independent variables.

Regression Coefficient—an estimate of a parameter that reflects the association between an independent variable and an outcome in a multiple regression setting; it is interpreted as the change in the outcome relative to a one-unit change in the independent variable, holding all other predictor or independent variables constant.

Rejection Region—the range of values in the probability distribution that leads to rejection of the null hypothesis.

Relative Frequency—the ratio of the number of participants in a particular group (frequency) to the total sample size.

Relative Hazard—in a Cox proportional hazards regression, relative hazard is the ratio of the hazard $h(t)$ at time t to the baseline hazard $h_0(t)$. The relative hazard is associated with potential risk factors $X_1, X_2, \ldots, X_p$ according to the following: $h(t)/h_0(t) = \exp(b_1 X_1 + b_2 X_2 + \cdots + b_p X_p)$.

Relative Risk—the ratio of prevalence or incidence in the exposed group to the prevalence or incidence in the unexposed group.

Retrospective—looking backward in time.

Retrospective Cohort Study—a study where exposures and outcomes have occurred at the start of the study; exposure or risk-factor status as well as outcome status is ascertained retrospectively.

Right Censoring—occurs when a participant does not have the event of interest during the study, and thus his or her last observed follow-up time is less than the time to event. This can occur when a participant drops out before a study ends or when a participant is event-free at the end of the observation period.

Risk Difference—also called excess risk. Measures the absolute effect of an exposure or risk factor on prevalence or incidence and is computed as the difference in prevalence or incidence between comparison groups.

Risk Factor—a characteristic or behavior that is hypothesized to be associated with the incidence of disease or the outcome of interest.

Sample—a subset of the population, with participants in the sample ideally selected at random from the population.

Sample Correlation Coefficient—also called the Pearson Product Moment correlation coefficient. Denoted r, ranges between -1 and $+1$, and quantifies the direction and strength of the linear association between two continuous variables.

Sample Mean—a measure of central tendency in a sample, computed as the ratio of the sum of the values in the sample to the sample size $\left(\overline{X} = \dfrac{\sum X}{n}\right)$.

Sample Median—a measure of central tendency in a sample reflecting the value that separates the top 50% of the values from the bottom 50%. Sometimes called the middle value or the 50th percentile of the sample.

Sample Mode—the most frequently occurring value in a sample. A sample can have no mode (when each value occurs as many times as every other) or more than one mode (e.g., a bimodal sample).

Sample Proportion—denoted $\hat{p}$ and computed by taking the ratio of the number of participants with a particular attribute in the sample to the sample size, $\hat{p} = \dfrac{x}{n}$.

Sample Range—a measure of variability in a sample, computed by taking the difference between the minimum and maximum values in the sample.

Sample Size—the number of participants in the study sample, denoted n.

Sample Standard Deviation—a measure of variability in a sample, denoted s and computed by taking the square root of the sample variance: $s = \sqrt{s^2} = \sqrt{\dfrac{\sum(X - \overline{X})^2}{n-1}}$.

Sample Variance—a measure of variability in a sample, denoted s^2 and computed by taking the ratio of the sum of the squared deviations from the mean to the sample size minus one: $s^2 = \dfrac{\sum(X - \overline{X})^2}{n-1}$.

Sampling—a procedure by which individuals are selected from a population into a sample.

Sampling Distribution—the probability distribution of a statistic produced by repeatedly selecting samples of the same size and computing the desired statistic (e.g., sampling distribution of the sample mean).

Sampling Frame—a complete list or enumeration of the population.

Scatter Diagram—a graphical display in which each point represents an (x, y) pair measured in a different participant. Scatter diagrams display the independent variable (X) on the horizontal axis and the dependent variable (Y) on the vertical axis.

Selection Bias—a distortion of the association (an over- or underestimation of the true association) between exposure and outcome status resulting from the process used to select participants into a study. Specifically, the relationship between exposure status and disease may be different in those individuals who participate in the study as compared to those who do not.

Sensitivity—the true positive fraction or the proportion of diseased participants who test positive, P(test positive | disease).

Sign Test—a nonparametric test to compare outcomes between two matched or paired groups, based on the numbers of positive and negative difference scores computed in matched pairs.

Simple Linear Regression Analysis—a technique used to estimate the equation of the line that best describes the association between one independent variable and a single continuous outcome or dependent variable.

Simple Logistic Regression Analysis—a technique used to estimate the equation that best describes the association between one independent variable and a single dichotomous outcome or dependent variable.

Simple Random Sample—a procedure whereby individuals are selected at random from a population into a sample.

Single-center Trial—a clinical trial conducted at a single site or clinical center (as opposed to a multicenter trial).

Slope—in linear regression analysis, the change in the outcome (Y) associated with a one-unit change in the predictor or independent variable (X).

Specificity—the true negative fraction or the proportion of disease-free participants who test negative, P(test negative | disease-free).

Spurious correlation—an apparent relationship that is likely due to a confounding variable.

Standard Deviation—the most commonly used measure of variability, computed as the square root of the variance.

Standard Error—the standard deviation of the sampling distribution of a statistic.

Standard Normal Distribution—a normal distribution with a mean of zero and a standard deviation of one.

Statistic—any summary number computed in a sample.

Statistical Adjustment—a procedure whereby we minimize the effect of a covariate or confounder using multivariable analysis to focus more specifically on the association or effect of the exposure or risk factor of interest and the outcome.

Statistical Inference—the procedures used to make generalizations about a population based on analysis of sample data.

Statistical Significance—a result is said to reach statistical significance if it is unlikely to be due to chance. In tests of hypothesis, we summarize statistical significance by the *p*-value.

Stratification—a process whereby participants are partitioned or separated into mutually exclusive or non-overlapping groups.

Stratified Sampling—a procedure in which the population is partitioned into non-overlapping groups or strata and whereby individuals are sampled within each stratum by simple random sampling or systematic sampling.

Study Design—the methodology used to collect the information from the study sample to address the research question.

Survival Analysis—statistical methods and procedures that handle outcomes that reflect times to event.

Survival Function—the probability that a person survives past a certain time point as a function of time.

Systematic Sample—a procedure whereby every *k*th person is selected from a sampling frame into the sample. The spacing or interval between selections is determined by the ratio of the population size to the sample size (N/n).

***t* Distribution**—a probability model for a continuous variable that is similar to the standard normal distribution Z, but takes a slightly different shape depending on the exact sample size. When the sample size is large, the *t* distribution is very similar to the standard normal distribution Z.

Test Statistic—a single number that summarizes important information in the sample and is used for hypothesis testing.

Third Quartile—denoted Q_3. The value in a dataset that holds 25% of the values above it, equivalent to the 75th percentile.

Time-to-Event Variable—a variable that reflects the time until a participant has an event of interest; used in survival analysis.

Two-factor ANOVA—a procedure used to test for differences in means due to one or both grouping variables (called factors) or the interaction between the two.

Two-sided or Two-tailed Test—a test in which the alternative or research hypothesis has investigators reject H_0 if the test statistic is extreme in either direction.

Type I Error—the situation in which we incorrectly reject H_0 when in fact it is true.

Type II Error—the situation in which we incorrectly fail to reject H_0 when in fact it is false.

Upper-tailed Test—a test in which the decision rule has investigators reject H_0 if the test statistic is larger than the critical value.

Variance—a popular measure of variability, interpreted as the squared deviation from the mean.

Wash-out Period—usually used in a crossover trial, represents the time during which no treatments are given. The wash-out period is included so that any therapeutic effects of the first treatment are removed before a subsequent treatment is administered.

Wilcoxon Signed Rank Test—a nonparametric test used to compare outcomes between two matched or paired groups which is based on the ranks of the absolute values of the difference scores computed in the matched pairs.

***Y*-intercept**—in linear regression analysis, the value of the dependent variable (Y) when the independent variable (X) is equal to zero.

Z score—also called a standardized score, computed as $Z = \dfrac{X - \mu}{\sigma}$, where μ is the mean and σ is the standard deviation of a variable X that is assumed to follow a normal distribution.

Index

A

acquired immune deficiency syndrome (AIDS), case series leading to discovery of, 9
active-controlled trial, 14
actuarial tables
    example studying time of death, 254–256
    time dependent covariates and, 271
adjusted analysis, phases of statistical analysis, 19
alpha (α), 179
alternative hypothesis. *See* research hypothesis
American Heart Association
    classification of blood pressure and cholesterol levels, 41
    statistics regarding cardiovascular disease, 1
analysis of variance (ANOVA)
    defined, 149
    steps in, 152–154
    sum of squares (SS), source of variation, and mean squares (MS), 150–152
    summary statistics for, 150
    table for two-factor ANOVA, 157
    when to use, 241
Anderson-Darling test, 224
ANOVA. *See* analysis of variance (ANOVA)
APGAR score, 224
attrition, 173, 176–178, 181, 185–187
axes (of graphs)
    honesty in visualization, 296–298
    normal distribution model, 78
    scaling to accommodate range of data, 41

B

bar charts
    for categorical variables, 46–50
    for dichotomous variables, 38–41
    histograms compared with, 44, 46
    with gridlines and data labels, 299–300
baseline comparisons, phases of statistical analysis, 19
Bayes' Rule. *See* Bayes' theorem
Bayes' Theorem
    for computing probability, 74
    example applying to predictive value of screening tests, 75
    formula for, 96
beta (β), 179
β (beta), probability of Type II error, 131
bias
    in case-control studies, 13
    defined, 8
    minimized in randomized studies, 14, 16
binomial distribution model
    computing probability of discrete outcomes, 75
    example applying to allergic reactions to medications, 76–77
    example applying to likelihood of heart attack, 77
    formula for, 97
    probability and, 235
biostatistics, introduction to
    defined, 2
    estimation of rate of development of new disease, 3
    identification of risk factors in development or progression of disease, 3

issues to address in applying, 2
quantification of extent of disease in group or region, 2–3
reasons for contradictory results between studies of same disease, 4
summary and references, 4–5
blind/double blind
　defined, 8
　ethics of, 14
　factors in clinical trials, 18
blinding (masking), of controls or placebos in clinical trials, 4
box-whisker plots
　of BMI of men and women in Framingham Offspring Study, 60
　for continuous variables, 57–58
　of diastolic blood pressures in Framingham Offspring Study, 59
　of sample mean, 87–88
　of total serum cholesterol levels in Framingham Offspring Study, 59, 309
　of weights and heights of men and women in Framingham Offspring Study, 60

C

case-control studies
　advantages and limitations of, 13
　issues that must be addressed in designing, 12–13
　nested case-control studies, 13
　overview of, 11–12
case reports, 9
case series, 9
cases, selection in case-control studies, 12
　categorical variables
　bar charts for, 46–50
　for blood pressure, total cholesterol, and BMI in Framingham Heart Study, 41–44
　defined, 36
　descriptive statistics for, 41
　overview of, 41
　statistics and graphical display for, 62
　CDC. See centers for disease control (CDC)
censored data, 250
censoring
　issues with person-time data in epidemiological studies, 30
　in study of myocardial infarction (MI), 251
　survival analysis and, 249–250
centers for disease control (CDC)
　demographic characteristics by state, 68
　guidelines for BMI, height, weight for infants and children, 87
　on trends in weight, height, BMI, 124
central limit theorem
　computing probability with, 125
　determining margin of error with, 103
　example applying to alpha fetoprotein (AFP) levels in fetus, 96
　example applying to HDL levels, 94
　example applying to LDL levels, 96
　example applying to Poisson distribution, 93
　example applying to population with dichotomous outcome, 90–93
　example of normal distribution, 89
　formula for, 97
　overview of, 89
　sample size and, 224
chart junk, 280, 300, 310
chi-square Goodness-of-Fit Test, 135–137, 139, 157, 160, 224
chi-square (χ^2) statistics, 217
chi-square test of independence, 159
choropleth maps, 324
CI (cumulative incidence). See cumulative incidence (CI)
CIs. See confidence intervals (CIs)
clinical trials. See also randomized controlled trials (RCTs)
　blinding controls or placebos in, 4
　defined, 8
　factors in, 18–20
　randomization of participants in, 14–16
Cochran-Mantel-Haenszel method
　example associating obesity with cardiovascular disease, 202, 216–217
　overview of, 202
cohort studies
　basing nested case-control study on, 13–14
　example of prospective cohort study, 26
　example relating BMI to cardiovascular disease, 270
　example studying time of death, 254–260
　Framingham Heart Study as, 17
　overview of, 9–11
　survival analysis and, 252–253
cohorts, 8
complementary events, 75
complex relationships. See multivariable methods
conclusions, of hypothesis tests
　ANOVA tests, 153
　example associating obesity with cardiovascular disease, 199
　matched samples, continuous outcome, 142, 145

Index 363

more than two independent samples, continuous outcome, 153, 155
null hypothesis, 127
one sample, continuous outcome, 132–133
one sample, dichotomous outcome, 134
one sample, discrete outcome, 138
research hypothesis, 130
two independent samples, continuous outcome, 141
two independent samples, dichotomous outcome, 147–148
two or more independent samples, discrete outcome, 161–162
conclusions, of nonparametric tests
example comparing anaerobic thresholds of athletes, 244
example evaluating interventions for preventing alcohol consumption during pregnancy, 267
example of anti-retroviral therapy for HIV patients, 232
example of chemotherapy treatment for breast cancer, 236
example of exercise reducing systolic blood pressure, 241
example of prenatal care via in-home visits, 230
overview of, 228
concurrent, 8
conditional probability
example of Down Syndrome screening test, 72
example of PSA screening test, 71–72
formula for, 97
overview of, 71
positive and negative predictive value of screening tests, 73
screening tests in, 71
sensitivity and specificity in screening tests, 72–73
confidence intervals (CIs)
key formulas related to, 119
matched samples, continuous outcome, 110–114
for odds ratio (OR), 117–119
one sample, continuos outcome, 104–105
one sample, dichotomous outcome, 105–106
for relative risk (RR), 115–117
for risk difference (RD), 114–115
two independent samples, continuous outcome, 107–110
two independent samples, dichotomous outcome, 114
types of estimates, 102
confidence intervals (CIs), sample size for
issues in, 172
matched samples, continuous outcome, 177
one sample, continuous outcome, 173
one sample, dichotomous outcome, 174–175
summary of formulas for, 189
two independent samples, continuos outcome, 175–177
two independent samples, dichotomous outcome, 177–178

confidence level, 102–105, 107, 111, 119, 164, 173–175, 177, 181, 336
confounding
analysis of, 194
Cochran-Mantel-Haenszel method in analysis of, 201
cohort studies complicated by, 11
defined, 8
example of drug trial for increasing HDL cholesterol, 199–201
example relating obesity with cardiovascular disease, 195–199
minimized in randomized studies, 14, 15
multiple linear regression analysis of, 212, 215
multiple logistic regression analysis of, 217
multivariable methods in assessing, 194
overview of, 195
confounding variable (confounder), 195
CONSORT (Consolidated Standards of Reporting Trials), 301–303
continuous outcome variables, confidence intervals
matched samples, 110–114, 119, 177
one sample, 104–105, 173
one sample with continuos outcome, 119
two independent samples, 175–177
two independent samples with continuous outcome, 107–110, 119
continuous outcome variables, hypothesis tests
formulas, 163
matched samples, 141–145, 185–187
more than two independent samples, 149–157
one sample, 131–133, 181–183
two independent samples, 139–142, 184–185
continuous outcome variables, sample size for
matched samples in confidence intervals, 177
matched samples in hypothesis tests, 185–187
one sample confidence interval estimates, 173
one sample hypothesis tests, 181–183
two independent samples in confidence interval estimates, 175–177
two independent samples in hypothesis tests, 184–185
continuous variables
box-whisker plots for, 57–61
correlation between, 203
defined, 36
descriptive statistics for, 50–57
estimating mean of, 87
nonparametric tests and, 227
normal distribution model and, 78, 79
overview of, 50

statistics and graphical display for, 62
t distribution probability model for, 103
controls
    factors in clinical trials, 18
    historical, 131
    selection in case-control studies, 12
convenience sampling, 70
CONVINCE principles, 280–281
correlation analysis
    example relating gestational age with birth weight, 204–210
    overview of, 203
correlation coefficient, 203–205, 207, 209
covariance, 205, 207
covariates, 19, 271
Cox proportional hazards regression analysis
    example investigating differences in mortality between men and women, 269–270
    example of cohort study assessing relationship of BMI to cardiovascular disease, 270–271
    overview of, 267–269
    time dependent covariates in, 271
critical values
    in choosing between null and research hypotheses, 125
    for *F* distribution, 150
crossover trials
    advantages and limitations of, 17
    overview of, 16–17
cross-sectional
    defined, 8
    studies made at point in time, 2
    survey, 9
crude analysis
    applied to risk factors, 202
    focus on single risk factor, 195
    phases of statistical analysis, 19
cumulative frequency, 42–43
cumulative incidence (CI)
    computing in cardiovascular disease example, 26
    formula for, 31
    person-time data and, 25
    problems estimating, 25
cumulative relative frequency, 42–43

D

data
    analyzing. *See* summarization of data
    interval data, 226
    managing in clinical trials, 19
    person-time data, 25, 30
    prospective data collection in clinical trials, 14
data visualization
    design principles, 280–281
    overview, 279–280
    presentation and statistical results in figures, 295–327
    presentation and statistical results in table, 285–295
    tips for effective use, 281–285
decision rules, in nonparametric tests
    example comparing anaerobic thresholds of athletes, 244
    example evaluating interventions for preventing alcohol consumption during pregnancy, 264–266
    example of anti-retroviral therapy for HIV patients, 231–232
    example of chemotherapy treatment for breast cancer, 236
    example of exercise reducing systolic blood pressure, 239
    example of prenatal care via in-home visits, 230
    overview of, 227
decision rules, in parametric tests
    ANOVA tests, 150, 152
    example associating obesity with cardiovascular disease, 197–199
    level of significance and, 125–126
    matched samples, continuous outcome, 142, 145
    more than two independent samples, continuous outcome, 152, 154
    one sample, continuous outcome, 132–133
    one sample, dichotomous outcome, 135
    one sample, discrete outcome, 136, 138
    setting up for null hypothesis, 126
    setting up for research hypothesis, 129
    two independent samples, continuous outcome, 141–142
    two independent samples, dichotomous outcome, 147–148
    two or more independent samples, discrete outcome, 160, 162
degrees of freedom (df), in ANOVA tests, 150–152, 154, 155, 157, 158
dependent samples, 219
dependent (response) variable, in regression analysis, 203
descriptive statistics
    for continuous variables, 50
    for dichotomous variables, 37
    example of continuous variables in Framingham Heart Study, 50–57
    example of dichotomous variables in Framingham Heart Study, 37–38
    for ordinal and cardinal variables, 41–44
    vs. inferential statistics, 23
    deviations, from mean, 53–57

df (Degrees of freedom), in ANOVA tests, 150–152, 154, 155–157
design principles, 280–281
dichotomous outcome variables, confidence intervals
    one sample, 105–106, 119, 174–175
    two independent samples, 114–118, 177–178
dichotomous outcome variables, hypothesis tests
    formulas, 163
    one sample, 133–135, 183–184
    two independent samples, 145–149, 187–188
dichotomous outcome variables, sample size
    one sample confidence interval estimates, 174–175
    one sample hypothesis tests, 183
    two independent samples in confidence interval estimates, 177–178
    two independent samples in hypothesis tests, 184–185
dichotomous variables
    bar charts for, 38–41
    defined, 36
    descriptive statistics for, 37–38
    logistic regression analysis applied to, 216–217
    overview of, 37
    statistics and graphical display for, 62
    survival analysis and, 250
difference scores
    for quantifying extent of disease, 27–28
    Sign Test and Wilcoxon Signed Rank test based on, 237
discrete outcome variables
    computing probability of, 75
    formula for, 163
    one sample, 135–139
    two or more independent samples, 157–163
disease
    estimation of rate of development of, 3
    identification of risk factors in development or progression of, 3
    incidence of. *See* incidence (of disease)
    prevalence of. *See* prevalence (of disease)
    quantification of extent of. *See* extent of disease, quantifying
    reasons for contradictory results between studies of same disease, 4
dispersion, 56–57, 61
distribution free tests. *See* nonparametric tests
dot plot, 303–304
double blind
    defined, 8
    ethics of, 15
    factors in clinical trials, 18
dummy variables, 215, 270

E

effect modification
    analysis of, 194
    example of drug trial for increasing HDL cholesterol, 199–201
    example of relationship between obesity and cardiovascular disease, 195–199
    multiple linear regression analysis of, 216
    multiple logistic regression analysis of, 217
    multivariable methods in assessing, 194
    overview of, 195
effect size, continuos outcome
    matched samples, 185–187
    one sample, 185
    two independent samples, 184–185
effect size, dichotomous outcome
    one sample, 183–184
    two independent samples, 187–188
efficacy
    in clinical trials, 19–21, 185, 199, 211, 213–214, 283
    multivariable methods, 194
    in randomized trials, 264
    test with matched samples, 143, 185
    test with one sample, 133
    test with two independent samples, 141–142, 175, 187
epidemiology, defined, 3–4
equally likely, 70, 109
errors
    margin of. *See* margin of error
    standard. *See* standard deviation
    Type I errors, 131, 179–181
    Type II errors, 131, 179–181
estimation
    confidence intervals for odds ratio, 117–119
    confidence intervals for relative risk, 115–117
    confidence intervals for risk difference, 114–115
    formulas, 119
    overview of, 102–104
    summary, practice problems, and references, 119–122
    techniques, 102
estimation, continuous outcome
    confidence intervals for matched samples, 110–114
    confidence intervals for one sample, 104–105
    confidence intervals for two independent samples, 107–110
estimation, dichotomous outcome
    confidence intervals for one sample, 105–106

confidence intervals for two independent samples, 114
ethics
    of blind/double blind, 14
    factors in clinical trials, 18
    vaccination example, 18
excess risk, difference measures for extent of disease, 27
expected frequency, 135, 159–162, 197–199
explanatory (predictors) variables
    correlation between, 203–204
    in regression analysis, 203
    varying over time, 271–272
exponential regression survival model
    for estimating survival function, 253
    hazard function in, 269
exposure. *See* risk factors
extent of disease, quantifying
    comparing between groups, 27
    difference measures, 27–28
    in group or region, 2–3
    incidence, 24–26
    issues with person-time data, 30
    overview of, 23
    prevalence, 24
    ratios, 28–30
    relationship between prevalence and incidence, 27
    summary, practice problems, and references, 30–34

F

F statistic, in ANOVA tests, 150, 152, 156–157
failure time, 250
false negative fraction, 97
false positive fraction, 97
FDA. *See* food and drug administration (FDA)
Few's principles, data visualization, 281
figures
    association among variables, 316–324
    bar charts for categorical variables, 46–50
    bar charts for dichotomous variables, 38–41
    box-whisker plots for continuous variables, 57–61
    components, 296–300
    data displays, 303–308
    data presentation, 295–296
    distribution displays, 303–308
    flow charts, 301–304
    geographical variations, 324–325
    group estimates, comparison, 308–313
    histograms for ordinal variables, 44–46
    line charts, 313–316
    participant flow charts, 301–303
    pie charts, 325–327
    proportionality assumptions in survival analysis, 271
    scaling axes to accommodate range of data, 41
    in summarizing data, 38
    tips for effective use, 281–285
first quartile, 54–55, 57, 86
fixed assignment, in crossover trials, 17
flow charts, 301–304
follow-up life table, in example studying time of death, 254–256
follow-up period
    in studies, 3
    survival analysis and, 253
food and Drug Administration (FDA)
    documentation of clinical trials required for applications to, 19
    New Drug Application (NDA), 20
    regulation of clinical trials by, 4
Framingham Heart Study
    confidence intervals for matched samples, continuous outcome, 110–114
    confidence intervals for matched samples, dichotomous outcome, 114–115
    confidence intervals for odds ratio, 117–119
    confidence intervals for one sample, continuous outcome, 104–105
    confidence intervals for one sample, dichotomous outcome, 105
    confidence intervals for relative risk, 115–117
    confidence intervals for two independent samples, continuous outcome, 107–110
    continuous variables in, 50–57
    Cox proportional hazards regression analysis, 268
    data on systolic blood pressure, 173, 185
    dichotomous variables in, 37–38
    examples of box-whisker plots, 58–60
    hypothesis tests for one sample, continuos outcome, 132
    hypothesis tests for one sample, dichotomous outcome, 134–135
    hypothesis tests for one sample, discrete outcome, 138–139
    hypothesis tests with matched samples, continuous outcome, 143
    hypothesis tests with more than two independent samples, continuous outcome, 149

hypothesis tests with two independent samples, continuous outcome, 140–141
incidence (of disease), 25–26
multiple linear regression analysis, 212
multiple logistic regression analysis, 217
multivariable methods applied in, 194
ordinal variables in, 41–44
overview of, 17
prevalence of cardiovascular disease, 24
sample size for one sample, dichotomous outcome, 183–184
study of risk factors in cardiovascular disease, 8
frequency bar charts
    for dichotomous variables, 39, 48–49
    uses of, 62
frequency distribution tables, for dichotomous variables, 37, 43–44
frequency histograms, 44–48, 62

G

Gaussian distribution, 78
gender, as example of dichotomous variables, 37
generalizability, 9, 14, 16, 18
geographical displays, 324–326
Gompertz method, for estimating survival function, 253, 269
goodness-of-fit test
    assessing normal distribution, 224
    one sample, discrete outcome, 137–139
    two or more independent samples, discrete outcome, 157
graphs
    association among variables, 316–324
    bar charts for categorical variables, 46–50
    bar charts for dichotomous variables, 38–41
    box-whisker plots for continuous variables, 57–61
    components, 296–300
    data displays, 303–308
    data presentation, 295–296
    distribution displays, 303–308
    flow charts, 301–304
    geographical variations, 324–325
    group estimates, comparison, 308–313
    histograms for ordinal variables, 44–46
    line charts, 313–316
    participant flow charts, 301–303
    pie charts, 325–327
    proportionality assumptions in survival analysis, 271
    scaling axes to accommodate range of data, 41
    in summarizing data, 38
    tips for effective use, 281–285

Greenwood's formula, for survival estimates, 258
groups
    comparing extent of disease between, 27
    quantification of extent of disease in, 2–3

H

hazard ratio, 268–271, 273
hazards
    Cox proportional hazards regression analysis, 267–271
    example investigating differences in mortality between men and women, 269–271
    proportionality assumptions, 271
histograms, 44–46, 62, 305–308
historical controls, in hypothesis testing, 133
hypothesis testing
    critical value and, 125
    errors in, 131
    matched samples, continuous outcome, 142–145
    more than two independent samples, continuous outcome, 149–157
    one sample, continuous outcome, 131–133
    one sample, dichotomous outcome, 133–135
    one sample, discrete outcome, 135–139
    overview of, 123–124
    steps in, 126–129
    summary, practice problems, and references, 163–170
    techniques for, 124
    two independent samples, continuous outcome, 139–142
    two independent samples, dichotomous outcome, 145–149
    two or more independent samples, discrete outcome, 157–163
hypothesis testing, sample size for
    formulas for, 189
    issues in, 179–181
    matched samples, continuous outcome, 185–187
    one sample, continuous outcome, 181–183
    one sample, dichotomous outcome, 183–184
    two independent samples, continuos outcome, 184–185
    two independent samples, dichotomous outcome, 187–188

I

ID (incidence density), 25–26
incidence (of disease)
    cumulative incidence (CI), 25
    defined, 8

example of Framingham Heart offspring study, 25–26
limitations of case-control studies, 13
person-time data and, 25
rate of, 25–26
relationship with prevalence, 27
incidence density (ID), 25–26
incidence rate (IR)
computing in cardiovascular disease example, 26
formula for, 31
overview of, 25–26
potential to change disease status and, 27
independence
defined, 159
in example of PSA screening tests, 74
in example of relationship between family history and cardiovascular disease, 74
formula for, 97
probability and, 74
independent events, 74
independent (predictors) variables
correlation between, 203
in regression analysis, 203
varying over time, 271–272
indicator variable, 210–211, 214–217, 252, 271. *See also* dichotomous variable
inferential statistics
broad areas in, 102
estimation. *See* estimation
hypothesis testing. *See* hypothesis testing
probability and, 67, 87
vs. descriptive statistics, 23
influential point, 320–321
intention-to-treat (ITT)
analysis sample, 19
defined, 8
nteraction, 194–195, 201. *See also* statistical interaction
intercept. *See* y-intercept
interquartile range (IQR), 54–57, 62
interrelationships, between variables. *See* multivariable methods
interval data, nonparametric tests and, 226
IQR (Interquartile range), 54–57, 62
IR. *See* incidence rate (IR)
ITT (Intention-to-treat)
analysis sample, 19
defined, 8

K

Kaplan-Meier (product-limit) approach
example comparing treatments for gastric cancer, 261
example studying time of death, 257–259
summary of, 272
Kolmogorov-Smirnov test, 224
Kruskal-Wallis test
example of albumin levels for adults on low protein diets, 241–243
overview of, 241
summary of, 246

L

least squares estimates, in simple linear regression analysis, 208–210
level of significance (α), in nonparametric tests
example comparing anaerobic thresholds of athletes, 243
example evaluating interventions for preventing alcohol consumption during pregnancy, 264
example of anti-retroviral therapy for HIV patients, 231
example of chemotherapy treatment for breast cancer, 236
example of exercise for reducing systolic blood pressure, 239
example of prenatal care via in-home visits, 230
overview of, 227
level of significance (α), in parametric tests
ANOVA tests, 150, 152
determining for null hypothesis, 126
determining for research hypothesis, 129
example relating obesity with cardiovascular disease, 197
matched samples, continuous outcome, 142–145
more than two independent samples, continuous outcome, 152, 154
one sample, continuous outcome, 131–133
one sample, dichotomous outcome, 133–135
one sample, discrete outcome, 136, 138
overview of, 125–126
two independent samples, continuous outcome, 140, 142
two independent samples, dichotomous outcome, 147, 148
two or more independent samples, discrete outcome, 160, 162
life tables
example comparing treatments for gastric cancer, 261
example studying time of death, 254–256
time dependent covariates and, 271
line plots, 323–324
linear regression analysis
multiple linear regression analysis, 212–216
simple linear regression, 207–211
log odds, 216
log-normal distribution, for estimating survival function, 269

log-rank test
　for comparing survival curves, 261, 262
　example comparing treatments for gastric cancer, 261–264
　time dependent covariates and, 271
logistic regression analysis
　example associating risk factors with infant weight, 217–219
　example relating obesity with cardiovascular disease, 216–217
　overview of, 216
logit. *See* log odds
lower-tailed test, 126, 132–134, 142

M

Mann-Whitney *U* (Mann-Whitney-Wilcoxon) test
　comparing outcomes between two independent groups, 228
　summary of, 245
margin of error
　in analysis of samples, 2
　effecting confidence intervals, 103
　for population mean, 172
matched samples
　confidence intervals for continuous outcomes, 110–114, 119, 177
　hypothesis tests for continuous outcomes, 142–145, 185–187
matched samples, nonparametric tests
　example assessing drug for repetitive behavior in autism, 232–235
　example of chemotherapy treatment for breast cancer, 235–239
　example of exercise for reducing systolic blood pressure, 239–241
　overview of, 232
matching, defined, 8
McNemar's Test, 149
mean
　box-whisker plots of, 87–89
　Central Limit Theorem and, 93–96
　computing average value of continuous variables, 52, 62
　confidence interval for, 104
　deviations from, 54–57
　estimating, 87
　normal distribution model and, 78, 80
　point estimates and margin of error for population mean, 172
　standard deviation of, 93
mean squares (MS), in ANOVA tests, 152, 154, 156–157

median
　computing average value of continuous variables, 52, 62
　normal distribution model and, 78, 79–80
　vs. mean when outliers are present, 54–57
misclassification bias, in case-control studies, 13
mode
　computing average value of continuous variables, 52–53
　normal distribution model and, 78, 79
　modeling probability binomial distribution model, 75–76
　example applying binomial distribution model to allergic reactions to medications, 76–77
　example applying binomial distribution model to likelihood of heart attack, 77–78
　example of assessing pain following hip replacement, 87–89
　normal distribution model. *See* normal distribution model
　overview of, 75
　sampling distributions using probability models, 87
　t distribution, 103
modeling, relationship of risk factors to outcomes, 3
monitoring, factors in clinical trials, 18
multicenter trails
　overview of, 14
　vs. single center trials, 18
multiple linear regression analysis
　example associating BMI with systolic blood pressure, 212–213
　example associating risk factors with infant weight, 215
　example of drug trial for increasing HDL cholesterol, 213–215
　overview of, 212
multiple logistic regression analysis
　example associating risk factors with infant weight, 217–219
　example relating obesity with cardiovascular disease, 217
　overview of, 216
multivariable methods
　Cochran-Mantel-Haenszel method, 201–203
　confounding and effect modification, 195–201
　correlation and regression analysis, 203–211
　multiple linear regression analysis, 212–216
　multiple logistic regression analysis, 216–219
　overview of, 193–195
　summary, practice problems, and references, 219–222

N

National Heart, Lung, and Blood Institute (NHLBI), 36, 41
NDA (New Drug Application), 20
negative predictive value
    example of Down Syndrome screening test, 73
    formula for, 96–97
    of screening tests, 73
nested case-control studies, 13
new drug application (NDA), 20
NHLBI (National Heart, Lung, and Blood Institute), 36, 41
nominal variables. *See* categorical variables
non-normal distribution, comparing with normal distribution, 78–79
non-probability sampling, 70
nonparametric tests
    for estimating survival function, 253
    matched samples, 232–241
    more than two independent samples, 241–245
    overview of, 223–226
    ranks and observed data, 227
    steps in, 227
    summary, practice problems, and references, 245–248
    two independent samples, 227–232
normal distribution model
    axes of, 78
    comparing with non-normal distribution, 78–79
    continuous variables and, 78, 79
    example applying to body mass index by gender and age, 80–87
    example applying to length of infant girls, 87
    example of normal distribution, 89–90
    formula for, 96–97
    Gaussian (bell-shaped) curve characteristic of, 78
    tests for assessing, 223–224
null hypothesis
    ANOVA tests, 151–156
    critical value and, 125
    defined, 123–124
    example associating obesity with cardiovascular disease, 197–198
    example of weight trends in men, 124–125
    more than two independent samples, continuous outcome, 151–154
    one sample, discrete outcome, 136–138
    steps in testing, 126–128
    two independent samples, continuous outcome, 141
    two independent samples, dichotomous outcome, 146–147
    two or more independent samples, discrete outcome, 160–161

O

observational studies
    case reports/ case series, 9
    case-control studies, 11–13
    cohort studies, 9–11
    compared with randomized studies, 7 8
    cross-sectional survey, 9
    Framingham Heart Study as, 17
    nested case-control studies, 13–14
odds ratio (OR)
    Cochran-Mantel-Haenszel method in analysis of, 201–203
    confidence interval for, 117–119
    formula for, 30
    ratios for comparing prevalence or incidence of disease, 28–30
one sample tests, confidence intervals
    continuos outcome, 104–105, 119
    continuous outcome, 173–174
    dichotomous outcome, 105–106, 119, 174–175
one sample tests, hypothesis tests
    continuous outcome, 131–133, 181–183
    dichotomous outcome, 133–135, 183–184
    discrete outcome, 135–139
one sample tests, sample size
    continuous outcome estimates for confidence intervals, 173–174
    continuous outcome hypothesis tests, 181–183
    dichotomous outcome estimates for confidence intervals, 174–175
    dichotomous outcome hypothesis tests, 183
OR. *See* odds ratio (OR)
ordinal variables
    defined, 36
    descriptive statistics for, 41
    in Framingham Heart Study, 41–44
    histograms for, 44–46
    nonparametric tests and, 227
    overview of, 41
    statistics and graphical display for, 62
outcome variables
    categorical. *See* categorical variables
    continuous. *See* continuous variables
    dichotomous. *See* dichotomous variables
    ordinal. *See* ordinal variables
    time-to-event variables, 249

outcomes
- assessing relationship to risk factors, 3
- binomial (success and failure), 75–78
- measuring as factor in clinical trials, 18
- ranking, 224

outcomes, confidence intervals
- key formulas related to, 107–110
- matched samples, continuous outcome, 177
- matched samples with continuous outcome, 110–114
- one sample, continuous outcome, 173–174
- one sample, dichotomous outcome, 174–175
- one sample with continuous outcome, 104–105
- one sample with dichotomous outcome, 105–106
- summary of formulas for, 189
- two independent samples, continuous outcome, 175–177
- two independent samples, dichotomous outcome, 177–178
- two independent samples with continuous outcome, 107–110
- two independent samples with dichotomous outcome, 114–119

outcomes, hypothesis testing
- matched samples, continuous outcome, 142–145, 185–187
- more than two independent samples, continuous outcome, 149–157
- one sample, continuous outcome, 131–133, 181–184
- one sample, dichotomous outcome, 133–135, 183–184
- one sample, discrete outcome, 135–139
- summary of formulas for, 189
- two independent samples, continuous outcome, 184–185
- two independent samples, continuous outcome, 139–142
- two independent samples, dichotomous outcome, 145–149, 187–188
- two or more independent samples, discrete outcome, 157–163

outcomes, sample size and
- matched samples, continuous outcome in confidence intervals, 177
- matched samples, continuous outcome in hypothesis tests, 185–187
- one sample, continuous outcome in confidence intervals, 173–174
- one sample, continuous outcome in hypothesis tests, 181–183
- one sample, dichotomous outcome in confidence intervals, 174–175
- one sample, dichotomous outcome in hypothesis tests, 183–184
- summary of formulas for, 189
- two independent samples, continuous outcome in confidence intervals, 175–177
- two independent samples, continuous outcome in hypothesis tests, 184–185
- two independent samples, dichotomous outcome in confidence intervals, 177–178
- two independent samples, dichotomous outcome in hypothesis tests, 187–188

outliers
- continuous variables and, 62
- median vs. mean and, 53–57

P

P-values
- ANOVA tests and, 152, 155
- determining if null hypothesis is true, 129–130
- level of significance and, 129
- Sign Test and, 234–235
- tabulating, 289–290

PAR (Population attributable risk)
- difference measures for extent of disease, 27–28
- formula for, 30

parametric tests, 224, 253. *See also* hypothesis testing

Pearson Product Moment correlation coefficient, 203–204

per protocol, 8, 19

percentiles, of normal distribution
- computing, 86
- formula for, 97

person-time data
- cumulative incidence (CI) and, 25
- issues with, 30

phase I: First Time in Humans Study, 19

phase II: Feasibility or Dose-Finding Study, 19–20

phase III: Confirmatory Clinical Trial, 20

phase IV Clinical Trial, 20

pie charts, 325–327

placebo
- in clinical trials, 14
- as controls, 4
- defined, 8

point estimates
- margin of error and, 102–103
- for population mean, 172–173
- types of estimates, 101–102

point prevalence (PP)
- difference measures for extent of disease, 27, 29–30
- formula for, 31
- overview of, 23–24

Poisson distribution, 93–94

population attributable risk (PAR)
- difference measures for extent of disease, 27–28
- formula for, 31

population mean, 172–173. *See also* mean

population parameters, 67–68
populations, of interest
    defining, 35–36
    factors in clinical trials, 18
    population standard deviation and sample standard deviation, 103
    in statistical studies, 2
positive predictive value
    example of Down Syndrome screening test, 74
    formula for, 97
    of screening tests, 73
power, in hypothesis testing. *See also* sample size
    matched samples, continuous outcome, 185–187
    one sample, continuous outcome, 183
    one sample, dichotomous outcome, 183–184
    probability and, 179–181
    two independent samples, continuos outcome, 184–185
    two independent samples, dichotomous outcome, 187–188
PP. *See* point prevalence (PP)
predictive value, of screening tests
    Bayes' Theorem in example of predictive value of screening tests, 74–75
    formulas for, 97
    positive and negative, 73
predictors (explanatory or independent) variables
    correlation between, 203–204
    in regression analysis, 203
    varying over time, 271–272
prevalence (of disease)
    computing in cardiovascular disease example, 26
    defined, 8
    example of cardiovascular disease, 24
    overview of, 23–24
    relationship with incidence, 27
probability
    Bayes' Theorem, 74–75
    binomial distribution model, 75–78, 235–236
    Central Limit Theorem and, 89–96
    conditional, 71–73
    confidence intervals compared with, 101–102
    defined, 70
    example applying simple random sampling, 70–71
    example of assessing pain following hip replacement, 87–89
    formula for, 97
    independence and, 74
    modeling. *See* modeling probability
    non-probability sampling, 70
    normal distribution model. *See* normal distribution model
    overview of, 67–68

parametric tests and, 223–224
power, in hypothesis testing, 179–181
summary, practice problems, and references, 96–100
probability sampling
    Central Limit Theorem, 89–96
    example of assessing pain following hip replacement, 87–89
    overview of, 67–69
    types of approaches to, 69
    using probability models, 87
product-limit approach. *See* Kaplan-Meier (product-limit) approach
prognostic factor, 8
prospective
    cohort studies, 9–10, 26
    data collection in clinical trials, 14
    defined, 8
protocols
    defined, 9
    factors in clinical trials, 18

Q

quantification of extent of disease. *See* extent of disease, quantifying
quantitative variables. *See* continuous variables
quartiles, continuous variables and, 54–55, 62
quasi-experimental design, 8
quota sampling, 70

R

race/ethnicity, as categorical variable, 36
randomization
    in clinical trials, 14–15, 18
    in crossover trials, 16
    defined, 9
    techniques for, 15
randomized controlled trials (RCTs)
    advantages and limitations of, 15
    compared with observational studies, 8
    crossover trials, 16–17
    details related to, 17–20
    as ideal for study design, 3–4
    overview of, 14–15
ranks
    example comparing anaerobic thresholds of athletes, 243
    example of albumin levels for adults on low protein diets, 243
    nonparametric tests and, 223–224
    outcomes and, 224

rate ratios, for comparing prevalence or incidence of disease, 28–29
ratios
   quantifying extent of disease, 28
   relationship between hypertension and cardiovascular disease, 28–29
RCTs. *See* randomized controlled trials (RCTs)
RD. *See* risk difference (RD)
recall Bias, 13
regions, quantifying extent of disease, 2–3
regression analysis
   Cox proportional hazards regression analysis, 268–271
   example of relationship between gestational age and birth weight, 203–207
   multiple linear regression analysis, 208–210
   multiple logistic regression analysis, 216–219
   overview of, 203–204
regression lines, 310–320, 316
relative frequency bar charts
   for dichotomous variables, 39–41
   for dominant hand, 49
   for smoking status, 50
   uses of, 62
relative frequency histograms, 45–47, 62
relative frequency, variable types and, 62
relative risk (RR)
   Cochran-Mantel-Haenszel method in analysis of, 201–202
   comparing prevalence or incidence of disease, 28
   confidence interval for, 114–119
   formula for, 30
research hypothesis
   ANOVA tests, 152–153
   critical value and, 125–126
   defined, 123
   example of association between obesity and cardiovascular disease, 194
   example of weight trends in men, 124
   more than two independent samples, continuous outcome, 156
   one sample, discrete outcome, 137–138
   two independent samples, continuous outcome, 142
   two independent samples, dichotomous outcome, 149
   two or more independent samples, discrete outcome, 160
residuals, in simple linear regression analysis, 208
response (dependent) variable, in regression analysis, 203
retrospective
   cohort studies, 10, 11
   defined, 9

right censoring
   issues with person-time data in epidemiological studies, 30
   survival analysis and, 249–250
risk difference (RD)
   confidence interval for, 114
   difference measures for extent of disease, 27
   formula for, 31
risk factors (exposure)
   addressing in case-control studies, 13
   assessing relationship to outcomes, 3
   for cardiovascular disease (Framingham Heart Study), 17
   Cochran-Mantel-Haenszel method in analysis of, 201
   competing risks in survival analysis, 271–272
   interrelationships between, 193–194
   multivariable methods in assessing, 194
   varying over time, 271
RR. *See* relative risk (RR)

S

safety analysis sample, 19
sample frame, in simple random sampling, 68–69
sample median, 52
sample range, 53
sample size
   factors in clinical trials, 18
   implications of, 20
   overview of, 171–172
   and representative nature, 2
   summary, practice problems, and references, 188–191
sample size, for confidence intervals
   issues in, 172
   matched samples, continuous outcome, 177
   one sample, continuous outcome, 173–174
   one sample, dichotomous outcome, 174–175
   summary of formulas for, 189
   two independent samples, continuous outcome, 175–177
   two independent samples, dichotomous outcome, 177–178
sample size, for hypothesis testing
   issues in, 179–181
   matched samples, continuous outcome, 185–187
   one sample, continuous outcome, 181–183
   one sample, dichotomous outcome, 183–184
   summary of formulas for, 189
   two independent samples, continuous outcome, 184–185
   two independent samples, dichotomous outcome, 187–188

samples
    of population in statistical studies, 2, 35–36
    size of and representative nature of, 2
    standard deviation and, 103
    summarizing data collected via. See summarization of data
sample variance, 54
sampling
    non-probability sampling, 70
    overview of, 68–69
    probability sampling, 69
    with/without replacement, 87
    Sampling variability, 67–68
sampling frame, 69
scatter plots, 316–322
screening tests
    assessing likelihood of medical condition, 71
    example of Down Syndrome screening test, 72
    example of PSA screening test, 71–72
    positive and negative predictive value of, 73
    sensitivity and specificity as measures in, 72–73
selection bias, challenges in case-control studies, 13
semiparametric models, 269
sensitivity
    example of Down Syndrome screening test, 73
    formula for, 97
    in screening tests, 72–73
Shapiro-Wilk test, 223–224
sign test
    compared with Wilcoxon Signed Rank test, 238–239
    example assessing drug for repetitive behavior in autism, 234–236
    for matched samples, 232
    summary of, 245
simple linear regression analysis
    example associating BMI with systolic blood pressure, 212
    example of drug trial for increasing HDL cholesterol, 214
    example relating gestational age with birth weight, 207–211
simple logistic regression analysis, 216–217
simple random sampling
    example applying, 70–71
    probability sampling, 69
source of variation, in ANOVA tests, 150–151, 153, 156–157
specificity
    example of Down Syndrome screening test, 73
    formula for, 97
    in screening tests, 72–73
spurious correlation, 323–324

squared deviations, from mean, 54–55
SS (sum of squares), in ANOVA tests, 150–151, 153, 156–157
standard deviation
    of binomial population, 78
    formula for, 68
    normal distribution model and, 78, 79–80, 83, 85
    of point estimate, 105
    population standard deviation and sample standard deviation, 103
    of sample mean, 93
    variability of continuous variables and, 50–53
standard error. See standard deviation
standard normal distribution, 82–83, 86, 97
statistical analysis, phases of, 18–19
    Statistical computing packages applied to multivariable analysis, 194
    applying Excel to survival probabilities, 256–257
    Excel as, 235
    implementing log-rank tests, 261
    time dependent covariates in, 271
statistical inference. See inferential statistics
statistical interaction. See effect modification
statistics, tools and techniques, 2
stratification, 9
stratified sampling, probability and, 69
study design
    implications of sample size, 20
    observational studies. See observational studies
    overview of, 7–8
    randomized controlled trials as ideal for, 3–4
    randomized studies. See randomization
    summary, practice problems, and references, 20–21
style guides, 286
subsets, of population in statistical studies, 2
sum of squares (SS), in ANOVA tests, 149–150, 152, 154–155
summarization of data
    bar charts for categorical variables, 46–49
    bar charts for dichotomous variables, 38–40
    box-whisker plots for continuous variables, 57–61
    continuous variables, 50
    conventions for reporting average value of continuous variables, 51–52
    descriptive statistics for continuous variables, 50–52
    descriptive statistics for dichotomous variables, 37–38
    descriptive statistics for ordinal and categorical variables, 41–44
    deviations from mean, 53–54
    dichotomous variables, 36
    histograms for ordinal variables, 43–45

ordinal and categorical variables, 41
overview of, 35–36
summary, practice problems, and references, 63–65
variability of continuous variables, 52–56
survival analysis
comparing survival curves, 261–267
Cox proportional hazards regression analysis, 267–271
estimating survival function, 253–261
example comparing treatments for gastric cancer, 261
example evaluating interventions for preventing alcohol consumption during pregnancy, 264–265
example studying myocardial infarction (MI), 250–251
extensions to consideration of, 271–272
overview of, 249–250
summary, practice problems, and references, 273–276
survival curves, 261–267
survival function, 253–261
survival time, 250
systematic sampling, probability and, 69

T

t distribution, probability model for continuous variables, 103
tables
components, 286–289
data presentation, 285–286
regression results, 292–295
statistical significance levels, 292
summary statistics, 289–292
tips for effective use, 281–285
test statistic, in nonparametric tests
example comparing anaerobic thresholds of athletes, 243–245
example evaluating interventions for preventing alcohol consumption during pregnancy, 264–265
example of anti-retroviral therapy for HIV patients, 231–232
example of chemotherapy treatment for breast cancer, 236
example of exercise reducing systolic blood pressure, 239–240
example of prenatal care via in-home visits, 230–231
overview of, 223
test statistic, in parametric tests
ANOVA tests, 149, 152
computing for null hypothesis, 129
example associating obesity with cardiovascular disease, 197–198
matched samples, continuous outcome, 142
more than two independent samples, continuous outcome, 150–153

more than two independent samples, discrete outcome, 160
one sample, continuous outcome, 131–133
one sample, dichotomous outcome, 133–134
one sample, discrete outcome, 136–138
selecting for null hypothesis, 125
selecting for research hypothesis, 129
two independent samples, continuous outcome, 142–145
two independent samples, dichotomous outcome, 145–149
χ^2 test of independence, 158
tests
nonparametric. *See* nonparametric tests
parametric. *See* hypothesis testing
text
tips for effective usage, 281–285
time, cross-sectional studies made at point in time, 9–10
time dependent covariates, in survival analysis, 271
time sequence, in study design, 8
time-to-event variables, 249. *See also* survival analysis
treatments, number of treatments as factor in clinical trials, 17
Tufte's principles, data visualization, 281. *See also* chart junk
two independent samples, confidence intervals
continuous outcome, 119
continuous outcome, 107–110
dichotomous outcome, 114–119
two independent samples, hypothesis tests
continuous outcome, 139–142
dichotomous outcome, 147–149
two independent samples, nonparametric tests
example of anti-retroviral therapy for HIV patients, 231–232
example of drug trial for asthma in children, 228–230
example of prenatal care via in-home visits, 230–231
overview of, 227–232
two independent samples, sample size
continuous outcome estimates for confidence intervals, 175–177
continuous outcome hypothesis tests, 184–187
dichotomous outcome estimates for confidence intervals, 177–178
dichotomous outcome hypothesis tests, 187–188
two or more independent samples, hypothesis tests
continuous outcome, 149–157
discrete outcome, 158–164
two or more independent samples, nonparametric tests
example comparing anaerobic thresholds of athletes, 241–243
example of albumin levels for adults on low protein diets, 241–242
overview of, 241–242

type I errors
    in hypothesis testing, 131
    sample size and, 179–181
type II errors
    in hypothesis testing, 131
    sample size and, 179–181

U

unadjusted analysis, 195
upper-tailed Test, 126

V

vaccination, ethics of blind studies, 18
variability of continuous variables, 53
variables. *See also* by individual type
    categorical. *See* categorical variables
    continuous. *See* continuous variables
    dichotomous. *See* dichotomous variables
    key statistics and graphical displays for, 61
    multivariables. *See* multivariable methods
    ordinal. *See* ordinal variables
    time-to-event. *See* survival analysis
    types of, 36

W

wash out period, in crossover trials, 17
Weibull method, 253, 269
Wilcoxon Rank-Sum test, 228
Wilcoxon Signed Rank Test
    example assessing drug for repetitive behavior in autism, 239–240
    example of chemotherapy treatment for breast cancer, 236
    for matched samples, 232
    summary of, 245–246

X

χ^2 goodness-of-fit test, 137–139, 157
χ^2 test of independence, 158

Y

y-intercept, 208–211

Z

z score, 83, 86, 96